Wireless Access and the Local Telephone Network

For a complete listing of the Artech House Telecommunciations Library,
turn to the back of this book

Wireless Access and the Local Telephone Network

George Calhoun

Artech House
Boston London

Library of Congress Cataloging-in-Publication Data

Calhoun, George, 1952-
Wireless access and the local telephone network / George Calhoun.
p. cm.
Includes bibliographical references and index.
ISBN 0-89006-394-X
1. Radio. 2. Telephone systems. 3. Computer networks.
I. Title.
TK6553.C248 1992 91-39436
621.385-dc20 CIP

685 Canton Street
Norwood, MA 02062

International Standard Book Number: 0-89006-394-X
Library of Congress Catalog Card Number: 91-7197

10 9 8 7 6 5 4 3 2 1

CONTENTS

PREFACE

This is a book about one of the most surprising and important trends in today's telecommunications industry: the emergence of advanced radio systems for providing access to the telephone network. *Wireless access,* as I use it here, is a generic term. It encompasses cellular mobile radio, personal communications concepts (PCN and PCS), the wireless local loop (BETRS and BEXR), wireless LANs, wireless PBXs, and other prospective applications, some of which appear to overlap one another. Collectively, these systems represent a vigorous challenge to the copper-based access facilities that have shaped telephony for more than 100 years. As but one indication of this challenge, consider that some forecasts now call for 40 million wireless access subscribers by 1999 *in the U.S. cellular radio market alone.* For comparison, the wireline telephone network did not reach 40 million subscribers in the U.S. until 1949—73 years after the invention of the telephone—and this forecast takes no account of the many other forms of wireless access now being discussed.

Actually, this volume really comprises *two* books, written for two rather different audiences. Parts I, II, and V are addressed to the "Big Picture"—what is happening to the telephone industry as wireless access muscles its surprising way towards center stage? These sections deal with issues of history, policy, and strategy, and may therefore be of interest to a general audience.

Sandwiched in between, in Parts III and IV, is a "textbook" for planners, managers, and applications-oriented engineers who are more concerned with the specifics of analyzing, evaluating, programming, and deploying wireless access systems in a telephony environment. Many of these readers may find the "Big Picture" sections in Parts I, II, and V to be of less immediate interest. One may certainly skip directly to the beginning of Part III for a more focused presentation of the technology, applications, economics, and engineering principles of wireless access.

To review briefly, Part I is an attempt to answer the questions: Why are we here? Why is wireless access, in its various guises (including cellular radio, PCN, and basic exchange radio), suddenly so prominent? Why is it "the most exciting development in the telephone industry since the break-up of AT&T" (in the words of a recent speaker at a standing-room-only session for the telephone industry)? In brief, the coming of wireless access has much to do with the coming of the Information Age itself. It portends fundamental and even unsettling changes in the nature of the telephone business. I may allude to those changes here by suggesting that there are basic parallels between the introduction of radio technology (i.e., "microwave") into the long-distance or interexchange business in the 1950s and 1960s (which led, I believe, in a direct causal chain to telecommunications competition, deregulation, and divestiture) and the imminent appearance of radio in the local exchange environment (which might be termed today the last bastion of the monopoly). Be forewarned, however, that in trying to encompass a number of very large issues, I find that the long first chapter has become itself a rather untethered dissertation, perhaps not every reader's cup of tea.

Part II begins a more linear exploration of three basic aspects of the "problem of access" and forms the background against which wireless access is emerging. Chapter 2 deals with the quantitative dimensions of the access problem in today's network. Chapter 3 addresses the costs and pricing of access—in short, the economics of access. Chapter 4 deals with the simultaneous "revolution" in telecommunication services that is redefining notions about "universal service" and the relation between enhanced access and enhanced economic and political opportunity.

Part III deals with the classification of the various applications of wireless access (Chapter 5), and provides a primer on the technologies that are making this all possible (Chapter 6).

Part IV is the most "nuts and bolts" section of the book, dealing with the methods for analyzing the economics of wireless (versus wireline) access facilities (Chapter 7) and with the issues raised during the actual engineering of a wireless access application, from radio propagation analysis to site preparation and generic installation procedures (Chapter 8). The material in Part IV is derived from practical experience with one of the pioneering companies in the entire field of wireless access and specifically from the problems encountered and solutions developed during the field deployment of more than 50 wireless access systems from 1986 through mid-1990 (serving several thousand users).

Finally, Part V returns to a consideration of some of the implications of the introduction of wireless access systems for the future of the local exchange environment. Chapter 9 discusses the complementarities of digital radio and fiber optics as the twin pillars of twenty-first-century access. Chapter 10 charts the course for the next few rounds of the PCN/PCS debate.

I have had the good fortune to play midwife to this new child of telephony. Beginning with IMM Corporation in 1980 (and with the assistance of M/A-COM and Hughes Network Systems), we developed a technological platform that brought advanced digital radio into the local exchange environment for the first time in the United States. We have worked through the arduous process of introducing a new approach to telephony to an industry that is steeped in traditional methods and procedures. I use the phrase "new approach" advisedly, for we have found that wireless access encompasses far more than just technology. It is a new way of looking at the problem of providing telephone service.

We have had to walk our ideas through the labyrinths of telephone company "product evaluation" and "standardization," develop multiple layers of documentation, prepare training cycles, and develop new planning tools. We have dealt with individuals at all levels and in all functions in the telephone business, from tariff experts and regulatory counsel to network planners and technologists, from chief executives to central office technicians and outside plant engineers. We have wandered the industry byways from the labs at Bellcore to rural "stepper" exchanges in Mississippi and Wyoming and Texas. In short, we had to confront the *culture of telephony*—still, overwhelmingly, a wireline-oriented culture where career paths often begin knee-deep in the mud, splicing cables. We have presented this culture with the first real alternative loop technology in 100 years. In doing so, we have learned that wireless access is both infinitely easier and, because it is so different, vastly more difficult than the comfortable "solution" of stringing more copper wire.

Nevertheless, wireless access is now well on its way. Almost every major telephone company has formed a wireless access group (under various names), and it is clear that many more will soon follow the paths that I have described here. It is my hope that in this book I've captured a significant number of lessons from our experience that may make this process a little easier for those who will follow.

Once again, as in my previous book, I must apologize for my ethnocentrism. The historical and political/regulatory context that has helped to create the "problem of access" is not entirely unique to the United States, but I must rely on the non-U.S. reader's own insight to see where the parallels apply and where they do not. Certainly, I believe that most of the technical "textbook" on wireless access that is presented in Parts III and IV will be relevant wherever such systems may be evaluated and deployed.

Philadelphia
August 13, 1991

ACKNOWLEDGMENTS

Some inventions are the work of an individual or a small team, but the invention of a whole industry is a collective accomplishment involving hundreds of enterprising people who are willing to go against the prevailing wisdom to bring forth something new under the sun. This book chronicles the invention of a new way of telecommunicating, and the ideas it contains are the coalescence of thousands of conversations and experiences contributed by the many coauthors of "wireless access." I only regret that the list is too long to recognize everyone by name.

I must, however, acknowledge the special roles played by some. In particular, I owe much to Jim Mullen, Dick Saunders, and Brian Kiernan, with whom I have stumped many a soapbox for the Cause. I have never known finer people. Others who stood on the front lines for "wireless" include Ted Clair, Mike Floyd, John Goetz, Tom Gorman, Bill Ronca, John Schaeffer, Moe Schuble, Kris Shelton, and Craig Startt. Those who helped create the analytical frameworks include Chris Chacona, Paul Clements, Jerry Fat, Jud Kenney, Craig Miller, Al Sacuta, and Jack Taylor. The guys who "made it work" include John Kaewell, Scott Kurtz, Ted Kunko, Mark Lemmo, Harry Leopold, Paul Levitsky, Dan Lin, Thad McGarrah, Mike Regensburg, Bill Vanderslice, and John Welch. I must also acknowledge, with deep gratitude, the steadfast support and the incredible ingenuity of the "Hughes team": Jack Shaw, Pradman Kaul, Pradeep Kaul, Jim Gandolfi, John Corrigan, Graham Avis, and many others (including, once upon a time, Carlos Roberts). And from SEL: Phil Benson, Manfred Böhm, Klaus Eckert, Erhart Ehlert, and Ken Giesken. As well, all my "tutors"—especially Ridge Bolgiano, Gil Lavean, and Eric Paneth.

Then there are the many others who have contributed assistance, material, ideas, or much-needed critiques: Art Biggs, Dale Brazale, Claude Buster, Hudson Barton, Don Beaupré, Tom Bernard, Peter Erb, Linda Garcia, Mike Hirsch, David Goodman, Kas Kalba, Nick Pagon, Peter Samek, Ray Simpson, George Squires, Michelle Suwala, and Richard Wolff.

Finally, I must again acknowledge Sherwin Seligsohn, whose intuitions have inspired this work, Steve Abramson for his patient encouragement, and Bill Hilsman who in many ways made it possible for me to write this book. And of course, again, Maria.

Part I
INTRODUCTION: BEYOND COPPER

Chapter 1
THE NOT-SO-ALL ENCOMPASSING NETWORK

1.1 LOOKING BACK: THE RIDDLE OF MODERNITY

The Copper Fabric

Our telecommunications infrastructure—the "network," as we acolytes usually refer to it—is, for Americans, a very special national resource. Indeed, until after World War II, mass-market telephony was a singularly American phenomenon. Foreign visitors often commented on our fondness for telecommunicating. "These people are forever sending off and receiving telegrams, messages, and cablegrams," wrote an English tourist in 1913. "They are continually telephoning." Moreover, this addiction seemed so severe that, when decoupled from these devices, the American specimen sometimes appeared to "relapse into a curious trance" [1]. Other outside observers were surprised to find that the telephone had even become a household appliance. "One thing which strikes one in America," wrote Sir William Preece, the Victorian-era head of the British Post Office (and the man responsible for bringing telephony to England), "is the enormous extent to which they apply the telegraph and the telephone for their own domestic purposes [i.e., in their homes]. . . . The reason they are driven to that is that it acts as a substitute for servants. Here [in England] we have a superabundance of messengers, errand boys, and things of that kind. In America they are wanted . . ." [2]. Imagine—a speaking telegraph in the scullery!

The anecdotes and impressions of American telephone addiction are borne out by the statistics. Through the 1920s, New York State had more phones than any European nation. And rural Iowa had more phones per capita than New York State [3]. In 1910, two-thirds of all the telephones in the world were in the United States, and as late as 1955 there were still more telephones in America than in all other countries combined [4]. This reflected something more than a merely quantitative comparison; the United States was the first (and for many years the only) major nation to make a societal commitment to an all-encompassing network. Figure 1.1 shows the penetration rates of telephony in Europe and America during this period, which portray what one European writer has referred to as "the almost abysmal gulf which . . . separated telephone development on each side of the Atlantic" [5]. Only in America (and, perhaps it might be admitted, in the showcase countries of Sweden and Switzerland) was there a commitment to anything like "one phone per household"—at least until well after World War II.

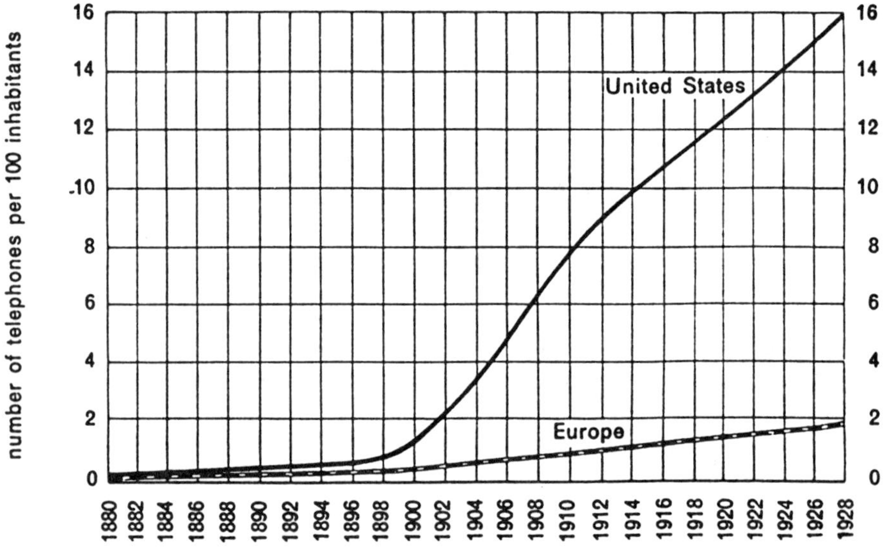

Figure 1.1 Comparative development of telephony in the United States and Europe. (From *Bell Telephone Quarterly*, 7(3), July 1928).

This social compact—"universal service"—was first articulated as part of Theodore Vail's public relations campaign for the antitrust-beleaguered Bell System, a *quid pro quo* for being permitted to retain a monopoly franchise. (Vail's full phrase was "One system, one policy, universal service.") Later embodied in the principles of the 1934 Communications Act, the availability of high-quality telephone service has come to be seen almost as one of the rights of American citizenship. Today, after more than a century of struggle, we have indeed managed to create something approximating "universal service"—and much else besides. This genial slogan—taken surprisingly seriously by later generations—has engendered a unique marriage of government policy and private enterprise, which, among other things, created the "biggest company on earth," spawned the largest private research and development program in history, gave birth to several new branches of science, produced something as close to successful socialism as the United States has ever seen, and, in the process, buried in the ground hundreds of billions of dollars of highly refined metallic copper.

This is a remarkable collective accomplishment. For the first and the last time in history, human beings wired an entire continent for the purpose of communications and administered it as a unified system of electrical signaling. Several miles of copper wire were laid down for every man, woman, and child. A dedicated copper circuit reached into each individual home and business, carrying the flow of conversation back toward the nearest switch, gathering with other wires from other homes into tributary cables, which in turn fed larger tributaries, flowing underneath the fields and the city streets, binding us together. Alexander Graham Bell's telephone instrument—a workbench collection of magnets and diaphragms—is the invention we celebrate, and we are in awe of the marvelous switching engines of today's high-tech networks. However, *it is in the fabric of copper wire laid down over a century of tremendous effort and expense that the network becomes a reality.* Other countries have telephones and switches, but only in the United States did the political determination exist, at such an early stage, to build the all-encompassing copper fabric.

Defeating Destiny

It bears asking, even to digress: Why should the telephone have flourished here as nowhere else? For some observers, the special circumstance of the American commitment to a wired society seems to be bound up with the problematic viability of the American nation. Historically, the young republic was confronted—as its founders realized—with the challenge of a continental geography, which greatly impeded the transportation of goods and people and the communication of information within the national framework. In 1799, it took a week for news of George Washington's death to travel from Virginia to the nation's capital (then in New York City). It normally required three to four weeks to communicate between New York or Philadelphia and cities on the periphery such as Charleston, Cincinnati, Detroit, and New Orleans [6]. The American economy was an unwieldy patchwork of isolated fragments, "a society of island communities" in the words of one historian [7].

Some even doubted whether a single polity could span such a vast area. "I shall refuse to believe," wrote the great French analyst Alexis De Toqueville in 1835, "in the duration of a government that is called upon to hold together forty different nations [i.e., the states] spread over a territory equal to one half of Europe" [8]. Twenty-five years later, on the brink of the Civil War, a sophisticated and not unsympathetic English visitor expressed the "general English feeling" that "this widespread nationality of the United States, with its enormous territorial possessions and increasing population, has fallen asunder, torn to pieces by the weight of its own discordant parts"—and foresaw not only the secession of South from North, but of the West also, driven apart by divergent economic interests, a disintegration not only inevitable, but "beneficial alike to both parties" [9]. Indeed, many important American leaders in this period had also questioned openly whether or not, as one put it, "the very idea of an American People, as constituting a single community, is a mere chimera?" [10].

Telecommunications has helped to defeat this geographic destiny. What economic historians often refer to as "a revolution in transportation and communications" began in the second half of the nineteenth century to create a truly cohesive national entity:

> The railroad, the telegraph, and the telephone were the transportation and communication links that by the beginning of the [twentieth] century bound the far-flung regions of the United States into a unified economy of unprecedented dimensions in world history. These innovations had . . . many impacts, but their greatest achievement was in giving operational meaning to the idea of a national economy. The size of the free-trade area within the borders of the United States in itself became a source of growth. . . . People in other regions of the world might possess resources and technical knowledge equal to that available to Americans, but no people had the opportunity to exploit a marketplace of comparable size. [11]

But the network is more than a scaffolding for the North American marketplace. It is the basis of modern civilization. It is part—a large part, I believe—of the answer to the riddle of modernity.

"The Continuity Snapped"

The riddle emerges from the observation, first made by thoughtful men about a hundred years ago—almost at the same moment as the transportation/communications revolution was taking hold—that a dramatic break had occurred in the course of human events. After eons of continuity and slow evolution of the basic forms of social and economic life, civilization (at least in Western Europe and North America) had leapt its tracks at some point in the nineteenth century and was now careening along a wholly new trajectory toward an uncertain future. The American social reformer Henry George, writing in 1879, evoked this Promethean feeling, as most have done, in terms of the cornucopia of recent mechanical inventions:

> Could a man of the last century—a Franklin or a Priestley—have seen, in a vision of the future, the steamship taking the place of the sailing vessel, the railroad train of the wagon, the reaping machine of the scythe, the threshing machine of the flail; could he have heard the throb of the engines that in obedience to the human will, and for the satisfaction of human desire, exert a power greater than that of all the men and all the beasts of burden of the earth combined; could he have seen the forest tree transformed into finished lumber—into doors, sashes, blinds, boxes or barrels, with hardly the touch of a human hand; the great workshops where boots and shoes are turned out by the

> case with less labor than the old-fashioned cobbler could have put on a sole; the factories where, under the eye of a girl, cotton becomes cloth faster than hundreds of stalwart weavers could have turned it out with their hand-looms; could he have seen the steam hammers shaping mammoth shafts and mighty anchors, and delicate machinery making tiny watches; the diamond drill cutting through the heart of the rocks, and coal oil sparing the whale . . . ? [12]

A generation later, during the time of Theodore Roosevelt when the trusts (including AT&T) were reorganizing the economy, the essayist and historian Henry Adams (a scion of the Massachusetts family that had produced two U.S. presidents, and himself a spokesman for the fading preindustrial aristocracy) cast his feelings succinctly in somewhat darker language: "In 1900, the continuity snapped" [13]. The symbolism of the machine was still, for Adams, the index of this break; after visiting an exhibit of the new technologies, where he was especially taken with the great electric "dynamos" silently coursing with power, he wrote that he "found himself lying in the Gallery of Machines at the Great Exposition of 1900, with his historical neck broken by the sudden irruption of forces totally new" [14].

Adams also sensed that these mechanical inventions were creating a completely new way of life, a "modern society"—but of this his vision was more clouded. He groped through metaphors for the substance of the changes, and saw little but blur and noise: "The typical American man had his hand on a lever and his eye on a curve in his road; his living depended upon keeping up an average speed of forty miles an hour, tending always to become sixty, eighty, or a hundred . . ." [15]. Others were equally dumbfounded. The aforementioned English tourist observed "a jerking, vital, *bizarre* 'rag-time' civilization" [16] and puzzled over its nature.

The search for the substance underlying these impressions and metaphors is not an easy one. Looking back over the past century in the hope of finding some explanation for, or simply a better definition of, the emergence of this "modern way of life," what is striking is how the modes of transportation, the patterns of commerce, the kinds of productive activities which "lead" the economy (however this may be defined) have all changed, and keep on changing, from generation to generation. In fact, the most important businesses of the period around 1900, which made the fortunes of the fabled "robber barons," have almost disappeared from the economic landscape. Some—such as meat-packing, leather, newspapers, sugar, street cars, even ice (for refrigeration)—have receded so far in importance that it is hard to imagine how prominent they once were [17]. And where once the steel mill was the very archetype of a "modern" industrial economy, today they make no more steel in Pittsburgh. The railroads, which seemed to nineteenth century observers to be the quintessential innovation of the modern era, now occupy a minor niche in our transportation eco-system. Social and economic preferences among alternative sources of power and energy—oil, natural gas, coal, nuclear, solar, gasoline/methanol/ethanol—shift nervously from decade to decade. Different nations today may make different choices as to how they move people around or keep them warm. It is now apparent that a "modern" society can be built with or without steel mills, railroads, or even gasoline. With conceivable technological breakthroughs in photoelectric power generation, it is even possible to imagine a modern society without the same central role for the electric power grid.

What is not possible to imagine is a modern society *without* a telecommunications network. The network is the enabler of so much else. From the first telegraph systems, which made large-scale railroad operations possible in the nineteenth century by permitting efficient scheduling and ensuring safe operations [18], to the powerful software environments that nourish the increasingly information-intensive service industries that employ the majority of the American work force at the close of the twentieth century, it is the increasing density of the communications grid that has fertilized the economic complexity perhaps most characteristic of modern life. A biological metaphor suggests itself. Very simple organisms can function without a formal nervous system; a preindustrial society can make do with word-of-mouth and hand-to-hand contact as the means of information transfer. More complex organisms and more complex societies must develop true nervous systems.

We have built our societal nervous system out of copper wire. We often equate "telecommunications" implicitly with a wireline connection. True, satellites and microwave radios exist somewhere "out there" in our long-distance circuits—but we still access them with a strand of copper—the limitations of which we only dimly appreciate. Yet if, as I am claiming, the character of modern civilization is bound up with our ability to telecommunicate, then the limits of copper-based telecommunications inevitably shape the kind of society we are.

The limits? We users are not accustomed to thinking of copper-based telephony in terms of its limits. The miracle of long-distance communication retains its power to amaze, regardless of other considerations. I remember a comment from a computer student recently who marveled that he could communicate so readily, at the speed of 1200 bits per second, with other computers across the country. The fact that he could also "communicate" (i.e., transfer information) with his printer, or his disk drive, or even with other computers on his local area network at speeds tens or hundreds of times faster had not yet awakened the thought that the copper-based public network might be the most constrictive gateway in his information-processing universe.

Nevertheless, copper-based telecommunications is indeed severely limited. We are beginning to be more aware of this as computer-driven applications push us into a new kind of economic environment, which some have dubbed the "Information Age." New types of telecommunications traffic are beginning to appear that the conventional copper fabric is not suited to handle. In particular, this new traffic is appearing in two forms:

1. *Nonvoice information:* Traffic that does not originate as speech, but as something else such as an image, which may require vastly greater capacity than the copper-based network was generally designed to handle.

2. *Information from mobile or movable nodes:* Traffic that does not always originate from a permanently fixed location, but instead appears from mobile or transportable or relocatable "terminals," as part of subnetworks that are characterized by continuous and rapid reconfiguration (as opposed to the stable, fixed star networks of wireline telephony).

1.2 LOOKING AHEAD: THE INFORMATION AGE

The concept of an information-based society is still a relatively new way of looking at the world, founded more on intuitive observations than rigorous statistics. The economist Fritz Machlup was the first to assay a quantitative measurement of the importance of information flows within the economy. In his pioneering 1962 book, *The Production and Distribution of Knowledge in the United States,* he offered a crude estimate that about 29% of the U.S. gross national product (GNP) was derived from the production and management of information [19]. He developed this figure by toting up the salaries of schoolteachers, the revenues of typewriter manufacturers, and so on—anyone or anything that appeared to be in the "knowledge business" in some fashion. Daniel Bell, in *The Coming of Post-Industrial Society,* observed that the relative importance of information-oriented activities is clearly on the rise, taking as one example the amount of the GNP devoted to education, which rose from about 3% in the 1930–1950 period to 7.5% by 1969 [20].

I believe, however, that these approaches to evaluating the importance of information transactions to modern economic activity miss the real point. What has happened in the last 40 years—accelerating especially in the last 15 or so years—is that almost all forms of economic activity have become more information intensive.

The Plumber's Bill: Information Intensity

Let us retell an old story (and, we may say, a true one) [21]. A family moves into a beautiful 100-year-old house, nicely restored to its Victorian splendor, only to find that the furnace in the basement is ailing. Winter is on the way and it will not keep them warm. But the cause of the problem is not easily diagnosed. The ancient heating system is a vast hodgepodge of disparate parts, from boiler to circulators to valves and pipes, originally installed in the days of coal, converted to oil, later to natural gas—a morass of patched-together pieces of mysterious hardware, mysteriously afflicted. The fuel is consumed, the furnace wheezes, but the flow of heat is somehow obstructed, although whether by a blocked pipe or a broken valve the homeowner (an amateur) cannot say. In an effort to fix it himself, he has read all the books, consulted his friends and relatives, but to no avail. He has called in several of the local plumbers, who proved to be good men for ordinary clogged drains and leaky faucets, but helpless before this monster of a heating system. They have applied their efforts—and submitted their bills—but the radiators are still cold.

Finally, and with some reluctance, the homeowner decides to consult a renowned master plumber, whose reputation is known throughout the city. The man is famous for his almost devilish skill with furnaces of all kinds, but he is also famous for his high rates. Nevertheless, the children's bedrooms are cold; the others have failed—what can the homeowner do? So the summons goes out.

The great craftsman arrives—empty-handed, without a single tool or instrument. He descends into the dark basement to inspect his adversary. For several minutes, he silently circumambulates the constipated beast, seeming to pay little attention to any particular detail, almost as if he were simply a curious bystander. The homeowner, already skeptical, begins to grow concerned. "How can this man do anything without tools?" he thinks. "Am I being conned?" Finally, in the most casual manner, the master plumber stoops down and picks up an ordinary brick that had been lying on the floor and gives one of the pipes a terrific bang. Immediately the sound of rushing water can be heard and the circulator starts to purr as its flow is finally released. Within a minute, as if by a miracle, the heat is returning to the cold pipes. The homeowner's wife appears at the top of the basement stairs, babe in arms, to say that the radiators are working again. The homeowner is of course delighted. The aura of domestic bliss returns to the household.

The plumber sits down and writes out a bill: "For Fixing Furnace—$500.00."

The homeowner's delight turns to dismay. "Five hundred dollars!" he cries, "But all you did was to bang it once with an old brick!"

"Yep, that's all," says the master plumber. (These are practically the first words to have escaped from his lips.)

"Well," replies the now indignant homeowner, "I demand an itemized bill."

"All right," says the master plumber. He patiently writes out another bill and hands it to the homeowner: "For Banging Pipe—$1.00, For Knowing Where to Bang—$499.00, No Charge for Tools. Total: $500.00."

"Now that," says the homeowner to himself as he pulls out his checkbook, "is information intensive."

The plumber's bill would still confound economists and accountants, who still struggle to capture the quantities and values of information transactions. Certainly, the plumber's actions would not have been recognized by Machlup's analysis. Information is not normally understood as an asset (except in the limited case of patents and copyrights). There is no easy way to quantify information intensity; most bills are not broken out as the master plumber's was. Yet it is clear to anyone who participates in the modern economy that the information intensity of almost all forms of economic activity is rapidly increasing, and the costs of this information are real (and often painful, as they were to this homeowner).

The field of health care is a good example, where the "plumber's bill" problem is acutely felt. Health-care costs in the United States have skyrocketed—from $13 billion (4.5% of the GNP) in 1950 to more than $500 billion (more than 10% of the GNP) in the early 1990s—in large part due to the higher information intensiveness of medical procedures [22]. Fifty years ago a doctor's tools (as he sat by the patient's bedside, in the patient's home) were a stethoscope, a thermometer, a tongue depressor, a knee reflex hammer, and a trained sense of touch and observation. Even so, medicine was already clearly a relatively information-intensive occupation. Today, the patient is often removed to a specialized diagnostic environment, where he is surrounded by high-tech (and high-cost) equipment, the purpose of most of which is to produce information to assist in the physician's decisions. In some cases, this information is produced in "real time," as in the delivery room or the intensive care unit. Fifty years ago most babies were still born at home. Today, the hospital delivery room is equipped with on-line fetal monitors, which deliver continuous printouts on various vital signs, ultrasound machines for direct visualization of the fetus, blood pressure and heart rate monitors for the mother (and, I suppose, if need be, for the father), with a myriad of other information-generating systems close at hand to help physicians and nurses respond to complications. Intensive care is virtually defined by the intensiveness of the information-gathering processes that are applied to monitor acutely ill individuals. More diagnostic systems are positioned off-line to run specialized tests or diagnostic procedures, from CAT scans to angiograms and fiber-optic probes.

Biomedical technology has been "blamed" for between 25 and 75% of the rise in medical care costs, depending on how the accounting is done [23] (a range of uncertainty suggests how difficult it still is to deal with information costs in a system of economic accounts still oriented toward toting up tangible merchandise). In any case, expenditures for diagnostic and other information-generating equipment are booming. Sales of diagnostic imaging devices (including techniques based on conventional x-ray, tomography, ultrasound, and nuclear magnetic resonance) grew at an annual rate of 24%, increasing more than 20 times between 1970 and 1987 to more than $8 billion, with the costs of "complementary inputs associated with the operation of the equipment" estimated at two or three times that amount [24]. Another major segment of the information-intensive medical equipment market is the field of patient-monitoring systems (including hospital-based intensive care units as well as the newer "ambulatory" and remote monitoring devices now becoming popular), which has grown almost as fast. The backlash against the high cost of these systems has led to complaints that "the modern medical education of the young physician has generated an addiction for diagnostic tests" [25]—yet few would advocate a return to the era of the medical black bag.

The physician today is very often in the position of the master plumber, mediating very "minor" interventions (when viewed in terms of the physical labor involved) with massive quantities of expensive-to-generate information. The cost for the labor involved in the physician's actual intervention is often minuscule compared to the cost of the information that guides this intervention, which includes not only the obvious diagnostic support but an amortized value for the physician's entire training. "Banging the pipe" is virtually a give-away, but "knowing where to bang" carries an increasingly steep price tag.

Of greater importance to the discussion here are the demands that such systems are already placing—and the much greater demands they will impose in the near future—on the telecommunications network. At least the master plumber was a self-contained diagnostic/surgical unit, not dependent on any form of telecommunicating (other than perhaps the phone call that brought him to the house in the first place). But the modern analog we have invoked—the practicing physician—is deeply entangled in the communications grid, and growing more dependent on it all the time. Let us elaborate with an example.

How Much Information Is Contained in a Heartbeat?

The human body, it might be said, produces its own diagnostic signal in digital form: the pulse. From the time of Hippocrates, medical practitioners have relied on information gained from taking the patient's pulse. Later the stethoscope amplified that information, by allowing the doctor to listen for more detailed characteristics. Fast or slow, weak or strong, regular or erratic, the heartbeat was an important indicator of many medical conditions.

In 1887, the electrical signature of the heartbeat (Figure 1.2) was first observed [26]. The name most associated with the development of the *electrocardiogram*

(a)

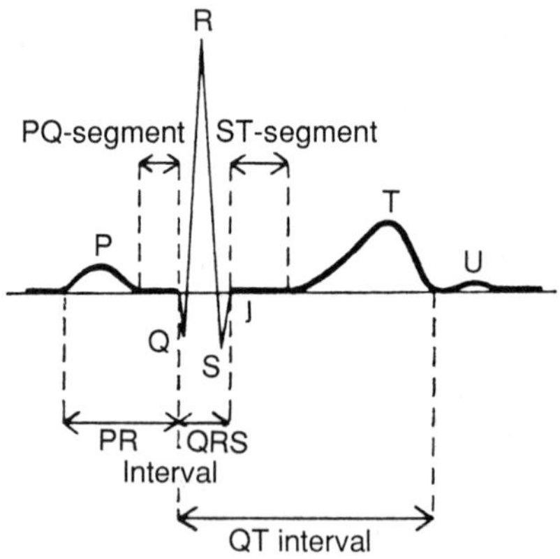

Figure 1.2 Electrocardiogram of a normal heart. (a) Standard nomenclature. (From J. Boutkan, *ABCs of the EKG,* Charles C. Thomas, Springfield IL, 1968.) (b) Electrical interpretation. (From Jules Constant, *Learning Electrocardiography,* Little Brown, 1981.) (c) Physiological interpretation. (From Michael Ritota, *Diagnostic Electrocardiography,* J.B. Lippincott, Philadelphia, 1969.)

(b)

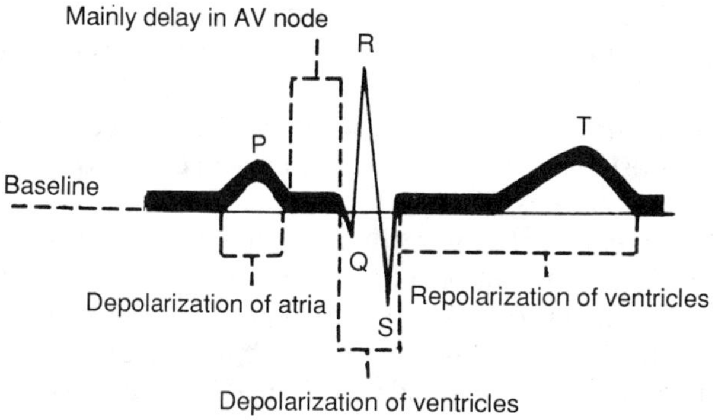

(c)

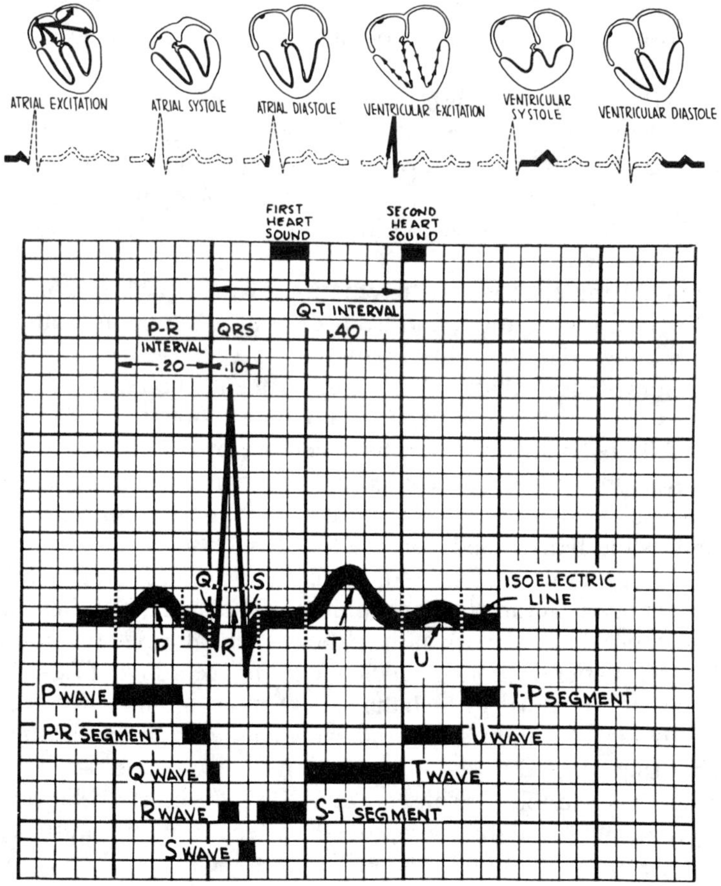

Figure 1.2 continued

(ECG) is Willem Einthoven of the Netherlands, who received the Nobel Prize for his work in 1924. Einthoven developed the first truly practical machinery for recording an ECG in about 1903, and, with others, over the next 20 years or so probed the relationship between the ECG tracings (at first thought to be simply a curiosity, of little diagnostic importance) and various pathological conditions of the heart, laying the foundation for modern cardiology.

It is interesting that one key to recognizing the value of the ECG was the ability to take an ECG from a large sample of hospitalized patients (since the cardiograms from normal people, taken over short periods of time, tend to reveal little). Yet Einthoven's machine, although effective, was a bulky laboratory device that could not be easily transported to or operated in a clinical hospital environment. The solution to this problem was expressed in the title of Einthoven's 1906 article, "Le télécardiogramme" [27]. He hooked up his machine to patients in the hospital more than a mile away over an ordinary telephone line. As one recent account has put it:

> Speculation [by others] that the electrocardiogram would not be of any clinical value might have remained uncontested longer if not for the practical idea of . . . linking the hospital to Einthoven's physiological laboratory with existing telephone lines . . . [which] . . . allowed electrocardiographic studies to be made in hospitalized patients, leading to Einthoven's many basic discoveries. [28]

It turned out that the electrical representation of the heartbeat contained a vast amount of relevant information, and from Einthoven's time until about 1960 the codification of knowledge about ECG patterns proceeded apace with the streamlining of the equipment such that, in the words of one textbook, "by 1950 good electrocardiograms were available to any doctor and any patient anywhere at low cost" [29]. The need for telephone hookups had been obviated.

However, a problem had always existed with conventional ECGs. The patient was normally hooked up for only a few minutes at a time, in a hospital or physician's office, usually while lying quietly on a table. Under such conditions, only gross pathologies can ordinarily be observed. Then, in 1961, N. J. Holter invented what has come to be known as the "Holter monitor." Essentially, a Holter monitor consists of a small tape recorder, worn at the patient's belt, attached to unobtrusive electrical leads on the patient's thorax, which allows the recording of ECG information for long periods of time during which the patient is not confined to a doctor's examination table, but is fully ambulatory and can go about his or her normal activities. Holter monitoring spurred a second revolution in electrocardiography, because it gave the cardiologist the ability to observe and record the heart's behavior all through a typical day (and night) and thus to capture the often rare or intermittent, but potentially life-threatening arrhythmias that elude the standard ECG setup, and to associate them with specific activities (sleeping, eating, exercising) that may play a causal role. Today, with many further advances in miniaturization, Holter monitoring is the preferred ECG technique in almost all situations other than intensive care, and the standard Holter "run" captures a full 24 hours of ECG information [30]. This has proved to be a dubious bounty, however; the Holter monitor records typically more than 100,000 heartbeats over a 24-hour period. Of these, only a very few are normally of clinical interest. Yet no human clinician can easily survey such a vast quantity of data "by eye." The Holter monitor collects a haystack along with the needle.

Therefore, from the very beginning, Holter monitoring has been associated with the exploration of various mathematical techniques for analyzing the data automatically. Of course, sifting the haystack to find the needle—that is, analyzing the huge amounts of cardiographic information—was clearly a task for a computer, and in those days it had to be a large one. Indeed, in that respect, the Holter approach represented a step backward toward Einthoven: The collected data could no longer be analyzed by the ordinary physician, but had to be sent to a specialized facility equipped with the requisite processing power. At first, this meant sending tapes by mail; but soon people had the idea of transmitting the information electrically over phone lines. This raised a new set of difficulties.

It soon became clear that although the original data were recorded as analog FM signals, the best way to transmit the information was to convert the data into digital form. The digitization of analog waveforms is something with which the communications industry is today quite familiar. The amount of data contained in a Holter recording, however, is staggering. Typically, frequencies up to at least 100 Hz are clinically relevant; this implies a minimum sampling rate of 200 samples per second. The American Heart Association has recommended standards that would require 500 samples per second [31]. Most quantization schemes would require between 8 and 12 bits per sample for the initial digitization of the waveform. This works out to between 138 and 518 Mb of data for a 24-hour run.

This presents a challenging communications problem. Even at the standard digital rate of an ISDN bearer channel—64 kb/s—the transmission of this information over a telephone line would require well over two hours, before allowances for error correction. (With the normal 2400-baud modem, which for practical purposes is still the limit today for many telephone lines, two and a half days would be required to transmit one Holter recording.)

Such unfavorable arithmetic has spurred an interest in data compression techniques. Since most heartbeats are quite similar—i.e., they contain very little new information compared to the previous beat, in information-theoretic terms—considerable compression should be possible. Anyone familiar with the approaches to low-bit-rate *voice* coding through the elimination of redundancies (see Chapter 6) can well imagine the variety of algorithmic strategies that have been applied to ECGs [32]. Over the past 25 years, significant progress has been made, with data compression ratios of 10 or more having been reported, which has brought the telephonic transmission of Holter ECGs to the threshold of feasibility.

Thus, it turns out that the simple diagnostic act of taking the pulse has—in the early dawn of the Information Age—ballooned into an information processing and communications task of huge proportions. With millions of ECGs being run now every year, and with the tremendous diagnostic powers of the Holter technique, it is easy to forecast that Holter ECGs alone will generate significant volumes of network traffic. This example suggests how the demands on the network are going to change as the Information Age unfolds. Yet ECGs are only the beginning; many other fields of medicine are candidates for the benefits of remote sensing and especially remote *in vivo* monitoring [33]. Many of the "intensive care" aspects of today's medical care could become decentralized (for economic reasons) by the judicious use of telecommunications. The modern master plumber (which may well be a computer in many cases) is vitally dependent on telecommunications to bring to him the vast quantities of information with which to make decisions.

The ECG example shows something else: A new kind of communicating node or message source has been created. And the traffic generated by this node is not only *not* human speech, it is most unlike human speech in terms of its:

1. *Information characteristics:* redundancy (information density), bandwidth, holding times, tolerance for errors
2. *Origination from a nonstationary node* (which may well become relevant as the Holter apparatus tends toward true on-line ambulatory monitoring capabilities, which would of course require a "tetherless" transmission means)
3. *Need for rapid reconfiguration* as new patients are fitted with the monitor.

Human-to-Machine, Machine-to-Machine

How are we to handle such traffic? For 100 years, most telecommunications traffic has consisted of real-time voice messages between human beings. Today's network is quite friendly to voice traffic—and distinctly unfriendly to other types of traffic. When we talk about telephone "service," we mean voice service. "Universal service" means that the voice network has nearly (if not quite) become all-encompassing. Most Americans are able to telephone—that is, to telecommunicate by voice with others—fairly much at will. At any moment in time, most of us are never more than a few seconds or minutes from a phone.

As we will examine in greater detail in Chapter 4, the conventional human-to-human voice communications business is a relatively "mature" market in the United States today. The amount of voice traffic per customer is increasing slightly, but it is not likely that Americans a generation or two from now will be spending, say, 10 times the number of minutes on the phone per day that we do. (Of course, many developing countries are still going through the growth curve in the evolution of their voice networks, and the telephone voice traffic per capita in, say, Mexico, may well be an order of magnitude greater 20 years from now.)

But the cutting edge of the Information Age lies beyond the domain of purely human interaction: a second dimension of Information Age communications traffic, with far greater implications for the network, is the growing importance of *man-machine communications.* Examples abound. More and more on-line database applications depend on the facilities of the shared network to channel information to users; the ordinary individual today interacts with such systems with increasing frequency, from point-of-sale validation of credit cards, to traces on Federal Express shipments, to obtaining airline schedule information from travel agents. These commonplace services are still relatively low bandwidth in nature; however, the next level of man-machine traffic now appearing in specialized work environments often involves the transport of images or video-oriented data of inherently higher bandwidth. Examples include computer-aided design and manufacturing (CAD/CAM), which often employs networking to share expensive equipment among many users, advanced avionics (including military), and transmission of data (including seismographic data) from oil exploration and production facilities. Even the ever-present FAX machine represents an important form of man-machine communication, or at least machine-mediated communication, which is generating annual traffic growth rates far higher than the growth rates for voice communications (see Chapter 4). From virtually zero traffic 30 years ago, man-machine communication is currently the fastest growing segment of the total traffic stream. Yet the network was not designed, and is largely unsuited, for much of this traffic. FAX is an example of how a machine-mediated communication service has had to adapt to the rather severe limits of the voice-friendly/data-unfriendly network.

Let us broaden the scope of our health-care example. Not only the Holter ECG, but many other diagnostic techniques of modern medicine generate huge quantities of information—so much that it threatens to utterly overwhelm today's communications facilities. The output from many diagnostic and analytical techniques consists of high-resolution images. This is true for conventional x-rays, as well as CAT scans, ultrasound, nuclear magnetic resonance, angiography, and, increasingly, video images produced from fiberoptic probes, microscopes (for pathology), and the like. In an increasing number of situations, the high cost of diagnostic instrumentation dictates that these resources be shared, which often requires that both input and output data be transmitted over some form of telecommunications network, whether a specialized local-area network or a long-distance (or wide-area) network based in whole or in part on facilities of the public telephone system. Figure 1.3 portrays the medical information network (largely driven by diagnostics) for one hospital (Georgetown University Hospital). This type of network is likely to expand to encompass other nearby facilities as well as offices for individual physicians and even, as a result of remote monitoring, the patients at home. Just as Einthoven coined his "télécardiologie," so we now have "teleradiology" [34] and "telepathology" [35] beginning to emerge as distinct fields of application. The challenge is not only that of sharing expensive hardware resources; it is also, and more basically, the

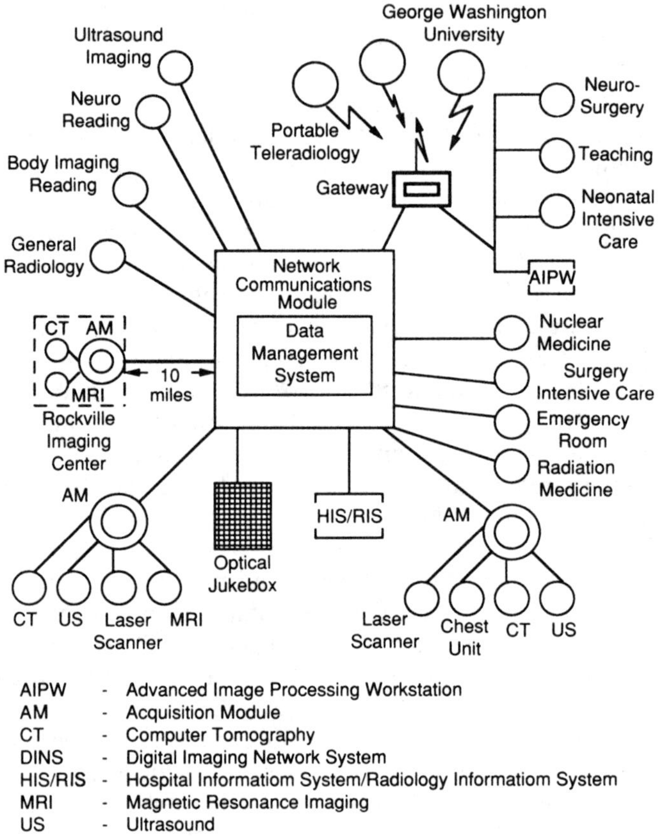

Figure 1.3 A modern broadbased medical information system. (From Jagdish Kohli, "Medical Imaging Applications of Emerging Broadband Networks," *IEEE Communications Magazine*, December 1989, p. 13.)

challenge of tapping into the vast information universe of medical knowledge, which is now and forever well beyond the compass of any single human physician. As one writer has put it:

> Expertise in pathology and its many sub-branches is at a premium. For many remote patients, the waiting time for diagnosis is often weeks or possibly months. A broadband pathology network allows the pathologist to be in two places at once, by communicating a high resolution video image of a patient specimen to a physician workstation located anywhere in the world. [36]

Where is this all heading? An article in the *Journal of the American Medical Association* has forecast that:

> The hospital of the future will be transformed into the critical care hub of a dispersed network of smaller clinical facilities, physicians' offices, and remote care sites that may stretch out as far as 200 miles from the core facility connected by air and ground critical care transport and *integrated* [my emphasis] by clinical information and patient monitoring systems. [37]

Yet a single detailed x-ray or CAT scan still-image may contain as much information as a 24-hour Holter ECG—up to several hundred megabits [38]. For "live" video applications, the bandwidth required may be hundreds of megabits *per second.* Clearly, while we may perhaps achieve sufficient compression of the Holter ECG to allow the use of today's voice-oriented telephone facilities, the transmission of diagnostic imaging outputs will swamp today's network.

There is more. Over the horizon, but approaching fast, is still another wave of communications traffic that will soon begin surging onto the overburdened network—*machine-to-machine communications,* not directly mediated by human translations. This is "pure data," which neither begins nor ends its life in a verbal or visual (i.e., human-adapted) form. Let us look briefly at three Information Age trends that are beginning to generate a different kind of telecommunications traffic consisting of messages prepared by machines, for machines:

1. The substitution of information for energy
2. The substitution of information for money
3. The substitution of information for physical goods.

Substituting Information for Energy

"Power" seemed so essential a feature of modernity to Henry Adams that he felt overwhelmed by the fact that the "coal-output of the world, speaking roughly, doubled every 10 years between 1840 and 1900" [39], and lapsed into awe at an exhibit of electric generators: "the dynamo became a symbol of infinity . . . [and I] . . . began to feel the forty-foot dynamos as a moral force, much as the early Christians felt the Cross" [40]. Certainly, the Industrial Revolution is first and foremost a revolution in powering the productive process. Indeed, the great boom in productive capacity of the last 150 years has been largely based on increasing the *energy intensity* of the production process—"mechanization" in the normal sense of the term. Machines—hardware-based innovations—amplified man's strength, speed, and endurance relative to the task at hand.

As we enter the postindustrial era, although we deploy vastly greater forces of energy than those of Adams's day, energy is becoming less central to the process of continuing to increase production efficiencies. In fact, since 1972 the energy intensity of the American economy (measured in terms of Btu's per dollar of economic output) has fallen by about 25% [41]. We are more concerned with the precise application of energy—which is increasingly achieved through *the substitution of information for physical force.* The Information Age is built on technologies that amplify man's decision-making powers, which even abstract those powers and embed them in "intelligent" tools, tools that adapt to the conditions of work, that optimize, that may even soon be allowed to "think." Many of the important innovations today are *software-based;* that is, they reside in the stored programs that control the hardware devices to which they are attached. For almost every type of machine product today, from automobiles to typewriters to VCRs to photocopiers (not to speak of the information engines, the computers, and communications systems themselves), the software component is increasing rapidly in importance, and in cost, with respect to the hardware. (Not long ago I had the opportunity to meet a group of engineers engaged in the development of intelligent washing machines; they were part of the same business unit working on a portable communications device, and were using some of the same technology.)

The real breakthrough has been the availability of commodity-like microprocessors, a development essentially of the last decade and a half. For a relatively small cost it is now possible when designing, say, a machine tool, to build in more computing power by adding a chip than would have been available from the average mainframe computer 25 years ago. The high-performance, real-time digital signal processors that will come with the next generation of digital cellular radios, for example, offer capabilities that were probably unavailable at any price two decades ago—and these marvelous miniature information engines are available in quantity for a few tens of dollars apiece [42].

As individual devices become "intelligent," the desire to link them together via communications networks will be driven by the advantages of coordinating at a higher system level. The "automated factory" is just one expression of this trend. Research in topics such as intelligent manufacturing systems and computer-integrated manufacturing proceeds in parallel with the elaboration of the necessary communications protocols, such as the manufacturing automation protocol, which are designed to support multimegabit data rates for short periods over coaxial cable or fiber optics [43]. Today, because of the centralized nature of most manufacturing processes, these communications developments are largely taking place within the framework of local-area networks. However, the obvious next step will occur as these local networks start to plug into the larger wide-area networks, including the PSTN. (We will look at this in greater detail in Chapter 5.)

Substituting Information for Money

Another fallen icon of pre-modern times is "hard" money. One hundred years ago, commerce was still bound to a tangible medium of exchange: gold. For most transactions to take place, a physical mass of gold metal had to be exchanged between the parties, or at least a certificate of ownership for the metal had to be tendered. Men had learned little of the art of creating money, an activity that was viewed in the same light as the alchemist's attempt to create gold itself. A few dire experiments, undertaken as extreme wartime measures (like Lincoln's "greenbacks"), had shown that it was possible to create a medium of exchange that was only loosely tied to gold metal, but these were viewed as grievous aberrations from sound policy. The commitment to gold as the physical reality of money anchored the economy and constrained every transaction. Throughout the nineteenth century, "delivering the payroll" (for example) meant physical labor (the gold was heavy) and armed security. And since money was inescapably linked to a tangible substance, debate was continual over which physical substance to use (gold, silver, or something else) and over whether or not the periodic shortages or surpluses of these substances (which had little to do with anything else in the economy) could hamper the proper management of growth.

Today we create money at will, in war or peace, as an instrument of public (or corporate) policy. The government runs on promises to pay that are no longer even fictionally linked to gold or any other tangible medium. Even private enterprises are involved in creating forms of money, through the issuance of stock, for example, the value of which may be only loosely related to the value of the tangible assets the corporation controls.

Whatever metaphysical shadow may cloud the "value" of our created money, the striking thing about the modern financial system is the almost total streamlining of the medium of exchange itself, the stripping away of all encumbering physical manifestations (metal, paper) that could impede rapid flows in the marketplace. "Electronic funds transfer" seems almost too prosaic a term for the distillation of money to its information-theoretic essentials, wherein we count neither coins nor certificates but electrical pulses or magnetic bubbles, with occasional printed statements to cater to our prehensile desire to hold "value" in our hands. Yet, the numbers in my bank account are no longer a pointer to something else (gold in a vault); those magnetic bubbles of information are my money. Even the title to these ephemeral possessions is also increasingly transferred electronically, rather than physically, when an economic transaction takes place. "Delivering the payroll" today is normally accomplished by "direct deposit" from one electronic account to another, without any human mediation or conversion into tangible form. And while most people still write checks, it seems more from emotional attachment to the myth of tangible money than from convenience or necessity. The trend surely points toward an increasingly checkless economy in the not too distant future (much as we operate in a largely cashless economy today). Any modern enterprise of any significant size—whether a corporation, a bank, or a government agency—is absolutely dependent on this new "informationalized" money—that is, money defined in terms of information rather than a physical substance. In our modern system, taxes could not be collected, bills issued, or payrolls met without the substitution of information for hard money. The costs of dealing in tangible money are simply too great.

The network of the Information Age is thus becoming the circulatory system of the economy, as well as its nervous system. This imposes a further set of demands on a communications grid that was designed for voice telephony between humans and is now being used to transfer units of economic value between computers.

Substituting Information for Physical Goods

Finally, it is becoming apparent that information can also be substituted for physical goods themselves. Perhaps the place where this is most obvious is in the management of inventories. During the early phases of the industrial era, maintaining a large inventory of finished goods (desired by sales and marketing personnel) and of raw materials (desired by manufacturing personnel) was viewed as a cost of doing business. It was the manager's way of coping with uncertainty—lack of precise information—about the needs of his customers and the availability of materials from his suppliers. The carrying cost of the capital required to maintain an inventory buffer was weighed against the cost of the lost order that would result if either finished goods or raw materials shortages developed. The rise of consumerism and mass merchandising in the modern sense placed if anything an even greater value on inventory as a competitive tool; large inventories guaranteed rapid response times to customers. In the consumer market, the cultivation of impulse buying patterns went hand in hand with the concept of department stores as vast consumer warehouses and plentiful goods located all through the distribution channel. The postwar marketplace was an era of bulging pipelines and rapidly gratified consumers.

The costs of these bulging pipelines were understood, at least intuitively, by the financial managers, who saw the cost of capital, the dangers of illiquidity, and the threat of write-downs for obsolete products. Clearly, reducing inventory without impairing the corporation's performance would be a positive accomplishment; yet inventory policy was so central to the corporation's performance that managers were loathe to tamper with it. As Peter Drucker has written, "the technical problems in inventory keeping—though quite considerable—pale into insignificance compared to the problems in other areas which any change in inventory-keeping will produce" [44]. However, some industrial engineers began to realize that the inventory management process, and particularly the reorder or restocking decision, was amenable to a formalized treatment. The rise of operations research (later dubbed "management science") in the 1950s feasted on the inventory management problem, producing ever more elegant and sophisticated mathematical models to allow managers to reduce inventories while maintaining response times to customers [45]. This first round of application of information to reduce the amount of goods carried in the pipeline produced important savings. However, these models shared a common *limiting* assumption: the firm was assumed to have the ability to estimate, *but not to control* the behavior of its customers and its suppliers.

Beginning in the 1970s, a second round of inventory-reduction techniques—this time largely pioneered by the Japanese—came to the fore. These new concepts, which go by many names, but are perhaps typified by the term "just-in-time" production, set a new standard for inventory reduction. The ideal was actually defined as zero inventory—or as close to zero inventory as possible. The aim was to squeeze simultaneously both physical inventories and schedule delays or buffers out of the production system. The reported results of these techniques are astonishing; a leading manufacturing consulting organization has estimated that:

> U.S. manufacturers can reduce inventory and production lead times by something like 90%, cut labor costs 10% to 30%, reduce set-up times by 75% and manufacturing and storage space by 50%, and improve the quality of manufacturing by 75% to 90%. [46]

Such low inventories have been achieved by in effect expanding the information network of the production facility to include both customers and suppliers. By linking customers and producers much more closely, uncertainty is reduced, which allows inventory to be reduced, and further changes the nature of the production planning process to allow a much more flexible, batch-oriented approach.

> Increasingly, large corporations are moving to bring their vendors and strategic partners onto their networks as Virtual Internal Departments (VIDs) to improve business efficiency. For example, auto manufacturers want to have their major component vendors "on-line" to accomplish a variety of functions such as placement of orders for just-in-time delivery, or downloading computer-aided design (CAD) files to a machine shop that can directly drive their computer-controlled machines. [47]

A recent article in *Scientific American* describes the application of these concepts to the textile industry (certainly one of the older offspring of the industrial age). I will quote at length because it illustrates how information can be applied intensively to actually replace physical inventories. Of course, the system is totally dependent on an Information Age communications system. The pioneering company in this case is the South Carolina-based Milliken & Co.

> The battle plan launched by Milliken . . . is called Quick Response. It is a way to knit all levels of the business—from fiber producers, to textile manufacturers, to apparel makers and finally to retail stores, in great part by exploiting electronic networks. . . .
>
> Typically, fabric and apparel producers have needed more than a year to fill retailers' orders for clothes. Retailers have had to forecast what their customers might be likely to buy and then order the garments accordingly. The result . . . has been unsold clothes at the end of the season that retailers must discount.
>
> Quick Response . . . opens a channel between the garment producers and the shelves in the store. In such a program, textile and garment makers agree to

> ship small batches of garments to retailers throughout the season. Retailers, who monitor which garments are selling, need only reorder those in demand. Fabric and apparel manufacturers need only make goods that are selling. Consequently, customers get what they really want [and] retailers avoid the nightmare of unsold stock. . . .
>
> Building such a system . . . relies heavily on an electronic network by which producers and retailers can swap data. Textile makers, accustomed to producing large batches of material, must retool plants to be able to quickly meet the demand for smaller quantities. . . . Before sending the cloth on to the apparel maker, the textile producer tags the fabric with standardized bar codes that describe the type of material, the color, and so on. The apparel maker, who scans these codes as the material arrives, then immediately puts the cloth on the proper production line. Finished garments, in turn, are labeled with specific bar codes—say, one identifying a woman's size-8 red shirt—so that the retailer can track precisely what goods are being sold. (These bar codes are scanned, in turn, by sales clerks when goods are sold.) The retail store then transmits data on sales back to the apparel maker, who electronically orders additional fabric from the textile mill. [48]

Consider what it will mean for the network as this example—reported as breaking news in 1990—becomes commonplace, and industry after industry develops its own network for tracing the distribution chain vertically from raw materials to final consumer. Every transaction, every sale, will generate a cascade of information events, all of which must be captured accurately and transmitted through a network that may encompass tens of thousands of nodes. As one study concludes, "a more meaningful measure of distance between corporate establishments, real or virtual, may be milliseconds of response time rather than statute miles . . ." [49]. The popularity of private networks seems to indicate that the telephone system may well have trouble keeping up with this surge in message traffic.

The Communication Needs of Intelligent Systems

In summary, the significance of the Information Age for the communications network is not so much that people will be communicating more with one another, but that we are constructing intelligent systems—factories, hospitals, banks, and dozens of other intelligent economic modules—that will consume and generate information on a vast scale. The next step will be for these modules to be linked, perhaps first along the lines drawn by their current distribution and financial channels, as in the textile example. Other communications grids will emerge around the sharing of expensive information resources, as in our health-care example. In short, intelligent systems, robots, databases, software-dominated machines, and point-of-sale terminals will discover *their* need to communicate, to pass information to other machines and human operators. And these resources will themselves be linked into subnetworks and supranetworks (overlays) that will allow the management of geographically dispersed systems.

The phenomenon is generic and can be approached from many perspectives. In the Milliken example, the distribution process is considered from the standpoint of making the right decisions about the kinds of products to produce. The same problem can be considered from the standpoint of reducing shipping costs and transit time. Consider, for example, a transport system for any common good, whether perishable (e.g., milk) or not (lumber), high value per pound (television sets) or low (cement). The physical transportation process today is increasingly managed as a total system, encompassing trucks, containers, trains, warehouses, inventories, production centers (factories), and end-users (customers). Ensuring the smooth and efficient flow of goods through such a system, with minimized inventories, delays, waste, theft, and expenditure of energy for transport, is recognized today as more a problem of information management than of brute hauling. Dispatching the trucks, routing the shipments, tracing individual orders, managing the inventories, and scheduling production are functions that are highly information intensive and dependent on the quality of the communications network(s). As we discuss in Chapter 5, the problems of managing transportation networks call for special solutions (e.g., mobile satellite terminals) that transcend the fixed-location, voice-oriented network of today.

Thus, I believe, the prodigious traffic streams of the Information Age will develop largely from man-machine and machine-machine communications, on top of the already enormous voice traffic base. The characteristics of that traffic will be quite different from the voice traffic now dominating the telecommunications network. Today's voice traffic is:

1. Highly *uniform* in character, overwhelmingly human voice communications
2. Relatively *indifferent to transmission quality* (largely because the human ear is so robust a receiver)
3. Sharply *bandwidth-limited*
4. Generated from predefined and not easily modified *fixed locations* (people can move to the phone more easily than the wire-bound phone can come to them)
5. Marked by stringent *real-time* requirements
6. Only *moderately concerned with reliability* ("acceptable" outages measured in minutes or hours).

By comparison, the Information Age traffic will be increasingly:

1. *Heterogeneous* in terms of the original form of the message, mixing modes (data, voice, video), bandwidths, and other characteristics in a common traffic stream
2. Subject to stringent *error control* to provide guaranteed quality of transmitted information

3. Rapidly *evolving in format* as the applications environment undergoes continuous technological change
4. Demanding of *flexible bandwidth*
5. *Semi-real-time* and *non-real-time* communications (including store and forward, and background process communications) will become much more important
6. Requiring *high reliability and redundancy* (acceptable outages measured in fractions of seconds).

Finally, above all—and this is one of the major themes of this book—*Information Age message traffic will increasingly be generated from "terminals" that are not permanently anchored to a given location.*

All in all, these trends indicate that our copper-based, industrial-age nervous system will not carry us far into the twenty-first century.

1.3 DIALTONE, A PARABLE

From an individual's standpoint, the network is rather less awesome than the foregoing discussion would suggest—indeed, it is normally out of sight, out of mind. The user is generally unaware of the processes of the core infrastructure. Its architecture matters little to him, and he is at best dimly conscious (from magazine articles and the like) of its sophisticated technologies. If all is in order, the network will be as nearly transparent as possible. It is, for most people, simply the mysterious home of dialtone.

Dialtone—mundane and unglamorous, yet reassuring and hope-giving—is the signal to the individual that he has *connected* to the network, that he has thus been afforded the opportunity to communicate. Once he is connected, the network will take care of the rest, rapidly, automatically, and transparently. What matters to most people, therefore, is how quickly, cheaply, and reliably they can obtain dialtone—in short, the quality of their access to the network. The distinction between *network* and *access* is crucial to an understanding of the technical and social challenges facing today's telecommunications systems. We will clarify these terms in Chapter 2. For the moment, we can think of access as the physical means by which an individual obtains a dialtone from the network.

Dialtone is very widely available in the United States. There is a phone in nearly every home, certainly every business, not to speak of the ubiquitous pay phones, and now millions of cellular phones proliferating—everywhere (it would seem) telephones are within easy reach. Another casual miracle is the fact that most of these telephones actually produce dialtone more or less on demand (which is not always true elsewhere). Thus it may be hard for Americans to visualize access as an problem. The ease with which we obtain our American dialtone renders us somewhat insensitive to the vast problems (and even more vast costs) involved in our commitment to creating and maintaining such an all-encompassing set of access facilities. It will come as something of a shock to many American readers to realize that building adequate access facilities is a far greater economic and technical burden than building the core network itself. In fact, the true value of the investment embedded in the access plant today is on the order of *ten times greater than the investment in the core network!* Constructing and supporting the access facilities of the twenty-first century for the United States alone will probably absorb something on the order of a trillion dollars. Yet for all this, even within the telephone industry a lack of awareness exists with regard to the scope and pervasiveness of the access problem. In Chapter 2 we will analyze in greater detail the nature and scope of the access problem. Here let us approach it through a sort of parable.

In 1982, I had an opportunity to visit the state of Alaska and to make something of a personal survey of a portion of the state's telecommunications infrastructure. In Alaska, telecommunications (like much else) is faced with unusual difficulties. The low population densities, the huge distances between settlements, and the severe environment present great obstacles to the implementation of ordinary telephone technologies. Long after universal service had become the norm in the "lower 48," Alaska was still trapped in the age of dog sleds and bush radios. In the late 1970s and early 1980s, however, a major program had been mounted to bring modern communications facilities to even the most remote villages, using a satellite backbone network, an innovative program that has received some attention [50].

By the time of my visit, in the major cities I found a telephone network and associated access facilities that were quite similar to those in any other part of the United States. In Anchorage or Fairbanks, I saw no evidence, as far as I could tell from casual use and observation, of any deficit in telecommunications opportunities. But we had come to see a different Alaska. Not far north of Fairbanks, the roads gave out. The only way to penetrate the Alaskan bush country beyond was by air. We chartered a small plane and flew north toward the Yukon River. Along the way our pilot was continually on the radio, chatting with various cabins and campsites as we passed over. Mostly the talk was about the weather. A few of our invisible communicants had specific messages to pass along to third parties, which the pilot seemed to record without the aid of pencil or paper. This crackly catch-as-catch-can radio link was their gateway to the outside world. After a while we ran out over the Yukon floodplain and the messages died away. My chief impression was of the vast empty flatness, uninhabited and visually uninteresting, punctuated at one point only by evidence of man—a rather large downed aircraft. This prompted a story from our pilot, who told how this craft had gone down in 1972 and it had taken many days to locate the occupants and bring them out, due to a heavy fog that lay over the ground and obstructed the aerial search. His final comment was "They went out with a bum radio." It was obvious that he considered this the height of folly.

An hour later we came down into Fort Yukon, a large Athabaskan village and administrative center for the region, located just above the Arctic Circle and comprising several hundred houses at that time. It was completely cut off from the world except by air, river, and satellite. From there we flew down river 100 miles to the somewhat smaller village of Beaver, Alaska. It was there that the problem of access first made a great impression on me.

Beaver was (and presumably still is) a fairly prosperous community. I was told that the annual household income at that time was in excess of $60,000, mostly from salmon fishing. The people lived in a tidy, compact enclave on the north bank of the Yukon. Many of the houses had racks for drying the fish, and a few had small gardens. They supplemented their diets and their income with local hunting and trapping. Beaver had a good gravel airstrip and a diesel generator that provided electricity for several hours each day. There were a few well-kept roads (all ended abruptly at the edge of town) and I saw an assortment of all-terrain vehicles. In short, the town seemed comfortable, albeit extremely remote.

We surveyed a number of points of interest to our mission, spoke to the local people, and then we decided to try to place a phone call. As indicated previously, the out-link was by satellite, which was in itself a marvel. The Alascom satellite network was put in place in the late 1970s and early 1980s by RCA, building on a project initiated by the U.S. Army. By the time I visited Beaver, the system had been turned over to Pacific Telecom, which was expanding the network. Beaver had a nine-meter satellite dish antenna—easily the largest and most impressive structure in town—which supported, as I remember, one voice telephone circuit and one television circuit.

To place our call, we first had to walk to a house out at the western edge of the village, where we were greeted by a somewhat doubting housewife who, after establishing our credentials, produced a key—the "telephone key" as she told us—which she entrusted to us only after repeated admonitions that we must return it promptly to her when we were finished, and in no event later than 30 minutes from that moment, since she was getting ready to leave on some errand down the river. She also reminded us to mark down the record of our call in the book she said we would find next to the phone. We thanked her, took the telephone key, and walked back to the community center, perhaps half a mile. Passing inside the community center, where the only occupants were an old man and a young boy watching a football game between the San Diego Chargers and the Miami Dolphins on a giant-screen TV, we entered a small, dark room at the back, where we came on an inner chamber that was locked. We unlocked it with the key, and went into the almost telephone-booth-like inner room, unfurnished and unlighted except for a small window, and there, on a plank, was a black dial telephone. I would say it was from the 1950s. It seemed quite worn from use. I remember thinking that it must have seen lengthy prior service somewhere quite different from Beaver, Alaska. Outside the window, the satellite dish loomed. It was tipped way over, pointing almost horizontally toward the south.

I reached curiously for the receiver, lifted it somewhat gently, and brought it to my ear. I heard the rushing sounds of the long-distance network plucking our request for service out of the ether, and after a couple seconds, there it was—dialtone. I slowly dialed a number back in Anchorage, and after a moment there was the sound of ringing, and then a voice answered. Except for the slightly bothersome satellite hop delay, the call was indeed clear and transparent. Next we called New York, then Philadelphia, then other cities in the lower 48, and every call was crisp and clear. The people we spoke to had difficulty believing that we were calling from an arctic village. "The call is too clear," one of them said. I think that at that moment, as jaded as we were by technology, we were all somewhat awed by the reach of the network. To place a call from the banks of the Yukon and talk over thousands of miles and though many time zones with the ease of placing a local call at home—that was amazing.

After many such calls, we suddenly remembered the woman's deadline. Thirty minutes had passed and then some; we had been so absorbed in the fresh miracle of telecommunicating that we had forgotten the time. She must be waiting for us impatiently. We still had a few more calls to make, but there was no way we could call her to let her know we might be late or to make other arrangements. So, as the youngest of that particular party, I was delegated to jog back down to her house and negotiate to keep the keys a while longer. I found her standing impatiently outside her cabin, fidgeting with the trappings on her ATV (which included a rifle). It was at least 20 minutes after she had said to be back. She was polite but clearly anxious to be on her way. I asked for a little more time, feeling a little uncomfortable even to ask. She said no, politely but firmly; she had to meet someone down the river, and she seemed rather agitated when she discovered I had not actually brought the key with me. To placate her I promised I would run back and bring it straight away. When I got back to the satellite station, my companions were still glued to the phone, and I had trouble convincing them to cut it short. In any case, 20 minutes later I was back at the woman's cabin. I put the key in her hand and tried to keep from wincing at the cramp in my side which all that running had produced.

As I walked back toward the community center for the third time that day, already a familiar of the village pathways, I could not help but reflect on the problem of access. The signals that the network could transport thousands of miles to and from this remote village could not be delivered the last half-mile to the woman's home. Indeed, no home in the village had a phone. I asked later and was told that the costs of installing and maintaining the wire-based outside plant in that environment would be simply prohibitive. (This statement came from a state official who, in effect, had just underwritten the quite substantial cost of the earth station.) The permafrost prevented burying the cable, and ice tore down most aerial installations nearly every winter. Telephone poles had to be shipped up the river by boat—they were too big and heavy for the airplanes that could land at Beaver's airstrip—and the river was frozen and impassable for eight months of the year. Setting the poles in ground that was either frozen or marshy was not easy. In some villages, I was told, people had even tried laying the access lines right on top of the ground, neither buried nor hung, but the snowmobiles chewed them up. The network could come right down into the heart of the village in the form of the advanced technology of the satellite earth station. But the antiquated art of copper wire and wood poles could not bridge the last mile. That single telephone set attached virtually to the base of the satellite dish—that was the sole means of access to the network for the people of Beaver.

I later realized that although this is an unusual situation in the United States, it probably typifies the access opportunities available to something like two-thirds of the population of the planet. A visit several years later to Argentina amplified this perspective. The decrepit state of telephony in Argentina is legendary. In downtown Buenos Aires I saw "pirate" telephone lines running between skyscrapers many stories above the busy streets. Not just a few lines, either, but thousands of them, from single bare wires to thick cables—to say that they "blackened the sky" would be an exaggeration, but it conveys the odd feeling of seeing so many wires hanging overhead in an otherwise modern-looking city. I climbed to rooftops of modern office buildings and watched as "linesmen" painstakingly traced working circuits through a mass of twisted copper wire dangling in the open air from an "illegal" connection board, searching for one or two lines that might still work. I sat in offices where receptionists steadily dialed the same telephone number again and again, sometimes for an hour or more, hoping to connect. I saw how people struggled to run businesses and pursue modern lives as best they could without adequate telephone access. I spoke to telecommunications authorities about the "official" 18-year waiting list for obtaining a telephone (and heard of course about the "short list," which, for a price, might bring a phone in only 5 years). I saw buildings advertised and sold on the basis of their precious telephone connections—sometimes two or three lines for a seven-story apartment building, with a huge sign on the front of the building proclaiming this prime selling point. I spoke to people living in a wealthy Buenos Aires suburb—which to outward appearances could have been a neighborhood in San Diego, California—who had been sufficiently desperate for dialtone that they had cut a deal with the state-run telephone company. All the neighbors chipped in money from their own pockets to buy a new telephone switch, a Siemens digital exchange as I recall, and the building to house it in their neighborhood. But apparently the phone company could not manage to install the wires to connect them up to their new switch, and two years later they still sat without telephones. Later I traveled to the pampas outside the capital, and saw dozens of huge ranches, with great mansions and herds of cattle valued at tens of millions of dollars, served by a single open-wire party-line circuit—iron wire strung on tree trunks, connecting to a dial-less telephone in a gas station (vintage 1913, manufactured by Western Electric). In another town I saw another new digital telephone switch sitting in crates; it had been there for more than a year and there was still no schedule for installing it; there was no hurry since there were no access lines to connect any of the customers.

Argentina is perhaps an extreme case—and yet almost any country in Latin America exhibits similar access deficits. I think again of driving through rural Mexico past mile after mile of bare and forlorn telephone poles where the wires had all been cut and removed by the peasants for the value of the copper, leaving only a few feet of unreachable wire dangling from the branches of each pole. Later still I realized how burdensome the access problem is even for many countries that are classified as "advanced"—countries like Spain and Italy, the countries of Eastern Europe, and even in some areas of western Europe, where to this day universal service is a still distant goal and the wait for obtaining telephone service can stretch to several years.

Finally, making the full circle of recognition, I came to realize that there are still *millions of households* in the United States that suffer from chronic poor or deficient access to the telephone network: open wire, party lines, noisy rural circuits incapable of supporting data modems or fax transmissions, lines in West Texas that expire whenever it rains. I realize that a large segment of American society, predominantly rural, does not succumb to the illusion of ubiquitous high-quality "easy" access. There is a division, even within our society, between the "information rich" and the "information poor," and the division is caused by, and defined in terms of, inequalities of access to the network. And as the Information Age pushes up the baseline, raising the standards for "acceptable service" far beyond simple voice telephony, these inequalities are bound to be exacerbated (as we shall examine in Chapter 4).

The paradox thus emerged for me of a network that can increasingly reach almost anywhere, to the most remote villages in Alaska or the Amazon, to connect any two points regardless of distance or terrain, even bring us live conversations and television from the moon—and of that same network saddled with the mysteriously intractable problem of bridging that "last mile" from the satellite dish or the microwave repeater or the digital switch to reach the cabin or home or factory or hospital where the end-user actually resides. As I learned about the ways in which the enormous costs of the last mile were displaced and shared within the web of "successful socialism," under private stewardship, that had been the old Bell System, and of the social and political compromises that had been engineered over many decades in order to support universal service—all driven by the enormous burden of the access problem—I saw that access is the key to understanding not only the problems at the fringe of the network, but also the diseconomies that underlie even my own ordinary, reassuring, hope-giving dialtone that I receive at my home in Philadelphia.

With the breakup of the Bell System and the introduction of competitive market policies in the long static world of monopoly telecommunications, the issue of access economics, access inequalities, and access technologies has moved directly to center stage. We examine in Chapter 3 how the issue of "access charges"—apportioning the costs of the access plant—has become a nexus of controversy for users and carriers alike. As competition has been permitted to flourish in the long-distance (interexchange) segment, the interexchange carriers have peered through the other end of the telescope, so to speak; for MCI and Sprint and others (and now even AT&T), the issue of access is an issue of how these carriers gain access to their customers over facilities they do not own or control. Meanwhile, the owners and controllers of those facilities—the local exchange carriers—are still regarded (by everyone from Judge Harold Greene to Bill McGowan) as a so-far-indissoluble lump of the old monopolistic worldview, a bottleneck to a wider deregulation of telecommunications.

In Chapter 4 we turn to the rising tide of new telecommunications services that is chewing away at the sand castles of telecommunications regulators. The same diseconomies that impeded the extension of voice-oriented access facilities to rural customers are poised to bedevil the Information Age. How long before ISDN reaches the smaller cities and the rural users? How long before FTTH ("fiber to the home") brings too-cheap-to-meter bandwidth to 50% of American users? To 80%? To 95%? (Every 1% is more than a million access lines—or a city the size of Philadelphia.) Even under the most optimistic scenarios for fiber deployment, the self-renewing problem of access will be with us well into the next century.

Thus, the problem that first struck my attention in a remote Athabaskan village on the Yukon River turns out to be the abiding heartache of today's network planners, managers, regulators, deregulators and—though we know it not, unless we live in Beaver, Alaska—of the users as well.

1.4 WIRELESS ACCESS: TO PCN AND BACK AGAIN

The Tyranny of Copper

Why is access such a difficult problem? The very short answer is that while the core network has evolved in wondrous ways, the facilities by which we gain our access to the network are antiquated, expensive, and technically primitive. Dialtone comes to my house (and undoubtedly to those of most readers) over a pair of copper wires, no different from the technology first used by Alexander Graham Bell more than a century ago. (It is interesting that although Bell is of course best known as the "father of the telephone"—meaning the telephone instrument itself; he had little to do with the network—perhaps his second most important invention in the field of telephony was the twisted-pair, the now-standard method of twisting the two strands of the metallic telephone circuit around one another to cancel inductive effects that would otherwise cause excessive crosstalk [51].)

The costs of bringing dialtone over copper wire are extraordinary. In my case, the "local loop," which connects my phone to the nearest switch, is just a little over a mile long; this is less than half the length of the average loop in the United States, and yet it still represents two miles of copper wire (for the pair that is needed) that is entirely dedicated to me and me alone. Moreover, the wire must be insulated, protected from moisture, and buried in conduits deep underneath the streets of Philadelphia; it must be protected from contacts with power lines and buffered against surges of lightning; it must be rat-proof, gopher-proof, and squirrel-proof; it must be maintained over decades in a hostile environment. If it should fail, I have the "reasonable expectation" that it will be repaired promptly, in a matter of hours or at most a few days. To support this service commitment, Bell of Pennsylvania maintains a huge squadron of armored infantry—trucks, cherry-pickers, backhoes, specialized machines to install new cable, specialized machines to remove and salvage old cable, pole-setters, trench-diggers, splicers, fans to exhaust the manholes, various forms of test equipment. One soon realizes that an enormous infrastructure is in place to support the wire loop. The "city under the city" is laced with hundreds of thousands of wire loops, connecting each individual home or business telephone to the local switch (the gateway to the network) with a dedicated wire-pair. Actually, each "wire-pair" is normally composed of several distinct physical segments, spliced together and connected through distribution frames, all administered separately. In fact, the telephone company plans and manages not loops *per se* but loop segments. The planning process is complex and difficult. (We review the loop planning methods in Chapter 8.) On most segments, the unit of analysis is the available wire-pair. On some segments, it may be necessary to administer "loop electronics" (part of the struggle to contain copper costs, as we shall see in Part IV). Failures take different forms on different sections of the loop. On some segments, the hazards are subterranean; seeping water primarily, but also the rodent kingdom and the wayward backhoes of street workers. To reconstitute the buried portions in an urban loop, the "outside plant engineer" delves through manholes and underground chambers. On other segments, such as the final or "drop" segment to the user's home, he emerges into the world of tree branches and squirrels. Some segments may be exposed to special hazards, such as the submarine cable often needed in low-lying areas and for islands.

An understanding of just how fragile and subtle are the conditions under which a copper filament can be made to carry a faithful electrical analog of the human voice, free of excessive noise or interference, impresses on us that the design of a "wire" or "cable" system capable of maintaining the necessary electrical environment over many miles and over many years is not a trivial problem. Historically, it took several decades after Alexander Graham Bell's invention of the telephone to perfect the design of a successful underground urban cable [52].

By the time all of this has been factored in, my little pair of wires has absorbed quantities of "overhead"—the labor to install it, the ducts and conduits to protect it, the maintenance it requires constantly, the allocations to reflect the trucks, the manpower, the heavy equipment, and the planning and management—which far exceed the value of the copper metal. The "true cost" of my dedicated loop facility is uncertain, composed as it is of various segments with different overheads, laid down at different times in the past (in some cases decades ago).

My own telephone situation is one of the most favorable, however; as an urban dweller, my loop is relatively short, the environment through which it runs is almost totally man-made, and the installation costs can be shared along almost the entire length by many other subscribers (through the use of multipair feeder-cables, as discussed in Part IV). In rural areas, it is not uncommon to find copper loops of 5 to 10 miles or more (the very longest loops in the United States can stretch in some areas to more than 100 kilofeet). The materials costs rise linearly with loop length. Moreover, the lower population densities mean that a much higher proportion of the total loop consists of totally dedicated installation (i.e., single-pair or at most two-pair cable). Thus, the installed cost rises at a rate faster than the increase in distance alone. Moreover, the terrain in many rural areas is much more hostile to the construction of buried cable facilities, and pole lines may be required, resulting in sharply increased annual maintenance costs. The per-unit-of-distance cost in rural areas is often higher, as well as the per-unit-of-time cost. The total life cycle cost of the loop to a country farmhouse located, say, 8 miles from the switch, may in actuality be 10 to 30 times higher than for my short urban loop (although, as we discuss in Chapter 4, traditional loop accounting methods tend to obscure the true costs of individual loops). Millions of such long loops exist in the United States today.

All in all, the investment required to fulfill the customer's expectation of instant dialtone is by far the largest component of the cost structure of his telecommunications service (much larger, as we shall see in the next chapter, than his share of the cost of the switch or the core network facilities). And yet it is the most grossly inefficient investment in the entire system. A typical household will use its loop for 15 or 20 minutes a day. The rest of the time it sits idle. The implications for the economics of access are severe (as we shall examine in Chapters 2 and 3). The distorted and inefficient economies of copper-based access are the very heart of the pricing conundrum that bedevils the telephone industry to this day. Indeed, the problematic contention that telephony constitutes a natural monopoly is based on the diseconomies of the access plant. The regulated-utility structure under which telephony has developed for decades, and which still holds sway in the local exchange environment (the regional Bell operating companies and other local exchange carriers), is rooted in the problems of copper-based access.

This dedicated facility is also, as we discuss in Chapter 2, enormously expensive to upgrade, and therefore subject to extremely slow technological evolution. It becomes the bottleneck in my communications link; it does not matter that the switch I connect to may be a prodigy of digital technology capable of supporting ISDN or some other marvelous new communications service—if my copper loop cannot support it, I cannot access it. Stated this way, it is obvious that the access facility is the anchor holding back the evolution of the entire telecommunications system. In fact, many of the advances in telecommunications that masquerade as network issues are really access issues. ISDN (to which we shall return in Chapter 4) is a good example; ISDN services have been mutually defined and constrained by the performance limitations of the embedded-loop plant. ISDN is really best seen as a strategy for alleviating the access bottleneck and saving copper. It is another chapter in the long struggle against the tyranny of the copper loop.

In fact, I believe that much of the history of research and development in telephony—exemplified in the United States by the rise to prominence of the Bell Laboratories, the largest private technology development program in history—can be seen as a series of variations on this theme: the war against copper, against the economics of copper plant, against the performance limitations of copper plant, against the inflexibility of copper plant. Whether it was the development of carrier techniques to multiplex many conversations on one physical wire (to save copper), the invention of digital transmission to extend the multiplexing principle into the time domain (to save copper), the fundamental research in electronics and materials science (to enable digital techniques to be implemented efficiently, again to save copper), the active pursuit of microwave radio (to avoid copper), or the development of lightwave transmission or fiber optics (to supersede copper), the underlying motivation was the same.

Yet for all the dollars invested and the scientific knowledge developed, the access plant in the United States today is still overwhelmingly based on twisted-pair copper wire. The fundamental feeder-cable designs are still based on those developed around the turn of the century [53]. In the past decade, an upsurge in the application of "loop electronics" (digital loop carrier) has occurred to derive additional efficiencies from the existing plant; however, these measures are still essentially band-aids for the basic problem.

It boils down to this: If we look at the three basic segments of communications plant—switching, long-distance transmission, and local access plant—we see that over the past 100 years enormous changes in technology have occurred in the first two segments (see Figure 1.4). One hundred years ago, switching was literally a form of manual labor, usually carried out by women who made each connection by hand, plugging in the lines on a switchboard. Today, a switch is a digital computer. One hundred years ago, "long-distance" lines were simply copper wires—open wire at that—strung on tens of thousands of telephone poles between the cities. (The first transcontinental line in the United States, opened in 1915, is said to have required 144,000 telephone poles to carry six pencil-thick wires to San Francisco [54].) Today, long-distance transmission is accomplished by microwave, satellite, or fiber optics—in most cases this means digital technology—and copper has been abandoned as a transmission facility in all but a small percentage of situations. Yet the technology of the local loop is fundamentally unchanged from the Victorian era.

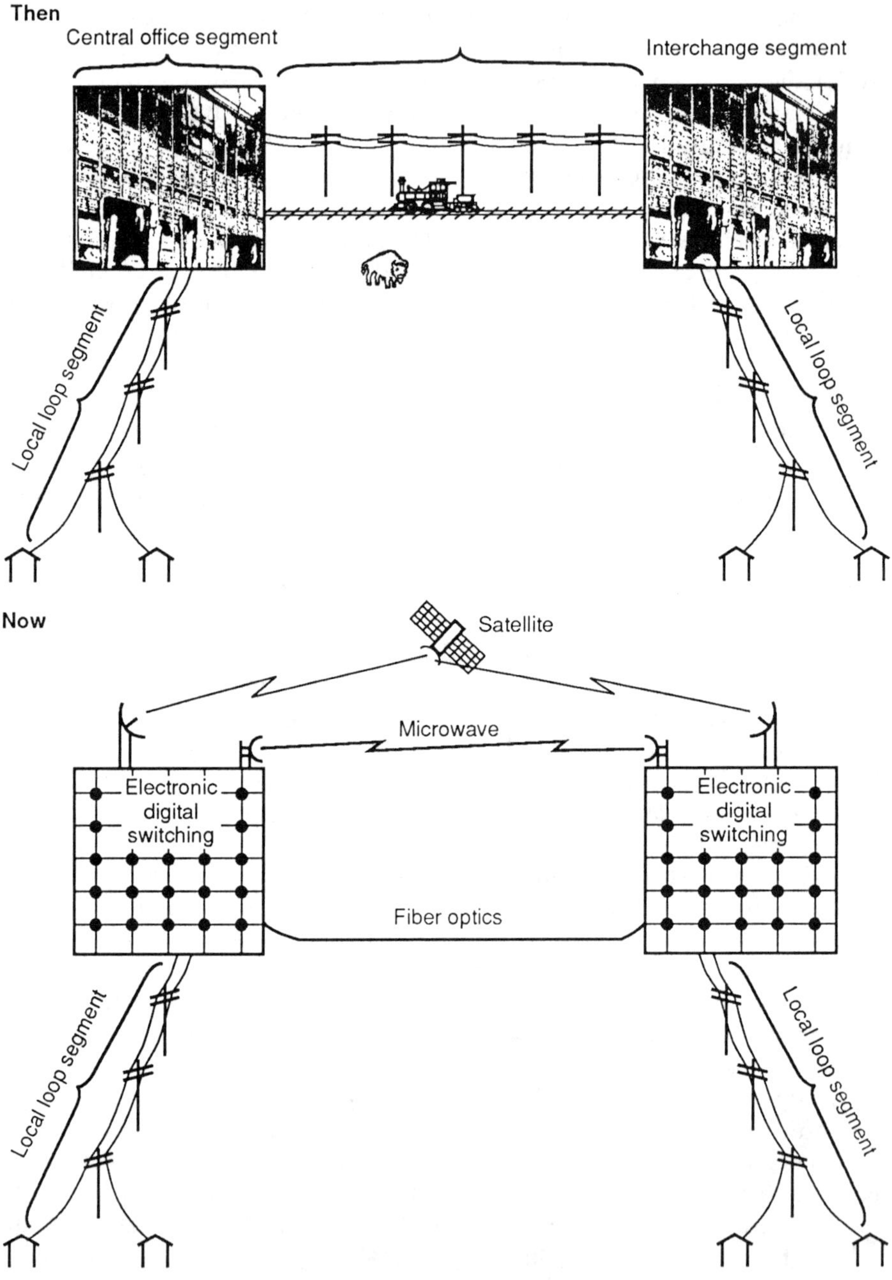

Figure 1.4 The three segments of the telephone network: then and now.

The Advent of Wireless Access

In the next 10 years or so, this situation will change at last. There is, I believe, a revolution in access technology already under way that will finally take us beyond copper wire in the access plant. Access in the next century will be based on two emergent technologies that are just beginning to penetrate the loop plant environment: fiber optics and digital radio.

In subsequent chapters, we will examine the prospective fiber/radio equilibrium and discuss the relative progress of both technologies. In general, however, we will observe that fiber optics, while bringing enormous bandwidth to support (potentially) a greatly expanded menu of basic services (up through and including video), still fits the same overall pattern of economics of deployment as copper. Both copper and fiber are physical media; both pose similar challenges as far as installation, access network design, maintenance, and so forth. This is not meant to minimize the differences between copper and fiber nor the importance of fiber as a new access technology; however, from the perspective of the third alternative—wireless access—the similarities are striking.

Digital radio, on the other hand, promises a far more radical change in the overall access equation. By avoiding the use of a physical medium altogether, radio breaks out of the linear, one-dimensional problem-space that copper and fiber alike must live within, and expands to a two-dimensional area coverage in which all the traditional problems and traditional solutions for designing and optimizing access plant networks are completely rewritten. The import of this brief statement will be made clear in Part IV; however, some idea of the transformation in thinking can be gathered from the observation that with radio-based access distance becomes essentially irrelevant as far as cost is concerned. In other words, *the costs of radio-based access for a subscriber 1 mile from the local switch and for a subscriber 20 miles from the switch will be practically the same.* Even before elaborating the analysis, it should be intuitively obvious that this will radically alter conventional access economics (and the basis of traditional access pricing).

In short, radio-based techniques are coming into the local exchange network. The initiatives are appearing in a number of different quarters, and include at least the following developments:

Cellular Radio. Cellular mobile radio is a boisterous and fast-growing communications market that until now has concentrated on upper-income and business users located in major metropolitan areas who need/want car phones. Hampered by an obsolescent radio technology (analog FM), cellular has nevertheless shown robust growth in its targeted segment and has generated surprising investor interest (and stock values) [55]. Because of the limitations of analog radio (in particular its gross spectrum inefficiency and lack of privacy) as well as regulatory uncertainties, cellular operators have shown scant interest in addressing the larger access questions I am raising here. But the cellular industry is in the process of transitioning to digital radio technology as the basis for a much more efficient and higher quality air interface. "Digital cellular" will constitute a technology that will be capable of supporting a broadbased radio-access system that could substitute for copper wire in many non-vehicular applications.

Basic Exchange Telecommunications Radio (Wireless Loop). In 1988, the Federal Communications Commission responded to a petition from the telephony community by authorizing a new class of radio service aimed at allowing telephone companies to use digital radio instead of copper wire in situations where it was more cost effective. Although "rural radio" had existed for some time prior to that, it had again been hampered by inadequate analog technology and high costs. Basic Exchange Telecommunications Radio Service (BETRS) represents an historic first step by the Commission toward recognizing radio as a fully equivalent alternative access technology for conventional telephone customers (i.e., not just mobile phones). As of the early 1990s, more than 50 BETRS systems in the United States served several thousand customers, which means that BETRS has achieved approximately the same level of development as FTTH access (see Chapter 9). BETRS, in fact, is based on time-division multiple-access radio techniques that are essentially similar to those being incorporated in the new digital air interface standard for digital cellular radio. Specifics of the BETRS technology and architecture are discussed in Chapter 5.

Thin-Route Microwave. Over the past 15 years another wireless access application has developed—largely outside the United States—based on so-called "thin-route microwave." These systems (discussed in Chapter 5) use a point-to-multipoint architecture to distribute telephone circuits in a time-division access architecture, and have found many applications, especially in remote areas and in developing countries, as an alternative to copper wire.

Satellite-Based Systems. The past decade has also witnessed the emergence of various communications systems that rely on geosynchronous satellites as the access facility to the network. Many of these technologies are finding their first application in what are essentially private networks, for example, very small aperture terminals (VSATs), which are gaining favor to link together corporate networks. Another fast-developing field is the use of satellites for mobile communications with interstate trucking. (These applications are described in greater detail in Chapter 5.)

Personal Communications Networks (PCNs). The PCN phenomenon is the latest star burst over the telecommunications world. As we will review in Chapter 5, the idea of employing wireless access on a large scale has been broached by the Department of Trade & Industry in the United Kingdom with their provocative concept of PCNs as the basis for competitive local exchange services. The idea has drifted across the Atlantic and is stimulating much discussion in the United States; at a recent conference, PCN was recently described as "the most exciting development in telecommunications since the break-up of AT&T"—which, judging by the standing-room-only audience, may not have been much of an overstatement [56]. The air is thick with bold new ideas, of which we may take as an example the recent proposal by Motorola to reallocate immediately several hundred megahertz of prime spectrum to PCN and similar services [57]. (Consider that the cellular industry fought for 15 years to win their initial allocation of 40 MHz away from the broadcasters [58].)

Yet PCN is still a largely undefined term; neither the application, nor the technology, nor the spectrum allocation, have been determined in the United States; nor do we have the political mechanisms to do so quickly or easily. Instead, it seems likely that a protracted public horse-trading is now beginning, which may carry well into the 1990s. Nevertheless, a psychological watershed seems to have been crossed; wireless access proponents are daring to dream the impossible dream—a parallel telephone system based on radio access—and suddenly the traditional political, regulatory, and spectrum constraints seem much more tractable.

The Many Portents of Wireless Access

The reason for the excitement is that some very startling conclusions can be readily drawn from the premise that radio is now poised to enter the local exchange segment of the telecommunications world and to become a mainstream access method. That is in large measure what this book is about. To whet the reader's appetite for what follows, let me briefly summarize what I believe will be some of the most important developments to follow from the spread of wireless access.

Redefinition of the Telephone Cost Equation. As I have already suggested, the basic cost structure of the telephone network is built around the economics (or dis-economics) of wireline-based access plant. Specifically, the utility-like cost structure (high fixed costs, lower variable costs), the use of price averaging, and the negative relationship between subscriber density and cost per subscriber, which places most rural areas at an inherent disadvantage, all derive from the nature of wireline plant.

Radio-based access will usher in a very different cost structure for many telecommunications services. Costs (on a per-line or per-subscriber basis) will become much more uniform across different service areas, varying much less as a function of distance, subscriber density, or other conventional factors. In particular, radio-based access will improve dramatically the economics of serving the "most expensive 5%" of today's subscribers, who place (as we shall see in Chapter 2) an inordinate burden on the entire system. Wireless access will greatly reduce the need for cost averaging and the regulatory superstructure that goes along with the administration of traditional tariffs.

Expansion of Universal Service. Again, as stated previously, the commitment to universal service is threatened by the transition to the true Information Age. Voice telephony is no longer an adequate definition of "service." Yet the many new service offerings now crowding into the communications marketplace are often impeded from being delivered to a wide range of customers by the performance limitations of the antiquated copper-based access plant. The danger (from a sociopolitical standpoint) is that twenty-first century enhanced access will be far less equitably distributed than voice service is today.

The advent of digital radio as an access technology will be an important weapon against this danger of information poverty. As we discuss in detail in Part IV, radio access is uniquely suited to allow early deployment of new services in free-market situations where demand for those services may not have reached the density to justify traditional (copper) or nontraditional (fiber) wireline access solutions.

Increased Flexibility in the Network. A customer today can obtain dialtone by means of wireless access in as little as half an hour (a typical time for installation of a BETRS subscriber station or a cellular car phone—obviously for truly portable applications the time-to-dialtone becomes virtually zero). No wireline solution (including fiber) can respond as rapidly, with as little overhead in terms of planning or preinstallation of common equipment. Moreover, if the customer relocates, or his service needs a change, the operator can reconfigure his wireless access plant very rapidly, avoiding the dreaded "stranded investment" so common with wireline plant. The Milliken example of the data network linking the textile industry vertically from fiber factory to consumer would obviously require a high degree of reconfigurability, especially at the retail end, as nodes are added, dropped, or moved. Of course, where true mobility or "tetherlessness" is required, only radio-based access can do the job.

I believe that much is encompassed in this innocuous word *flexibility,* which will transform the nature of the planning and administration of access plant. Not overnight certainly, but gradually and inexorably, the vast overhead associated with the care and feeding of the copper plant will begin to wither away. Trucks and cherry-pickers and backhoes will still be found in the arsenals of the telephone operators of the twenty-first century, but there will be far fewer of them.

The End of the Natural Monopoly. The introduction of radio technologies into the long-distance segment of the telecommunications market in the 1950s led directly to the breakup of the monopoly for long-distance services. Freed from the cost constraints of wireline plant in the long-distance application, operators found that they could effectively compete, and tolerate competition, without destroying either themselves or the market. It took the regulatory structures some time to catch up; however, competition was eventually embraced, and today we have AT&T arguing that it has come to the point where they should no longer be saddled with the assumption that they are the dominant carrier (and with the handicapping measures that were designed to allow their "weak" competitors to grow).

Wireless access will work the same transformation in the local exchange market. Make no mistake about it, we are witnessing the beginning of the end of the natural monopoly. In the twenty-first century access environment, there will be no reason why there cannot be as much competition in telephony as there is in, say, the airline business or the trucking business.

The Problem of Methods and Procedures

The advent of wireless access is greeted with a paradox of emotions by the telephone industry. On the one hand, the idea of wireless access can be simple and compelling: for someone who has spent a career in the wet, muddy, dry, dusty, hornet-infested, iced-up, lightning-tormented, middle-of-the-night world of the *outside* plant, pulling cable through dark places, splicing circuits in the rain, hauling and climbing and staring through the windshield for hours on end, the appeal of replacing all that with a small black box called a digital radio that goes into service in minutes is tangible and immediate.

On the other hand, as anyone who has ever tried to sell a piece of equipment to the telephone companies is painfully aware, the industry is hidebound with respect to its "methods and procedures" of *deploying* anything. Before the telephone company can commit itself to a new technology—even one that has been assessed and validated from a technical performance standpoint—techniques must be developed for economic analysis, site engineering, installation; drawings and documentation must be prepared for almost every conceivable (and not a few inconceivable) contingencies; training for phone company personnel must be programmed, policies developed for repairs, tariffing, etc. This is entirely understandable; the telephone industry is supported, at the local exchange level by, literally hundreds of thousands of technicians and managers throughout the country, most of whom are steeped in the culture of wireline plant and very experienced in installing and maintaining the old technology, following procedures that have evolved almost Darwinistically over many decades, but who are, as a rule, completely unfamiliar with radio as an access technology. I have found a degree of apprehension about wireless access, based on this lack of familiarity and perhaps exacerbated by the mysterious nature of radio transmission.

Thus, to bring radio-based access techniques to their full potential will require not only the development of the hardware itself, but the elaboration of methods and procedures for actually deploying it in the field. This is something that the euphoric advocates of PCN sometimes overlook. For example, the evaluation of a prospective radio link—the calculations for path loss, foliage attenuation, and reflections and diffractions from the intervening terrain—is a task that would be alien to most outside plant engineers today. The fact that such propagation analysis is simple and can be easily automated is of no immediate help; the methods and procedures for this analysis must be codified and reduced to an easily trainable and easily operated vehicle with which today's copper-oriented engineers can feel comfortable. As we discuss at length in Part IV, there is really nothing that difficult about engineering radio-based access—indeed in many respects it is much simpler than engineering wireline access facilities—but it is still *new,* and the industry has yet to learn it. The second half of this book is largely concerned with the elaboration of these methods and procedures.

1.5 THEORIES OF HISTORY

The patient reader will have perceived the author's weakness for tugging on the threads of history that bind us and our contemporary problems to the past and on which we rappel toward the future. In this introductory chapter, I have taken the liberty of trying to circumambulate the topic of wireless access, as a prelude to wrestling with it directly. We have wandered through the pre-modern past and into the Information Age future, and laterally from Alaska to Philadelphia. We have sketched the issues that will form the core of the remainder of the book. It remains only to reset the narrative on its track, which is done in Part II, where we examine the trends in network evolution in a more linear manner, focusing on three parallel themes that frame the issue of wireless access:

1. The emergence of access as the defining technical challenge for telephony in the 1990s (Chapter 2)
2. The unraveling of the long-standing compromise on telephone service pricing, under the impetus of "divestiture" and "competition" trends around the world (Chapter 3)
3. The rapid evolution in the menu of telecommunications services offered on a common carrier basis (Chapter 4).

For the reader who wishes to delve immediately into the topics of *wireless* access—including applications, technology, economics, and system engineering—Part II can be skipped or reviewed later. However, for those who choose to follow the discussion of historical trends in the next three chapters, I wish to comment briefly on the nature of the historical outlook adopted herein.

I believe that we are entering what I call the "Era of Access," the third era of the telephone industry (following the "Era of Interconnection" and the "Era of the Network"). This periodization of telecommunications history is, so far as I know, original. These historical epochs are not defined by regulatory trends (such as the tolerance or suppression of competition) nor by industry structure, nor by technology. Rather technology and regulation respond to the one central problem of each age. Each era is preoccupied with its defining challenge, which seems to dominate all other issues and to determine (for that era) the nature of telephony itself. The central problem calls out the research agenda and determines which technologies are developed most rapidly. It guides the visible hand of regulation.

The defining challenge of each age is perceived, particularly at the beginning of the period, as a task so awesome that it is hard for those at the time to imagine that it will ever be overcome. For example, if we place ourselves at the beginning of the first era, at the beginning of telephony, we would have confronted the staggering challenge of interconnecting millions and millions of people with copper wire. One can imagine the response of some time-transplanted latter-day venture capitalist to Mr. Bell's scheme:

"You mean that for this "telephone," to work, you will have to run a wire to every home in the country?"

"Actually, a pair of wires, that is, to do it right."

"A pair of wires? Pure copper wires? Do you know how expensive that would be?"

"And poles, of course. You need poles to hold up the wires. Or maybe in the cities, underground cables, if we can figure out how to make them work."

"And you need a separate pair of wires to every single user?"

"Yes, to every home, and every business."

"Not just from city to city, like the telegraph?"

"No, the telephone must be universal."

"Well, Mr. Bell, we'll have to think about this and get back to you . . . ."

Next let us imagine ourselves transported forward to the year 1940 or so. The brute force efforts of several generations has brought the goal of universal interconnection within reach. Yet now a new challenge begins to loom. The system is so large that it is becoming clear that its operations, particularly switching, must be automated. Indeed, an entire hierarchy of automated switching and signaling must be created, unless we are to draft half the work force to sit at switchboards. Yet looking out at the tens of thousands of manual and inflexible electromechanical switches in the system in 1940, the idea of replacing them all with programmable automatic devices, and upgrading them several times as technology moves forward, must have appeared as a daunting prospect. Some of us, the most farsighted, will have begun to realize that we are contemplating a qualitative leap forward—the creation of the integrated network. (As we make clear in Chapter 2, there was no such thing as an *integrated* network, in reality or even in conception, much before this time.) From the perspective of 1940, the program for developing and building this automated nationwide system would have appeared to stretch decades into the future.

A problem on such a scale shapes (one might say, *distorts*) the outlook of successive generations of telecommunications managers, engineers, and regulators. It outlasts careers and administrations. The central problem is so obvious and apparently so overriding that it merges into the background of inherited and unquestioned assumptions against which we see the development of all the characteristic policies and institutional arrangements of that period.

Overcoming this kind of challenge is an ambition that transcends the scale of any individual plan or effort. It calls for a program on a societal scale, rather like the construction of the great cathedrals of the Middle Ages. Our attitude toward such a challenge is naturally one of some humility; we can sometimes feel for ourselves the strange exhilaration of laying down another row of bricks on an edifice that we did not begin and which we shall not finish. Of course, another common reaction is a sense of unfulfillment and constraint. We scale back our sense of the possible and we strive to adapt to our particular defining challenge, to build up partial solutions, to ameliorate slowly its oppressive presence.

Yet the problem of the first era finally gave way to—or engendered—the superseding problem of the second era. And the second era is now, I believe, drawing to its close. Even as the transformation of the network appears to build to its climax, a new challenge has appeared: the challenge of obtaining access to the transformed network. We are facing the apparently awesome prospect of uprooting several hundred million circuit miles of copper and replacing it. The mind rebels at the idea, much as it must have rebelled at the thought of wiring every home in America, or automating every switch. After all, it may seem as though the problem is really still "in the switch." As Robert Lucky of Bell Labs recently put it: "I've never understood why communications cost so much. Transmission is the easy part. . . . The real challenge is building a switch that can accommodate the bandwidth" [59]. In other words, in his mind we are still in the Era of the Network (see Chapter 2). The reader can already anticipate that we shall take issue with the notion that "transmission is the easy part." Yet in the same interview, Lucky admitted that he saw the outlines of something else materializing in the fog: "Ubiquitous personal access is a theme that caught us all by surprise" [60]. So there you have it.

In fact, I contend that many of the classic problems of network-building have been overcome. All that remains, as they say, is the implementation. Standing on the brink of the third era, this may not be entirely clear. So many of the grand plans of the network builders have just begun to come to fruition. ISDN, broadband services, intelligent networks—all of these have only begun to make their mark. But the deployment of these services will be not so much a problem of upgrading network facilities (although tens of billions of dollars will be spent in this area), but of upgrading the access facilities. Indeed, as we will see in Chapter 4, the matter of ISDN (to pick a celebrated example) is really best seen as a problem of access. Without a revolution in access, there can be no revolution in telecommunications service.

Finally, let it be understood that this view of telecommunications history is not a claim to truth but to usefulness. I believe that the framework of thinking about access as our central problem will clarify much about our current situation and the alternatives we face. However, I can already see some embryonic developments that may render the access issue far less urgent in the next century. The problem will be "dis- solved," as a former teacher of mine used to say; the framework will at some point lose its relevance. Such is progress, at least by one definition—the loss of relevance of the old problems.

NOTES

1. Rupert Brooke, *Letters from America,* New York: Scribners, 1916, p. 68.
2. Cited in J.E. Kingsbury, *The Telephone and Telephone Exchanges,* London: Longmans, Green, and Co., 1915, pp. 208–209. (Reprinted by the Arno Press, New York, 1972.)
3. Department of Commerce, Bureau of the Census, *Telephones & Telegraphs, and Municipal Electric Fire-Alarm and Police Patrol Signaling Systems—1912,* Washington, D.C.: U.S. Government Printing Office, 1915; Department of Commerce, Bureau of the Census, *Census of Electrical Industries: Telephones—1922,* Washington, D.C.: U.S. Government Printing Office, 1924; Department of Commerce, Bureau of the Census, *Census of Electrical Industries: Telephones and Telegraphs—1937,* Washington, D.C.: U.S. Government Printing Office, 1939.
4. Robert J. Chapuis, *100 Years of Telephone Switching (1878–1978),* Amsterdam: North-Holland Publishing Co., 1982, pp. 138, 281.
5. *Ibid.*, p. 138.
6. Sidney Ratner, James H. Soltow, and Richard Sylla, *The Evolution of the American Economy,* New York: Basic Books, 1979, p. 124.
7. Robert H. Wiebe, *The Search for Order, 1877–1920,* New York: Hill & Wang, 1967, p. xiii.
8. Alexis de Tocqueville, *Democracy in America,* New York: Alfred Knopf, 1980 (orig. 1835), Vol. I, p. 397.
9. Anthony Trollope, *North America,* New York: St. Martin's Press, 1986 (orig. 1862), Vol. 1. p. 6.
10. The comment here is from a benighted ancestor, John C. Calhoun, cited in James A. Ward, *Railroads and the Character of America, 1820–1887,* Knoxville, Tenn.: University of Tennessee Press, 1986, p. 13.
11. Sidney Ratner, James H. Soltow, and Richard Sylla, *The Evolution of the American Economy,* New York: Basic Books, 1979, pp. 341–342.
12. Henry George, *Progress and Poverty,* New York: Walter J. Black, 1942 (orig. 1879), p. 3.
13. Henry Adams, *The Education of Henry Adams: An Autobiography,* Boston: Houghton Mifflin, 1974 (1918), p. 457.
14. *Ibid.*, p. 382.
15. *Ibid.*, p. 445.
16. Rupert Brooke, *Letters from America,* New York: Scribners, 1916, p. 45.
17. See the introduction to Alfred D. Chandler, *Strategy and Structure: Chapters in the History of American Industrial Enterprise,* Cambridge, Mass.: MIT Press, 1962.
18. See, for example, George Rogers Taylor, *The Transportation Revolution, 1815–1860,* New York: Harper & Row, 1951, p. 151, who despite the focus of his own study on methods of transportation, notes that "in an age of revolutionary developments in transportation and communication, perhaps the most drastic change resulted from the magnetic telegraph."
19. Fritz Machlup, *The Production and Distribution of Knowledge in the United States,* Princeton, N.J.: Princeton University Press, 1962.
20. Daniel Bell, *The Coming of Post-Industrial Society,* New York: Basic Books, 1973, p. 213.
21. My apologies to the fans of Mark Twain who was, I believe, the first to tell this story; I have embellished it from personal experience.
22. E. O. Attinger, "Impacts of the Technological Revolution on Health Care," *IEEE Transactions on Biomedical Engineering,* Vol. BME-31, No. 12, December 1984, pp. 736–743.

23. *Ibid.*, p. 741.
24. *Ibid.*, p. 739.
25. *Ibid.*, p. 740.
26. Dennis M. Krikler, "Historical Aspects of Electrocardiography," *Cardiology Clinics,* Vol. 5, No. 3, August 1987, pp. 349–355; James K. Cooper, "Electrocardiography 100 Years Ago," *New England Journal of Medicine,* Vol. 315, No. 7, August 14, 1986, pp. 461–464.
27. W. Einthoven, "Le Télécardiogramme," *Archives of Int. Physiology,* Vol. 4, 1906, pp. 132–164.
28. James K. Cooper, "Electrocardiography 100 Years Ago," *New England Journal of Medicine,* Vol. 315, No. 7, August 14, 1986, p. 463.
29. Thomas M. Blake, *The Practice of Electrocardiography,* 2nd ed., Medical Examination Publishing Co., 1986, p. 12.
30. Nitish V. Thakor, "From Holter Monitors to Automatic Defibrillators: Developments in Ambulatory Arrhythmia Monitoring," *IEEE Transactions on Biomedical Engineering,* Vol. BME-31, No. 12, December 1984, pp. 770–778.
31. Urs E. Ruttimann and Hubert V. Pipberger, "Compression of the ECG by Prediction or Interpolation and Entropy Encoding," *IEEE Transactions on Biomedical Engineering,* Vol. BME-26, No. 11, November 1979, pp. 613–623.
32. A sampling of useful references, in case anyone may actually want to follow up on this tangent, would include the following (in historical sequence): J.R. Cox, F.M Nolle, H.A. Fozzard, and G.C. Oliver, "AZTEC, a Preprocessing Program for Real-Time ECG Rhythm Analysis," *IEEE Transactions in Biomedical Engineering,* Vol. BME-15, April 1968, pp. 128–129; Urs E. Ruttimann and Hubert V. Pipberger, "Compression of the ECG by Prediction or Interpolation and Entropy Encoding," *IEEE Transactions in Biomedical Engineering,* Vol. BME-26, No. 11, November 1979, pp. 613–623; John P. Abenstein and Willis J. Tompkins, "A New Data-Reduction Algorithm for Real-Time ECG Analysis," *IEEE Transactions in Biomedical Engineering,* Vol. BME-29, No. 1, January 1982, pp. 43–48; B.R. Shankara Reddy and I.S.N. Murthy, "ECG Data Compression Using Fourier Descriptors," *IEEE Transactions in Biomedical Engineering,* Vol. BME-33, No. 4, April 1986, pp. 428–433; Borivoje Furht and Alex Perez, "An Adaptive Real-Time ECG Compression Algorithm with Variable Threshold," *IEEE Transactions in Biomedical Engineering,* Vol. 35, No. 6, June 1988, pp. 489–494.
33. For example, Brian D. McKean and David Gough, "A Telemetry-Instrumentation System for Chronically Implanted Glucose and Oxygen Sensors," *IEEE Transactions in Biomedical Engineering,* Vol. 35, No. 7, July 1988, pp. 526–532.
34. For example, J.N. Gitlin, "Teleradiology," *Radiological Clinics of North America,* Vol. 24, No. 1, March 1986.
35. For example, R.S. Weinstein, "Telepathology and the Networking of Pathology Diagnostic Services," *Archives of Pathology and Laboratory Medicine,* Vol. 111, July 1987; E. F. Brown, "Telepathology Allows Long-Distance Consultation," *American Medical News,* October 24–31, 1986.
36. Jagdish Kohli, "Medical Imaging Applications of Emerging Broadband Networks," *IEEE Communications Magazine,* December 1989, p 8.
37. Cited by Robert A. Schoellhorn, "Medical Technology: The Unseen Crisis," *Vital Speeches of the Day,* December 1, 1987, pp. 122–125.
38. Jagdish Kohli, "Medical Imaging Applications of Emerging Broadband Networks," *IEEE Communications Magazine,* December 1989.
39. Henry Adams, *The Education of Henry Adams: An Autobiography,* Boston: Houghton Mifflin, 1974 (orig. 1918), p. 490.
40. *Ibid.*, p. 380.
41. See "Growth without New Energy," *Science,* June 22, 1990, p. 1486; this article summarizes a report from the Office of Technology Assessment of the U.S. Congress, entitled "Energy Use and the U.S. Economy" (1990).

42. George Calhoun, *Digital Cellular Radio,* Boston: Artech House, 1988.
43. See Michael A. Kaminski, "Protocols for Communicating in the Factory," *IEEE Spectrum,* April 1986, pp. 56–62; and Theodore S. Rappaport, "Indoor Radio Communications for Factories of the Future," *IEEE Communications Magazine,* May 1989, pp. 15–24.
44. Peter F. Drucker, *Management: Tasks, Responsibilities, Practices,* New York: Harper & Row, 1973, p. 544.
45. One operations research textbook put it this way: "More O.R. has been directed towards inventory control than towards any other problem area in business and industry." C. West Churchman, Russell L. Ackoff, and E. Leonard Arnoff, *Introduction to Operations Research,* New York: Wiley, 1957, p. 195.
46. Figures from Arthur Andersen & Co., quoted in James Cook, "Kanban, American-style," *Forbes,* October 8, 1984, p. 66.
47. P.D. Lattner, R.L. Fike, and G.A. Nelson, "Business and Residential Services for the Evolving Subscriber Loop," *IEEE Communications Magazine,* March 1991, p. 110.
48. Elizabeth Corcoran, "Milliken & Co.: Managing the Quality of a Textile Revolution," *Scientific American,* April 1990, pp. 74–76.
49. P.D. Lattner, R.L. Fike, and G.A. Nelson, "Business and Residential Services for the Evolving Subscriber Loop," *IEEE Communications Magazine,* March 1991, p. 110.
50. Alex Hills and M. Granger Morgan, "Telecommunications in Alaskan Villages," *Science,* Vol. 211, January 16, 1981, pp. 241–248.
51. R.V. Bruce, *Bell: Alexander Graham Bell and the Conquest of Solitude,* Boston: Little, Brown and Company, 1973.
52. J.E. Kingsbury, *The Telephone and Telephone Exchanges: Their Invention and Development,* London: Longmans, Green, and Co., 1915, Chap. 22. (Reprinted by the Arno Press, New York, 1972.)
53. *Ibid.*
54. J. Brooks, *Telephone: The First Hundred Years,* New York: Harper & Row, 1976.
55. George Calhoun, *Digital Cellular Radio,* Boston: Artech, 1988, Chap. 1–4.
56. Neal Hightower, presentation on "PCN" at Supercomm, in Atlanta, April 17, 1990.
57. This has been a staple of Motorola presentations beginning about 1989.
58. George Calhoun, *Digital Cellular Radio,* Boston: Artech, 1988, Chap. 3 and 4.
59. "Broadband is Big," *Communications Week,* June 4, 1990, p. 34.
60. *Ibid.*

Part II
THE PROBLEM OF ACCESS

The telephone system in the United States has passed through two phases in its evolution and is now on the threshold of a third. The first phase—from 1876 until roughly 1950—focused on interconnection, wiring up the homes and businesses across the country. The second phase—from 1950 to 1990—dealt with the technological transformation of the core network, to expand, automate, and expedite the switching, call processing, and transport functions, and to begin the conversion to an intelligent network. The third phase—now beginning—will be the Era of Access.

The defining challenge for the industry today is the need for an overhaul of the access facilities, to replace the antiquated copper wire plant with fiber optics and digital radio, the twin pillars of enhanced access for the twenty-first century. This will present an enormous economic challenge to the industry, since the cost of access is approximately 10 times greater than the cost of the core network on a per customer basis.

In this section, we will formulate the problem of access. We will examine the historical development of this problem (Chapter 2), the economic and political origins of today's access bottleneck and the regulatory problems it creates (Chapter 3), and the impending explosion of new digital services, which threatens the social compact of universal service and heightens the urgency of the access problem for the industry and the entire society (Chapter 4).

Chapter 2

THE EVOLUTION OF THE TELEPHONE SYSTEM AND THE ERA OF ACCESS

2.1 THE ERA OF INTERCONNECTION, 1876–1950

From the time of Alexander Graham Bell until about the middle of the twentieth century, American telephony was defined by the physical challenge of wiring the continent.

The industry started out with primitive technology, without any master plan. The principle of the telephone instrument itself was so elementary that anyone could build one, and once the Bell patents expired (in the mid-1890s) hundreds of workshops stood ready to meet the demand. "Switches" were slow to evolve from their beginnings as manual pegboards. Indeed, the first large mechanical switches offering customer-dialing capabilities were not installed by the Bell System until the 1920s [1], and did not become prevalent until the 1940s. The dominant factor in shaping the "system" during this long period was neither the terminal technology nor the switchboard technology. The challenge lay in building the infrastructure to connect one to the other. This infrastructure was also based on extremely low-level technology—iron wires were the early rule (for strength), sometimes copper or iron clad with copper for better conductivity; ground circuits were prevalent (to save the cost of a second wire) until the rise of general electrification (especially the electric trolleys in the cities) rendered the earth too noisy to use as a return circuit; wooden poles supported the wire—preferably rot-resistant chestnut or cedar until these species were exterminated from the forests—or fence posts, or sometimes wires strung on living trees; in some urban areas where the scrub of poles and wires grew too dense, subterranean cable was laid down.

First and foremost, telephony was a construction business. Its arts were those of the setting of poles, the opening of cobblestone streets, the laying of conduits for proper drainage, the lacing of wires, the manufacture of soldering fluxes, glass insulator design, and rot-inhibiting wood treatments. The early issues of the *Bell System*

Technical Journal are peppered with articles about such dirty-fingernails science [2].

Over decades and generations, a nationwide hodgepodge of telephone systems was slowly pieced together by entrepreneurs, fast-buck artists, and populists. Exchanges often began as isolated community development projects, particularly in rural areas and in states outside of Bell's strongholds in the East. Franchises were awarded, repealed, competed, sold, consolidated, and every year hundreds or, in some years, thousands of new companies and cooperatives were formed. Even after the "Kingsbury Agreement" in 1913 (in which AT&T agreed to play by a particular set of rules in order to avoid the fate it would suffer 70 years later), the patchwork of American telephony remained remarkably heterogeneous and technologically diverse. Exchanges were patched up to neighboring systems in a haphazard fashion, and often the operators themselves were uncertain as to quite how they tied into the larger network. Different companies were often mutually hostile. Customers on one system often could not call other people who subscribed to a different system in the same town. In 1912, the Census Bureau collected the testimony of a large number of independent telephone company operators about the status of their integration with the larger system. This material helps to convey just how unintegrated the "network" really was in this period. In the following quotes, the "Bell Company" stands, in effect, for the national network, or the next closest thing:

> "So far as I have any knowledge of the company's affairs, there is no direct connection with the Bell Telephone Co. or with any company operating for profit [!]. The *** company operates in this county, and we have indirect connection with them through some of their country switches and through private families who have phones on both lines; but we have no connection between centrals.
>
> Our company does not directly connect with Bell lines, but can get these connections through the *** Telephone Co. at ***."
>
> "The *** Telephone Co., which operates in this county with an exchange and headquarters at ***, has connection with the Bell Co. The toll lines of the Bell Co. connect with the *** switchboard of the above-mentioned mutual company, and a number of the rural lines at this company connect with the Bell switchboard at ***."
>
> "Our company has no connection whatever with the Bell Co., either directly or indirectly, in this vicinity, and are in direct competition . . . . There are some points, however, at distances where we formerly had exclusive connection that have taken on Bell toll lines. . . . In a few cases where the Bell Co. have bought out independent exchanges our connection has not been disturbed, and subscribers from those exchanges can reach our subscribers over independent toll lines and similarly Bell subscribers over Bell toll lines. [3]"

Of course, some of the fuzziness over interconnection was deliberate, Kingsbury notwithstanding. Competing telephone systems existed in some large urban areas well into the 1920s and 1930s [4], and the granting or withholding of high-quality interconnection was still used as a competitive weapon. This did not make for a seamless technological fabric. Small independents (like those quoted above, persisting right up to the present day) often followed their own seat-of-the-pants approach to telephony. The larger independents followed stridently divergent technical programs, and often developed their own equipment. Western Electric did represent a source of conformity for many types of equipment within the Bell System; however, the methods of the Bell managers differed widely from one area to another [5]. The idea of standards had been discussed for many years, but in the field there were few indications of anything that could be called a uniform body of technical or operational telephone practice [6].

During this period, the industry (viewed as a whole) struggled against daunting logistical and economic obstacles in pursuit of rather limited objectives, the most important of which was, quite simply, to provide as many people as possible with the means "to make a call." Sometimes this meant what a modern reader would generally assume it to mean: a private, single-party access line connecting an individual home to the network. Most of the time it meant access to a shared circuit, or party line, often serving eight or more households and perhaps 50 or more individuals. Indeed, for large parts of the country, "to make a call" meant digging out a nickel for a pay phone at the corner gas station or drugstore [7]. As late as 1950, two-thirds of all the residential lines in the United States were still party-line circuits [8].

The only "network service" provided during this epoch was call-switching for voice circuits—in other words, the interconnection of two parties for a voice telephone call. For most people during much of this period, "telephone service" meant strictly local service. Long-distance lines (where they were available) required special conditioning and often could not be interconnected to ordinary local exchange lines [9]. Indeed, few people needed or wanted to make long-distance calls on any regular basis. The Census takers in 1912 found that only 2% or so of the telephone traffic went over toll lines [10] and even this figure excluded the so-called "interior" or "private-branch calls," which probably made up an absolute majority of telephone communications during that period. During the first era, for the vast majority, telephony was construed very much in terms of what we would today call a *lifeline service.* It was not the convenience that mattered as much as the simple opportunity to make a call.

To reiterate, the network was not even fully interconnected for most of this period. The Bell System—the very idea of which represented an important concept in competitive marketing—did provide a modest degree of integration among urban telephone systems in the East. And, of course, Bell monopolized long-distance service [11]. But for nearly 50 years, telephony in the far West developed in physical isolation. In the South, vast segments of the population were bypassed by the system builders altogether [12]. The West and the South were after all politically and economically marginal. Even in the Midwest, however, as late as World War I most of the telephones were operated by non-Bell companies. Independent subscribers outnumbered Bell customers in Ohio, Indiana, Wisconsin, Minnesota, Iowa, Missouri, the Dakotas, Kansas, Kentucky, North Carolina, Florida, Arkansas, Nevada, and Texas (and served more than one-third of the total subscribers in 13 other states, including key industrial states such as Illinois, Michigan, and Pennsylvania) [13]. Although AT&T's formal competitive policy of refusing to interconnect with these "foreign" operators was ended by the Kingsbury Agreement in 1913, it can be assumed that until the 1920s most of these non-Bell customers enjoyed something considerably less than a "comparably efficient interconnection" to the Bell-controlled long-distance facilities. Yet there were literally thousands of independent telephone companies, and tens of thousands of "private" networks, many operated by rural farmers' cooperatives, and possessing no long-distance calling capabilities whatsoever. For decades many telephone systems were essentially isolated islands serving only their own communities (Figure 2.1). There were few common standards for equipment. There was (until 1925) no Bell Labs. There was no network.

It was only toward the end of this period—which lasted 75 years—that the many political, technical, and economic obstacles to interconnection were substantially overcome, and it began to be possible to speak realistically about a single network serving the entire country. The change was more than simply a quantitative accumulation of telephone lines; it reflected qualitative technical and administrative changes in the nature of telephone service. By the end of World War II, millions of households were still without phones. But urban America was fully interconnected and the road toward true universal service—even for rural areas—was at last clear [14]. Those with telephones could now reach out from their local exchange islands and interconnect to other customers and systems across the country on a fairly routine basis. True, call connection for long-distance service lagged behind local service in automation, speed, and convenience. Certainly it was expensive. Even in 1950, *direct distance dialing* (as it came to be called) was essentially unknown; long-distance calls had to be arranged by human operators who would notify the caller when the circuit was set up. But it was now possible for someone to call a distant city from his own home—something that had not been possible for most customers a generation earlier.

As the system became fully interconnected, a new problem emerged. The structures and technologies that had worked well enough on a small scale were increasingly inadequate for a nationwide system serving tens of millions of people. It became necessary to evolve hierarchical structures, adding what eventually became five to seven layers above the local exchange level [15]. Automation became a necessity for growth. The qualitatively new capabilities of an automated, unified network transformed the telecommunications agenda quite suddenly after World War II. The

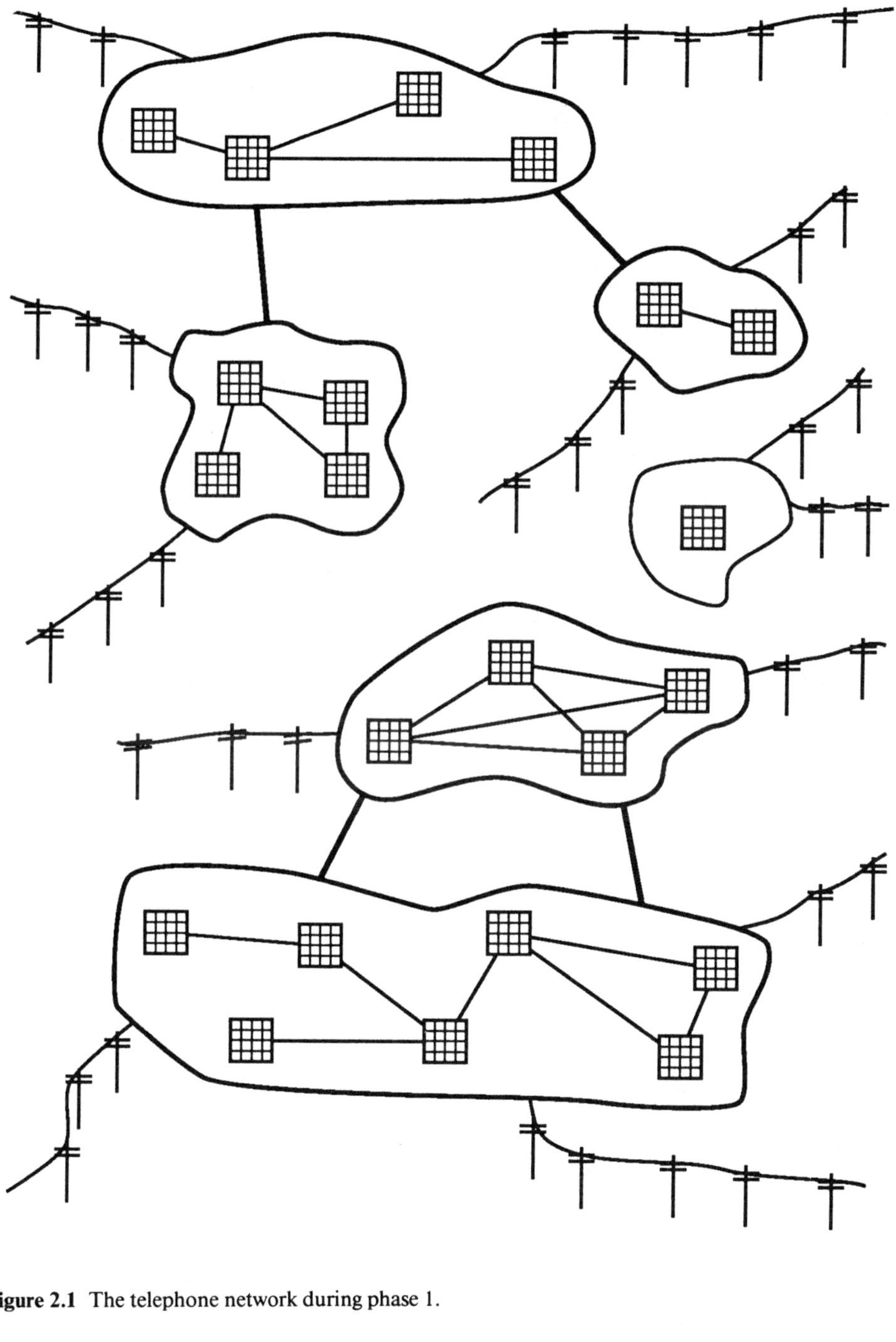

Figure 2.1 The telephone network during phase 1.

telephone network—a phrase that hardly held any profound significance (beyond the notion of interconnection) for the first 75 years of telephony—became *the Network*. In this subtle shift in terminology, we see reflected an astonishing technological metamorphosis spanning the years from 1950 to 1990.

2.2 THE ERA OF THE NETWORK, 1950–1990

The emergence of the unified network signaled the beginning of the second era of American telephony. This phase, which has lasted from 1950 until approximately the present day, has coincided with the completion of the universal service commitment through the extension of telephone service to over 90% of the population. But this belated achievement of universality may be regarded as a simple continuation of the trends in subscriber growth established during the first era. The defining characteristic of this second era has been the revolutionary transformation of the network core infrastructure by advanced technology.

To appreciate this transformation, consider the telephony landscape in 1950. There were approximately 30 million telephone access lines—and perhaps twice that many customers, accounting for party lines—served by something on the order of 120 million miles of local loop facilities [16]. There were also about 27 million circuit-miles of toll line (i.e., long-distance) facilities [17]. In other words, the local exchange was almost five times larger than the long-distance network.

The network in 1950 was 100% analog (the word *digital* was itself a neologism that as yet had no meaning in telephony). The voice network was entirely based on copper wire plant [18]. The switching infrastructure was based on electromechanical machinery; indeed, a significant portion of the subscriber base was still being interconnected through manually operated switchboards [19]. The transistor was unknown to the working engineers in the industry [20]. Long-distance connections were still made over largely open wire pole lines. Microwave radio did not yet exist as a practical tool [21]. Despite some pregnant developments in the labs at Bell and elsewhere, which in hindsight loom large in historical significance, in 1950 the fundamental technology on which the telephone system was built—the basics of transmission and switching—had changed very little in half a century.

Forty years later, the network has been utterly transformed. At first glance, we are impressed by the quantitative aspect of the transformation. The sheer size of the toll network has increased by a factor of approximately 100—to something on the order of 2.5 billion circuit miles [22]. Another way of looking at this is to realize that 99% of today's core network has been built in the past 40 years. But these statistics perhaps mask a more important aspect of the transformation. Most of the growth occurred not in the subscriber plant (although the number of subscribers did increase by a factor of 2 to 3 [23]) but in both the short-haul and long-haul interexchange networks. The mileage in the local exchange plant increased to approximately 300 million circuit-miles [24]—an increase of only about two and a half times in 40 years. The ratio of subscriber plant to network plant has been drastically shifted. In 1950, the exchange plant was about five times larger than the toll network. By 1990 the relationship was reversed: The interexchange circuit-mileage is six to eight times greater than the local exchange (see Figure 2.2). In less abstract terms, this meant more long-distance circuits per capita served, and a denser network of interexchange trunks in the large urban areas. In short, the network was becoming denser, more interconnected, and capable of bearing much higher traffic per capita. This was reflected in the more rapid growth of interexchange traffic and revenues (Figure 2.3).

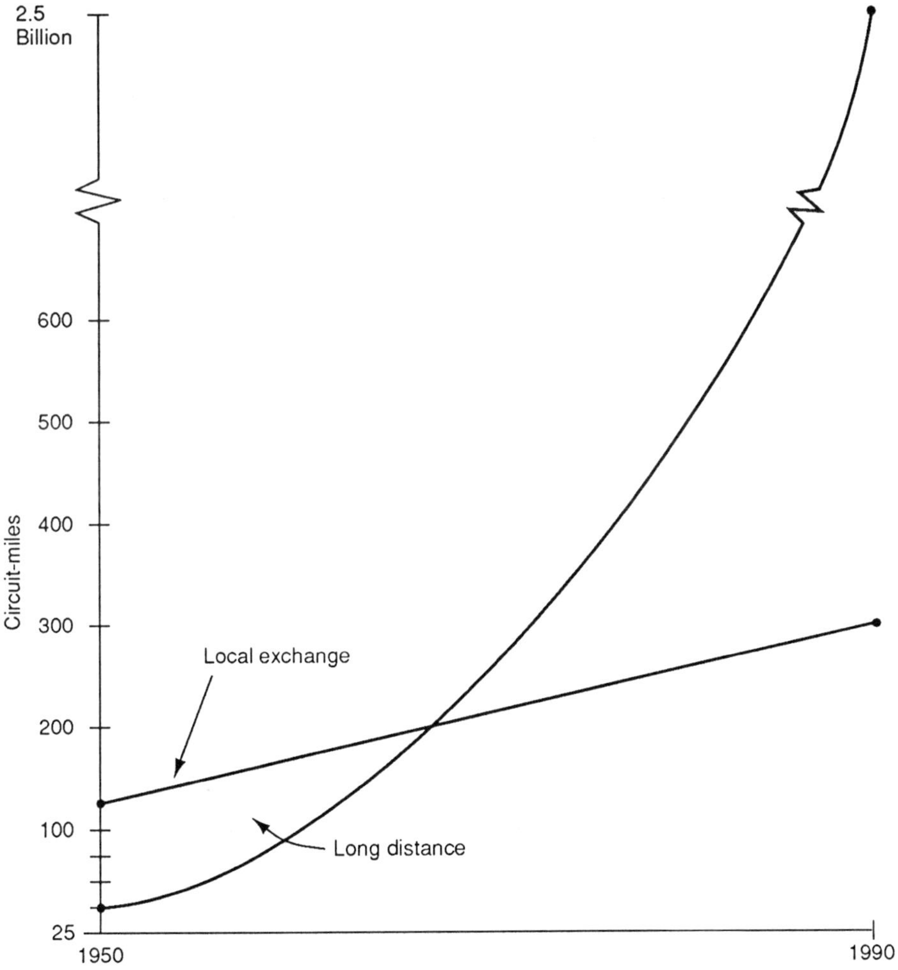

Figure 2.2 Relative size of local exchange *versus* toll networks. (Adapted from FCC data).

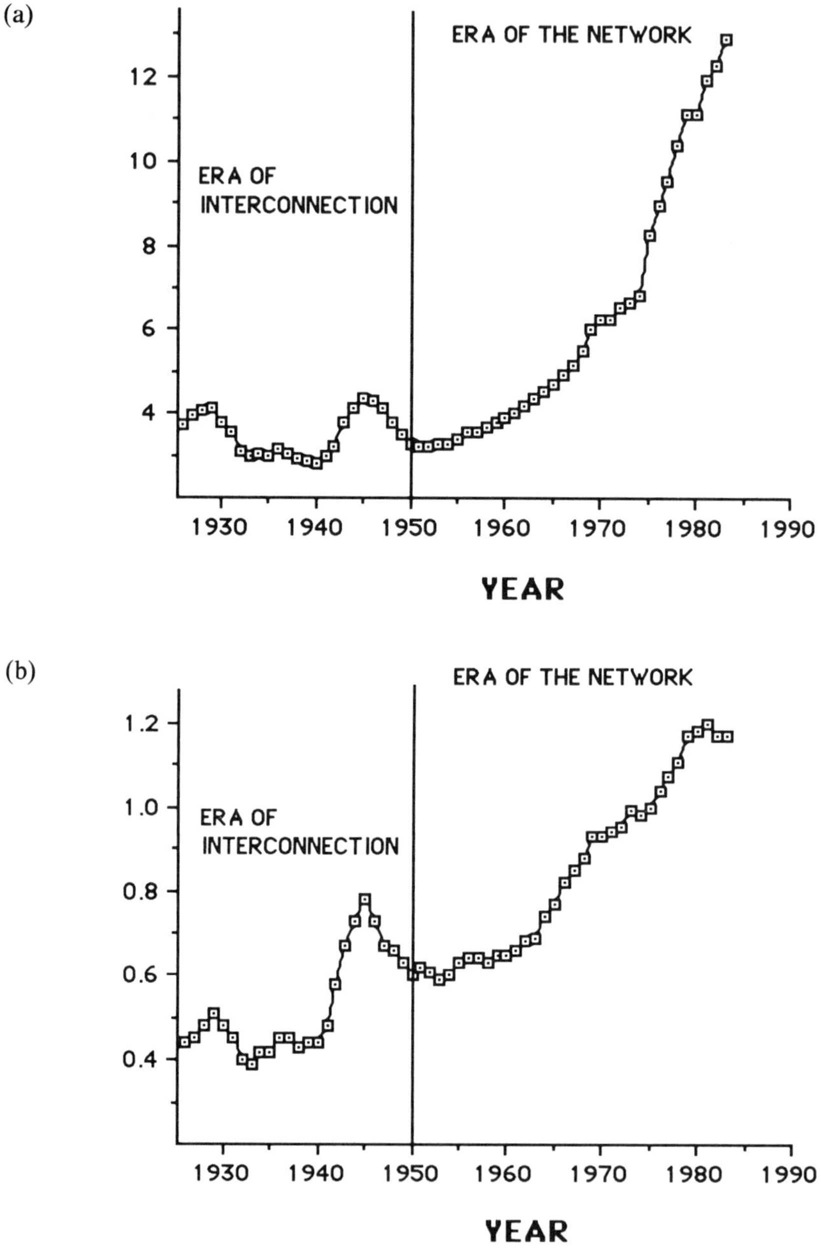

Figure 2.3 Relative growth in local exchange *versus* toll traffic; (a) toll calls as a percentage of local calls, (b) ratio of toll revenues to local service revenues. (Adapted from FCC data.)

Over the same period, the real (inflation-adjusted) costs of long-distance service have fallen by a probable factor of 100, enabling long-distance to become the ubiquitous service that it is today [25]. The technological driver for this economic aspect of the transformation has been microwave radio (see Chapter 3), which still accounts for more circuit-miles than all other technologies combined—although fiberoptics transmission (another revolutionary interexchange technology) is perhaps poised to overtake radio as the dominant long-distance vehicle. Satellites have revolutionized the international communications market.

The network has become highly automated. Call-routing is performed by computers. Operator-assisted call setup has been replaced by direct customer dialing even for most overseas calls. Many routine information services are partially or wholly performed by machines. Credit card calls have been automated. Many diagnostic and even maintenance functions are now accomplished without human intervention. Trouble reports are aggregated to regional network management centers, where decisions are made that often need not involve local personnel. Even operations such as circuit provisioning and telephone number changes are increasingly performed from terminals located at distant, centralized facilities rather than locally. The work environment of the network world has been transformed; where 40 years ago the typical telephone facility was a large building filled with huge volumes of clattering machinery and the din of busy people, most of them talking to customers or each other, today's central office is pervaded by the quiet presence of electronics and perhaps a technician or two monitoring a screen of diagnostic messages.

The network is now well along the road to a complete conversion to digital technology. Digital switches now serve most of the lines in the United States, and will be essentially universal by the end of the 1990s. Digital T-carrier predominates in the interexchange trunking segment; soon nearly all of the interoffice short-haul circuits will be digital [26]. Although a large residual base of analog hardware exists within the network, the standards for new equipment that have been developed by Bellcore and other standards groups within the major carriers now specify digital technologies.

Network services encompass much more than voice telephony. Data communications have grown at a tremendous rate and are expected to equal the voice traffic within several years. Facsimile is another nonvoice service that has exploded on the telecommunications scene in a very short period of time. And with the major telephone carriers agitating in Congress and before the Federal Communications Commission to be permitted to enter the television business—presumably over optical fiber—video services loom just over the visible horizon.

Digital networks are becoming intelligent. Network elements are increasingly linked in signaling and control architectures that allow operators to manage the network much more efficiently. Software has become the dominant resource in these advanced systems. These architectures begin to anticipate the more thorough integration of communications with computing. The modern network can already support many new services that will transform telephony from a voice-oriented communications system into something much, much broader—a system for processing digital signals. The underlying content of these signals—for example, voice, data, video—will be less important than the processing parameters, such as bandwidth and tolerances for delay or bit errors. System architects are beginning to take advantage of the pliability of digital information in order to envision a portfolio of basic operations, such as synchronization, regenerative transmission (at several standard

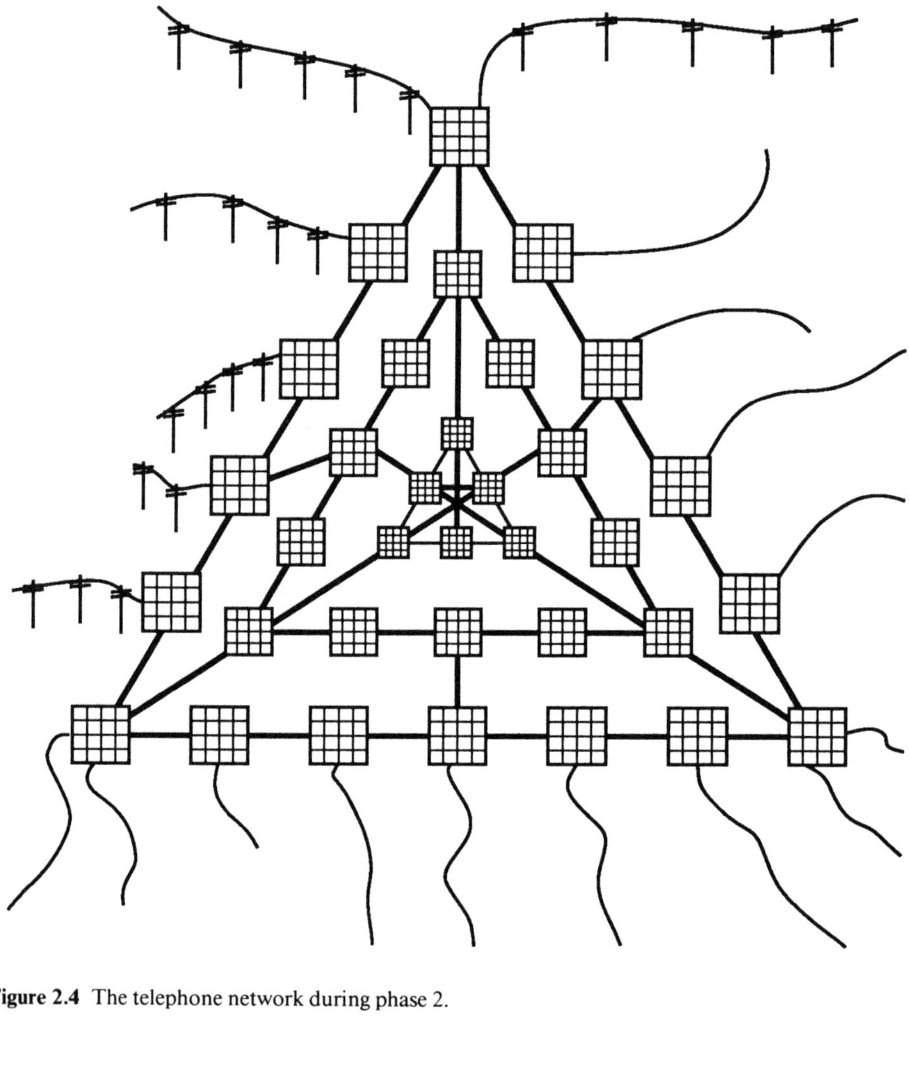

Figure 2.4 The telephone network during phase 2.

rates), equalization, error correction, time-division switching, packet switching, formatting, protocol conversion, compression, storage, encryption, and so forth, built into multilayered digital formats that, in principle, should maximize interoperability. These basic operations will be applied to create a wide variety of new services for voice communications, data communications, private networks, video—all capable of being supported by a common network infrastructure. Although many of these service concepts are still in their infancy, and many standards remain to be sufficiently defined, the trend is already clear and all major carriers and equipment manufacturers are aligning their plans in this general direction.

Thus, toward the end of this second era, an intoxicating vision emerges of a seamless network, transporting trillions of uniform molecules of information every second, completely transparent to all services, voice, data, and video (see Figure 2.4).

Yet, an obstacle remains to all of these plans. The network infrastructure—telephone exchanges and interexchange trunking facilities—has been and is being transformed by technology, creating the possibility for a host of new services; however, the ability to deliver those services to the users is hampered by an obsolete system of telephone access lines still based on copper wire-pairs—the same technology used by Alexander Graham Bell more than a century ago. This copper-based access technology is extremely costly (as we examine in the Part III) and it was not designed to handle the emerging services. The local loop plant has become, in effect, a bottleneck to the continued evolution of the entire network. The third era of American telephony will be defined by the problems of overcoming the access bottleneck created by the limitations of traditional copper-based loop transmission.

2.3 THE ERA OF ACCESS: 1990 AND BEYOND

Conventional twisted-pair copper plant will not disappear overnight; the installed investment of some 130 million access lines will be "milked" for decades to come. It will be a deadweight pulling against every forward thrust of technology unless there is a particularly far-sighted shift in regulatory attitudes, which seems unlikely. Nevertheless, for reasons we examine further in Chapters 3 and 4, it is becoming imperative to break the access bottleneck. To do this, a new set of physical transmission technologies must be developed. The creation and deployment of these new access media will constitute the defining challenge for the telecommunications industry for the next several decades.

Why do we insist on the preeminence of access as the defining problem of our era? What about the nascent marvels of photonic switching, asynchronous transfer mode (packet voice) switching, video services, broadband ISDN, and the like? We often look to the digital switch as the "engine" of the information age, and the metaphor is apt; like the old locomotives of the nineteenth century railroads, a large modern digital telephone switch is indeed an impressive technological ensemble, one not fully appreciated by those outside the industry. Consider that today a new generation of digital switching hardware and software can represent several billion dollars of research and development industry-wide, and sales of perhaps $50 to 100 billion worldwide, and it is clear why switching technology so dominates the thinking of industry planners, managers, technologists, and regulators [27].

The parallel with railroading is apt, as well, to point out the relative insignificance of the locomotive as a capital item in the construction of a railroad. The real driving economic force in the railroad industry was of course the tremendous effort and expense of constructing the railroads themselves. The value of the rolling stock was never but a small fraction of the value of the land, the rails, the roadbeds, and the capitalized costs of construction of the tracks. Today the telephone industry is faced with a problem that, to continue our analogy, is as though the railroaders had invented a new type of locomotive that promised tremendous advantages in performance and economics, but required a new rail technology to replace the existing infrastructure. Indeed, the analogy hits the mark, because today the railroad industry is faced with just such a dilemma; the new super-high-speed trains being developed in Europe and elsewhere cannot be used on today's rails, but would require an expensive reconstruction of roadbeds and replacement of rails with a completely new system based on welded tracks.

Although not yet fully recognized, this is the situation in which the telephone industry finds itself. It is developing new "information engines" that cannot run on the existing tracks (the wire access loops). The industry is really only beginning to consider the scope of the challenge involved in overhauling these "tracks" and replacing them with new ones that can carry the freight of the true Information Age. Indeed, an unpleasant surprise lies in store for the planners of the twenty-first century network: They will discover that as expensive as switching systems seem to be, the cost of upgrading the "existing tracks" will be at least 10 times greater than the cost of the engines themselves (see Section 2.5).

In fact, as we will see throughout this book, the preponderant importance of the copper access plant in all matters of telephony economics (hence, also of technical performance, at least as far as commercial deployment is concerned) is something that the industry has really always understood implicitly. Yet we are often startled by the extent of the copper bias built into many cost analyses and planning models. We are perhaps even less aware of how the historical evolution of telephone technology has been driven by the problem of copper costs. From the earliest analyses of electrical transmission (mainly in support of cable design) [28], to the elaboration of frequency-division carrier techniques (for more capacity per pound of copper) [29], to the first digital hierarchies (to reduce the wire-pairs needed for trunk circuits) [30], and even to the first plans for ISDN [31], the prime objective (often so obvious that it was often left unstated) was always to find some way to mitigate the economic burden of a copper-based twisted-pair access network.

In the third era of telephony, which we are now entering, the supremacy of copper will be overcome. In the next decade, two "new" access technologies will begin to play larger roles in the local loop, and in the twenty-first century they will become the twin pillars of the access network (see Figure 2.5). The first of these two technologies—*fiber optics*—is well known because of its spectacular success during the 1980s in ousting other technologies from large parts of the long-distance transmission segment. Fiber is now pushing into the interexchange trunk network, and even into the feeder portion of the loop plant. Over the past several years, a number of industry planners have extrapolated this trend to encompass a vision of *fiber to the home* (FTTH)—the extension of a fiber link all the way to the subscriber's premises. FTTH presages a qualitative change in telephone service; fiber loops provide such an enormous communications bandwidth to the end-user that the technical "bottleneck" disappears for almost any currently conceivable digital service. The revolutionary potential for fiber optics as an access technology is being explored today in a number of FTTH projects, serving collectively several hundred subscribers as of the beginning of 1990. Although a number of serious political roadblocks exist with regard to the delivery of broadband nonvoice services, the future for fiber in the access network seems full of promise.

The second new access technology—digital radio—has attracted much less attention from the traditional (i.e., regulated) telephone industry. Like fiber optics, digital radio techniques have a prior history of rather extensive application in the long-distance and short-haul interexchange segments of the network. Unlike fiber, digital radio is still stereotyped within mainstream telephony as a long-distance or trunking product. *Radio* still means point-to-point microwave (hardly suitable for access applications) in the lexicon of traditional telephone engineering.

Recent technical breakthroughs have created a new form of digital radio, suited in economics and performance to local loop applications. For one thing, the cellular mobile radio industry is currently in the throes of converting its most basic transmission standard to digital radio technology [32]. In the past few years, moreover, a number of digital loop radio systems have also emerged that have begun to address the integration and standardization questions that regulated telephone planners will pose to any wireless access option [33]. In fact, there were more wireless access trials under way and more subscribers using wireless access systems as of the beginning of 1990 than there were FTTH trials and subscribers (see Chapter 9).

Clearly, however, wireless access based on digital radio is still very much in its infancy. Planning methods, cost comparisons, deployment methods and procedures, and equipment performance standards are all being developed. Yet there is reason to believe that the impact of integrated digital radio on the problems of network access may be surprisingly swift. That is what the explosive success of cellular radio portends. Whether in the long run radio plays a subordinate role to fiber optics (as most believe), the importance of wireless access is certainly going to be very great as the network moves into the twenty-first century.

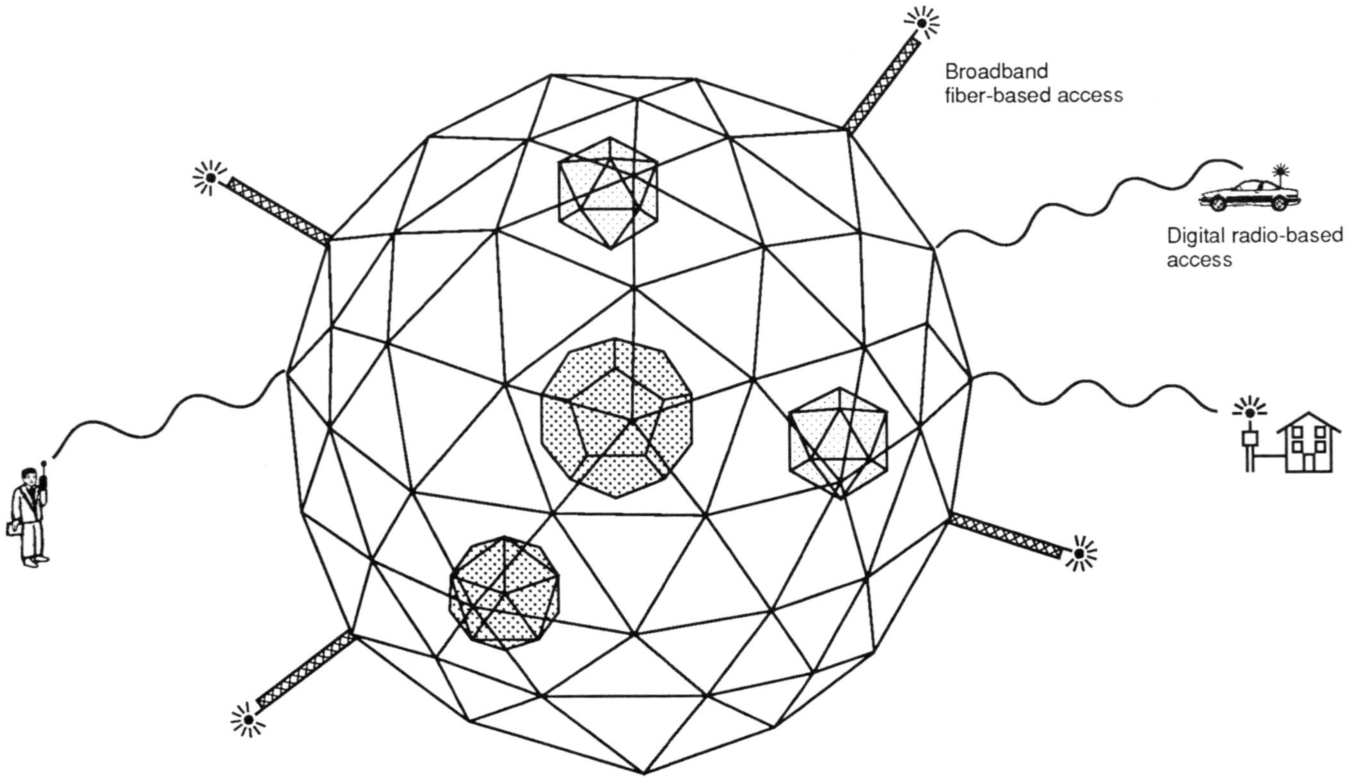

Figure 2.5 The telephone network during phase 3.

2.4 TERMINOLOGY: NETWORK AND ACCESS

The foregoing discussion has introduced, implicitly, a somewhat novel terminology—network versus access—that merits a more formal presentation. First we consider the more common nomenclatures:

Switching versus *Transmission.* A distinction is often drawn between switching and transmission as the two fundamental hemispheres of the network [34] (see Figure 2.6). It is an analytical cliché to point out the alleged engineering trade-off between

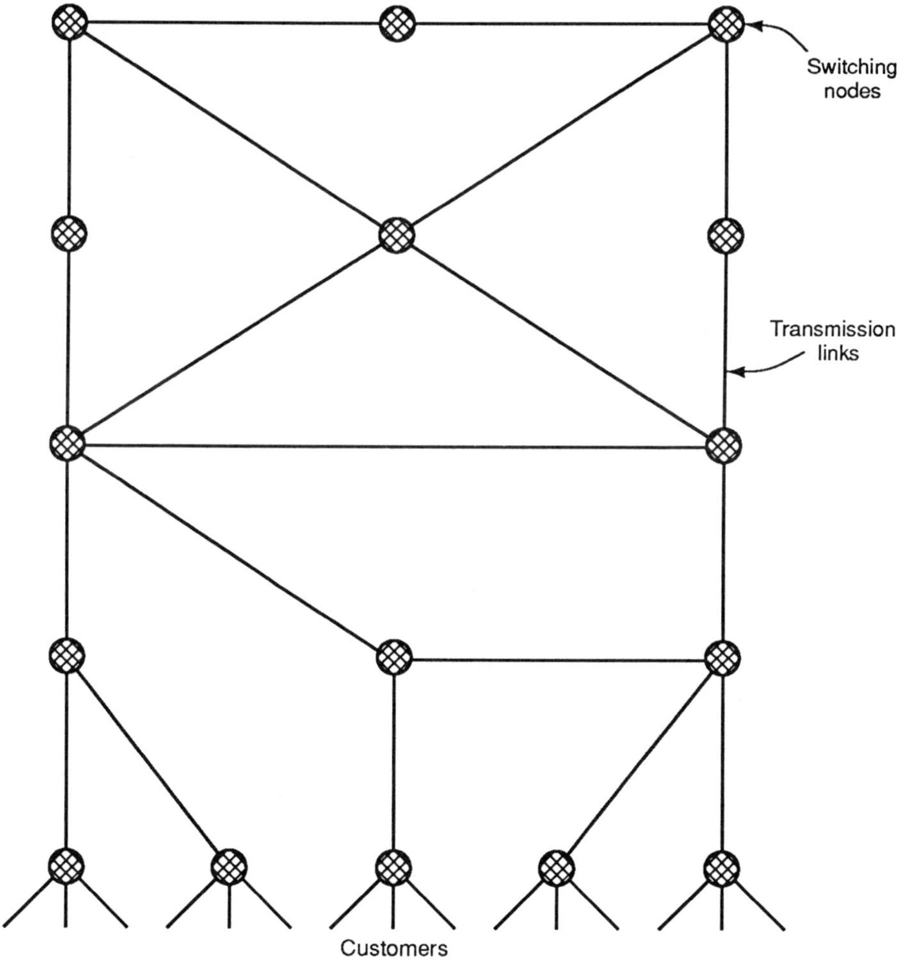

Figure 2.6 Switching and transmission: nodes and links.

switching and transmission plant [35]. In simple terms, it is said that a network can be built from many smaller switches located closer to the end-users, which means less wire and cable is needed, or with a few large switches, which are naturally farther, on average, from the customers, necessitating more wire and cable. Switches are, in this view, essentially a means of reducing the mileage of transmission plant [36]. A crude equilibrium is assumed, which is altered slowly over time as the relative costs of switching versus transmission equipment change. Some analysts draw weighty conclusions from this simple formula. The influential "Huber report" on telephone competition prepared in 1987 for the Justice Department in connection with a review of Judge Greene's Modified Final Judgment (the decision that broke up the Bell System) contains several typical pronouncements:

> When switching is expensive and transmission is cheap, the efficient network looks like a pyramid. . . . By contrast, when switching is cheap and transmission expensive, the efficient network is a ring . . . network design used to be driven by a single overpowering reality: switching was very slow and expensive, enormously so in the days when a human operator in front of a plug board provided the switching intelligence. Transmission over copper wire was, by comparison, quite fast and cheap. In these circumstances it made engineering sense to channel traffic through a minimum number of switches [. . . etc. and so forth. . .]. [37]

We shall have occasion to reexamine the so-called "geodesic network" idea at the end of this book. For the moment, we simply note that although transmission over copper wire is indeed fast, it has never been "cheap" in relative terms at any time in the history of the network.

Moreover, the growing use of intelligent systems in all parts of the telecommunications grid—including the subscriber's own terminal equipment—renders this switch/transmission dichotomy obsolescent and often makes it difficult to apply rigorously. Switch-like capabilities are being incorporated into more and more network elements. Today a *time-slot interchanger* (the basic subsystem for a digital switching function) can be purchased on a single chip. From the standpoint of an equipment designer, the decision to incorporate certain rudimentary switching functions into a particular piece of hardware has nearly reached the point of "Why not?" [38].

A more serious objection is that "transmission" is a very heterogeneous category. A point-to-point high-capacity digital microwave system or an undersea fiber-optics cable exhibits few meaningful similarities to an open-wire loop serving a country farmhouse. The wide disparity in technologies and economics of different types of "transmission" has misled at least one recent analyst into concluding that "transmission costs have dropped much faster than switching costs" [39], when (as we shall see in Section 2.5) the truth is quite the opposite. Indeed, this points to the anachronism of these terms. During the first era, there were for all practical purposes no real alternatives to copper wire-pairs for transmission, whether for local loops, interoffice trunks, or long-distance circuits. The transmission plant, on the whole, was fairly homogeneous technologically. Transmission and switching were serviceable categories of analysis. But the startling qualitative changes in long-distance and interoffice transmission methods during the second era (radio, satellites, fiber optics, high-rate digital multiplexers) have drained this tired dichotomy of much of its original analytical usefulness.

Central Office versus *Local Loop.* Another terminology that shows its age draws the distinction between the central office (roughly, the local switch and its associated CO trunks out to other exchanges) and the local loop (comprising the bulk of the local within-the-exchange transmission plant) [40] (see Figure 2.7). The loop plant is further distinguished into a feeder segment and distribution or drop line segments (sometimes additional distinctions are drawn—see Chapters 5 and 7).

This could be termed a *naturalistic* conception of telephony. The underlying topological metaphor is that of a river gathering in its tributaries. Individual lines from every home are brought together into multiline cables, which merge with other cables to create larger cables (the feeders), which may continue merging into ever-increasing gauges until they terminate at the main distribution frame in the central office. The following passage is taken from a famous early pamphlet by Alexander Graham Bell:

> At the present time we have a perfect network of gas-pipes and water-pipes throughout our large cities. We have the main pipes laid under the streets communicating by side-pipes with the various dwellings, enabling the members to draw their supplies of gas and water from a common source. In a similar manner, it is conceivable that cables of telephone wires could be laid underground, or suspended overhead, communicating by branch wires with private dwellings, country houses, shops, manufactories, &c., &c., united them through the main cable with a central office where the wire could be connected as desired, establishing direct communications between any two places in the city. [41]

In the autumn of 1877 when these words were drafted, Bell's vision of a network covering an entire city was quite bold considering that at the time there were in all the world only a few hundred telephones (and as yet no switches) tucked into a corner of New England.

This terminology addresses only the microcosm of the local exchange and does not encompass—except by reference to gateways or points of presence—the larger network (in the more evolved sense of the word) of long-distance transmission and toll switching. It is clearly a "field" language of working-level telephone engineers and planners, rather than technologists, telecommunications policy analysts, or network architects. It has served the industry well enough through the first and second eras. The terms are sometimes difficult to apply as advanced technologies begin to reach out into the loop plant in the form of digital loop carrier systems and remote

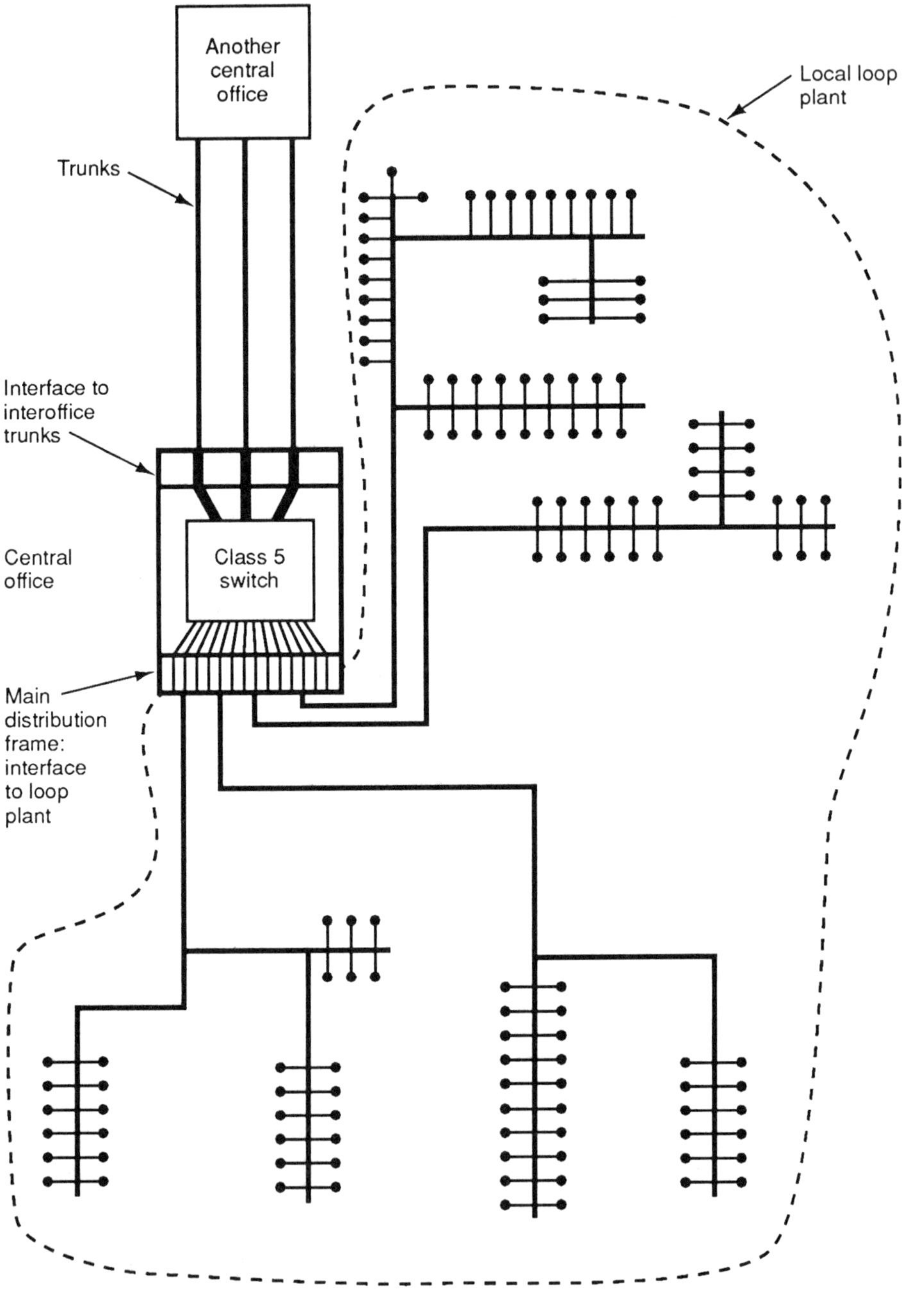

Figure 2.7 Central office and local loop.

switches. As a concrete and easily habituated terminology, however, these terms are still quite useful; they describe fairly accurately the environment of the contemporary installed base of local exchange plant, and we shall refer to this nomenclature frequently in this book.

In the third era, however, the terms *local loop* and the like will lose much of their meaning. Newer concepts such as *multiple access, demand assignment, local-area networks* (as well as *metropolitan-area networks* and *wide-area networks*), *rings, servers, packet nodes,* and other less familiar words are already working their way into the industry lexicon. The century-old metaphor of a stepped hierarchy of tributaries draining traffic from the exchange into the network will be less relevant 10 years from now.

Network and Access. I wish to promote a more contemporary construct. Unlike *switching* and *transmission,* this terminology reflects the changes that have taken place in the telecommunications industry during the second era, and I think it is more amenable to the broader policy discussions than the field terms.

In brief, the *network* comprises all the elements of the telecommunications system that are *shared* by many users during the process of communication. These elements include switches, short-haul interoffice trunks, and long-distance trunks.

For historical reasons, we still tend to see the network in terms of its principal function: switching. I believe this is too narrow. Although the primary function of the public network is still to provide a switchable interconnection between individual users, a host of other functions is now emerging. Also, many private and semi-private networks supplement the public network in various ways and occasionally interface to it. In many of these networks, conventional circuit switching for direct real-time user-to-user communication plays little role and may not even be offered—consider an electronic mail service, for example. Most computer networks are not oriented to this type of traffic, but to communications between the user and some central resource such as a processor or a database. In the not too distant future, switching may no longer be the predominant characteristic of the public network, either. For example, it seems likely that in terms of traffic bandwidth, the broadcast delivery of prepackaged, standardized information services to users (e.g., videotext or cable TV) may soon outpace the traffic on switched circuits. The traffic patterns envisioned by proponents of FTTH often assign conventional circuit-switched telephony a relatively low position in the gamut of broadband services. Thus, rather than defining the network in terms of a specific service or capability, I prefer to define it—for reasons I hope will be clear below—as the collection of elements shared by many users.

Access, by comparison, refers to the means by which individual subscribers are connected to the network. Today, access is normally achieved by means of copper-wire local loops, which physically (i.e., electrically) link each subscriber to his or her local telephone switch. By and large, access elements are dedicated to individual users.

Some of the important distinctions that can be drawn between network and access, as categories, include the following.

Fundamental Economics. Network elements are optimized for heavy traffic and a high duty cycle. For example, a trunk circuit connecting two central offices and shared during the course of the day by hundreds or thousands of callers can be designed for as much as 80 to 90% occupancy. In contrast, access elements (e.g., local loops) are idle most of the time. A typical residential telephone and its local loop circuit from the home to the central office switch will be in use for perhaps 20 minutes a day, sitting completely unused for the other 1420 minutes in each 24 hours.

This creates an overwhelming distortion in access economics compared to network economics. Consider an idealized example. Two telecommunications facilities are to be compared. Technically, physically, the two facilities are identical; each is a pair of copper wires of, say, two miles in length. One is an interoffice trunk circuit and the other a subscriber loop circuit. They carry exactly the same type of traffic—voice telephone calls. The facilities have essentially the same capital requirement per foot installed. The difference between the cost of operating the circuit (i.e., using it to actually carry voice traffic) and the cost of leaving the circuit idle is negligible.

In short, the costs of the two facilities are virtually identical. The network trunk facility, however, is in use 50% of the time because it is shared by many users on a demand-assigned basis. The access facility is in use only 1% of the time; it is a dedicated loop to a farmhouse.

If we now insert into our example a "man from Mars" and we given him the appropriate instruments and tell him to examine for a short defined period (say 10 seconds) the activity on each of the two circuits—without telling him ahead of time which is which—he will not be able to tell whether he is looking at a trunk circuit or a loop circuit. In principle, our Martian would conclude, the rate charged for such traffic ought in principle to be identical on both circuits. We would then allow him to observe these two circuits for a longer period, and he would discover that one of the two circuits is carrying 50 times the traffic of the other. Told that the cost of the two circuits is the same, the Martian will politely insist that, given the opportunity, he would much prefer to own the circuit with 50 times the revenue-potential. (QED Divestiture!)

Relatively speaking, the access plant is extremely underutilized and unproductive. The *fundamental* economics of the access plant are poor, which discourages investment. In the scenario described above, access investments are a necessary evil for the telephone engineer. In many areas, extensive tampering with telephone pricing mechanisms is the only reason the access plant attracts any investment at all (see Chapter 3). But the fact that the operator can recoup his investment by pricing to cover the inefficiencies of the local loop access plant should not obscure the real burden that the system carries because of these inefficiencies.

Economies of Scale. Costly, complex facilities can often be deployed economically in the network because their burden is shared by so many users. The cost per user for upgrading shared network elements is low, almost by definition. Significant economies of scale exist for the deployment of many network elements, especially advanced network transmission systems and the software "overhead" that embodies the intelligence in modern switches. The access plant shows fewer economies of scale since, by definition, the access plant is largely dedicated. It is true that there are some opportunities for economies of scale in the use of, for example, multipair cable to save on installation and labor costs, but these are not as readily engineered. The access plant engineer must serve his customers where he finds them and this often results in a substantial portion of each loop being totally dedicated. (We examine this in greater detail in Chapter 7.)

Cost Trends. The economics in the network tend to improve, on a per-user basis, because sharing and centralization of many network facilities can occur across larger and larger subscriber bases. An example is the trend toward sophisticated network management systems that can perform a wide array of functions (e.g., telephone number changes) from remote centers, functions that used to require local capabilities that were duplicated in every exchange. Another example is in the interoffice trunking segment, where the same physical facilities (wire-pairs) can today carry far more traffic than 20 or 30 years ago as more sophisticated multiplexing systems are applied. On the other hand, the economics in the access plant tend to be intractable and slow to improve, because of the dispersed, duplicative, and nonshared nature of the traditional local loop.

Cost Inequalities. Access is a "good" that is distributed very unequally. Typically, for example, those living in urban centers have much better and cheaper access to the network (and usually a much broader range of services) than people in rural areas. The inequalities in access costs are inherent for any wireline access system. Wireline plant economics are largely a function of the distance of the subscriber from the central office. The cost to provide a rural customer with access may be easily 10 times the cost of access for a city dweller (see Chapters 3 and 7).

On the other hand, the per-customer cost for network facilities is much more uniform. In fact, within a given exchange, the cost of network facilities should normally be exactly the same for all customers. Differences between exchanges range over as much as a factor of 3 or 4; very small switches have a higher per-line cost than large switches. But the overall distribution of network costs per customer is much less skewed than the costs of access (see Figure 2.8).

Access Pricing versus *Network Pricing.* The charges for use of shared network elements can generally be priced in a straightforward manner because the total cost of many network facilities is traffic-sensitive. Shared systems can be optimized for a given level of traffic (and given service objectives, such as blocking); once thus optimized, a small increase in traffic will necessitate adding additional equipment. The relationship between cost of capacity and throughput occurs in small steps and can be easily plotted and priced.

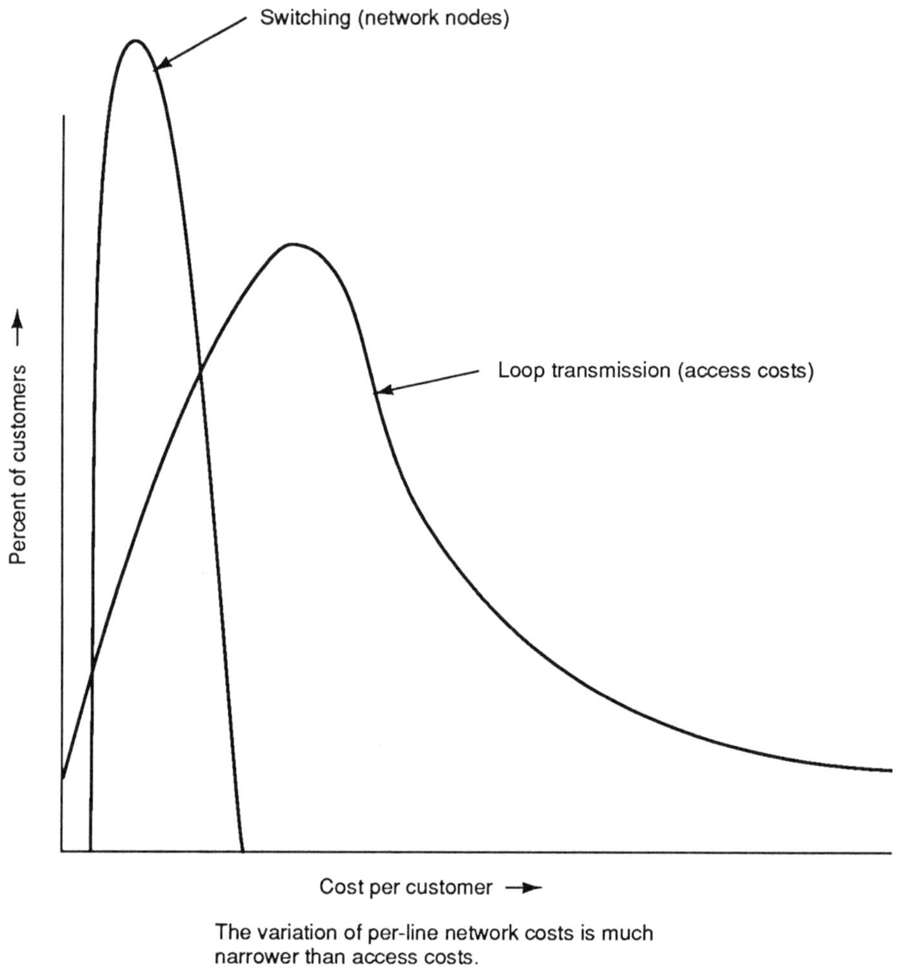

Figure 2.8 Log-normal distribution of access costs *versus* more concentrated distribution of per-line switching costs (idealized).

On the other hand, access circuits operate at a very small fraction of their capacity. The cost of a residential local loop used for an average of 5 minutes of calling per day is essentially the same as one connecting to a busy store that generates 5 hours of calls each day. The traffic on a local loop circuit can fluctuate widely with little or no impact on the cost of the facility. Access plant is largely nontraffic-sensitive.

Operating Environment. Network elements (such as telephone switches) are usually housed in a controlled environment and protected from the weather and extremes of temperature as is reflected in their design. In contrast, access equipment must often be located out-of-doors, where it is exposed to sun, rain, ice, wind-loads, lightning, insects, woodpeckers, rodents, and man-made hazards such as wayward backhoes.

In my experience, the cost of "ruggedizing" access facilities to survive for years or even decades in the outdoor environment is a severe constraint on the design of access equipment. It can become the largest single cost factor, especially as new technology comes into the access segment. Imagine, for example, the difference in cost between a television set—approximately equivalent in complexity to some of the radio equipment we will discuss in this book—designed to operate for about 10 years in a heated, protected living room, and (hypothetically) the cost of a TV designed to sit outdoors on a wooden pole for 10 years. As one Bellcore paper notes with respect to the difficulties of placing fiber optics in the loop:

> The range of temperature and humidity in the OSP [outside plant] is significantly wider than that of a telephone central office. . . . Since degradation of electronic and optoelectronic components increases exponentially with temperature, degradation can be as much as 10 times faster in the OSP placing very stringent requirements on device reliability. [42]

In short, the operating environment becomes the dominant factor in the design of new access facilities.

Integration. The public network is itself (at the end of the second era) a system whose elements are highly integrated and interconnected. By contrast, the access system is not really an integrated entity, but a vast agglomeration of feeder routes. The characteristics of each feeder—its performance, its cost, its ability to handle advanced services—is independent of any other feeder route. Each must be treated as a separate entity. And of course 30 to 50% of the total access investment is in the drop wire, which is a dedicated pair connecting the subscriber to the nearest feeder cable [43].

The Divergent Meanings of "Progress." In the world of the network, *progress* (i.e., "improved service") often leads directly or indirectly to lower costs—for example, through reduced maintenance for electronic (as opposed to mechanical) switches. On the other hand, in the access environment of the first and second eras—where voice was the main form of traffic—the improvement of service has generally come about through the elimination of ground return circuits, and the upgrading of party lines to single-party service. Yet a single-party access circuit carries much less traffic than the same facility used as a party line. In other words, the fundamental economics of access have deteriorated as the traffic per circuit-mile of access plant has dropped.

The Pace of Technological Change. The network has undergone a revolution in technology in the past 40 years. Digital switches, microwave radio, and fiberoptics systems have transformed the network and brought down costs by orders of magnitude in many segments. Yet access technology has been static for more than a century.

The overwhelming majority of access lines in the United States is still based on the technology from the Civil War era.

In short, network and access are very different environments, physically, technologically, and economically (see Table 2.1). The network is a dazzling hive of modern technology, soon to be almost all digital, software-dominated, intelligent, integrated, with rapidly falling per-unit-of-traffic costs. The access environment is still literally bound in its harness of copper wire, primitive, stagnant, still almost entirely analog, an incongruous patchwork of antiquated elements.

Table 2.1
Network and Access: Basic Comparisons.

Network	*Access*
Shared facilities	Dedicated facilities
Integrated	Fragmented, not integrated above feeder level
High Usage	Low usage
Controlled environment	Field environment
Advanced technology	Primitive technology
Rapid technological evolution	Technological stagnation
Software-dominated	Hardware-dominated
Costs falling rapidly	Costs rising
99% digital (by Year 2000)	99% analog (to end customer)

Like all terminologies, the network/access distinction is imperfect. For one thing, the division between the categories is not sharp. Line cards, for example, are elements of the switch that are usually not shared but are entirely dedicated to individual subscribers. Are they actually access elements? They suffer the same kind of cost penalty for idleness as other forms of access plant, and yet in most other respects they are more similar to network facilities. I prefer to view line cards as an element of the network. I know that a tomato is really a vegetable (even if the botanist says it is a fruit).

I also find that the term *network* is so protean, and so handy, that it cannot be entirely confined within a narrow definition. The phrase *access network* is sometimes useful to designate the totality of the access plant viewed collectively.

This terminology will undoubtedly lose its usefulness someday. Yet I believe that for today's telecommunications industry, the older dichotomies are outmoded and misleading. These are better categories for talking about today's issues.

2.5 THE DIMENSIONS OF THE PROBLEM OF ACCESS

Estimating true access costs is difficult. Historical costs, book value, and the like—the type of numbers available from telephone accountants—are often quite misleading. For example, a telephone exchange will typically carry assets on its books identified as "central office equipment"—a significant portion of the network plant—which usually constitute 25 to 35% of the telephone company's total assets. "Outside plant"—principally local loops associated with access—can account for 30 to 60% of its book assets and is usually the largest item [44]. Another deceptive indicator is the annual construction budget for the telephone industry, which consistently shows about 60% more money invested each year in switching upgrades than for outside plant construction [45].

These figures would seem to suggest that network and access costs (per subscriber) are of a similar order of magnitude. The problem is that the numbers are distorted by regulatory policies and accounting fictions associated with the rate-making process, as well as the tendency for PUCs and the Rural Electrification Administration (REA) to look favorably on network investments (e.g., digital switches). Also, investment programs are heavily weighted toward switch upgrades where the technology is.

A better approach is to look at replacement cost. This is now being done more frequently in connection with a scenario that is much discussed these days among telephone planners: the implementation of FTTH. In this scenario, fiber optics replaces copper in the local loop all the way to the customer's premises. An important and interesting question that often comes up in these analyses is: What would be the cost to rewire America with optical fiber? In more generic terms: *What is the replacement cost of the entire access plant nationwide?*

We can make some rough estimates. There are approximately 130 million access lines in the United States today. The average loop length is approximately 12,000 feet (12 kilofeet in telephone terms). The total loop plant in the U.S. telephone system is on the order of 1.5 trillion circuit-feet, or about 300 million circuit-miles. As large a number as this seems, it accounts for only 10 to 15% of the 2.5 billion or so circuit-miles in the entire system.

Unfortunately, the cost of new loop plant per access line is quite variable, depending on three principal factors—the gauge of the cable used, the method of construction (aerial, buried cable, conduit), and the terrain and type of soil. Nevertheless, for purposes of a rough estimate, I have found that a round number of $1 per installed foot—or $5000 per route-mile (which is not the same as a circuit-mile) holds fairly well as an average for new copper-wire loop facilities, tending to be slightly higher in rural areas and somewhat lower in urban areas [46]. For example, the cost per route-mile of plant in the REA database is nearly $7000 [47].

The next question is: How many circuits are installed per route, on the average? In other words, how much of the total circuit-mileage is carried in single-pair lines versus multipair cables? Of course, electronically derived circuits (e.g., digital loop carriers) also play a role in raising the ratio of circuit-miles to route-miles. This is another difficult answer to derive from the readily available data (see Chapter 7 for further analysis of the problem). Many factors need to be considered. For example, there is the problem of unused, spare, or out-of-service pairs. A given six-pair cable may support only three actual circuits, with one circuit in reserve and two no longer usable. In this case, the circuit mile–to–route mile ratio would be properly counted as somewhat lower than 3:1, since the cost of the unused circuits must be borne by the active users. There is always some portion of the loop that is single-pair—the so-called *drop* portion of the loop, which connects the subscriber's premises to the nearest feeder or distribution cable. So the typical loop may start out as part of a 500-pair feeder cable for the first third of its length, be spliced to a 25-pair distribution cable for the second third, with the remaining one-third as single-pair drop line. Based on very rough analyses of certain access networks, I have assumed that the average number of circuits per route-mile is about five, which rounds the cost to $1000 per circuit-mile.

Based on these rough assumptions, the replacement of the entire access plant nationwide would require some $300 billion. (Compare this with the book value of outside plant, estimated variously at $50 billion [48] to $70 billion [49]—not all of which qualifies as "access plant" by our definition.)

This figure is roughly in agreement with recent FTTH studies by Bellcore, which estimated the cost at some $250 billion [50]. Other Bellcore studies have shown higher figures. One Bellcore analyst recently estimated that to rewire the country for fiber optics to the end-user, it would be necessary to spend "$15 to $20 billion per year for several decades, peaking at about the year 2015 but continuing through at least 2030 . . . in all, several hundred billion dollars for loop construction." The analyst further acknowledged that, in the current scope of the industry's construction activities, this program would require that "virtually all construction funds go into the loop" [51]. By the way, the cost of completely replacing all the network switching engines with broadband ISDN switches would appear as a mere bubble in the chart (see Figure 2.9). Another well-informed observer has bracketed the cost for a fiber-based loop at "150 billion dollars up to 400 billion dollars" [52]. Based on a number of studies, I could well imagine the actual replacement cost might run as high as $500 billion. The logistics of replacing a network that has taken more than a century to build would be staggering.

Now consider what it would cost to replace every central office in the United States. This number is easier to develop because switching costs per line are much more uniform. Current prices for new digital switching systems suggest that $200 per line would be defensible as an assumption for such a program [53]. (Of course, the logistics of a large program of switch replacement would favor lower prices.) At this figure, the replacement of 130 million lines of switches with new digital equipment would cost $26 billion. Even if we pad our estimates to allow for higher per line costs in small rural exchanges, a figure of $40 to $50 billion should be adequate. This is in general accord with commercial market assessments.

In other words, *to replace the access plant entirely in the United States would cost 10 times more than the cost to replace every telephone switch.*

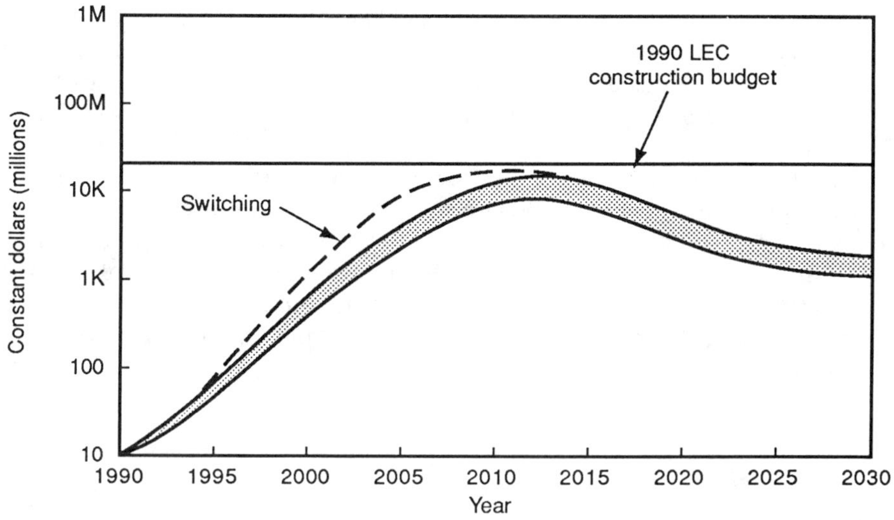

Figure 2.9 Projected annual construction expenditures for broadband access (from Bellcore Studies). (From Bellcore, Network Systems and Services Research Division, 445 South Street, Morristown, NJ, 07962-1910.)

We can also look at per-line costs for new facilities (1990 pricing). As noted, the cost to modernize a local telephone exchange with state-of-the-art switching is about $200 per line [54]. To build new access plant (copper only) in even the most densely populated urban setting costs several times as much. One article recently cited that "in Southern Bell, the cost to provide copper in the distribution plant is slightly more than $1200 per living unit" [55]. It is not clear whether this includes the feeder portion of the loop. Elsewhere the study notes that costs may range from $400 to $1600. Another study cites a range of $900 to $1400 for the "cost of copper-based telephony" [56]. Other Bellcore studies use an "average installed first cost" of $1750 to $1800 per copper loop [57]. In short, when we look at costs on a per-line basis, with today's pricing, we find that it costs anywhere from 6 to 10 times more to upgrade (replace) the access plant as it does to replace the central office switch—and this is not even fiber for broadband services, which "could run between $5000 and $10,000 per residence today" [58].

The actual range of loop costs is even larger; the figures cited above implicitly eliminate the highest cost loops, mainly in rural areas, where loop replacement can cost thousands or tens of thousands of dollars per customer. The skewed nature of loop costs produces extreme diseconomies for rural operations. One Bell operating company has indicated to me that some 30% of their annual operating, maintenance, and upgrade costs are associated with only 3% of their most rural loops! In my experience, rural areas often experience costs of $3000 to $6000 per loop; one large independent telephone company has cited a figure of 25% of their loops costing more than $3000. Thus, the disparity between network upgrade and access upgrade costs (per line) in rural areas may be more on the order of 20:1 or even 30:1—a staggering differential.

This should give some idea of the dead weight of the access plant. Changing out facilities to introduce new technology costs money. The local exchange manager could replace his switch 10 times over before it would equal the cost of replacing his loop facilities. This is at the heart of the technological backwardness of the access plant. New technology has a much harder time penetrating the access environment. This is the challenge facing American telephony as it moves into the third phase of its evolution.

NOTES

1. M.D. Fagen, ed., *A History of Engineering and Science in the Bell System: The Early Years (1875–1925),* Bell Telephone Laboratories, 1975, p. 585.
2. Some representative titles from the *Bell System Technical Journal:* "Creosoting Plants for Treating Chestnut Poles" (1925), "Recent Developments in the Process of Manufacturing Lead-covered Telephone Cable" (1928), "A Study of Telephone Line Insulators" (1930), "Cellulose Acetate Treatment for Textile Insulation" (1932), "Some Physical Properties of Wiping Solders" (1932), "Forces of Oblique Winds on Telephone Wires" (1936), "The Corrosion of Metals: Lead and Lead-alloy Cable Sheathing" (1936), "Determination of the Corrosion Behavior of Painted Iron and the Inhibitive Action of Paints" (1936), and "The Relation between Penetration and Decay in Creosoted Southern Pine Poles" (1936).
3. Department of Commerce, Bureau of the Census, *Telephones & Telegraphs, and Municipal Electric Fire–Alarm and Police Patrol Signaling Systems–1912,* Washington, D.C.: U.S. Government Printing Office, 1915, p. 37.
4. For example, the Keystone Company provided alternative local exchange service in Philadelphia until 1945; see J. Brooks, *Telephone: The First Hundred Years,* New York: Harper & Row, 1976, p. 160.
5. See A. Hibbard, *Hello, Goodbye: My Story of Telephone Pioneering,* Chicago, Ill.: A.C. McClurg & Co., 1941.
6. J.E. Kingsbury, *The Telephone and Telephone Exchanges: Their Invention and Development,* London: Longmans, Green, and Co., 1915, Chap. 22. (Reprinted by the Arno Press, New York, 1972.)
7. Readers may call to mind the scene toward the end of the vintage 1941 film *Sergeant York* when Gary Cooper (as York) phones his family back in Tennessee. The war hero, much decorated and acclaimed, and yet at root a simple man from a preindustrial world, is overwhelmed by his reception in Manhattan, the ticker tape parade, and the luxurious suite in the Waldorf Astoria that is made available to him. He is further taken aback to discover a phone, a private line, *actually in the hotel room.* On the other end of the line, his family and friends are gathered around a community phone in Walter Brennan's dry goods store. The scene is meant to convey the social distance between the high life of Manhattan and the hero's commonplace origins in Tennessee.
8. See telephone company statistics in Department of Commerce, Bureau of the Census, *Historical Statistics of the United States: Colonial Times to 1970,* Washington, D.C.: U.S. Government Printing Office, 1976.

9. See A. Hibbard, *Hello, Goodbye: My Story of Telephone Pioneering,* Chicago, Ill.: A.C. McClurg & Co., 1941.
10. Department of Commerce, Bureau of the Census, *Telephones & Telegraphs, and Municipal Electric Fire-Alarm and Police Patrol Signaling Systems–1912,* Washington, D.C.: U.S. Government Printing Office, 1915, p. 31. The marvel of making a long-distance call at all was the dramatic framework for that scene in *Sergeant York,* and yet we should remember that the scene is supposed to take place in 1918, four decades after the invention of the telephone.
11. For example, through the control of Pupin's loading coil patents; see M.D. Fagen, ed., *A History of Engineering and Science in the Bell System: The Early Years (1875–1925),* AT&T Bell Laboratories, 1975.
12. See, for example, Department of Commerce, Bureau of the Census, *Census of Electrical Industries: Telephones–1922,* Washington, D.C.: U.S. Government Printing Office, 1924.
13. *Ibid.*, pp. 36–37.
14. The Rural Electrification Administration was a creature of the Roosevelt era, focusing initially on fostering the spread of electric power to rural areas. After World War II, the REA's mandate was broadened to include rural telephony, and it has been one of the major factors responsible for the promotion of rural telephone service.
15. The classical switching hierarchy cited in all standard telephony references includes five levels, hence the designation of a "standard" local exchange switch as a Class 5 central office. However, the use of tandems at several levels, as well as various specialized subsidiary switching systems, from PBXs to remote switching units (RSUs) to, more recently, specialized switches for (among other things) cellular radio, add complexity to the basic five-level scheme.
16. These figures are based on FCC data, slightly massaged to eliminate extension phones and to separate local from toll plant. In general, although many historical statistics are available for telephones, the master work of collation and standardization of the data remains to be done. As anyone who has worked in the numbers will acknowledge, there is a terrible mixing of categories in the historical data, which can result in gross errors in interpretation. I hope I have been careful enough to avoid the deepest pitfalls.
17. E.F. O'Neill, ed., *A History of Engineering and Science in the Bell System: Transmission Technology (1925–1975),* AT&T Bell Laboratories, 1985, p. 784.
18. There was some use of coaxial cable in the interexchange network, amounting to perhaps 20% of the circuit–miles by 1950; *ibid.*
19. See annual FCC industry surveys.
20. The transistor was invented in 1947 by Bell Labs scientists, but was not used in commercial telephony until the mid-1950s. Interestingly, the first application was in a rural subscriber-line carrier system, called *P-carrier,* which was developed and tested in 1954 and 1955. As the AT&T technical history comments, "Type P carrier was not very successful, as much perhaps because of the system concept as because of the embryonic devices, but, by the late 1950s, highly reliable junction transistors were becoming available. . . . "; E.F. O'Neill, ed., *A History of Engineering and Science in the Bell System: Transmission Technology (1925–1975),* AT&T Bell Laboratories, 1985, p. 489.
21. Microwave had been used a little for quasiexperimental transmission of television programming in lieu of coaxial cable. The first telephony use of microwave came in 1950–51; in fact, the first use of microwave may be seen as a symbolic delimiter between the first and second phases of the evolution of the network; E.F. O'Neill, ed., *A History of Engineering and Science in the Bell System: Transmission Technology (1925–1975),* AT&T Bell Laboratories, 1985, pp. 291 ff.
22. Once again, a precise figure is difficult to develop. FCC data would seem to indicate a slightly lower figure—perhaps on the order of 1.7 to 1.8 billion circuit miles by the end of the 1980s. However, I believe significant underreporting or undercounting exists. There is, of course, the ever-present problem of how to count the plant-miles and circuit-miles associated with large PBX systems. I do not include this element in my estimate. The AT&T history also provides a higher figure—more

than 1 billion circuit miles in 1980, at a time when the FCC figure is only about 800 to 900 billion circuit-miles after adjustment to eliminate local exchange mileage. In any case, the basic orders of magnitude, on which the argument rests, are okay.

23. Again, the relationship between the number of subscribers and the number of access lines is not the same in 1950 as in 1990. In 1950, most subscribers shared a line. By 1990, the two numbers are close to equal. Thus, the growth in access lines—a factor of about four—is somewhat greater than the growth in subscribers.
24. I derive this figure, roughly, by multiplying the average loop length, which has stood at about 12 kilofeet for several decades, by the 130 million access lines.
25. The authors of the AT&T history are, if anything, more aggressive in their calculations: "The period from 1925 to 1975 saw a spectacular cost reduction in the line cost of [interexchange] transmission circuits, approaching, in normalized dollars, a factor of 1000"; E. F. O'Neill, ed., *A History of Engineering and Science in the Bell System: Transmission Technology (1925–1975),* AT&T Bell Laboratories, 1985, p. 782. See Chapter 3 of this volume for further discussion of these cost trends.
26. Different sources offer slightly different figures. According to L.K. Vanston, R.C. Lenz, and R.S. Wolff, "How Fast is New Technology Coming?" *Telephony,* September 18, 1989, p. 50: "In 1989, 84% of interoffice equivalent voice circuits in use are digital."
27. See, for example, P.W. Huber, *The Geodesic Network: 1987 Report on Competition in the Telephone Industry,* Washington, D.C.: U.S. Government Printing Office, 1987, Chap. 14.
28. J. E. Kingsbury, *The Telephone and Telephone Exchanges: Their Invention and Development,* London: Longmans, Green, and Co., 1915, Chap. 22. (Reprinted by the Arno Press, New York, 1972.)
29. See E. F. O'Neill, ed., *A History of Engineering and Science in the Bell System: Transmission Technology (1925–1975),* AT&T Bell Laboratories, Chaps. 5, 9, and 10.
30. *Ibid.*, Chap. 18.
31. See Chapter 4 of this volume for references on the motivation of ISDN.
32. By 1990, standards had been set for digital cellular in both Europe and North America.
33. I am referring here to *basic exchange radio,* a precursor for more general wireless local loop applications, which is covered in some detail in Chapter 5.
34. For example, the elaborate, multivolume technological history of the Bell System published by AT&T during the 1970s and 1980s consigns switching and transmission to separate volumes.
35. For example, A.E. Joel, "Circuit-Switching Fundamentals," in J.C. McDonald, ed., *Fundamentals of Digital Switching,* New York: Plenum Press, 1983, pp. 1–39.
36. P.W. Huber, *The Geodesic Network: 1987 Report on Competition in the Telephone Industry,* Washington, D.C.: U.S. Government Printing Office, 1987, p. 1.14: *"It has always been true that equipment and transmission are substitutes at the margin: a national network can be assembled with a billion wires and one switch, or a billion switches and one wire, or any mix in between."* This idea, which smacks of graph theory, has always struck me as too transparent a pedagogical device. Switching and transmission are not opposite ends of a continuum, as the analogy implies, but qualitatively different functions that are interchangeable to a limited degree within a large network. Moreover, the cliché blurs the important differences between long-haul transmission and access transmission, which Huber elsewhere recognizes.
37. Peter W. Huber, *The Geodesic Network: 1987 Report on Competition in the Telephone Industry* Washington, D.C.: U.S. Government Printing Office, January 1987, p. 1.3.
38. There are good reasons, such as greater system redundancy, to build in certain switch-like capabilities into many systems, which, viewed from a higher level in the hierarchy, would be clearly classified as transmission elements (such as digital loop carrier systems).
39. K. Flamm, "Technological Advance and Costs: Computers versus Communications," in R.W Crandall and K. Flamm, ed., *Changing the Rules: Technological Change, International Competi*

tion, and Regulation in Communications, Washington, D.C.: The Brookings Institution, 1989, pp. 59–60. Flamm really duffs it: *"The otherwise sober Huber Report makes* [*an improper*] *leap when it argues that 'in the past fifteen years . . . revolutionary developments in electronics have slashed the costs of switching and other forms of network intelligence.' Huber argues that the fundamental network trade-off between trunking and switching has been tilted toward switching by this revolutionary cost decline. His observation that local, low-density transmission costs have not much changed and are not likely to do so leads him to conclude that as switch costs decline, switching will move out toward the user so that communications firms can economize on relatively more expensive transmission. But for the network as a whole, the historical record refutes these assumptions. Transmission costs have dropped much faster than switching costs and, with the advent of fiber optics, have continued to do so."* This dissertation is fraught with confusion, introduced primarily, I would argue, by the obsolete switch/transmission dichotomy. The cost per circuit-mile for interexchange transmission has fallen dramatically—by a factor of hundreds—in the past 50 years. On the other hand, the cost of local exchange transmission (per access line or circuit-mile) has probably not fallen appreciably, if at all, over that time frame. And fiber optics will not change that basic loop cost equation very much. See Chapter 9.

40. The *local loop* refers of course to the metallic electrical closed circuit, conceived by early telephone engineers as a "loop" of wire from the central office out to the customer and back. The colloquial character of many of the terms in the telephony planner's lexicon belies the relatively unsophisticated perspective that most early telephone technicians brought to their craft.
41. John E. Kingsbury, *The Telephone and Telephone Exchanges: Their Invention and Development,* London: Longmans, Green, and Co., 1915, p. 9. (Reprinted by the Arno Press, New York, 1972.)
42. P.W. Shumate, "Special Broadband Needs of the Outside Plant," *Proceedings of the First Broadband Exposition and the Thirteenth Fiber Optic Communications and Local Area Networks Exposition,* San Francisco, October 30, 1989, p. 4.
43. One study found that 20% of the cable investment was in interoffice trunking—part of the core network in our terminology—and 35% in the feeder segment of the loop, leaving 45% in the distribution segment, of which at least two-thirds and probably more is associated with the dedicated or drop segment. See R.S. Wolff, "What's Ahead for Copper?," *Telephone Engineer & Management,* October 1, 1988.
44. See annual equipment and assets surveys published by the REA and by *Telephony* magazine. 45 P.W. Shumate, "Cost Projections for Fiber in the Loop," *IEEE LCS Magazine,* February 1990, p. 73; see also annual surveys of plant expenditures in *Telephony* magazine.
46. See details in Chapter 7.
47. U.S. Department of Agriculture, Rural Electrification Administration, *1988 Statistical Report, Rural Telephone Borrowers,* REA Bulletin 300–4, 1989. This figure is "contaminated" by some radio route-miles, but it is probably not off by more than 10%.
48. D.L. Waring, J.W. Lechleider, and T.R. Hsing, "Digital Subscriber Line Technology Facilitates a Graceful Transition from Copper to Fiber," *IEEE Communications Magazine,* March 1991, p. 96.
49. R.S. Wolff, "What's Ahead for Copper?," *Telephone Engineer & Management,* October 1, 1988.
50. *A Compilation of Results from Recent Bellcore Studies on Fiber to the Home (FTTH),* SR-TSY-001468, Issue 1, December 1989, Bellcore, 1989; see also the related references cited in Chapter 9 on this book.
51. Richard Wolff, District Manager–Local Distribution Networks, Bellcore, in a presentation entitled "Local Loop in Transition—The Next Decade," given at Supercomm, Atlanta, April 17, 1990.
52. J.F. McLaughlin, Executive Director of the Program on Information Resources Policy, Harvard University, "Making Money as a Telephone Company: Risks, Opportunities, and Likely Failures," prepared for the United States Telephone Association Capital Recovery Seminar, October 23, 1989.

53. P.W. Huber, *The Geodesic Network: 1987 Report on Competition in the Telephone Industry,* Washington, D.C.: U.S. Government Printing Office, 1987, p. 14.11, note 50.
54. See discussion of switching costs in K. Flamm, "Technological Advance and Costs: Computers versus Communications," in R.W. Crandall and K. Flamm, ed., *Changing the Rules: Technological Change, International Competition, and Regulation in Communications,* Washington, D.C.: The Brookings Institution, 1989, pp. 13–61.
55. P.W. Shumate and R.K. Snelling, "Evolution of Fiber in the Residential Loop Plant," *IEEE Communications Magazine,* March 1991, p. 73.
56. P.W. Shumate, "Optical Fibers Reach into Homes," *IEEE Spectrum,* February 1989, p. 44.
57. K. Lu, R. Wolff, and F. Gratzer, "Installed First Cost Economics of Fiber/Broadband Access to the Home," *Proceedings of the IEEE Global Telecommunications Conference,* November 28 1988, pp. 48.4.1–48.4.7.
58. P.W. Shumate, "Optical Fibers Reach into Homes," *IEEE Spectrum,* February 1989, p. 44.

Chapter 3
THE COSTS OF ACCESS

3.1 A PEARL IN THE OYSTER

Telephony has been the subject of comprehensive state and federal regulation in the United States for nearly 80 years [1]. The major business decisions that in most industries are determined solely by corporate management in light of market forces—decisions about pricing, selection of lines of business, defining standards of service, the timing of investments and upgrade programs, even the choice of technologies in the infrastructure—are instead negotiated with politicians and their appointees, as well as, occasionally, the courts, and are continuously supervised by civil-service bureaucracies [2]. The rationale for this regulation is often lost in the sediment of history. The standard cliché has been that telecommunications is a business that is "clothed in the public interest" [3], or a "public utility"—although, in the words of one skeptic, "to say that a public utility is a business affected with a public interest is to include piggeries and mortuary parlors" [4]. Nevertheless, we tend to accept the proposition, out of convenience: The telephone system becomes a public asset, which in most countries is administered directly by the government. In the United States, where we prefer private enterprise, the right to operate this utility is assigned to a franchisee, whose tincture of self-interest is thought to be a protection against the inefficiencies that would result were the government to operate it directly. Pricing is regulated to ensure fairness; the private franchisee is guaranteed a reasonable rate of return on its investments and is protected against competition. Finally, there is the social compact: the franchisee agrees to promote universal service—which we may describe as the devout wish, inscribed but vaguely in the law and with much interpretive latitude, that all American citizens should enjoy the opportunity to have a telephone, and at a reasonable price [5].

More and more, we question these arrangements. Some of the questions are political. For example, Americans are suspicious of monopoly powers, even those endowed by the government. The antitrust tradition in the United States will always resist the conclusion that any business—even a public utility—must be exempted from competition. There is also the question of state versus federal jurisdiction. Is local telephone service truly a local business, with predominantly local interests and responsibilities like, say, the water company? Or is it an integral and inseparable part of the nationwide network, that unique national resource we extolled in Chapter 1? Disagreements between state and federal regulators are not uncommon, and the parties often seek recourse with the third and most powerful set of supervisory agencies, the courts, which follow their own idiosyncratic inspirations.

But more and more, the questions about the public utility framework are inspired by the recognition that the technological revolution has undone so many of our long-held assumptions about the nature and particularly the economics of the telephone business. It is clear that the advanced segments of the telecommunications business today do not at all fit the traditional public utility models. And by fits and starts the regulators have responded to these changes. "Competition," which was once unimaginable, now seems imminent everywhere. It is almost a fad. The long-distance segment, the customer-premises equipment segment, the network equipment segment, data networks, transoceanic fiberoptic cables, value-added services—we are advancing rapidly toward free-market conditions on all fronts.

Except one:

> *Local telephone service remains a regulated monopoly with no alternative for the foreseeable future.* [Emphasis added] So-called bypass, or alternative, connections are for all practical purposes nonexistent. Certainly, for purposes of planning telecommunications networks, the operating assumption for investment decisions must be that alternative technologies and suppliers do not yet provide a meaningful option to local exchange carrier-provided facilities.
>
> William G. McGowan, 1989 [6]

The monopolistic character of the local exchange is widely perceived as intractable, and the boldest prophets of change grow pessimistic in the face of it.

> When you look at the overall picture, the Local Exchange Carrier remains the powerful, ubiquitous local network. And . . . *I see nothing . . . that will undermine that monopoly.* [Emphasis added]
>
> Henry Geller [7]

> *The current roadblock to competitive conditions is local distribution.* [Emphasis added]
>
> Gerald Brock [8]

Copper Economics

This persistence of the local monopoly is based on the economics of electrical transmission facilities—copper wire. Simply stated, these economics, left to their own natural tendencies, would have promoted the development of a network that would have been, as with many things in our economy, skewed in favor of the urban centers and the major corridors of trade. The reason is simple: in a copper-based access network, the cost-basis for a given customer is essentially a function of distance from the nearest exchange (Figure 3.1). The more wire to run, the higher the cost. A customer who lives 10 miles from the center of town (and the local switch) may cost 10 times more to serve than someone who lives a half-mile from the switch. As we shall examine in Chapter 7, this means that the costs of copper-based local access rise rapidly as a function of declining population density [9].

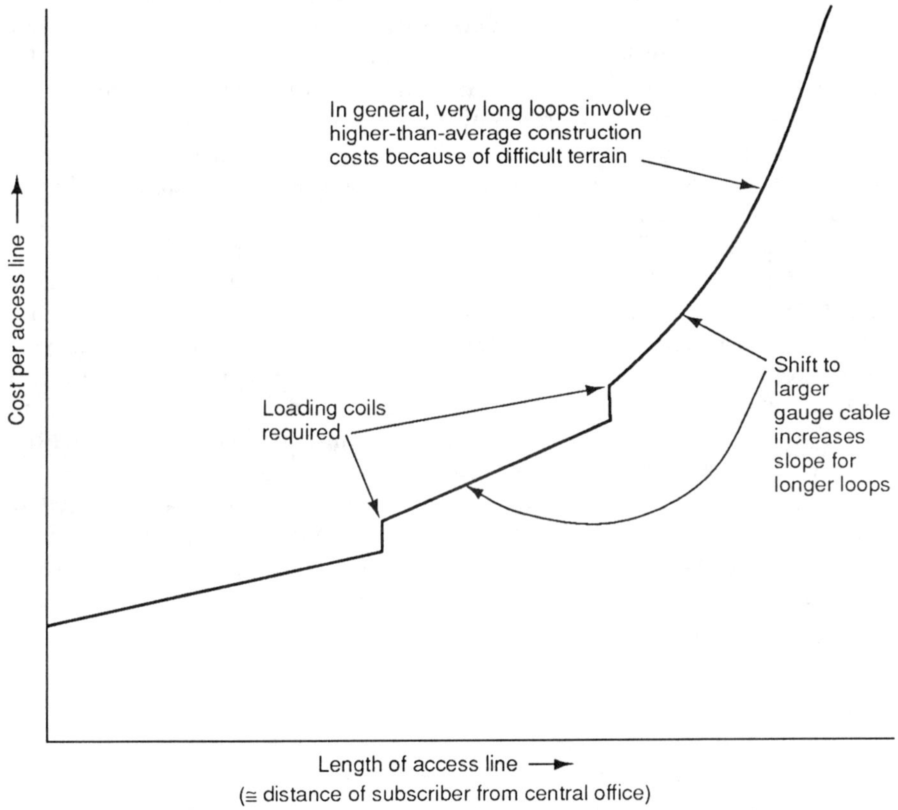

Figure 3.1 Cost/distance relationship for copper plant.

Indeed, in most countries of the world, this is precisely how the telecommunications networks have grown during the copper wire era: The major cities enjoy fair-to-good service, but rural areas and even some suburbs remain woefully underserved, even cut off. This is not true only for the Amazon jungle. Travel 30 miles outside Buenos Aires and you are reduced to a few CB radio links and an occasional open-wire party line that may serve several dozen ranches. In Mexico City there are 10 times more telephones per capita than in rural areas of the country. Vast regions of the Third World are completely without telephones. In India, with 850 million people, there are about 5 million telephones, all located in a few large cities. Even in Europe, rural areas are often quite poorly served.

That this fate was avoided in the United States was due to an imaginative political arrangement: essentially, the telephone company was conceded the power to tax and redistribute, to construct a private welfare system for rural (i.e., nonurban) telecommunications. The regulators exercised oversight, but essentially at an arm's length. In time, they became cheerleaders, because the system allowed them to reap political benefits without forcing them to be tax collectors. This arrangement developed into a byzantine system of financial transfers, administered cooperatively by AT&T and its regulators, and with some added support from the federal government (e.g., low-interest financing for the rural non-Bell carriers). The local monopoly was converted through political alchemy from a problem into a solution, a solution to the social problem of guaranteeing universal service. In the end, it became an icon, a "natural" monopoly, untouchable in the minds of many state regulators.

Indeed, over many decades, a complicated body of regulatory theory and practice has grown up around the gritty problem of the copper-based local exchange monopoly, and there are still those who would argue that it is a kind of pearl conceived within the tightly closed oyster of the predivestiture, pre-Carterphone, pre-competition telephone industry of the good old days. In our misguided enthusiasm for trustbusting (so they believe), we may have thrown away the oyster. But we should keep the pearl. Other observers and participants, bedeviled by the challenge of fostering competition through half-measures in an industry still blighted with monopoly abuses and bottlenecks, would just as soon be rid of it—but they cannot figure out how, since without these subsidies so carefully constructed, most rural telephone companies would shrivel up like a Nebraska cornfield when the Oglala aquifer runs dry.

The Bottleneck

But even beyond the question of universal service as such, other troublesome ramifications of the local exchange monopoly are becoming clearer as we move into the Information Age.

The Equal Access Problem. As long as the local exchange constitutes a monopoly, the local exchange carrier (LEC) in effect "owns" the end customers and their traffic from the standpoint of competing interexchange carriers (IXCs) and other providers of value-added information services who need to reach those customers to do their business. In other words, for an IXC to reach from his point-of-presence in a local access and transport area (LATA) to his customers' premises, he must rely on the only facilities that are available (in most cases): the copper "last mile" of the local loop (Figure 3.2). This monopoly power raises many problems for the public and private administrators of the competitive IXCs. The same is true for providers of value-added services.

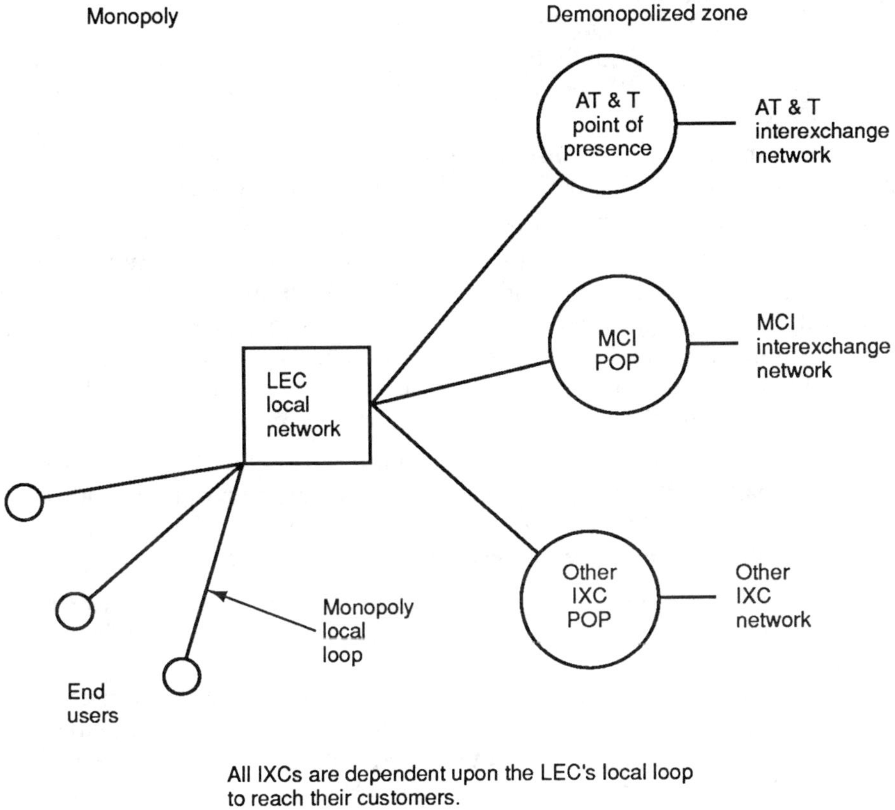

Figure 3.2 The issue of access from the interexchange carriers' perspective.

The Rural Quality-of-Service Problem. From the telephone customer's standpoint, the local exchange monopoly means that he has none of the traditional marketplace leverage to obtain any particular "type, quality, or price" of access to the network. Instead, he must rely on a diffuse political process, applied through state public utility commissions, to secure the best bargain possible; often, for customers in rural areas, this bargain is not particularly good, nor is the recourse particularly swift. In my experience, service deficiencies go back years, decades, even to the beginning of telephonic history. Statistically, although more than 90% of American households enjoy a reasonable grade of voice service, there are still millions of households that do not. For example, several million party lines still exist, and hundreds of thousands of households are without any service at all (see Chapter 5 for statistical estimates). Beyond such obvious deficiencies, it is clear to anyone with experience in the rural telephony environment that there is often a *de facto* second class of service in rural areas, associated with deteriorating plant, high maintenance costs, slower response times, longer outages, and qualitatively poorer transmission quality. In the words of one West Texas rancher, "Whenever it rains, the phones go out. Thank God it don't rain much." And finally, new services that depend on a higher quality channel, such as facsimile, work well on most urban loops, but may not be reliably deliverable on many rural circuits. Yet what options do these rural customers have?

The Regulatory Stalemate. Some argue that, like partners in a marriage gone sour, the local telephone carriers and their regulators have settled into a dysfunctional pattern of mutual provocation, misunderstanding, and ritualized bickering that has become self-defeating. The often adversarial rate-of-return approach to pricing seeks to cope with the absence of any market-like discipline by imposing elaborate and artificial tests of profitability that often encourage and even sanction substandard service and inefficient investments. As one author has commented:

> Cost-of-service regulation provides little incentive to be efficient. . . . Bad management is bailed out while good management receives no reward. In addition, reducing costs might involve the utility in innovation and risk-taking. If the measures fail, the company might even be penalized by the regulators. [10]

Enforced Technological Obsolescence. To keep subscriber costs down, extremely long depreciation lifetimes have often been imposed on key categories of equipment [11]. The result has been to keep large amounts of equipment in service for years, even decades after it has been rendered obsolete by technological progress. Aside from the impact on the customer, the consequences of this underinvestment in local exchange technology are felt throughout the system of equipment suppliers, as well, discouraging R&D for loop-oriented technology. This is, in effect, one more type of subsidy—borrowing from the future to subsidize the present.

The Wireless Millennium for the Local Exchange

The advent of wireless access will, in time, permanently transcend these difficulties. By reducing the vast disparities between the cost bases of telephone service for urban and rural customers—from say 10 or 20 to 1 with copper, to perhaps 2 to 1 or less with radio—the true costs of local service will fall more squarely on those who create them. The need for gross subsidies will diminish. Also, by making possible a much more competitive environment, it will allow regulation to be supplemented by the discipline of the free market. Radio is quite obviously *not* a natural monopoly; indeed, the introduction of radio in the long-distance business 40 years ago provides a vivid precedent for the coming of radio-based competition in the local exchange.

It should be obvious that radio will spawn a competitive industry in the local exchange. The broadcasting industry already shows how multiple operators can farm the same geographical franchise, using different layers of the electromagnetic spectrum. Imagine, technically, adding a two-way communications capability to today's television receivers—voilà, competition in local telecommunications. Any local radio-based access network will face many challenges—not the least of which is to compete with the entrenched, often largely depreciated investment in wireline plant—but it does not inherently abhor competition.

But the most difficult roadblock toward local exchange competition based on wireless access is likely to be the political fragmentation of the regulatory process itself.

Three Parallel Traditions of Regulation

The fabric of telephone regulation is far from homogeneous. In the American system, perhaps uniquely in the world, there is no single center of supervisory authority, no coordinated national plan or policy for telecommunications. This will complicate the transition to a competitive local exchange environment, and may delay it for some considerable time.

Through the years there have been at least three broad strands of regulatory policy, often at odds not only with the carriers themselves but with one another.

In the first place, the courts and the Department of Justice have waged a continuing campaign under the banner of antitrust, marked by long periods of latent strife, punctuated by epochal lawsuits, which are in turn followed by truces (e.g., Kingsbury, the 1956 Consent Decree), that mark great turning points but never seem to provide a definitive release from further contention. The latest watershed is the Modified Final Judgment (MFJ) of Judge Greene's federal court, which in the 1980s propelled the entire industry into a new orbit, and which still sits in judgment on the fate of dozens of large and small competitive thrusts and parries from the affected carriers and others [12]. We discuss below the theory of the Bell breakup in more detail. Yet, despite its titling, the MFJ has provided little finality and certainly no rest for the lawyers.

The second strand of regulation comes down from the Communications Act of 1934 and the creation of the Federal Communications Commission (FCC), charged with the regulation of interstate telecommunications and the uses of the radio spectrum. Since the 1930s, the role of the FCC has evolved and grown more sophisticated. It, too, has grappled with the enigma of competition. Hampered by its now-obsolescent charter and often overmatched by the politics of managing a transition to the Information Age, the FCC has struggled to provide that which it is perhaps uniquely suited to provide—a vision of the whole. After all, the network is always tending toward integration. One of the greatest dangers in making telecommunications policy is the temptation to make do with piecemeal solutions to particular problems, which nevertheless make their unpredicted consequences felt throughout the network.

Third is the welter of rate cases and other regulatory actions issuing continuously from the 50 state public utility commissions. These range up and down the scale of enlightenment, but they are almost invariably at odds with much of the "bigger picture." Because of the network's role in spurring the economy toward national and international integration, the province of intrastate telecommerce becomes more difficult to define. Eighty years ago, or even 40 years ago, local telephone service might have been accurately construed as primarily a local business facilitating transactions within a defined community, with but an occasional Mother's Day call to breach the constitutional boundaries. But in this day of 800 numbers, competing long-distance carriers, voice mail, fax machines, and computer networks, the term *local service* bears less and less relation to any geographically integrated community. The very term *access*, the more modern usage, connotes this changing perspective on the function of the local loop as first and foremost a gateway to the wider world. The temptation is to announce the death of local service as a concept. As Alfred Kahn has written:

> Access costs are incurred when the subscriber subscribes—not to interstate service, not to intrastate service, not to local service, but to the availability of any and all of these. [13]

The boundaries of their jurisdiction are disappearing, for telecommunications purposes, yet we can hardly expect the state regulators to disestablish themselves serenely. They would rightly point to the many ways in which the network is still not complete (e.g., the rural service problem) and for which their vigilance may still be required.

Telephone regulation has followed, thus, these three themes: antitrust, states rights, and the vision—three traditions not exactly at odds, but rarely in harmony. We observe the development of wireless access and local exchange competition through these diverse lenses.

Add to these three constants the occasional attentions of Congress, its Office of Technology Assessment, the executive branch (although telecommunications has rarely been a focal concern for presidential-level policy), lobbies, trade associations, and public interest groups and there are many hands grasping for the steering wheel of telecommunications policy in this country. Taken as a whole, the regulation of telecommunications is probably more pervasive and complex than for any other industry, the developments of the last decade notwithstanding.

Moving Forward: More or Less?

The telecommunications industry would seem to provide the textbook case both for and against regulation.

On the one hand, the United States has created universal service or something close to it, and is still one of only a few countries that can truly make that statement. This is undoubtedly the triumph of the regulatory framework. Also, the industry structure is so much the creature of the regulatory environments within which it operates that it is difficult to conceive of treating it as an ordinary business. If we smash the hothouse windows, will the orchids die?

Perhaps; therefore, say the optimists about the possibilities of modern government, let us step up to the challenge. We have been building up sophisticated regulatory mechanisms for the past 40 years; we can continue to do so. We should not shy away from complexity *per se*—the whole economy is so much more complex, the government is so much more complex, than it was 40 years ago. And indeed, even after Judge Greene, this is hardly a moot argument: The United States has not yet committed itself to actual deregulation. For all that we may imagine that we have fostered deregulation during the 1980s, the quantitative intensity of government regulation is if anything higher than it was 20 years ago, especially since we are now in one of the proactive phases of the antitrust cycle. There is little argument that telecommunications regulation is a difficult and tricky business, but it has often been viewed as inescapable. As one person has put it: "Regulation is like growing old; we would rather not do it, but consider the alternative" [14]. And in the end, since we are fond of self-congratulation, we often tell ourselves that we have the "best telephone system in the world" and many will readily attribute this to the success of the regulatory framework. Besides, remember those Nebraska cornfields.

On the other hand, there is a growing feeling that the framework may be unraveling beyond repair. The traditional roles of the three main regulatory agents are all tinged with obsolescence. The antitrust courts are still operating under the framework of the Sherman Act (1890) and the Clayton Act (1914), and they have often fumbled the technological issues. On states' rights, there has been a recent backsliding toward court rulings based on clearly antiquated principles, such as the recent case in Louisiana that blocked the FCC's proposals to update depreciation schedules to help spur technological upgrades—a decision almost everyone in the modern industry found distressing [15]. The FCC, for its part, is bound by an unamended charter from the Depression era (1934), which (as the Louisiana case has shown) has left it without sufficient powers to manage the technological dimension of the network evolution. The states are for the most part still fighting the battles of the Progressive Era when the idea of a "public utility" was born—in an age when it is finally dying. Meanwhile, the network has grown by a factor of 100 in scope and has undergone a technological metamorphosis resulting in new capabilities that we are really unprepared to manage in the traditional manner. We have been watching the caterpillar for so long, we are not ready to deal with the dazzling butterfly it has now become.

In the not-so-distant past, there was hope of containment as a viable long-term solution. The philosophers of public utility administration (including both regulators and the managers of the regulated carriers) often tried to restrain and even suppress these technical developments in ways that now seem ludicrously misguided. The long battle over the interconnection of "foreign" devices to the network—from Hush-A-Phone in the 1950s to Carterfone in the 1960s to the protective coupling arrangements of the 1970s—is a strange enough story [16]. Can it have been so recently that it was a *crime* to own an answering machine? A far more important suppression of innovation has been the long delay in the introduction of digital systems into the network. For all its vaunted technical sophistication, the Bell System was still introducing new analog switching and transmission systems well into the 1970s [17]. Meanwhile, the development of data communications has taken place largely outside of the traditional network, and has provided by now a compelling alternative model for the management of a complex, technologically sophisticated industry that is based on networking principles. As we stand back just a bit from the past, we now see fairly clearly how many similarities there are between the telecommunications industry—which is highly regulated—and the computer industry, which is virtually unregulated.

I believe we have reached a stalemate. Various complex technological trends are undermining the regulatory framework. Diagnosticians have pointed to the diverse symptoms, and the commentators have bestowed new names on the emerging realities of the Information Age and the geodesic network that clearly do not fit the traditional regulatory models [18]. The death of the public utility concept has been announced many times [19]; but earlier "obituaries" have usually reflected an American's political discomfort with the idea of a government-protected, privately owned monopoly. Its present and final demise in telecommunications has to do with the long-delayed but now inescapable consequences of a technological revolution.

A new paradigm is emerging, of course. It is based on the new icons of competition and market mechanisms. So stated, the deregulator may feel his plan is straightforward; he has but to remove the scaffolding and to emulate the market structures he finds in other segments of the economy. But deregulating telecommunications (if it were to occur) would not be simply a matter of waving away the cobwebs of history. There is still the problem of ensuring universal service. It is still true, I believe, that universal service could not survive the full deregulation of the network, given the current dependence on copper-based local access. At the heart of this paradox is a hard knot of diseconomy that is, I believe, central to understanding all regulatory solutions past, present, and future.

We do not need to recapitulate the vast subject of telephone regulation in its entirety. Many fine works are available in this field [20]. In any case, our focus is much narrower: We need to understand how the copper-based access plant still confounds the regulators and still blocks the path to the technicolor networks of the next century, and how the advent of radio-based access technologies will release the process from these disadvantages.

In the following sections, we will review some of the milestones in the theory and history of telephony regulation, as it bears on our main topic.

3.2 SUCCESSFUL SOCIALISM

The local exchange has been called a *natural monopoly*—a troublesome concept, "both static and dubious" [21], a "fiction" [22]. Is there such a thing? Are there forms of economic activity that are by their nature best accomplished by a single firm, and is telecommunications one of them?

I prefer historical answers to questions of this type, rather than answers given by theory. Whether a telephone monopoly is like a redwood tree, a thing of Nature, or whether it is a most unnatural artifice of cunning businessmen and colluding regulators, one thing is certain: The idea of a natural monopoly is a powerful political slogan. It captures a complex set of economic arguments in a phrase that seems somehow transparent even to the layman. Theodore Vail perceived this in his successful deflection of the federal antitrust suit against AT&T in 1913; he offered his "One system, one policy, universal service" as a blueprint for what became a unique structural solution to the problems of competition and conglomeration: a nationwide corporation, a privately owned enterprise, of massive proportions (the "biggest company on earth"), upon which was conferred a government-protected monopoly franchise. In fact, some eight years after the Kingsbury truce with the executive branch, Congress specifically emancipated the telephone company from antitrust worries by passing the Willis-Graham Act [23]—which allowed AT&T to grow from about a 50% share of the telephone access market in 1910 to about 84% by 1950. In return for the monopoly, Vail promised to support universal service (although it would be at least another half century before this promise could be deemed to have been kept) and to submit to financial regulation of his rates of return (although this, too, was often loosely applied).

How did the telephone companies pursue their promise of universal service? Essentially, the framework that evolved was based on two pricing concepts—*averaging* and *separations*—both of which were forced on the industry by the economics of copper-based access plant. Both are inherently subsidy mechanisms. (The term *subsidy* is used here to mean simply that some users pay substantially more than others for the same service and/or that some users pay substantially less than the actual costs of the facilities used to provide them with service.) As such, they require protection from cherry-picking competitors who could attack the subsidized markets at their weakest points. Furthermore, the whole arrangement was enormously aided by a series of happy accidents of technology that allowed for an even more far-reaching subsidy system to be implemented with relatively little political or customer resistance.

Averaging

In 1937, the counsel for the new-born Federal Communications Commission wrote that:

> [It is] universally recognized that telephone toll rates cannot successfully be based upon the particular costs involved . . . an averaging process is therefore essential. [24]

Cost averaging is practiced, to a degree, by almost every business. The cost of shipping a can of Coke from the bottler to a store 100 miles away is greater than the cost of shipping the same can to a store 5 miles away. Yet the price is the same; thus, the customers of the nearby store could be said to "subsidize" (however slightly) the customers in the outlying town. The price is based not on the microscopically accurate actual costs of each shipment, but on the average cost for all shipments. (Of course, in extreme situations, even the cost of a soft drink may be affected; I remember paying $3 for a can of Coke in a northern Alaskan village, where all such supplies were brought in by plane.)

In short, averaging means that instead of basing the price to a given customer on the actual costs incurred to provide service to that customer, the total cost to serve all customers is divided by the number of customers served to yield an average cost. The average cost is then marked up to achieve a fixed rate of return, and that average price becomes (with some exceptions, mainly for extremely remote customers) the price for all.

Classical economic theory teaches that such a system should present potential competitive opportunities. The customers who provide the subsidy—that is, whose true costs are less than the average cost—can become targets for a competitor who does not have to address the distant market. Nevertheless, the system works well, even in a competitive market, as long as the variation in the cost base is not too great. Probably the difference in the shipping costs of the two cans of soda is less than 1%. The subsidy is so tiny that there is little room for a competitor to gain a market edge. If significant cost disparities do develop, the free market allows tiered pricing, so that prices more closely match actual costs and eliminate these competitive vulnerabilities.

This is not true for telephone service. First of all, the costs of serving different customers, even within a single exchange, can vary by at least a factor of 2 to 5. A recent study of fiber-in-the-loop identifies this as a difficulty for planning the conversion to fiber optics:

> The key target for first-cost parity with copper ranges from $400 to $1600. The wide variation is due to . . . the very nature of the marketing application of the service, which includes the provision of one, two, three, four, or more voice-grade lines to the living unit. The second primary reason is the demographics of the application. It is very difficult to arrive at a meaningful average cost, since the density of living units . . . drastically affects the cost study comparison. [25]

Another study indicates that the density of lines "varies by several orders of magnitude between rural, suburban, urban and super-urban population areas . . . [and] the line cost is found to be inversely proportional to customer density" [26]. In short, cost disparities across different exchanges—say, from the urban center to an adjacent rural area—can vary by a factor of 10 or more, ranging today from several hundred to several thousand dollars per access line [27].

Second, the normal competitive response—tiered pricing—is not an available option (except in the most extreme situations). It is not politically acceptable to charge a farmer 5 or 10 times what the city dweller pays for basic service, even if his cost base is 10 times greater. Moreover, if this were done, the result would likely be a decrease of rural service on a massive scale. The 1934 Communications Act specifically provides for "reasonable charges," and glosses this by pronouncing "any unjust or unreasonable discrimination in charges" to be "unlawful" [28]. Whether the pricing of local rates on the basis of individual costs, with the resulting wide range of variation, would be held to be discriminatory is not easy to say. "In the end," writes one author, "undue discrimination is whatever the regulatory commissions or the courts say is undue discrimination" [29]. The determination would be political, undoubtedly; but ever since the earliest days of telephony, price averaging has been practiced. Indeed, the scope of this averaging has tended to expand over time, such that today the use of averaging extends nationwide, for certain components of access cost [30].

The result is that urban customers tend, on the whole, to subsidize rural customers. It has been argued that this is not simply a tax on urban telephone subscribers, since the quality of the service provided by the network is enhanced by having more people gain access to it. Nevertheless, it is clear that most rural loops are *per se* money-losing propositions, or would be if the accounting were done on an individual basis. And many urban and especially business users are paying not only for their own costs but for a majority of the costs of these rural customers as well.

This unusual situation is a straightforward consequence of the cost structure of copper-based electrical transmission facilities. It could not survive a competitive environment, where the competitor could skim the cream by eliminating this subsidy from his prices. But it made the system work, and allowed the extension of service into large segments of the population where telephony would otherwise never have reached.

Separations

The other important pricing principle for telephone access is rather more esoteric. As this regulatory framework began to take shape, the telephone company (we shall refer here to AT&T as the principal actor) found itself under pressure from local regulators to moderate its rates for basic local service. This pressure took the form of court challenges to AT&T's pricing policies, which ultimately resulted in the landmark case of *Smith v. Illinois Bell* coming before the United States Supreme Court in 1930 [31]. In essence, this case involved an argument over the allocation of costs between local service and long-distance service. Since local access facilities carried both types of traffic, the answer was never completely straightforward (see Figure 3.3). It seemed unfair to base local service rates on the assumption that 100% of the costs of the local exchange (the local switch and local loop facilities) should be allocated to local service, which would result in higher local rates. The court, obviously struggling with difficult technical questions, came to the conclusion that some apportionment of the costs of the local access facilities to the two types of service was necessary.

> It is obvious that, unless an apportionment is made, the intrastate service to which the exchange property is allocated will bear an undue burden—to what extent is a matter of controversy. [32]

This decision became the basis for the financial compromise that probably made universal service possible—and which now holds the industry in thrall.

Happy Accidents of Technology

The compromise was, at first, a logical one. The relative percentages of local traffic and interexchange traffic could be used to separate the expenses of the local exchange into two accounts, the first to be used for establishing local service rates and the second to be paid for out of long-distance or toll revenues. As one history comments:

> The whole process was called the "separations" process. Its procedures were embodied by 1947 in a complex Separations Manual that—as revised—has been the bible of this private-public process ever since. [33]

For anyone who has ever been involved even on the periphery of this process, the religious metaphor will seem apt.

(a)

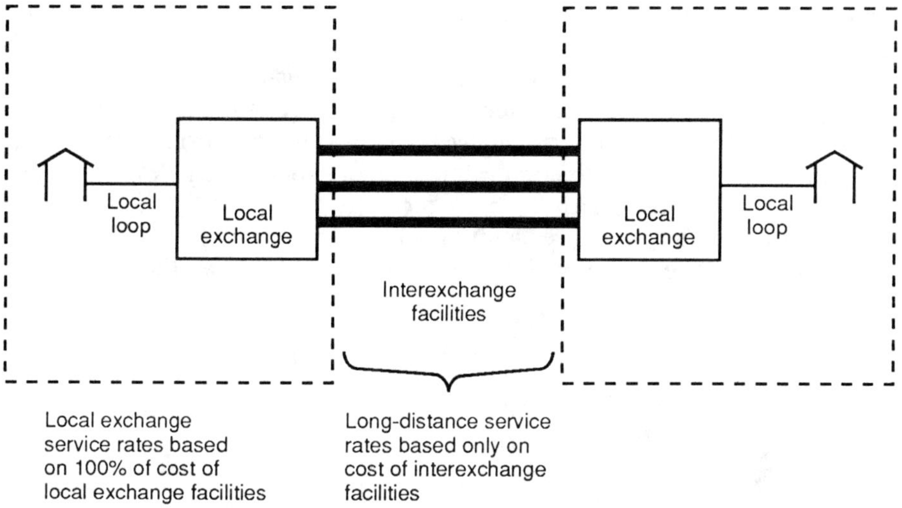

(b)

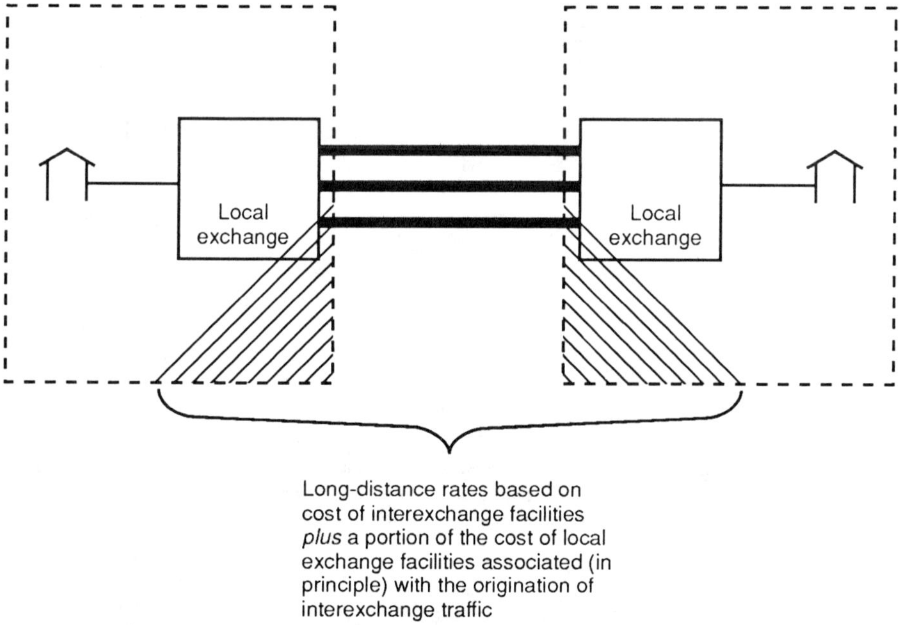

Figure 3.3 (a) Board-to-board *versus* (b) station-to-station accounting of network costs.

The initial allocation of exchange costs to interstate traffic was based on actual minutes of interstate traffic, and was therefore quite small—less than 5%. But the imposition of this new arrangement coincided with the beginnings of the technological transformation of the core network, as we have discussed in the previous chapter. Carrier systems, microwave, and finally fiber optics had a tremendous impact on the economics of the interstate, interexchange segment of the network. In particular, the costs of long-distance transmission declined by a factor variously estimated (as we have seen) at between 100 and 1000 in real terms [34]. Specifically, the net book cost per circuit-mile for long-distance facilities fell from $217 in 1930 to $11 in 1980, a tremendous decrease even before inflation adjustments [35] (see Figure 3.4).

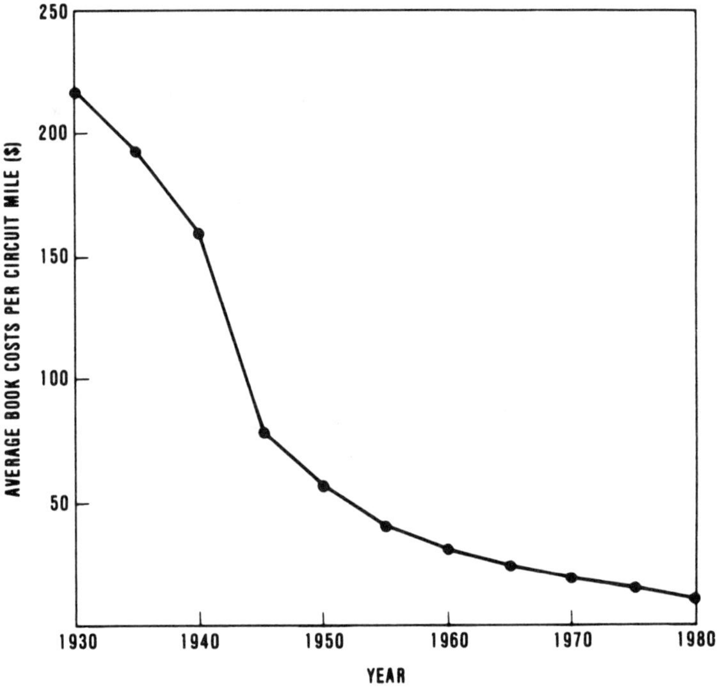

Figure 3.4 AT&T long lines department average embedded cost per circuit mile, 1930–1980. (From E.F. O'Neill, ed., *A History of Engineering and Science in the Bell System: Transmission Technology (1925–1975)*, AT&T Bell Laboratories, 1985, p. 781.)

This created an opportunity for a painless expansion of the access subsidies. The long-distance business was becoming immensely profitable, even after dramatic rate reductions for toll calls. In fact, the burgeoning profits of the interexchange sector were an embarrassment for AT&T, which was under continuing antitrust pressure. Meanwhile, the per-line costs of local exchange operations—bypassed by the technological revolution—were not declining, and the company was expanding service to higher cost segments of the population beyond the major cities. The obvious and generally acceptable solution was to allow AT&T to begin to skew the separations process so as to shift a portion of the long-distance profits to support the local loop. The percentage of local exchange costs that was allocated to the interexchange sector began to rise, far beyond the much slower increase in actual interexchange traffic. By 1980, the subsidy from the long-distance business had reached nearly 30% of local exchange costs (see Figure 3.5).

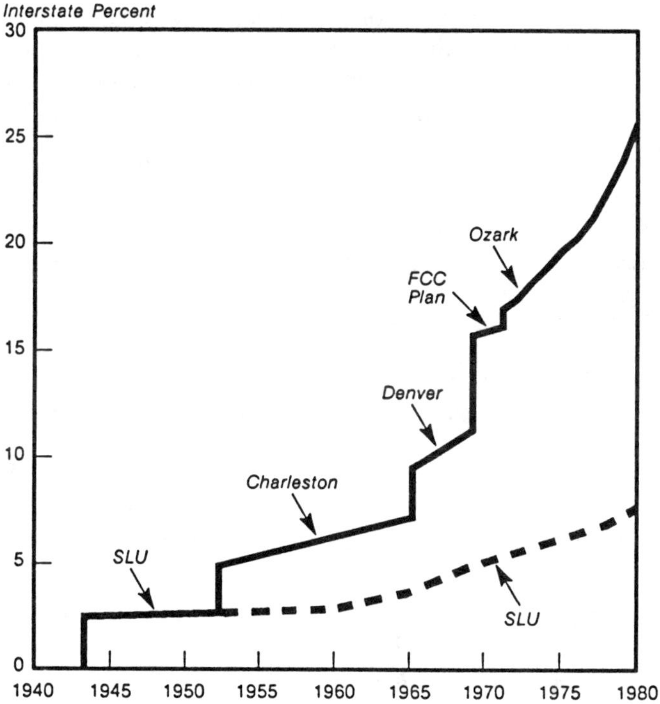

Figure 3.5 Exchange plant allocated to interstate service, 1940–1980. (From P. Temin, *The Fall of the Bell System,* Cambridge University Press, 1987, p. 26.)

Once again, it is striking how the economics of copper-based electrical transmission facilities skew the system; had the long-distance network remained copper-based, these subsidies would not have been nearly as easy or painless. Local access pricing would have remained closer to actual costs (assuming no direct government subsidies). Universal service would undoubtedly be not quite so universal.

Successful Socialism

This, then, was the formula for successful socialism. A private monopoly was allowed to operate its own welfare system for telecommunications (supplemented to a degree in extreme situations by more explicit government subsidies such as the REA funding). The "Haves"—businesses (who always paid more than residential customers, regardless of the fact that costs did not greatly differ), long-distance users (who were enjoying cheaper and cheaper rates anyway), and urban customers—were levied, taxed, overcharged, tithed, whatever word applies, to support the "Have Nots," mainly residential and nonurban customers whose per-line costs were so much higher. To make it workable, the private operator had to be allowed (guaranteed) a profit on the overall operation of the network. He had to be protected against competition from others who might try to skim the cream (underprice to win away the lucrative urban and business customers on whom the subsidies depended). As one recent history has commented:

> Regulators could channel the benefits of Bell Labs' new technology to local rates in pursuit of universal service because the Bell System functioned as a unit. It did not matter—or at least it did not seem to matter in the 1950s—that prices and costs for individual services were increasingly diverging. [36]

Therein lay the instability, or the vulnerability, of the system. To work, the monopoly had to become all-encompassing. And given the American antipathy to any monopoly, the persistence of these arrangements depended on the continuing credibility of the courts and others for the natural monopoly argument. Alas, the same technological breakthroughs that had made this welfare system economically feasible were also beginning to undermine the viability of the natural monopoly argument, and as the courts and the FCC began to recognize and act on this fact, the careful arrangements of this successful socialism began to unravel.

3.3 THE UNRAVELING OF THE NATURAL MONOPOLY

The Genie Escapes: Microwave Radio

The economics of copper plant had forced the creation of this system. Beginning in the 1950s it was radio—with its very different cost structures—that began to undo it. At that time, of course, it was not radio in the local loop, but point-to-point microwave radio for long-distance transmission.

Microwave was developed in the 1940s and first applied commercially for long-distance telephony in 1950–51 [37]. It is perhaps surprising to look back now on the tremendous impact it had. In 1950, there were about 25 million circuit-miles in the U.S. toll network. By 1980, there were more than one billion circuit-miles in the long-distance network—an increase by a factor of 40 [38]. Microwave technology was the driver for most of this expansion. From the mid-1960s on, there were more circuit-miles on microwave radio systems than on all other technologies combined (Figure 3.6).

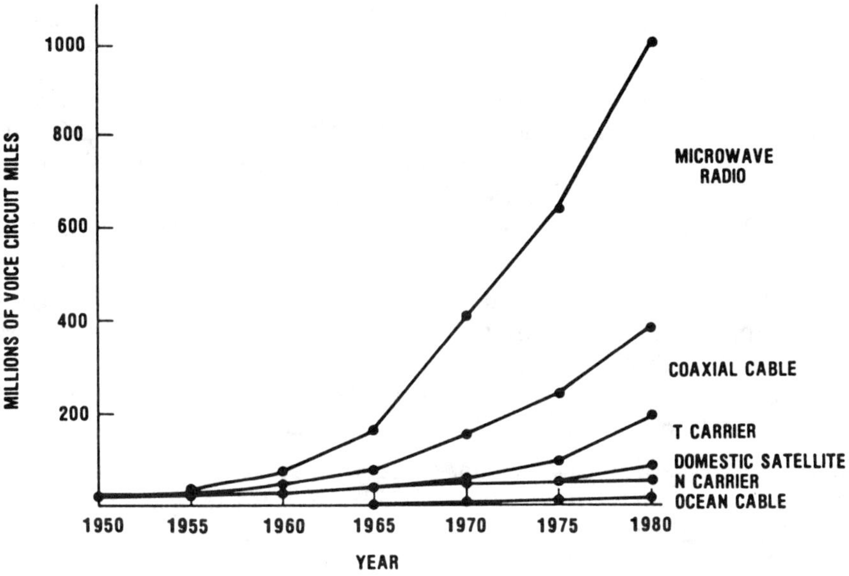

Figure 3.6 Transmission plant mileage, 1950–1980. (From E.F. O'Neill, ed., *A History of Engineering and Science in the Bell System: Transmission Technology (1925–1975),* AT&T Bell Laboratories, 1985, p. 781.)

It was clear almost from the outset, however, that radio did not fit the mold of natural monopoly. Other radio-based services, such as broadcast radio and TV, as well as the nascent land mobile radio business, were marked by vigorous competitive market structures that, if anything, tended toward fragmentation rather than oligopoly. From the beginning, the FCC had sanctioned the sale of microwave equipment to certain classes of users, such as railroads and electric utilities, who operated these systems themselves, for their own purposes, in areas where there were no adequate common carrier facilities [39]. But now the question became more pointed. If AT&T might build a microwave link between two cities, or even an entire microwave network, there was no obvious reason why someone else—anyone else—should not be able to do the same. Many large communications users—including, for example, the growing TV networks who needed a lot of bandwidth for distribution of their programming to affiliates—saw in microwave radio the opportunity to build their own private networks and free themselves from dependence on AT&T. Abetted by opportunistic suppliers such as Motorola (who needed to foster a non-Bell market for their radio equipment), these large users ultimately brought the issue before the FCC [40], where in 1959 the landmark "Above 890" decision was issued.

Above 890 and Beyond

The matter before the FCC had to do with the allocation of new spectrum above 890 MHz, which the advances in microwave technology had recently made available. Essentially, the FCC decided to allow large users to build their own radio networks. There were certain conditions imposed, but no one was deceived. The issue was competition. Henry Boettinger, heading an AT&T planning group that was charged with assessing the impact,

> saw the *Above 890* decision as a decisive turning point in FCC policy: "Competitive supply would be substituted for the principle of common carrier provision of an important and growing sector of communications." There in microcosm was the central issue that would lead AT&T and the federal government into deeper and deeper conflict over the next two decades. [41]

This much was perhaps consciously understood by the FCC and the industry. But it seemed a rather abstract proposition. The decision enabled a few large companies to buy microwave radio hardware for their own use. The managers of the telephone monopoly, by nature, resisted any loss of territory, any hint of competition. The regulators, with their simmering antitrust inclinations, sought at least to restrain the more exaggerated ambitions of the monopolists. These were the natural positions of the parties. As such, the outcry seemed but one more cliché of the regulatory process. For all the dire proclamations, there appeared to be no real cause for alarm. The FCC had allowed competition in mobile radio, for example, without precipitating a catastrophe. And after all, private microwave was such a tiny niche, far outside the mainstream of the telephone business. It seemed to provide little precedent for any drastic alterations in the basic telecommunications monopoly.

But the implications for the successful socialism, however obscure at that moment, were dire indeed. As the authors of *The Fall of the Bell System* wrote in 1987, after it had all come unraveled:

> The FCC's decision in *Above 890* broke with the basic principle of nationwide averaging and cross-subsidization in the telephone network. It marked the first step down the distressing path. . . . [42]

"Above 890," as it turned out, was one of those seemingly insignificant acts that became the first falling domino in a great historical process. The chain of reactions and escalations—the AT&T predatory TELPAK tariff, the FCC's MCI decision, the creation of the specialized common carriers, the Execunet decisions, and many other regulatory milestones through the 1960s and 1970s—can all be traced back to the decisive impact of radio technology on the thinking of the industry and its regulators [43]. Here was a transmission technology with a cost structure that was clearly compatible with a competitive environment. Essentially, radio made competition possible in the long-distance segment of the industry, which led to the creation of companies such as MCI and the acceptance of market mechanisms instead of public utility regulatory theories for supervising interstate telecommunications. In time, even AT&T acknowledged and embraced the competitive implications in the interexchange segment [44], realizing that its greatest advantage in the interexchange business accrued if it was fully free to compete (a status it has still not entirely achieved). Eventually and inevitably, competition put pressure on long-distance prices. As pricing became competitive (i.e., cost-based), the internal surpluses that AT&T Long Lines had been generating began to evaporate. Whither the telecommunications welfare system? Whither the local loop?

During the technological metamorphosis of the core network, the local access system remained copper-bound, relatively untouched by high technology. It had endured in its larval form, arguably still a natural monopoly. As the 1960s gave way to the 1970s, the differences between the uncompetitive, anticompetitive, subsidy-ridden local access environment, and the highly competitive market for interexchange services began to stand out more and more. The AT&T monolith began to look more and more like a pair of nonidentical Siamese twins: a still-massive public utility providing basic low-tech telephone service as a strict monopoly, joined to a rapidly evolving long-distance network operating in a ferocious competitive jungle. The antitrust watchdogs picked up the scent again. Whether it was ever AT&T's intent to use its monopoly power to bash its long-distance competitors, it soon became clear that its ownership of the monopoly facilities of the loop provided AT&T with so many opportunities for inadvertent anticompetitive behavior that supervision, however vigilant, and mere sticks and carrots, would be insufficient to prevent abuses.

Shedding the Copper Husk

The solution was radical surgery, called *divestiture,* incorporated as the Modified Final Judgment of Judge Harold Greene's federal court, handed down in 1982 [45]. AT&T decided to shed 80% of its facilities investment—the obsolescent copper-based loop networks—and to keep the competitive, high-tech service and equipment segments of Western Electric and Long Lines (and, of course, Bell Labs).

I can remember at the time being taken aback by what seemed the drastic nature of AT&T's dismemberment [46]. Back then, most observers (and many insiders) expected some type of horizontal amputation, the separation of some of the local

Bell operating companies and probably Western Electric as well [47]. The spin-off of all the operating companies, and two-thirds of the assets, was breathtaking, stunning. It was debated immediately as to whether the divestiture was a fiasco or a brilliant coup, and the interpretations of the MFJ outcomes continue to puzzle many observers.

Whether we judge it a success, a disaster, or simply an inevitability depends I think on what we believe the MFJ was intended to accomplish. Many theories have been put forth. The AT&T theory, and one that appears sincerely held, is often said to be based on the perceived convergence of computers and communications and the company's desire to find some way to free AT&T to compete in new equipment markets [48]. That may be one of the outcomes, but I think the real objective was to solve the problems created by the intractable local exchange monopoly, which came to be called by a new term conveying the frustrations of the new era—the *bottleneck:*

> The theory that underlies the modified final judgment is straightforward. AT&T owned bottleneck facilities in a regulated industry. This control gave it the power to exclude competitors in upstream and downstream markets through denial of access or patronage and allowed it to subsidize unregulated activities from its regulated markets. Because its operating companies controlled access to more than 80% of the country's telephones, it could use its franchised monopoly of local exchange and access services to monopolize equipment supply and interexchange. [49]

The Huber Report, three years after, concurred:

> The MFJ's central assumption is that local exchange service . . . is a natural monopoly. [50]

The divestiture of AT&T did not eliminate this natural monopoly. It alleviated one rather narrow subset of antitrust concerns. To a degree, it did isolate the monopoly elements from the rest of the system—to what end? It surely failed to simplify the regulation of the local exchange. Policy-makers are still scrambling to substitute for the old, simple, effective averaging procedures that allowed the integrated AT&T to function. The local monopoly still distorts the structure of the industry. It defines the boundary between "regulated" and "unregulated" services—the most fundamental strategic consideration for investment decisions by common carriers today. It has even redrawn our map of the network. According to the U.S. Department of Justice:

> The primary objective of the exchange area concept [i.e., the LATA] is to separate competitive . . . intercity services, from the bottleneck local distribution function. [Emphasis added] [51]

But perhaps it did bring us a large step closer to the ultimate solution. Forty years ago, even 20 years ago, the stark diseconomies and technological shortcomings of the local access system were hidden from clear sight, shrouded in subsidies and specialized accounting. Even the managers of the system, I am convinced, did not understand their access costs particularly well. They had little incentive to do so. They made their return on assets, not on operating efficiencies. The oyster was closed shut.

Today, the copper loop stands out like a sore thumb, in all its baroque inefficiency. We are, I believe, on the downslope of the life cycle, where with every passing year the obsolescence of the copper-based local exchange monopoly becomes more and more dangerous, like an old house that threatens to collapse. The underlying threat is the potential for a painful retrenchment of rural telephone service, which becomes more real as competition advances further into the profitable segments of the network and eliminates the various sources of subsidy one by one. In every industry where we have unleashed competition into formerly regulated markets—whether airlines, bus service, even medical services—we have seen rural dieback. For example, it is a poorly kept secret in the industry that many of the major LECs have already concluded that they do not want to serve rural customers. There is too little revenue potential and too much cost exposure. Also the long depreciation cycles make for extremely unattractive commitments of capital in an era when the technology life cycle is probably less than 10 years. In many cases, the LECs are actively investigating ways of shedding these money-losing customers at the first opportunity. The minute they can, some of them will begin to do so. Are we prepared to maintain the local monopoly indefinitely, along with all of its props and prosthetics? Can Judge Greene live forever? The social compact is in danger.

We are just beginning to understand the many penalties we pay—technical and economic—for the neglect of the local access system. *Neglect* may seem a harsh word. I mean not so much the failure to invest, but the failure to modernize. As we enter the 1990s, the vast majority of telephone switches in the former Bell System are still analog switches. End-to-end digital transmission on the loop side of the switch is still extremely rare. The laboratories of the Bell System are on the cutting edge, but the deployment of new technologies into the access environment has been appallingly slow. To put it in perspective, if the same principles were applied in other industries, we would still mostly be driving cars designed (if not actually built) in the 1950s, and using computers from the punch card era. Sounds like Cuba, or East Germany.

This metaphor, too, may prove apt. Many have commented on the broad organizational and cultural similarities between state-run monopoly enterprises and the benign soviet of the Bell System. The MFJ was the counterpart of glasnost for the Eastern Bloc. It did not in itself eliminate the local monopoly, but it made it possible for us to lay clear plans to do so. From a technical standpoint, those plans will be vitally dependent on wireless access. Just as radio "liberated" the long-distance business, both physically to expand by a hundredfold and economically to become a competitive market, so will radio now liberate the local exchange. Just in time, too, because the services revolution we shall discuss in Chapter 4 is about to raise the stakes.

NOTES

1. We may arbitrarily date the supersession of market forces by government regulation and the beginning of the public utility theory of telecommunications to 1913, when the federal government concluded a truce in its antitrust proceeding against AT&T, known as the Kingsbury Commitment. It is true that regulation by some state utility commissions predates this by a few years, and it is also true that, Kingsbury notwithstanding, it was not until the 1930s and the creation of the Federal Communications Commission that we find the mechanisms for active (as opposed to reactive) regulation actually in place.
2. *"The essence of regulation is the explicit replacement of competition with governmental orders as the principal institutional device for assuring good performance."*—Alfred E. Kahn, *The Economics of Regulation: Principles and Institutions,* New York: John Wiley & Sons, 1970, Vol. I, p. 20. The broad scope of this regulation may be surprising to those not directly involved with it. Consider the following excerpts from a state PUC annual report (California, 1989), in which the PUC describes its own regulatory agenda with respect to the telephone company carrier: "[*Two*] *issues remain unresolved and the Commission's review continues in separate proceedings. The first issue concerns a continuing audit of affiliate transactions in three areas: joint ventures, strategic alliances, and research and development. The second issue . . . is a determination of whether the plant modernization decisions of Pacific Bell's managers were prudent . . .* [*Further comments on the second point included the following description of the "paper war" between Pacific Bell and the PUC over the "prudence" of management decisions.*] *Pacific Bell introduced a lengthy report by A. D. Little (an accounting firm) that concluded approximately 2% of Pacific Bell's investments were partially unreasonable. In response* [*the PUC*] *contracted with SRI (a consulting firm) to examine Pacific Bell's investment. SRI produced a lengthy study in March 1988 that concluded approximately 24% of the projects examined fell into a category in which SRI deemed investment justifications were deficient. Furthermore, SRI estimated that approximately $206 million of the $3.02 billion of investment that it examined were uneconomic . . .* [*The PUC*] *has subsequently issued a report calling for a revenue requirement adjustment of $440 million to penalize Pacific Bell for this poor performance, and an annual reduction of $130 million in its revenue requirement until Pacific Bell demonstrates that it has adopted procedures that yield prudent investments. . . ."* From *The 1989 Workplan,* published by the California Public Utilities Commission, pp. 54–55.
3. From the famous *Munn v. Illinois* decision of the U.S. Supreme Court in 1877, the case which established the basis for government regulation of certain industries as public utilities; see L.S. Hyman, R.C. Toole,, and R.M. Avellis, *The New Telecommunications Industry: Evolution and Organization,* Arlington, Va.: Public Utilities Reports, Inc., 1987, p. 196.
4. Arthur Stone Dewing, *The Financial Policy of Corporations,* New York: The Ronald Press, 1953, p. 309.
5. J.C. Bonbright, A.L. Danielsen, and D.R. Kamerschen, *Principles of Public Utility Rates,* 2nd ed., Arlington, Va.: Public Utilities Reports, Inc., 1988, pp. 35 ff.
6. William G. McGowan, "Investment in Telecommunications: Opportunities or Pitfalls in Today's Environment," *IEEE Communications Magazine,* January 1989, p. 30.
7. Henry Geller, "U.S. Domestic Telecommunications Policy in the Next Five Years," *IEEE Communications Magazine,* April 1989, p. 20.
8. G.W. Brock, *The Telecommunications Industry: The Dynamics of Market Structure,* Cambridge, Mass.: Harvard University Press, 1981, p. 303.
9. Notwithstanding the ability of some carriers to "prove" the opposite, in connection with esoteric and often specious rate case arguments; see Congressional Budget Office, U.S. Congress, "The Changing Telephone Industry: Access Charges, Universal Service, and Local Rates," Washington, D.C. U.S. Government Printing Office, June 1984, p. 73, Note 9.

10. L.S. Hyman, R.C. Toole, and R.M. Avellis, *The New Telecommunications Industry: Evolution and Organization,* Arlington, Va.: Public Utilities Reports, Inc., 1987, p. 206.
11. K. Flamm, "Technological Advance and Costs: Computers vs. Communications," in R.W. Crandall and K. Flamm, ed., *Changing the Rules: Technological Change, International Competition, and Regulation in Communications,* Washington, D.C.: The Brookings Institution, 1989, pp. 55–56: *"Historically, rates of capital equipment depreciation in the telephone industry have been very low. In the computer industry, new products have typically had a life expectancy of six or seven years. By contrast, in the late 1960s the FCC-approved service life for most central office communications equipment was about twenty years, and forty for electronic switching systems."*
12. *United States v. Western Electric Company, Inc., et al.,* Civil Action No. 82-0192; Notice of Entry of Final Judgment, *Federal Register,* Vol. 47, No. 177, September 13, 1982, pp. 40392–40396.
13. Alfred E. Kahn, "Some Thoughts on Telephone Access Pricing," based on comments delivered at Workshop on Local Access: Strategies for Public Policy, September 14–17, 1982, St. Louis, Missouri, p. 12.
14. Cited in J.C. Bonbright, A.L. Danielsen, and D.R. Kamerschen, *Principles of Public Utility Rates,* 2nd ed., Arlington, Va.: Public Utilities Reports, Inc., 1988, p. 29.
15. *Louisiana Public Service Commission v. FCC 476,* US 355 (1986).
16. See P. Temin and L. Galambos, *The Fall of the Bell System: A Study in Prices and Politics,* Cambridge, U.K.: Cambridge University Press, 1987, p. 47.
17. R.W. Crandall, "The Role of the U.S. Local Operating Companies," in R.W. Crandall and K. Flamm, ed., *Changing the Rules: Technological Change, International Competition, and Regulation in Communications,* Washington, D.C.: The Brookings Institution, 1989, pp. 114–146.
18. P.W. Huber, *The Geodesic Network: 1987 Report on Competition in the Telephone Industry,* Washington, D.C.: U. S. Government Printing Office, 1987.
19. For example, Horace Gray's famous article, first published in 1940, entitled "The Passing of the Public Utility Concept," *Journal of Land & Public Utility Economics,* February 1940. (Reprinted in American Economic Association, *Readings in the Social Control of Industry,* Philadelphia: Blakiston, 1942, pp. 280–303; this was paid further homage by Alfred E. Kahn, "The Passing of the Public Utility Concept: A Reprise," in Eli M. Noam, ed., *Telecommunications Regulation Today and Tomorrow,* New York: Harcourt Brace Jovanovich, 1983, pp. 3–37.
20. See the following representative texts: J.C. Bonbright, A.L. Danielsen, and D.R. Kamerschen, *Principles of Public Utility Rates,* 2nd ed., Arlington, Va.: Public Utilities Reports, Inc., 1988; L.S. Hyman, R.C. Toole, and R.M. Avellis, *The New Telecommunications Industry: Evolution and Organization,* Arlington, Va.: Public Utilities Reports, Inc., 1987; C. F. Phillips, *The Regulation of Public Utilities: Theory and Practice,* 2nd ed., Arlington, Va.: Public Utilities Reports, Inc., 1988.
21. L.S. Hyman, R.C. Toole and R.M. Avellis, *The New Telecommunications Industry: Evolution and Organization,* Arlington, Va.: Public Utilities Reports, Inc., 1987, p. 264.
22. Horace Gray, "The Passing of the Public Utility Concept," *op. cit.,* p. 284.
23. P. Temin and L. Galambos, *The Fall of the Bell System: A Study in Prices and Politics,* Cambridge, U.K.: Cambridge University Press, 1987, p. 11.
24. Cited in R.H.K. Vietor, "AT&T and the Public Good: Regulation and Competition in Telecommunications, 1910– 1987," in S.P. Bradley and J.A. Hausman, ed., *Future Competition in Telecommunications,* Boston: Harvard Business School Press, 1989, p. 46.
25. P.W. Shumate and R.K. Snelling, "Evolution of Fiber in the Residential Loop Plant," *IEEE Communications Magazine,* March 1991, p. 72.
26. T. R. Rowbotham, "Local Loop Developments in the U.K.," *IEEE Communications Magazine,* March 1991, p. 57.
27. This does not include the "extreme cases" in which remote rural customers may sometimes experience costs of as much as $10,000 to $40,000 for an access line. In most such situations, the costs are felt to be prohibitive and the facilities are not actually constructed.

28. See P. Temin and L. Galambos, *The Fall of the Bell System: A Study in Prices and Politics,* Cambridge, U.K.: Cambridge University Press, 1987, p. 11.
29. H. Craig Peterson, *Business and Government,* New York: Harper & Row, 1985, p. 184.
30. Congressional Budget Office, U.S. Congress, "The Changing Telephone Industry: Access Charges, Universal Service, and Local Rates," Washington, D.C.: U.S. Government Printing Office, June 1984.
31. P. Temin and L. Galambos, *The Fall of the Bell System: A Study in Prices and Politics,* Cambridge, U.K.: Cambridge University Press, 1987, p. 22.
32. *Ibid.*, p. 22.
33. *Ibid.*, p. 23.
34. E.F. O'Neill, ed., *A History of Engineering and Science in the Bell System: Transmission Technology (1925–1975),* AT&T Bell Laboratories, 1985, pp. 781 ff.
35. *Ibid.*, p. 782.
36. P. Temin and L. Galambos, *The Fall of the Bell System: A Study in Prices and Politics,* Cambridge, U.K.: Cambridge University Press, 1987, p. 27.
37. E.F. O'Neill, ed., *A History of Engineering and Science in the Bell System: Transmission Technology (1925–1975),* AT&T Bell Laboratories, 1985, pp. 291 ff.
38. *Ibid.*, pp. 783–4.
39. P. Temin and L. Galambos, *The Fall of the Bell System: A Study in Prices and Politics,* Cambridge, U.K.: Cambridge University Press, 1987, p. 29.
40. *Ibid.*, pp. 28 ff; see also L. S. Hyman, R. C. Toole, and R. M. Avellis, *The New Telecommunications Industry: Evolution and Organization,* Arlington, Va.: Public Utilities Reports, Inc., 1987, pp. 124 ff.
41. P. Temin and L. Galambos, *The Fall of the Bell System: A Study in Prices and Politics,* Cambridge, U.K.: Cambridge University Press, 1987, pp. 31–32.
42. *Ibid.*, p. 31.
43. The details are recounted in a number of recent studies, including P. Temin and L. Galambos, *The Fall of the Bell System: A Study in Prices and Politics,* Cambridge, U.K.: Cambridge University Press, 1987, and G. W. Brock, *The Telecommunications Industry: The Dynamics of Market Structure,* Cambridge, Mass.: Harvard University Press, 1981.
44. P. Temin and L. Galambos, *The Fall of the Bell System: A Study in Prices and Politics,* Cambridge, U.K.: Cambridge University Press, 1987, pp. 62 ff.
45. *United States v. Western Electric Company, Inc., et al.,* Civil Action No. 82-0192; Notice of Entry of Final Judgment, *Federal Register,* Vol. 47, No. 177, September 13, 1982, pp. 40392–40396.
46. *"The agreement was startling in its severity and for its dogmatic view of the world."*—L. S. Hyman, R. C. Toole, and R. M. Avellis, *The New Telecommunications Industry: Evolution and Organization,* Arlington, Va.: Public Utilities Reports, Inc., 1987, p. 157.
47. See the many "reactions" described in Steve Coll, *The Deal of the Century: The Breakup of AT&T,* New York: Atheneum, 1986.
48. *Ibid.*, p. 332.
49. R.W. Crandall, "The Role of the U.S. Local Operating Companies," in R.W. Crandall and K. Flamm, ed., *Changing the Rules: Technological Change, International Competition, and Regulation in Communications,* Washington, D.C.: The Brookings Institution, 1989, p. 118.
50. P.W. Huber, *The Geodesic Network: 1987 Report on Competition in the Telephone Industry,* Washington, D.C.: U.S. Government Printing Office, 1987.
51. Response of the United States to Public Comments on Proposed Modification of Final Judgment, 47 *Federal Register* 23320, 23341 (1982).

Chapter 4
THE SERVICE REVOLUTION AND THE UNCOMPLETED NETWORK

4.1 THE VOICE-OPTIMIZED NETWORK

The telephone network is still essentially a single-purpose instrument. It is so thoroughly optimized for real-time voice communications between human beings that in many ways it is but an electrical extension of ear and mouth—which is precisely how Alexander Graham Bell came to his invention in the first place. As he groped toward his conceptual breakthrough, Bell dimly glimpsed the telephone as a kind of artificial ear, with diaphragm and a magnet in place of the eardrum and the tiny bones of the inner ear, to translate the vibrations of the air into magnetically induced fluctuations in a variable electrical circuit. He went so far as to procure the ear fróm a cadaver, which he hooked up to a stylus to produce a visual record of spoken sounds [1].

Many mechanical inventions begin as prosthetic devices. In the case of the telephone, the imprint of human physiology is still pervasive. Analog telephone channels (99.9% of all local loops in the United States) are engineered to match the frequency response of the ear [2]. The standard 64 kb/s digital circuit used in the core network (and planned for the loop in the future) is derived from the same model; for example, the 8-kHz sampling rate is designed to capture (by means of Nyquist's theorem) only the information below 4 kHz—the frequency "weighting" of the human ear. The voice network mirrors the auditory filter that each of us carries in his or her head. It further takes for granted the traffic patterns of human conversation—the characteristic holding times, the symmetrical flows, the diurnal fluctuations—and optimizes everything in the system around those parameters. This presents no problem as long as human beings are the only ones trying to communicate over the network. Once we begin to hook up machines to talk to people and to other machines, we need a different kind of network, and a different way of accessing it.

4.2 THE EXPLOSION IN DIGITAL SERVICES

There always was, of course, an entirely separate network for data communications: the telegraph system built and operated by Western Union. It was engineered for different traffic characteristics, with different transmission objectives. For the most part, telegraph traffic was not real time; nor was it truly two-way in the sense of a duplex telephone circuit (with some exceptions, such as military telegraphy). It normally comprised short messages transmitted in bursts; and the messages consisted not of electrical analogs of speech sounds, but of a stream of digital symbols borne on short pulses of direct current.

Telegraphy led in an entirely different direction scientifically. Instead of studying acoustics and auditory and articulatory physiology, telegraph engineers were concerned with how to increase the rate of transmission of pulses through the electrical circuit and how to invent codes that could capture the information as efficiently as possible. These investigations laid the groundwork for modern information theory [3], but they lay almost entirely outside the compass of telephony, however. The two original networks—telephony for voice, telegraphy for data—were separate and distinct, carried on separate facilities.

About 30 years ago, that began to change. The emergence of time-shared computer systems led to the requirement for a new kind of data communications network based on a form of data traffic that was very different from telegraphy. Computers spoke to one another and to their users not in short, one-way messages, but in long interactive "sessions"; the volume of data transferred in a typical hook-up leapt a thousandfold or more and an interest in timeliness and error control began to predominate.

At first, time-sharing systems were implemented as closed networks, linking terminals to a mainframe over hardwired connections within a limited geographical area (such as a computer center) without any direct connection with the outside world. But almost in parallel, the desire for intercomputer communications arose. Moreover, the need for true remote terminals meant that dedicated hardwire connections were increasingly expensive. By the early 1970s, the need for a more advanced form of data communications had become acute. Solutions were developed along two quite different paths, neither of which has proven entirely satisfactory.

4.2.1 Voiceband Services

The first approach, not surprisingly, was to try to utilize the existing voice network—the telephone network—to handle data traffic. The voice network had many obvious advantages. It was ubiquitous; it was already "paid for" by the voice traffic it carried—adding data would be a way to increase revenue without substantially increasing costs. For the most part, it was well-engineered. It was also a switched network, which meant that a data communications overlay would inherit the convenient dial-up capabilities that telephone users already enjoyed.

The voice network, however, was essentially an acoustical system. While there were mixed analog and digital facilities throughout the system, the point of equivalence was the transparency of all facilities to a 3-kHz acoustical bandwidth. Whatever fit within that bandwidth—the "voiceband," roughly 200 to 3200 Hz or so—could be transmitted over the telephone system. Therefore, beginning in the 1960s, manufacturers began to develop equipment to translate the digital electrical signals of computers into acoustical signals—usually, at first, in the form of tones of specified frequencies—that could be passed through the voice network. It soon became routine to transmit digital information between computers over ordinary telephone lines. This is known as *voiceband data communications.*

One problem was speed, or the lack of it. Direct connections within a closed computer network (e.g., from the processor to a printer) operate at very high rates over specially designed facilities (e.g., ribbon cables). The first modems designed to work over ordinary telephone loops worked at 300 b/s or less. This was the equivalent of about 40 alphabetic characters per second. That was a lot faster than most telegraphs, and for some applications, such as the transmission of text to "live" operator terminals to be read by a human being, 300 b/s is (barely) adequate. For other applications, however, it proved far too slow and expensive. For host-terminal communications over long distances, the costs of the hook-up vastly exceeded all other computing costs. Nevertheless, the voiceband approach did allow the telephone network to be used for some modern data communications. Over time, the standard rate for computer modems inched up to 1200 b/s, 2400 b/s, and even higher. By ingenious digital signal processing, rates of 9600 b/s are possible on most circuits and higher rates (up to 19.2 kb/s) are possible on certain lines [4].

In the 1980s, an even more important form of voiceband data communications began to surge onto the network: *facsimile* transmission of images, or *fax.* The same technical strategy of adapting the digital information to the voice frequency characteristics was employed. Once again, the challenge was to speed up the data rate as much as possible while staying within the voice bandwidth. Advanced signal processing techniques helped raise the effective rate to the 2400 to 9600 b/s range—and the result was an explosion in fax traffic. I am told by Bell System insiders that today fax traffic accounts for several billion dollars per year in revenue, and is growing far more rapidly than any other telecommunications service. It is said to account for as much as 30% of trans-Atlantic traffic and up to 50% of trans-Pacific traffic [5]. Clearly, fax is one of the great success stories in the recent history of telecommunications. It was also largely unplanned from the standpoint of the network operators; facsimile has been a grassroots phenomenon.

The future of voiceband data communications services faces two serious problems, however. First, as noted, the voiceband channel is fundamentally limited as to the rates of information that it can transmit. Today's equipment has probably just about reached the practical limits of this approach. More importantly, as we push the acoustical channel to its limits, we are discovering that not all voice circuits are created equal. We fall back on the underlying problem—the economics and cost/performance equations for copper wire access facilities. Especially in rural areas or, more generally, for longer loops, the actual transmission characteristics may deviate from the norm. For voice communications, where we can rely on the marvelous "built-in" filter of the human ear to prevail, even under relatively adverse electrical conditions, these deviations may be acceptable in the sense that rural customers may still enjoy telephone service (albeit of a somewhat lesser quality); they can still "make a phone call." But for voiceband data systems, which rely on extremely sophisticated techniques to milk the acoustic circuit of every drop of carrying capacity, these small degradations loom large. In fact, many rural telephone circuits cannot reliably handle fax and data traffic.

All in all, the exploitation of the possibilities for carrying data traffic on the voice-oriented, essentially acoustically engineered telephone network is probably reaching the point of diminishing returns, at least in terms of bit rates and technical performance.

4.2.2 Packet Data Networks

The second strategy that evolved during the 1970s to handle computer traffic was very different. Instead of trying to utilize the existing voice network, an alternative network was developed—again separate and distinct from telephony, just as the telegraph system had been. This network was based on a very different architecture, which was better suited to the special characteristics of computer traffic. Unlike voice communications, computer-mediated communications tend to be somewhat delay tolerant (i.e., not quite real time), error intolerant, asymmetrical (flows in one direction did not always match flows in the other direction), and very bursty. Rather than force this traffic to fit the voice network, which meant incurring higher costs, lower data rates, and higher error rates, computer network designers began to evolve a different approach, which was based not on switched circuits, but on the routing of message units through a shared, robust, and high-speed network. This approach came to be called *packet communications*.

The technical differences between data-oriented packet communications and voice-oriented circuit communications are discussed in Section 6.4.6.1. The relevant point here is that packet networks have evolved as separate networks, optimized for data. They now offer much higher bit rates for file transfer and other data operations within computer networks. Packet networks may be configured either as local-area networks (LANs), within an office or factory, or, in the future, on a larger scale—either metropolitan-area networks (MANs) or wide-area networks (WANs). Today, there are many efficient new architectures for such networks, and this is one of the most active fields for standards development.

These new networks are based on transmission facilities that are usually quite different from those traditionally used to provide telephone access. Although some LANs can operate over conventional twisted-pair copper wires, the more common backbone is coaxial cable, optical fiber or some enhanced form of electrical transmission (shielded wire-pairs). The high bit rates these systems require for efficient throughput mean that conventional copper alternatives cannot be used. It also means that in general today's local loop telephone plant cannot be used to access such networks.

4.2.3 Narrowband Integrated Digital Services

To counter this trend toward separate networks for voice and data, the telephone industry has developed its own concept for an advanced integrated voice and data network that will be based not on the acoustical properties of the analog telephone circuit (and its transparent digital counterpart), but on an entirely digital transmission format, implemented either electrically or optically—the Integrated Services Digital Network (ISDN).

The first version of ISDN will be based on the integration of voice and what we may call *narrowband* digital services. In practical terms, narrowband digital services are those that would be needed to support most computer networking applications involving numerical, text, or text-like data, as well as conventional low- to medium-resolution facsimile. The data rates involved range up to 144 kb/s for a so-called "basic rate" ISDN interface. This is felt to be suitable for an individual household or small business, i.e., roughly comparable to an analog access line in today's network. The ISDN basic rate would provide services based on either standard voice circuits (up to two), a 64 kb/s digital bearer circuit, or a 16 kb/s packet channel (see Chapter 8).

The advantages of ISDN, from a network planner's perspective, are straightforward. Like the voiceband overlay, both voice and data services would share common facilities, which should be much cheaper than building separate networks. Moreover, the ability to integrate simultaneous voice and data transmissions could allow for innovative service offerings, such as the ability to link computers during a voice phone call, which could allow for interesting "electronic" Yellow Pages-type services, for example. Any customer service application where one end of a telephone call involves someone looking at a computer screen full of information (e.g., travel agents, airlines, hospitals, department stores, credit card companies) could become bilateral, with the same screen of information available at both ends.

And, of course, from a transmission engineer's perspective, ISDN provides a true digital format at the electrical (or optical) level. This allows higher bit rates by a factor of 10 or so compared to voiceband systems, which in turn facilitates a range of new services, such as the next-generation facsimile, based on 64 kb/s. It should also allow for better control over the error performance of the channel. The tempting conclusion is that 144 kb/s ought to be enough (for nonimaging applications) into the foreseeable future.

Alas, this will probably not be the case. Already, dedicated LANs can offer data transfer rates at megabit speeds to computer users, at least over short distances, and most new workstations are being designed around the assumed availability of communications channels capable of much higher data rates than 144 kb/s. We must be careful of assuming that there is a "natural" cut-off in data rates for important applications at 64 kb/s, 144 kb/s, or any other (low) level. It may be true that today the vast majority of all nonvideo applications would comfortably fit within the ISDN narrowband channel—but this is simply because higher rate channels have never really been available on a commercial basis to end users.

In fact, once again, if we peel back the layers of the ISDN architectural philosophy, we find the reptilian spoor of the copper-based access plant. The 64 kb/s bearer channel is obviously a relic of the voiceband network. Who said that our data requirements would come forth in 64 kb/s quanta? The 2B+D concept (see Section 8.1.1.2) is clearly derived from the two-voice-circuit-per-user serving area concepts developed in the 1950s and 1960s—entirely within the framework of voice telephony [6]. Indeed, the upper limit of 144 kb/s is based on the presumed limitations of the existing twisted-pair copper wire plant. Consider the very latest "application rules" for ISDN published by Bellcore in 1990:

> The ISDN-DSL [i.e., digital subscriber line] will use pairs in the [existing] loop cable plant. In the same cable sheath there will be pairs carrying baseband voice, data, and other transmission services. . . . A goal of the application is to allow easy selection and assignment of pairs suitable for DSL operation without special conditioning or field tests. . . . In brief, the rules will indicate that a "clean" pair is needed to avoid excess loss (i.e., no load coils or bridge lifters, limited bridged taps, etc.). These rules will permit operating ranges up to 18 kilofeet with restrictions . . . and less with bridged taps depending upon their length and location. . . . In general it is expected that standard DSL cannot operate in any binder group with early or present-vintage single-channel analog subscriber carrier systems, in the same or adjacent binder groups with multichannel analog subscriber carrier, or in the same binder group with data-over-voice or long 15-kHz program services. . . . [7]

Anyone who has ever examined the cross section of a large feeder cable, its binder groups and transpositions, and its mathematically prescribed twists can imagine the difficulties of implementing ISDN under field conditions.

In short, the limitations of ISDN are not market-determined limitations; they are the digitalized ghost image of the age-old limitations of the copper plant. Bellcore's new manuals admit as much:

> The standardization of this access facility [i.e., ISDN] proved to be a formidable technical task because of the variable high-frequency properties of the existing loop plant (such as impedances, frequency response, crosstalk coupling, and noise) and the data processing required to overcome these impairments. [8]

These descriptions evoke for me a certain sense of horror. *Those who do not learn the lessons of history are doomed to repeat them.* ISDN has been advertised as the veritable Millennium of telecommunications, but it begins to seem but the same old flawed, cuprified system compounded, the old mule decked out in feathers. The installed base of 130 million or so copper loops overshadows our thinking. We are laying down more new copper loops in the United States and Canada every year than the entire base of cellular radio over the past seven years. With the ISDN architecture, we would bind ourselves to how many more generations of copper wire in the access network? ISDN would improve the performance of some of this loop plant—at the cost of massive new investments in terminal electronics—but it is still tied to the copper access network, with all its limitations, economic and technical.

Moreover, just as the ability to receive high-quality voiceband services is limited by the underlying quality of the copper loop, so the availability of ISDN in any given area, to any given customer, will be determined (for quite some time, at least) by the state of the existing loop plant. The Bellcore applications rules would restrict ISDN to customers on loops shorter than 18 kilofeet—which leaves out some 15 to 20 million largely rural access lines in the United States. How many urban dwellers enjoy "clean" loops is somewhat harder to quantify. And even those of us lucky enough to live at the end of a short "clean" loop will find the very definition of "digital services" limited to those that can fit over an average wire-pair—and not even a new wire-pair at that, but one that in all likelihood was buried or hung years before anyone could spell ISDN. Will this be good enough to supplant the dedicated packet networks? Will this train take us to Information City? I think not. Most of us are already transporting data around at much higher rates, at least in our local environment, and what we are looking for, it seems to me, is the ability to extend the range of our megabit capabilities, not to reduce the bit rate to a relative trot because of limitations in cable design that were standardized in the 1890s [9].

4.2.4 Broadband Integrated Digital Services

A variety of architectures are under consideration for a future network that would support much higher bit rate services to end-users.

Wideband ISDN would deliver bit rates in the 1.5 to 45 Mb/s range (corresponding to the DS1 and DS3 rates in the current core network [10]). One might point out that these are simply standard "rungs" in the traditional digital hierarchy for voice transmission, but that would in some sense be just a quibble. These are bit rates that would accommodate many (though not all) of the new computer networking requirements and even some video applications. The program for such services is also rather unimaginative (which need not be interpreted as a criticism), since it is simply an extension of current core network transmission formats out into the loop environment as an access service.

More revolutionary concepts gather under the heading of *broadband ISDN,* or B-ISDN, which anticipates bit rates up to 1 Gb or more, sufficient for even full-motion, high-resolution video. Some of these concepts are combined with so-called "fast-packet" architectures in which voice, data, and video would all be intermixed on common facilities (switching and transmission) with the greatest of ease (see Section 6.4.6.2 for a very brief discussion).

Such transmission rates, if they become available to the end-customer as a form of enhanced access, would no doubt dispel the misgivings expressed in the previous sections of this chapter. A transmission rate of 600 Mb/s would supersede all of today's real or imagined service constraints. Yet there is a roadblock on the way.

All of these broadband concepts require a complete rewiring of the United States. The existing plant is totally incapable of carrying such Niagaran information flows. To even think about such advanced services, we must posit the replacement of the copper fabric by a nationwide network of optical fiber reaching all the way to the home. And this raises at least three extremely challenging problems:

1. *Cost:* The costs of such a rewiring were touched on in Chapter 2 and will be analyzed further in Chapter 9. It will undoubtedly require hundreds of billions of dollars in new investment.
2. *Time:* In any reasonable scenario for deployment of new plant, many decades would pass before even a majority of the access lines in America had been upgraded to fiber. I do not know whether this will mean 30 or 50 years to reach the 80% penetration point, but what is certain is that tens of millions of today's subscribers will not live to see the broadband era.
3. *Cost structure:* The most inescapable of the difficulties with fiber is that its cost structures—as an access vehicle—are essentially the same as the copper it replaces (see Chapter 9.) For all the talk about crossover points and falling costs of fiber electronics, the real economic savings of ISDN and B-ISDN is in the cost-per-bit-delivered, not in the cost-per-customer-served. There is no question that fiber can deliver more information for the same cost; this creates new revenue opportunities, which could be very important in driving the penetration of fiber. But the baseline cost of an access line—and, as we shall discuss in

Chapter 7, the relationships between categories of costs, material, labor, maintenance, etc., and population patterns are fundamentally similar. It is still outside plant.

4.3 THE INFORMATION RICH AND POOR

Imagine an electrification program in some fictional developing country. The planners move in, set poles and wires, install a small generator, and soon the village households are equipped with electricity—enough for a single 40-W light bulb per household. The engineers move out across the country, wiring it up the same way—one bare bulb per family. That is all the network can handle; the generators are sized around this requirement, the gauge of the wires fixed by the assumption that a 40-W bulb is all anyone will ever need. All the other possibilities, from refrigeration to electric stoves to vacuum cleaners, are excluded by the nature of the network design. The thinking is that everyone should have electric light, but the luxuries are beyond reach. (This is a socialist fiction.)

Such is, by analogy, the nature of the telephone network we have built [11]. Successful telecommunications socialism (see Chapter 3) resulted in a leveling of access opportunities and telecommunications services. Nearly everyone was able to obtain "basic" voice telephone service, and few resources were wasted on the production of "luxuries" (such as efficient computer communications). The unhappy consequence, however, is that we have an access network that is almost completely unsuited for the arrival of the Information Age and competitive markets for telecommunications services.

It is not difficult to see where this will lead us. As we dismantle our telephone socialism and allow competition into the industry, the gap between the Information-Rich—those who have high-quality access to the most modern services—and the Information-Poor—those who are shut out from the new services, for long periods of time—must grow. We have spent 100 years building a network around the policy of universal service, which until now has always meant simply voice telephony. Now, in a short decade or so, this equality of opportunity will be shattered. By the end of this century, the service options for the Information Rich will be dazzling. Those of us who live in metropolitan areas, centers of modern industry, or along the "fiber highways," will enjoy a widening advantage over those in the hinterland, many of whom will be struggling simply to pay their unsubsidized basic telephone bills. In the major metropolitan areas, the combination of competition and technological revolution will fuel economic growth and improved quality of life. In rural areas, it is quite likely that telephone companies will be looking to abandon even basic service, where possible. The rich ranchers will get by with satellite dishes, but the majority of the rural population will fall further and further behind.

It is a cruel paradox. Just as we have achieved universal service, the name of the game has changed. Voice services are no longer enough to compete. A business that does not have access to fax and data services is already severely handicapped. In 10 years, even higher bit rate services (whether ISDN, B-ISDN, or some other architecture) will be essential for economic competitiveness.

This is the final legacy of the copper access network. It may be clear by now just how constrained we are. Existing services are limited to what can pass through the needle's eye of the 3-kHz acoustical voiceband. Even our dreams of an integrated digital future betray themselves as products of our copper-obsessed imagination. We are facing a multi-hundred billion dollar "rewiring" before we can really think freely about twenty-first century telecommunications. Meanwhile, the Information Rich will surge ahead, free of the obligation to subsidize their backward cousins.

Of course, we could anticipate that some new social compact might be pieced together. Just as Theodore Vail and the Communications Act set the scene for universal access to voice telephony, supported by decades of government funding and even more far-reaching intrasystem subsidies, so we might now mandate a new twenty-first century version of *universal digital service* or some such idea. Alas, it is unworkable. For one thing, we have already unbottled the genie of competition; competition abhors subsidies, and without subsidies, the enormous disparities in the costs of wireline access (e.g., fiber) for urban and rural customers will defeat us. For another, how would we define the new baseline? Some telephone companies are now including voiceband data (up to 2400 b/s and Group 3 fax) in their definition of POTS (Plain Old Telephone Service). But will we accept the proposition that every American citizen has a right to ISDN access? Will we pay the enormous tax that would be required to achieve this?

Given the economics of wireline facilities (including fiber), the possibility for a new kind of poverty in the twenty-first century is very real. We can already see its outlines. People in major cities have more options for broadcast television, radio, and cable TV. They will soon have much better telecommunications capabilities. Perhaps this will be the more relevant form of "poverty" in our age of affluence—not the inability to obtain the material means of life, but the inability to access the modern networks.

We shall discuss the details of this economic equation in the following chapters, especially Chapter 7. The solution, of course, is radio-based access. I use the word *solution* hesitantly because it is not as painless or inexpensive or easily constructed an answer as that simple word connotes. But the economics of radio-based access, as we shall examine in the following chapters, transforms the equation that relates facilities cost and population density. By transcending the limits of physical plant, wireless access can help us to equalize the access opportunities available to people in different geographical areas and different socioeconomic segments.

NOTES

1. R.V. Bruce, *Bell: Alexander Graham Bell and the Conquest of Solitude,* Boston: Little, Brown and Company, 1973, p. 121.
2. See J.C. Bellamy, *Digital Telephony,* 1st ed., New York: John Wiley & Sons, 1982, p. 33.
3. See, for example, R.V.L. Hartley, "Transmission of Information," *Bell System Technical Journal,* July 1928; H. Nyquist, "Certain Factors Affecting Telegraph Speed," *Bell System Technical Journal,* April 1924; and H. Nyquist, "Certain Topics in Telegraph Transmission Theory," *AIEE Transactions,* April 1928.
4. Bellcore, *Telecommunications Transmission Engineering, Volume 2—Facilities,* 3rd ed., Bellcore Technical Publications, 1990, Chap. 5.
5. I have these numbers informally from several sources and informal presentations by Bell and AT&T engineers.
6. Bellcore, *Telecommunications Transmission Engineering, Volume 2—Facilities,* 3rd ed., Bellcore Technical Publications, 1990, pp. 93 ff; see also G. T. Hawley, "Historical Perspectives on the U.S. Telephone Loop," *IEEE Communications Magazine,* March 1991, pp. 24–28, for a discussion of the historical origins of the serving area concept.
7. Bellcore, *Telecommunications Transmission Engineering, Volume 2—Facilities,* 3rd ed., Bellcore Technical Publications, 1990, pp. 116–118.
8. *Ibid.,* p. 111.
9. J.E. Kingsbury, *The Telephone and Telephone Exchanges: Their Invention and Development,* London: Longmans, Green, and Co., 1915, p. 9. (Reprinted by the Arno Press, New York, 1972.)
10. Bellcore, *Telecommunications Transmission Engineering, Volume 2—Facilities,* 3rd ed., Bellcore Technical Publications, 1990, Chap. 6.
11. The analogy is appropriated from an article by Bob LeBlanc, formerly of Contel.

Part III
WIRELESS ACCESS: TECHNOLOGIES AND APPLICATIONS

Radio is as old as Marconi (or Hertz). But wireless access—the use of radio to provide access to the telephone network—is still in its embryo. For all the excitement it has engendered, the ultimate scope and direction of its development remain uncertain. Unlike optical fiber, which replaces specific links in the current network, wireless access will involve the creation of a totally new communications link and capabilities. To understand its future, we must answer two very broad questions:

1. What can we accomplish (i.e., what are the technological options)?
2. What do we want to accomplish (i.e., what are the applications)?

At best, the answers are preliminary. New technologies and applications concepts for wireless access are multiplying rapidly, yet few of these have suffered the reality test of the marketplace. In fact, however, the digital radio techniques now emerging are so robust and flexible that the technology *per se* will not be a limiting factor, except for short periods of time in the initial phases of commercializing new applications. The greater difficulty lies in identifying which applications will succeed in the market, what economic, technical, and operating requirements must be met, and what sort of regulatory framework is needed. Therefore, we begin, in Chapter 5, with the question of applications. Chapter 6 will survey some of the broad technological trends.

Chapter 5
APPLICATIONS OF WIRELESS ACCESS

The past few years have seen a proliferation of new ideas for radio-based telecommunications services, and an alphabet soup of new acronyms to designate them. Those for which a standard either exists or has been proposed include the following (with sample references given in brackets):

AMPS—analog cellular radio, North America [1]

TACS—analog cellular radio, United Kingdom [2]

NMT—analog cellular, Scandinavia [3]

AMPS-D—digital cellular, North America [4]

Japanese digital cellular [5]

GSM Cellular (Group Speciale Mobile; digital, European) [6]

CT-2—cordless telephone, second generation [7]

CT-3 [8]

DECT—digital European cordless telephone [9]

PCN—personal communications network [10]

PCS—personal communications service [11]

UDPC—universal digital portable communications [12]

FPLMTS—future public land mobile telephone system [13]

BETRS—basic exchange telecommunications radio service [14]

BEXR—basic exchange radio [15]

Mobile RACE—research on advanced European communications [16].

There are other wireless access applications for which standards have not yet been proposed (particularly in the PCN theater), or which, because of the nature of the application, may develop without a formal standard, using proprietary formats, such as mobile satellite services, wireless LANs, wireless PBXs, and thin-route microwave. (In addition, of course, there are the traditional radio services such as paging, citizens band radio, analog cordless phones, and "private" analog radio systems, some of which provide at least some of the elements of wireless access, and which still account for well over 90% of all radio devices in use today.)

The first impression is thus one of diversity, perhaps even confusion. It is not always clear which of these proposals may compete with, overlap, replace, or leapfrog one another. Some are little more than speculative concepts, which face long years of wrangling over spectrum and/or technical standards before they can reach their intended markets, while others are fully operational systems serving large numbers of subscribers with commercial "off-the-shelf" hardware.

Underlying this diversity, however, is a common conceptual core, inspired by a shared vision of the potential for a broader role for radio in the network. The systems concepts now being conjured in so many quarters are responses to this opportunity, which is driven by several enabling developments:

1. New technology is becoming available to expand the performance horizons for radio-based systems and to enable the exploitation of much more broad-based applications.
2. A new regulatory climate favors the expansion of radio-based services, to a degree that would have scarcely seemed possible only a decade ago (particularly in the United States).
3. The lure of competitive opportunities in the local exchange environment (which radio-based access will make possible) is a powerful stimulus for the entrepreneurial spirits, large and small, who have moved into the telecommunications industry in the past decade.
4. The demand exists on a very large scale for radio's unique capabilities, such as mobile and portable communications.

The goal of this chapter is to organize this diversity, at least provisionally [17]. Unfortunately, the target is moving. More than 50 applications for experimental licenses for various wireless services were filed with the Federal Communications Commission (FCC) between the first and final drafts of this chapter. A simple resumé of the proposals on the table would not be of much help; the book would be out of date before it reached its readers. Instead, the aim is to categorize the applications concepts that underlie the diverse set of wireless access proposals. Figure 5.1 is an "applications tree" for wireless access, which may serve as a road map for this long chapter. The "branches" of the tree correspond to the subsequent sections and subsections of the chapter.

We are tempted to conclude that, in part, the confusion is intentional, as different interest groups try to jockey their own proposals ahead of their competitors'. Each group understandably tries to define the terms of the debate in a manner congenial to its objectives. Hence, many overlapping programs and competing architectures revolve around the same basic applications. Nevertheless, radio does lend itself

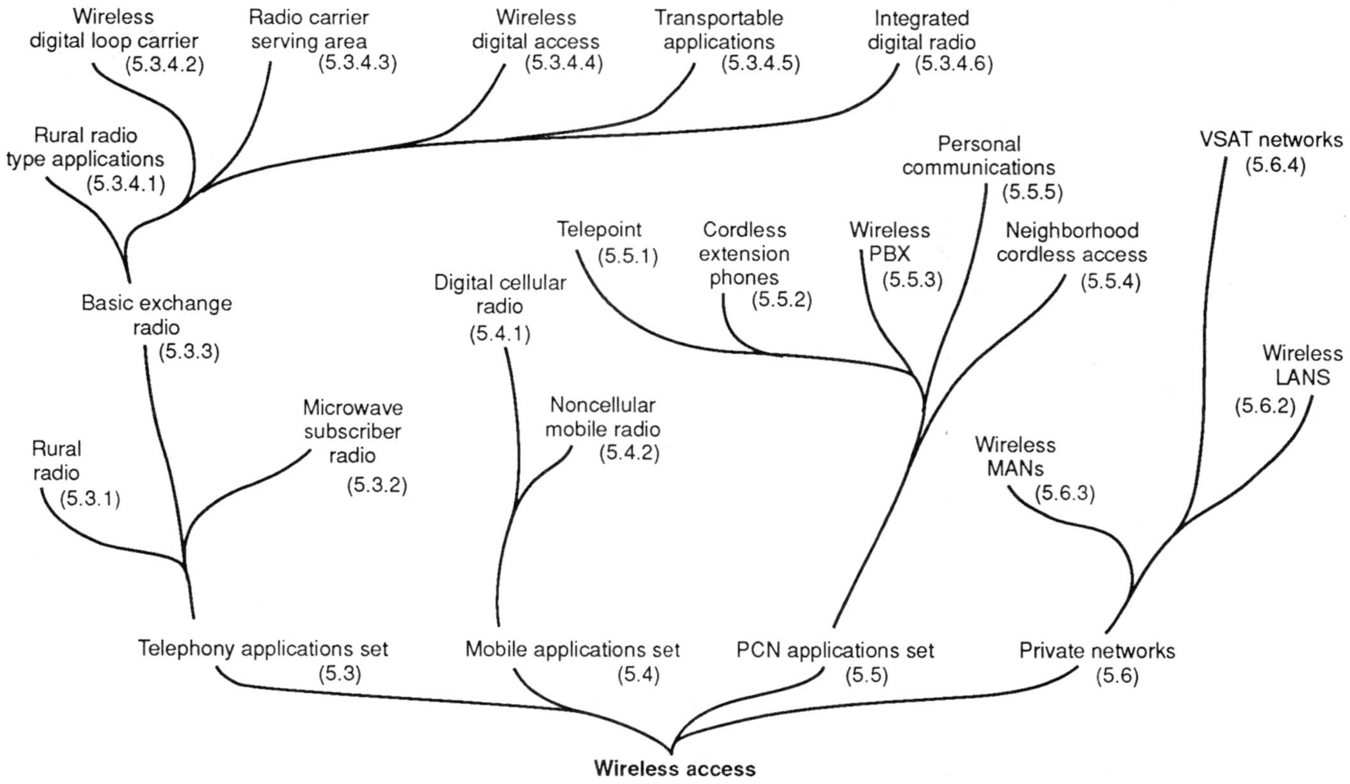

Figure 5.1 A tree diagram of major wireless access applications.

to an amazing number of uses, and there is a danger of falling into a parochial perspective, in which we take one particular application as the paradigm for all the others. For example, while portability is an important aspect of radio's promise, I would hazard the guess that the majority of wireless access devices over the next two decades or so (and who can really see further than that?) will not be shirt-pocket portable. Nor is "mobile radio" necessarily the central paradigm. Those who come from a mobile radio background may be surprised to learn that wireless access systems based on sophisticated digital radio technology (of the kind just now entering the cellular arena) have been in use for years by the wireline telephone companies, who have begun to lay the foundation for a mass-scale wireless network that *per se* has nothing to do with cellular car phones. Many new applications that cellular business planners now eye eagerly for their own are already staked out by the local exchange carriers. And there are many other players and other plans as well. The interexchange carriers are also becoming involved, as are some of the cable TV operators, and other dark and darker horses. The coming battle for the soul of PCN will clarify for all of us just how many different interests are involved.

5.1 A BASIC MODEL OF WIRELESS ACCESS

Radio is used for an incredible number of communications needs, everything from opening garage doors to transmitting digitized photographs from deep-space probes. Very few of these qualify as wireless access. Therefore, we must define the subset of radio applications that fall within the scope of this book.

System Building Blocks

We begin with a generic model of a wireless access system (Figure 5.2). The four basic building blocks are:

1. A radio terminal, or subscriber station
2. A radio base station
3. A switch or network control device for the wireless access system itself
4. The network, to which the wireless system provides access.

The radio terminal or subscriber station is often the most visible element of the system. It is a transceiver device carried by, or located near, the user. It may be fixed, mobile (i.e., vehicular), or portable—indeed, a given wireless access application may encompass all three basic types of subscriber terminals. In mobile applications, the terminal is the mobile radio unit in the car. In portable applications, it is the handheld telephone. In fixed applications, it may be a rather large box mounted on a pole outside a building.

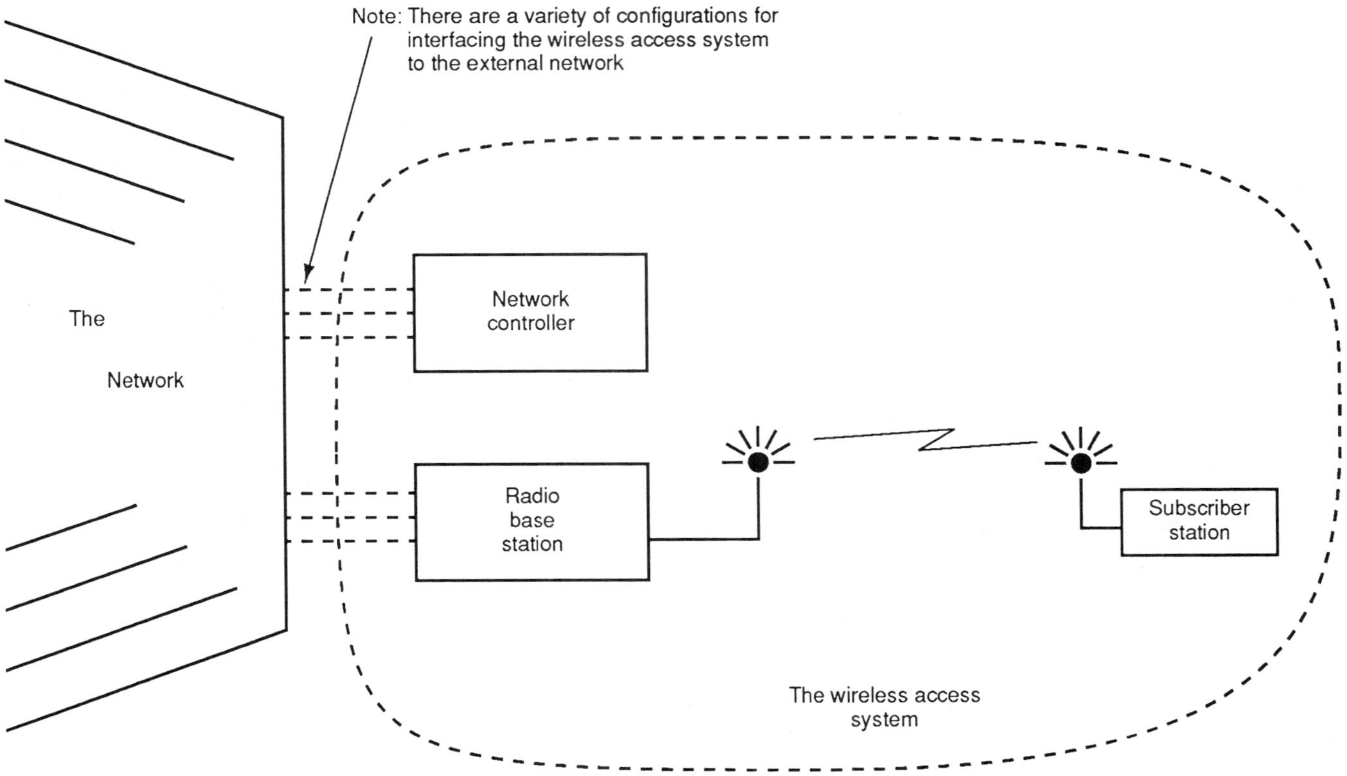

Figure 5.2 The generic building blocks of a wireless access system.

The subscriber station provides, generally, a single circuit accessing the network (although multicircuit configurations are easily arranged). The functions of the subscriber station are to convert the user's message (speech, data, images, etc.) from its original form into an appropriate digital signal, to further translate this signal into a form suitable for radio transmission, to establish radio access to the network though the base station, and to effect the transmission of the message to the base station over a designated radio channel (Figure 5.3). This is generally a two-way (duplex) process.

The subscriber station encompasses two other subsystems besides the transceiver itself: (1) a *power supply* and (2) a *user interface.* Conceptually these are separate functions, and in some applications they may be engineered as separate physical devices. For example, in some applications the user interface device may be an ordinary telephone set, which is connected by indoor wiring and an outdoor drop line to the transceiver, which may be located hundreds of feet or even several miles distant. The power supply subsystem in such applications may be a standard ac connection backed up by a car battery, with a separate housing, separate electric meter, and extensive grounding and circuit protection—in all, a "package" considerably larger than the transceiver unit itself. Of course, at the other end of the spectrum, a portable cellular phone, for example, integrates all three subsystems—transceiver, power supply, and user interface—in a single physical unit.

These ancillary subsystems may also be subject to very different regulatory treatment. For example, the user interface in some applications will be defined as *customer-owned equipment* (CPE) while the subscriber transceiver will be owned by the carrier (see Section 5.2.1).

In some configurations, the subscriber station may be connected by either wire or radio, or both, to other subscriber stations (Figure 5.4). In some applications this may evolve to a pure radio repeater. A full net-like architecture in which subscribers form a radio-fabric among themselves and the base station assumes a less central role is also possible and may be regarded as a variation on the generic wireless access architecture. This type of system has been fielded in a military environment [18]. However, since most of the systems that have been proposed or deployed involve a point-to-multipoint architecture with a centralized base station, that is the structure on which we concentrate in this book.

The *radio base station* is a centrally located, generally multicircuit transceiver, which radiates over an area called a *cell* (if generated by omnidirectional transmission) or a *sector* (if generated by a sectorized transmission from a wide-beam directional antenna). The size of the area covered may range from hundreds of feet to hundreds of miles across. (Wireless access base stations do not normally employ highly directional point-to-point transmitters such as those used in long-haul microwave systems.)

The base station comprises a number of functional subsystems: (1) radio modules, including transmitters, receivers, power amplifiers, combiners, filters, antennas,

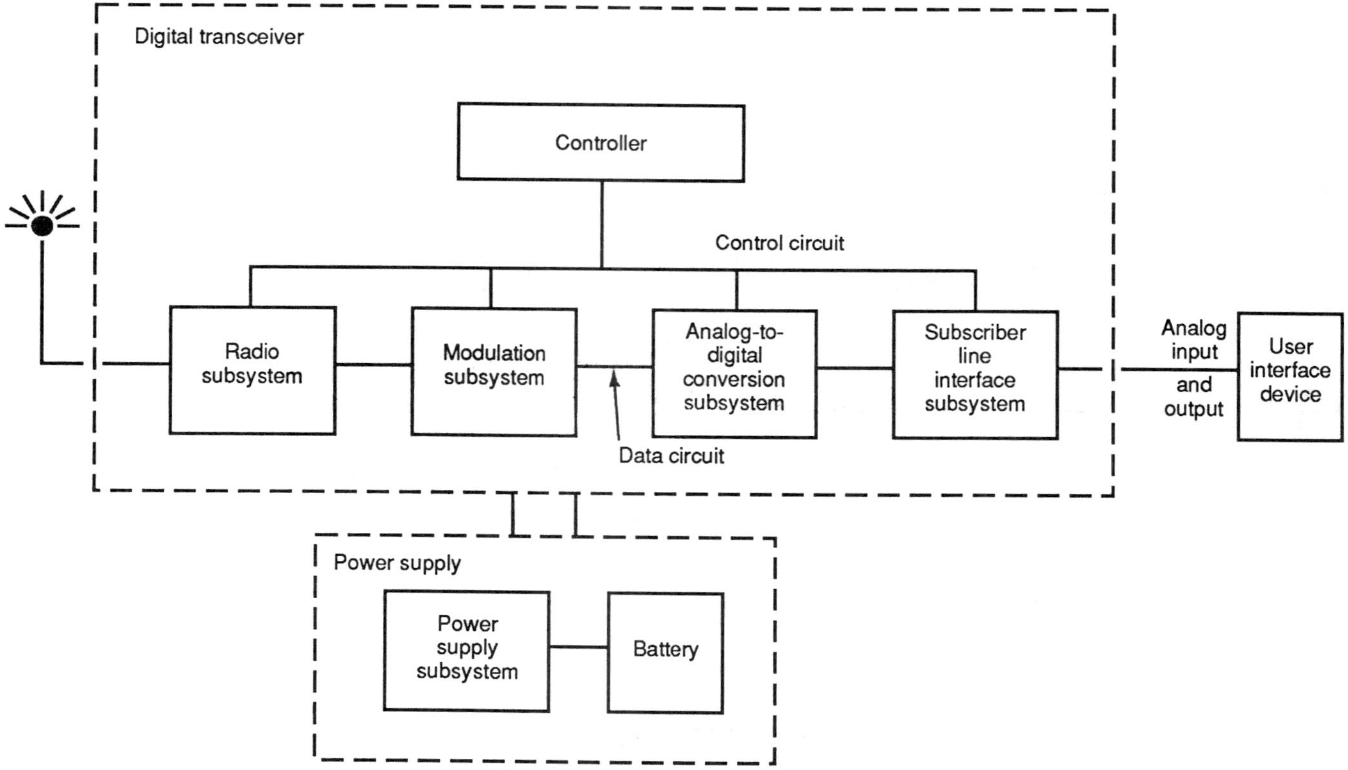

Figure 5.3 A functional diagram of a subscriber station.

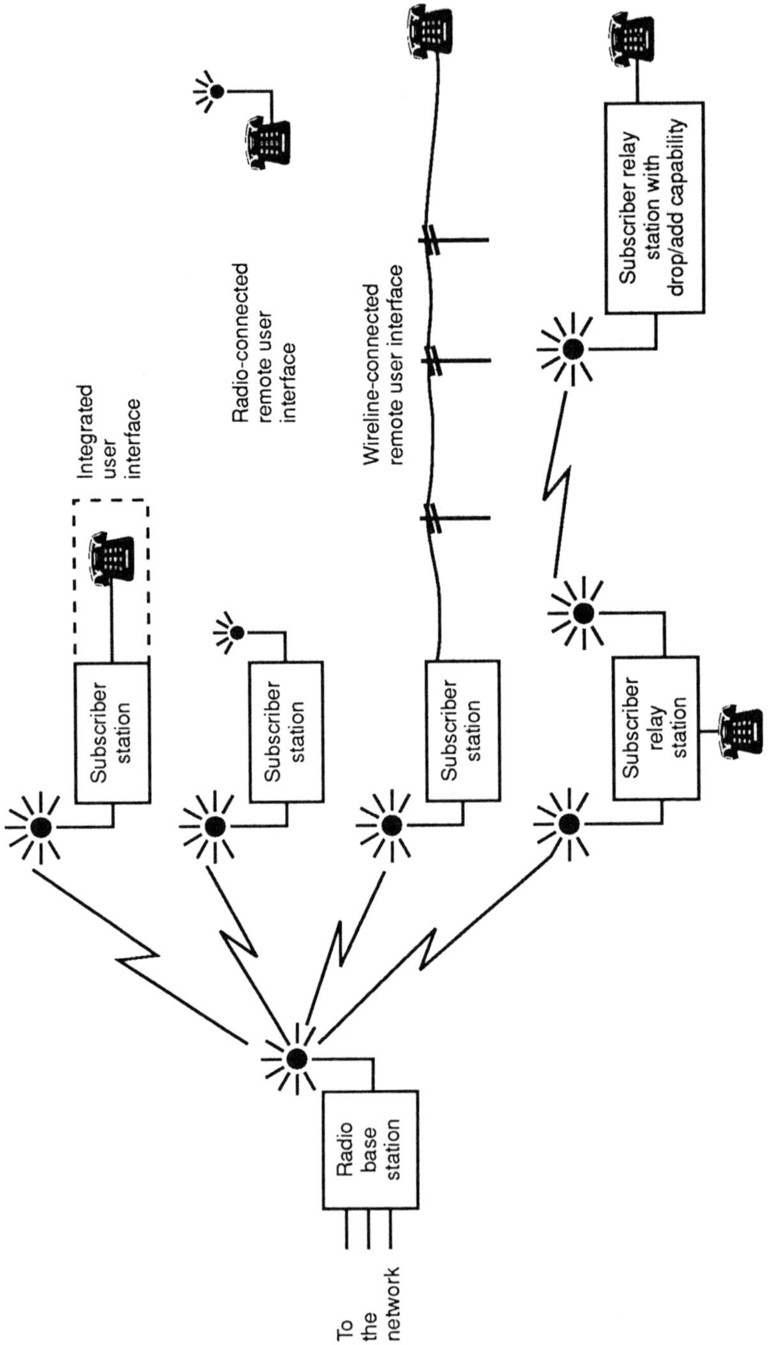

Figure 5.4 Terminal connection alternatives.

and so forth, some of which may be duplicated for every channel and some of which may be common or shared by a large number of channels; (2) baseband or "digital signal processor" modules; (3) network-interface modules; and (4) common equipment, including power supply, system control, and so forth (Figure 5.5). These modules may be engineered as separate pieces of hardware or they may be highly integrated.

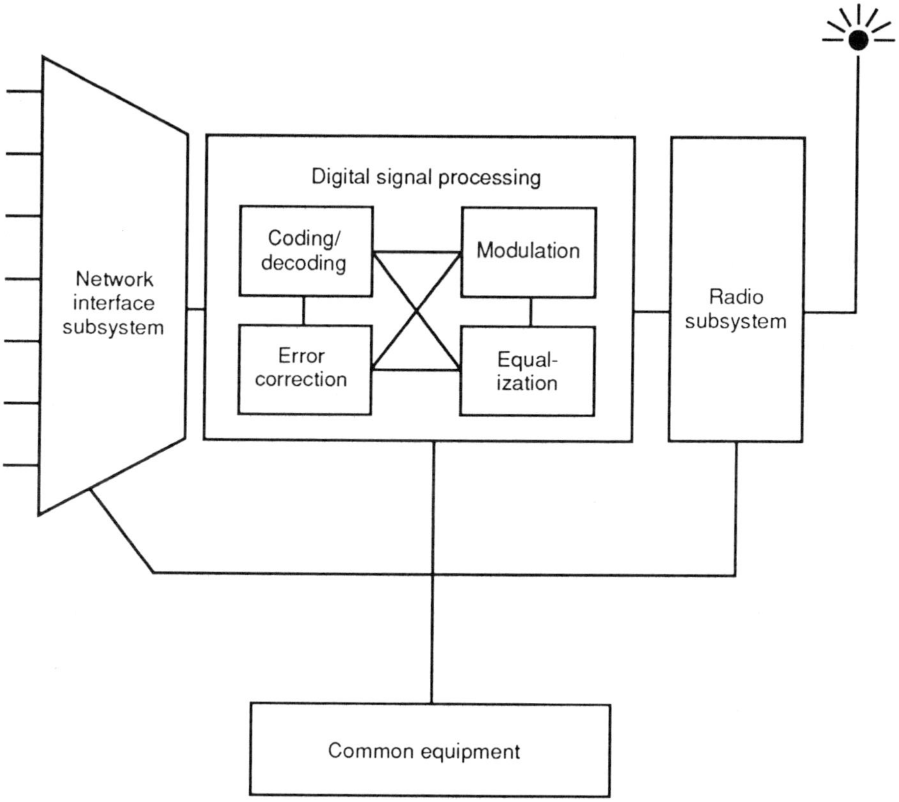

Figure 5.5 A functional diagram of a base station.

The *network controller* is the subsystem that controls the operation of the entire wireless access system. It determines the assignment of circuits to individual users. It monitors and controls system performance. It provides and controls the interface between the wireless access system and the external network. It may provide other functions such as hand-off or vehicle location (for mobile systems). These functions can be implemented in many ways. For example, a cellular system incorporates its own switching system, which may include large central switches connected to intermediate distributed switching elements, sometimes referred to as *base station controllers* (Figure 5.6). At the other extreme, some telephony-oriented wireless access systems interface to the wireline network by means of a standard line interface to an existing switch, with a small adjunct processor to handle radio channel assignments (Figure 5.7). Yet at the generic level, both configurations realize the network control functions. We will return to these differences in the following subsections of this chapter.

Finally, the *external network* to which the wireless access system connects may be:

1. The public switched telephone network (PSTN) of today
2. The integrated digital network (IDN or ISDN) of tomorrow
3. A private network (MAN, LAN, WAN, etc.).

In general, we will be talking about PSTN-type interconnections, since the PSTN is the ubiquitous "network" of contemporary experience.

The Wireless Access Service Offering

Different assumptions about the type of service being offered define the main "branches" in our taxonomical tree of wireless access applications, as we discuss in Section 5.2. However, there are certain commonalities to most true wireless access applications, which can be summarized here as the generic service assumptions of wireless access. These include:

1. Wireless access provides a circuit-like communications channel.
2. Wireless access is broad-based and high capacity, capable of serving very large numbers of users.
3. Wireless access is fully interconnected to the public network (albeit perhaps via a private network in some applications).
4. Wireless access service is highly integrated with wireline service concepts.

Circuit-like communications channels may be actual or virtual circuits, but in either case they conform to a number of functional requirements. Circuit-like channels are real time (i.e., low delays), suitable for voice communications. They are user-to-user, or point-to-point channels, rather than broadcast or net channels. They are private and they are modular—i.e., circuits can be added, subtracted, or switched like lines.

Broad-based wireless access systems are designed to serve large numbers of users. But how large is large?

In 1980 there were probably fewer than 250,000 users of radio-based interconnected telephone systems in the United States (mostly IMTS mobile systems and a

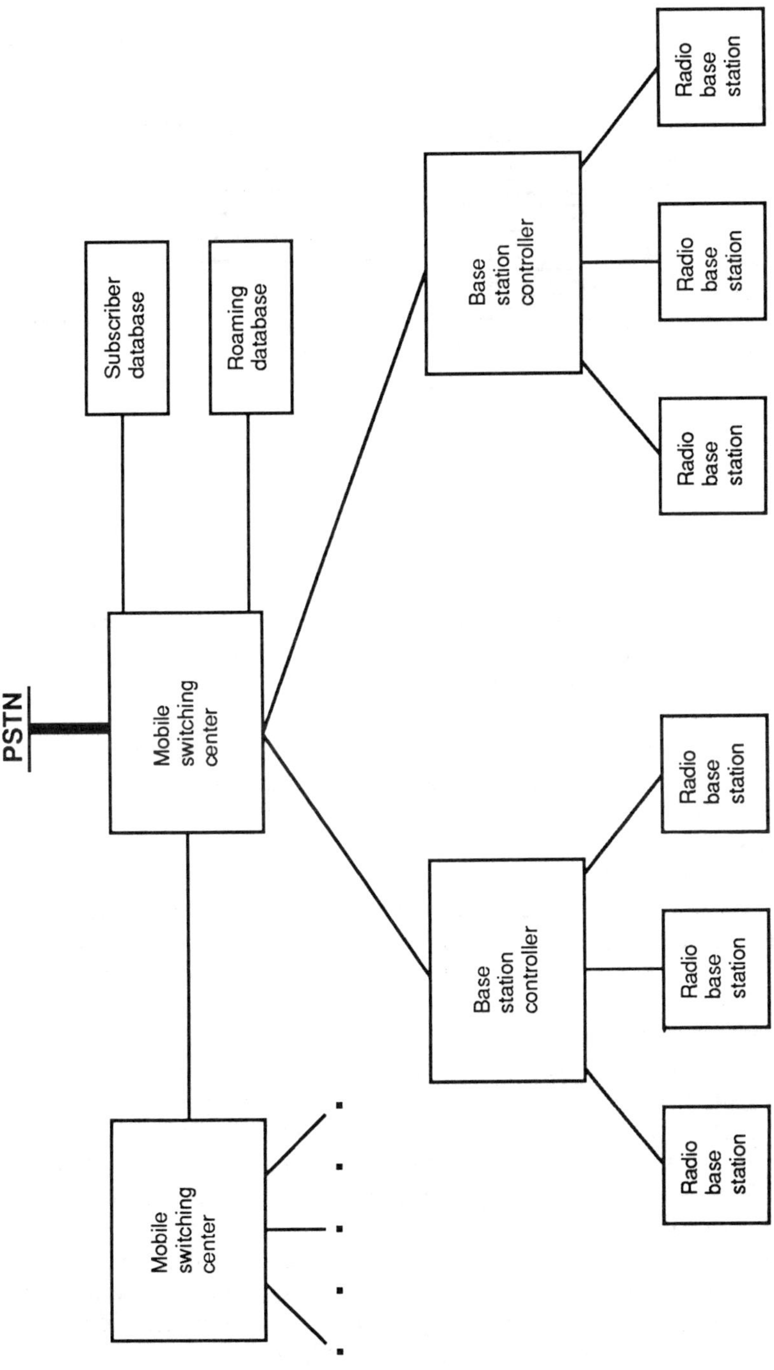

Figure 5.6 A cellular switching infrastructure.

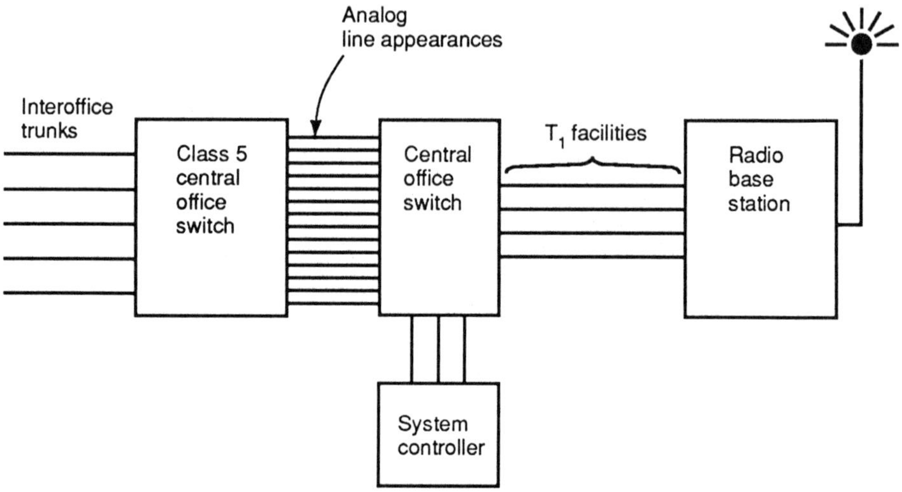

Figure 5.7 A line interface for BETRS-type wireless access systems.

very small number of rural radio users), and a few individual systems encompassed more than 1000 users. The largest radio access system in the world was probably the unique Aurora network in the province of Alberta, Canada, which served about 20,000 people [19].

By 1990, the largest integrated radio access systems were cellular systems in Los Angeles, London, or Tokyo, reaching customer bases perhaps 10 times larger than Aurora—but still only a tiny fraction of the size of the coterminous wireline networks in those same cities. Moreover, these systems were based on analog radio transmission, which, for reasons discussed in the following chapter, cannot be regarded as true wireless access systems. The number of wireless digital access customers in the United States in 1990 was less than 5000 [20], and worldwide probably less than 100,000 [21]. Even if we accept analog cellular as a forerunner of true wireless access, in the United States overall there were probably only about three million active cellular subscribers in 1990 [22] and about the same number in the rest of the world, with only a handful of individual systems yet exceeding 100,000 active customers. Compared to 130 million wireline access customers in the United States, and perhaps 500 million worldwide, it is clear that radio-based access is still embryonic.

Nevertheless, the market's enthusiasm for cellular service, despite its current limitations, has inspired some heady forecasts. Some now project 20 to 40 million users of cellular radio in the United States by the year 2000, and up to 100 million worldwide [23]. The vast majority will be using digital cellular technology in one of several forms. Precise numbers are not important; the point is that many observers have begun to believe that millions, and perhaps tens of millions, of customers will be using wireless access systems in just a few years time.

Full interconnection means that wireless access customers are assumed to be able to tap into the large public networks for worldwide voice (and some data) communication—i.e., they have the ability to "make a phone call." In the past, most radio systems did not offer this capability, and today there are still many more non-interconnected radio users in the United States than there are interconnected cellular subscribers. Some of the new emerging applications—such as mobile satellite services for interstate trucking—may not offer interconnection. However, all the system concepts listed at the beginning of the chapter call for full interconnection, either directly or through a private network that interfaces to the PSTN.

The user is normally not very concerned with exactly how the interconnection is achieved. The user is quite concerned, on the other hand, with the quality of his interconnection. *Integrated wireless access systems* are those in which the communications capabilities available to the user are not altered or constrained by the wireless access channel itself, where the quality of wireless access approximates the quality of wireline access.

One aspect of interconnection quality is the grade of service. True wireless access systems should approximate wireline blocking rates (we return to this issue in detail below). Another aspect of quality is the system's transparency to new services. If the wireline system supports facsimile transmission, then so should the wireless system. Wireless customers will demand as many of the services enjoyed by their wireline cousins as they can.

These few generic service goals are far from sufficient to describe fully any particular wireless access concept. In my view, the way in which the service concept is fleshed out constitutes the most important factor in shaping subsequent decisions about the architecture, performance objectives, and operational details of any particular species of wireless access. In the following section we will examine the major variants in the service model for wireless access, as it is viewed by different telecommunications constituencies.

5.2 ALTERNATIVE SERVICE MODELS

We can start with the very basic question: What kind of telecommunications service is wireless access seen to be? Is it, for example, an outgrowth of the existing wireline telephone network, plain old telephone service by another means? Or is it an expansion of mobile radio—a "better" car phone, which may incidentally encroach on nonmobile applications? Or is it something else—a "personal" communications service—and what does that really mean? The detailed answer to this question defines what I will call the *service model of wireless access.*

A succinct rephrasing might be: Who owns wireless access? In the current discussions within the industry, four alternative answers have been offered:

1. The existing wireline common carriers—i.e., the telephone companies
2. The existing radio common carriers—i.e., the cellular operators

3. Neither of the above—i.e., an open field for all comers
4. Private carriers, generally large corporations or service companies.

Each answer leads to a different set of conclusions as to how the wireless access service would be established, operated, and regulated.

5.2.1 Wireless Access as an Extension of the Network

(In beginning with the telephony service model, I am counting on its shock value. When they hear the word radio, most people think more readily of cellular or PCN than of wireless local loops in rural areas. Nevertheless, the telephony model is considerably further advanced than either cellular or PCN, as this section will, I hope, make clear.)

One vision of wireless access considers it simply as an extension of the regulated, common carrier telephone network. Wireless access is, in this view, simply another means of delivering dialtone, another transmission technology in the local exchange carrier's tool box. In support of this view, the telephone engineer points to the use of alternative technologies in the long-distance business, from microwave to fiber to satellites, all of which produce the same "product"—long-distance telephone service. Why should local exchange service be regarded any differently? Moreover, if we accept this premise, then the frameworks for evaluating and prescribing answers to many questions about economics, service standards, engineering practices, and methods of operation pertaining to wireless access are already in place. We inherit and adapt them from the wireline model.

The reader may anticipate that things will not turn out to be quite so simple. Nevertheless, the telephony-oriented view of wireless access is in most respects the strongest construct, which is not surprising since it has the longest service tradition on which to draw. (Mobile radio, by comparison, is only just becoming a mass market service.)

Although we discuss specific architectures in the following sections, it is useful at this point to refer to a generic architecture for wireless access, as viewed by the telephony service model (Figure 5.8). In this model, note that wireless access is freely mixed with wireline access, and both access modes interconnect to a standard (normally preexisting) telephone central office switch (Class 5). The salient characteristics of this service model are:

1. A tendency to view wireless access as merely an alternative technical vehicle for accomplishing the same basic service objectives, but at a lower cost (with less delay, etc.).

   This principle may appear to be a negative one because it tends to discount the unique capabilities of radio-based access (such as mobility). How-

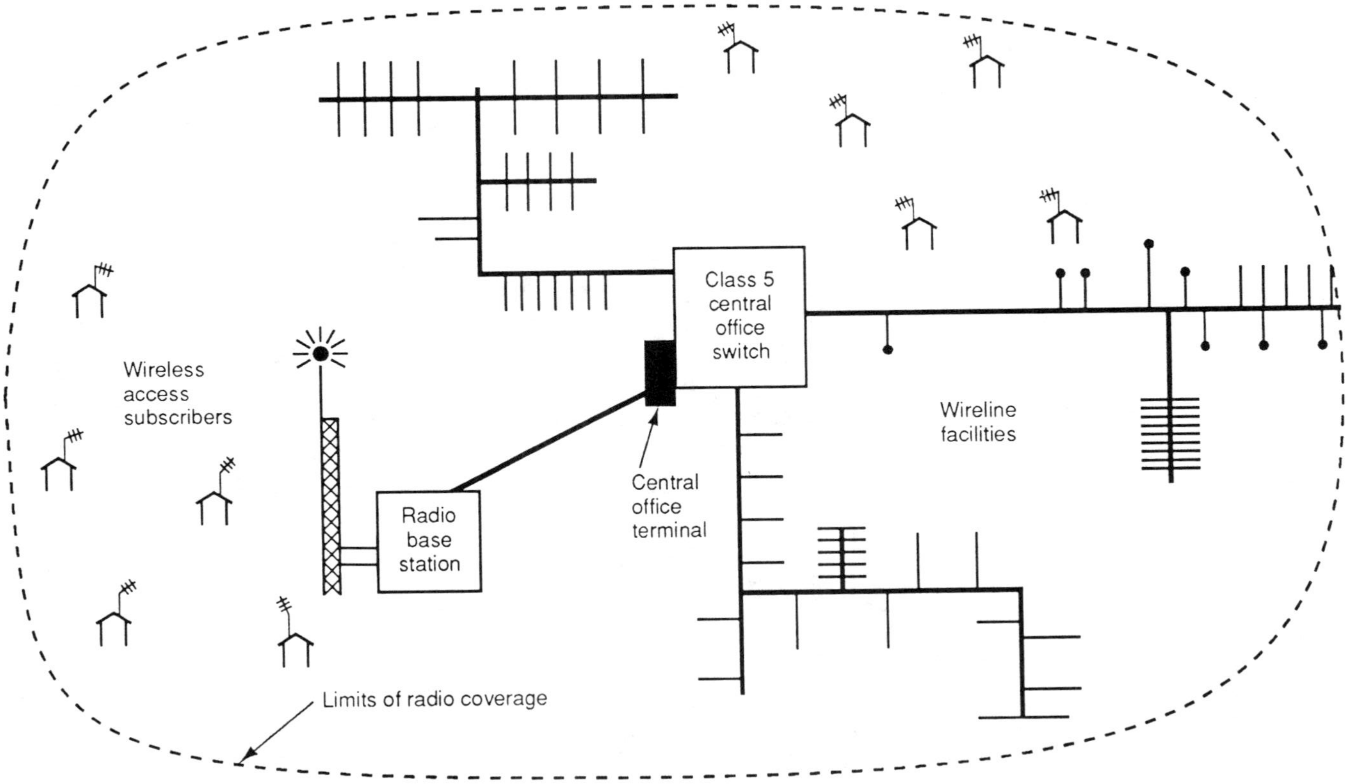

Figure 5.8 The basic architecutre of wireless access in the telephony service model.

ever, it turns out to have many positive and challenging ramifications for technical standards and systems deployment. The most important of these is this: *All of the performance standards for wireline service are carried over and applied to wireless access.* This "doctrine of equivalence" is expressed in several ways.

Transparency. Telephone-oriented views of wireless access stress that a wireless loop should "look like" a wire loop, as much as possible, from both ends of the circuit.

For the *subscriber,* functions such as call setup ("dialing") should mimic the wireline practices. This requirement imposes significant constraints on systems design. For example, the use of a "send" button to transmit a call request will generally be rejected in favor of conventional dialtone and on-line dialing. Transparency also implies that a wireless access system will be forced to adapt to many existing forms of telephone signaling (such as dc "hook-flash" or pulse dialing), which may also impose constraints on even the basic air interface technology [24]. For example, the requirement that a wireless local loop handle a FAX transmission without any special user intervention (i.e., the same as a copper loop) may be difficult to meet for systems that have optimized spectrum efficiency through voice compression algorithms. Yet in a telephony-oriented application, fax transmission is likely to be viewed as a basic requirement. In general, realizing the design principle that the wireless access circuit must mimic as much as possible the operational and functional characteristics of copper-based local loops is a major technical challenge.

From the *central office end,* transparency means that the transmission link must support the operation of automated loop test equipment now coming into use (despite the fact that the designers of such equipment never anticipated finding anything in the loop but a metallic circuit). The base station equipment must conform to the operations, maintenance, and network management requirements flowing from the wireline network [25]. Circuit provisioning and administration of wireless loops follow the same general procedures as for wireline loops. (This may change in time; as wireless access comes to play a larger role in the total local exchange picture, new procedures may be developed that will be optimized for radio circuits.)

Equivalent quality of service. In practice, this means *equivalent voice quality.* In other words, the wireless loop should "sound the same" as a wireline loop. Yet this is difficult to measure. In part, equivalent quality will be defined in terms of noise and other degradations that originate externally to the transmitted signal. Here it would seem there are standards that can, in principle, be imported from copper telephony. For example, idle channel noise is one measure for which any telephone engineer can usually cite a standard. As we will discuss in Chapter 8, however, even such "obvious" measures must be applied with caution in a wireless access application.

However, the assurance of equivalent quality is complicated by the use of new voice processing technologies. Some wireless access systems today employ coding algorithms that use less than 25% of the digital bandwidth of wireline digital transmission. In the next several years, coding techniques that operate with bit rates more than *10 times lower* than standard *wireline* digital standards will come into widespread commercial use. Moreover, whereas wireline digital transmission is based on waveform coding methods in which the loss of sound quality is in some sense uniform and natural (see Chapter 6), the low-bit-rate coders now coming on the scene are *synthetic*—they analyze, model, and recreate the voice waveforms in ways that can introduce subtle colorations of the transmitted speech. Thus, whether or not something "sounds the same" has become much more difficult to measure, and radio-based access systems are in many ways pioneering the measurement of voice quality [26].

Equivalent grade of service. Grade of service is another simple concept that becomes complicated once we strip out the wireline-oriented assumptions, as we will do in Chapter 8. Most engineers are familiar with the basic concepts of blocking, trunking efficiency, and the principles of traffic analysis. A more sophisticated understanding of the true grade of service takes into account the measure of equipment availability. Yet here again, even the simplest measures may cause problems.

As a practical matter, radio systems are almost always designed as trunking systems. By contrast, most segments of most wireline loops are completely dedicated, from the line card to the drop wire. Even where electronic pair-gain is used, there is a tendency to refrain from implementing concentration. Loop plant engineering does not depend greatly on statistical traffic analysis today. Radio engineering, on the other hand, is all statistics. In practice, the result is that an engineer who has previously dealt almost entirely with nontrunked loop facilities may be uncertain how to engineer a trunked radio system to provide an equivalent grade of service.

2. A tendency to view the decision to deploy wireless access as the operator's prerogative, rather than the customer's.

   Under the telephony model, if the operator (the telephone company) can indeed assure that a wireless loop is fully the equivalent of a wireline loop, then whether the company deploys a wireless loop or a wireline loop in a particular case is its decision, to be based on its costs and the available alternatives. The customer buys access, period. Whether it is delivered by radio or copper is not the customer's concern. After all, today the telephone customer does not ordinarily know or care whether he is served by a dedicated copper twisted-pair, an electronically derived "pair" (and dedicated drop) from a loop carrier system, or even a fiberoptic loop carrier (and dedicated drop). Why should transparent, equivalent wireless access be any different? That, at least, is the way the telephone world works today. It affords the operator with a great deal of flexibility in configuring and managing a plant.

A corollary of this is that the telephony service model tends to view radio-based access as *fully integrated with the rest of the (wireline) network.* A wireless access system should connect to the same central office and the same switch that handles the local wireline loop traffic. If the local switch happens to be a 40-year-old step-by-step electromechanical switch, so be it. The sophisticated radio systems in the loop will have to adapt to their immediate network environment.

Another implication—and I raise it here only to suggest how pervasive this assumption of integration of wireless and wireline facilities can be—is that the same craftspeople who today keep the copper plant working should be capable of installing and servicing the wireless loop equipment. The radio systems must be designed for the common denominator; we have found, for example, that even the need for an additional item of field maintenance equipment (e.g., a spectrum analyzer) may be perceived as a burden, not so much because of the cost *per se,* but because of the need to provision local central offices with a new item to administer.

3. A tendency to assume that wireless access must be priced (and tariffed) in the same way as wireline access.

The telephony service model accepts price regulation by government agencies as a fact of life. In the United States, the rates for local telephone services require the approval of state public utilities commissions. The approved *tariffs* specify the type of service and the methods for calculating customer charges, often in considerable detail. As a practical matter, the filing of a new tariff often entails an adversarial process between the telephone company and the regulatory body (who represent the consumers' interests). It can become a difficult and costly proceeding, highly political, often resulting in unlikely horse trades between rate increases in one area and service commitments in another. A tariff proceeding is not something the telephone company enters into lightly.

In several respects, wireless access differs sufficiently from conventional copper-based access to raise the specter of tariff problems. The following examples are drawn from actual cases:

Power. Radio-based access requires electric power to be supplied to a radio transceiver located at or near the subscriber's premises. The easiest way to deal with this requirement is often to use the subscriber's own electric service (i.e., plug it into the nearest outlet). However, this means that the subscriber pays an additional amount for his telephone service delivered by means of wireless access—an amount he would not pay for an ordinary copper loop circuit. The cost may be nominal (less than $1 per month in most situations), but it can create a potential tariffing problem—tariffs designed for copper-based loops do not provide for subscriber-supplied power [27].

This can hamper the deployment of wireless access systems. One U.S. telephone company, rather than use subscriber-supplied power for wireless access terminals, decided to install separate power lines and meters for every subscriber station, billed back to the phone company at bulk rates. This substantially, and needlessly, increased the cost of each installation. In another case, the phone company decided it was willing to use wireless access—with subscribers supplying the power—but only for customers who had never had wireline service. It was felt that customers who had previously been served by wireline circuits (and had not paid for power) would find it objectionable to now pay extra for "the same service." As a result, people in rural areas who were experiencing poor service on deteriorating cable plant could not be upgraded using wireless access.

The subscriber-power issue also hinders the deployment of the fiber-to-the-home technology because fiber-connected terminals will need local power for ringing and other functions. Nevertheless, some state regulators have opposed the idea of asking the subscriber to pay for power, however nominal the charges may actually be [28].

Reason must eventually prevail. Both fiber optics and radio-based access terminals will require small amounts of power, which is best supplied by the customer. The cost of the power will be literally only a few pennies a month [29], and can be handled either as a *de minimis* matter (i.e., ignored for tariffing purposes) or credited in some fashion in future rate adjustments. State regulators need to send the proper signals to the industry to help remove this roadblock.

Party-line service. A similar problem arises from the fact that wireless access does not easily lend itself to party-line service. Indeed, radio is uniquely suited to help operators eliminate party lines. Nevertheless, party-line service is tariffed in some states as a "last resort" class of service for rural customers, who may pay as little as $5 a month. It was originally viewed as a reduced quality of access for people who otherwise might have had no access at all. In some states, this has been turned on its head; it is viewed by some regulators as the customer's "right" to request a cheaper rate. In the course of our early experiments with wireless access, I was once told by a state regulator that there was little chance that wireless access would be permitted in his state, because it would take away the "opportunity" of remote rural customers to choose party-line service and pay party-line rates. The fact that without wireless access the customers in question would have no service at all was a point which I found difficult for some reason to communicate.

CPE. Another regulatory and tariffing issue that is potentially even more ominous (because it may invoke federal restrictions) is the question of whether a wireless access transceiver terminal should be considered as *customer-premises equipment* or as network equipment. The telephone companies now dread CPE because in many cases they are legally prohibited from owning it. Therefore, if a wireless access terminal is deemed to be CPE, it must be sold, like a cordless phone or an answering machine, to the end-user. This interpretation would hamstring wireless access in the telephony environment by forcing the often poor or rural customer to lay out significant sums up front to purchase a piece of transmission hardware that looks more like, say, a power transformer bolted to a utility pole than a consumer product (see Section 5.3).

Fortunately, the regulators generally seem to realize that classifying wireless access subscriber equipment as CPE would be self-defeating. The FCC has mandated that wireless access terminals be treated as network equipment, equivalent for accounting and ownership purposes to the wire and cable and poles that it replaces [30].

Distance charges. Still another example of the difficulties caused by the "Don't touch the tariffs" principle of the telephony service model arises from the fact that in many states a special provision exists to allow companies to recover their extremely high costs of rural copper facilities by applying a monthly surcharge to customers living outside the basic service area (remote ranches, for example). The surcharge is often based on the distance in miles beyond the boundary of the basic serving area. One typical rate calls for $12 per month plus $0.90 per quarter mile (per month) beyond the edge of the serving area. For a customer 15 miles outside of town, this could amount to $60 to $80 per month for basic service!

One of the unique advantages of radio-based access is that the costs are essentially distance-insensitive (see Chapter 7). Yet, in a state where mileage charges are applied, the carrier may be required to charge a customer outside the serving area a mileage surcharge, regardless of whether wireless or wireline access is used. The result is to price wireless access as though it were based on copper wire! Since the copper-based pricing is prohibitive for remote customers, it completely eliminates the rationale for the wireless alternative. This scenario might seem contrived, were it not for the fact that it actually occurred, holding up deployment of wireless access by a major telephone company in one state for more than a year, and eventually leading to a formal tariff action on wireless access (one of the first I am aware of that has dealt with wireless access explicitly) to free it from mileage charges.

I have participated in many straight-faced discussions with telephone engineers who were struggling against their own preconceptions to draw such obvious and advantageous distinctions between copper and radio as modes of access. One of the disadvantages of the telephony service model is the conceptual baggage that must be disposed of in order to free radio-based access for more general deployment. These are the kinds of issues that will control the pacing during the early phases (read: the first 10 years) of extending wireless access into the regulated common carrier telephone network. Based on my experience, they should not be underestimated.

4. A tendency to assume that wireless access will be deployed as a part of the monopoly franchise.

   Radio does not lend itself to a natural monopoly. Nevertheless, local telephone service is a monopoly franchise today in most jurisdictions. Adherents of the telephony service model tend to assume, therefore, that any access vehicle used in the local exchange environment will be controlled and administered by the monopoly operator. The radical potential of radio to upset the equilibrium and bring full-blown competition to the local exchange is largely ignored.

In summary, generally, what we may call the *wireline-derived* view of wireless access tends to be cautious and pragmatic. Its principal concern is often to ensure that radio can be used as a fully equivalent access vehicle to copper. It holds up high standards. It tends to ignore or downplay the unique aspects of radio (e.g., mobility). It tends to be very conservative (and not very innovative) in application engineering. It plans on long, slow deployment cycles, and tends to assume a stable, controlled, centrally administered evolution of technology. It tries earnestly to fit radio into the mold of established telephony. And it underestimates the implications of wireless access for increasing competition.

5.2.2 Wireless Access as an Outgrowth of Cellular Radio

Proponents of cellular radio exhibit a different psychology from their wireline cousins. Theirs is a young industry, characterized by rapid growth, generally free (so far) from state-level pricing regulation, and free also from many of the assumptions that comprise the universal service commitment of telephony. The cellular carriers tend to look at wireless access from a rather simplified, if slightly apprehensive, viewpoint. They ask, in effect, what is all the fuss about? Wireless access is here already, and it's called *cellular radio.* All that is really necessary (in their view) is to free cellular operators to expand into new applications and provide them with spectrum to do it.

Their apprehensiveness derives from the uneasy sense that new competitors could emerge, under the guise of some vague new service concept (such as PCN), which could diminish the value of today's semi-monopolistic cellular franchises [31]. And unfortunately, in the United States at least, cellular radio has suffered a black eye on the crucial issues of spectrum efficiency and technological leadership and has lost some of its leverage in the spectrum allocation process. In the mid-1980s, the FCC articulated a rather severe stand against allocating any new spectrum to cellular services, for the time being [32]. This position appears to be rooted in the commission's frustration with the fact that initial petitions for cellular spectrum relief were filed less than two years after cellular operations commenced [33]—an unexpected and unsettling development from the FCC's standpoint, which had been led to believe that the cellular concept would solve the mobile radio spectrum shortage for the foreseeable future [34]. This hard line carried over in the 1990 Notice of Inquiry on PCN-type services, in which the FCC indicated that cellular operators might be banned from seeking such licenses, in the interests of promoting competition [35].

Nevertheless, the cellular operators in the United States today do enjoy considerable latitude as to what services they offer and what technologies they employ. While PCN may be years away in the United States (given the exigencies of the spectrum allocation process), cellular operators are already beginning to experiment with "microcellular" concepts and other cellularized versions of various wireless access applications. Thus, a cellular service model for wireless access is emerging, comprising the following assumptions:

1. A willingness to embrace somewhat more innovative service objectives.

   Whereas the telephony service model views wireless access as a different means to the age-old end of cheap, universal telephone service, the cellular service model starts with the assumption that wireless access offers a novel communications opportunity that exploits radio's unique capabilities as a transmission medium. Specifically, cellular assumes that wireless access begins with mobile, i.e., vehicular, communications as the basic model. In many ways, therefore, the cellular model is not constrained by the weight of telephone experience, and traditional telephony service objectives are only obliquely relevant.

   For example, transparency *per se* is not a meaningful objective for the cellular operator. Cellular systems do not try to mimic the procedures used on wireline loops. Hook-flash signaling, for instance, is irrelevant if the cellular network is designed around DTMF (touch-tone) from the beginning. *Equivalent quality of service* is replaced by a much looser notion that might be termed acceptable quality of service, or even *salable quality of service.* And the *equivalent grade of service* sought by the telephony model is replaced by, in the first instance, the rather hopeful FCC criterion of 2% blocking [36], and in practice by a much looser standard in which blocking may reach 20% or more in some cells at the busy hour [37].

   Cellular phone calls today are poorer in quality than wireline calls. From the standpoint of the cellular service model, however, the point of reference (if there is one) is to pre-cellular mobile radio, where blocking and service outages were even more severe. In their minds, cellular operators market a *mobile* communications service with which the wireline network cannot compete, by definition. Therefore, the proposition that car telephones should provide the same quality and grade of service as the wireline network represents, at best, an ideal of sorts; as a practical matter, it is the competitive environment of the cellular duopoly that provides the real discipline. If one operator's grade of service is markedly inferior to his competitor's, in time he will lose market share. The real measure of service for this model is perhaps best expressed as *good enough to stay competitive.*

The shortcomings of this service concept are readily apparent to those who subscribe to the telephony model. However, the positive side of the cellular service model is that it tends to spur thinking about innovative types of service offerings. The operator is much more willing to exploit the uniqueness of radio as a nonphysical access medium. The cellular model also tends to adopt a more experimental approach; a new service offering may start out with somewhat inferior but market-acceptable performance and then improve over time. This lowers the barriers (psychological and economic) to service experimentation, at least by comparison with the telephony model, in which any new wireless access offering tends to be held up to the standards of mature wireline telephony right from the beginning.

However, the innovativeness within the cellular model is still constrained by the dominant, and highly successful, car telephone application. If a cellular operator has a choice between allocating a portion of his spectrum to some new type of service—say, telepoint (see next section)—and loading on more car telephone customers, the established segment will tend to be preferred.

2. A tendency to view wireless access as a service/product that is market-driven.

The cellular model sees wireless access as a consumer-oriented business. Many of the key decisions that drive the technological cycles are customer-driven. The features that sell are those (such as portability) that offer attractive possibilities to the end-user, rather than the essentially invisible benefits (operator cost-savings, network flexibility) that wireless access connotes to the telephony model.

Wireless access terminal equipment is viewed within this model as CPE. This has manifold implications within today's regulatory environment. It also means that the consumer bears a substantial up-front cost to initiate service, as well as the risk of loss, theft, or damage. Moreover, as consumer equipment, wireless access terminals will tend to be designed for shorter product cycles, *planned obsolescence* may even come into play. Certainly, it can be expected that a wireless access customer with a vehicularly mounted terminal would probably replace that equipment at least as often as he replaces his automobile, which in the United today averages about 4–5 years. This implies a different perspective on product design than, say, the 20-year traditional service lives associated with telephone outside plant.

Also, the cellular model does not posit a fully integrated role for wireless access, but instead follows the assumption of parallel or overlay network development. Wireless and wireline access are not mixed as they are in the telephony model.

3. Wireless access is seen as an unregulated service.

   The cellular model assumes that the operator of wireless access will have the ability to set and modify prices in response to market conditions, including competition, untrammeled by state regulators. In short, the cellular model tends to view wireless access as a business, rather than a utility.

   This ideology has been only partly thought through. In fact, cellular licenses are semi-monopolies; indeed, this is the basis for much of their value. To the extent that wireless access is viewed as a monopoly, it invites regulation. Some state regulators are beginning to take up the question of their role in ensuring adequate competition, fair pricing, service quality, and so forth [38]. The apparent lack of price competition in many cities suggests that cellular operators may be exposed in the future to greater scrutiny.

   Surprisingly, some proponents of the cellular model have advocated expanding the cellular franchise to cover certain applications—such as basic exchange service or wireless local loops to fixed locations (see Section 5.3)—which are today fully regulated at the state level. The exposure to full-fledged state-level price regulation would seem to be something that the cellular operators would generally seek to avoid.

4. Wireless access is seen as a semicompetitive service.

   The cellular service model is uncertain as to how much competition to embrace. The North American framework creates two licenses for each franchise. The decision to limit it to two was historically based on arguments that have lost much of their technical merit [39]. Moreover, as some state regulators have pointed out, the flourishing cellular reseller market tends to undercut the notion that cellular services can support only a limited amount of competition. In California, for example, there are more than 50 cellular resellers; yet only two entities are allowed to own facilities in any market in any given serving area [40].

In summary, the cellular service model tends to take a somewhat simplistic approach to the broader questions of wireless access. Wireless access is seen as a series of applications offshoots of today's mobile and portable cellular telephone services, to be pursued in due time by the cellular carriers. Service standards are loose. Serious conflict exists on the issues of competition and regulation.

The chief merit of the cellular model, in my opinion, is that it combines a modicum of quality-of-service consciousness (perhaps reflecting its telephony parentage) with a greater readiness to exploit the unique attributes of radio. It is less confined by the sometimes irrelevant aspects of the telephony service model and is more free to innovate.

5.2.3 Wireless Access as PCN

The third view of wireless access services is frankly revolutionary. (If this means that it is also at times rather overblown, so be it.) Radio is welcomed in its full potential as a technology of the next millennium, the communications medium of the future, the realization of science fiction dreams from *Dick Tracy* to *Star Trek.* Radio is not to be shackled to antiquated service ideas rooted in the saurian swamp of wireline telephony—on the contrary, the PCN revolutionaries rhapsodize about radio communications as a great technological imperative that will redefine our expectations about what communications services should do for us. Just as the airlines killed the railroads, so will personal, portable communications networks reduce the wireline network to a vestigial remnant, a backup system at best, a pastime for antiquarians.

It may prove a bit too topical to label this the *PCN service model,* especially since *PCN* (i.e., personal communications networks, or *PCS* for personal communications service) is still an undefined term in the United States. However, the use of a neologism of uncertain significance is perhaps appropriate. PCN models have no history (outside of science fiction genres), and PCN almost certainly faces a protracted birthing process. In North America, there are no PCN spectrum allocations, no PCN service definition, no licensing procedures, no technical standards, no pioneering manufacturers. Yet we can discern some of the implicit assumptions of the PCN service model.

1. A strong emphasis on the unique and "revolutionary" capabilities of radio technology, as contrasted with wireline systems.

   PCN focuses on what radio can do that wireline communications can never do. Specifically, PCN stresses portable communications. The idea of a completely portable communications device is not new, and many presentations on PCN have an odd, almost dated quality about them. As far back as the turn of the century, when "Hertzian waves" first came to the attention of the commercial world, grandiose extrapolations to a wireless, in-touch-anywhere network were commonplace. Yet despite the availability of radio communications systems from an early date, it was radio broadcasting that took off. In the meantime, several hundred million wireline access lines were installed around the world. Whither portability?

   The extent to which completely portable communications will actually be embraced by the mass market is one of the uncertainties hovering over the topic. Nevertheless, the PCN service model leaps to the end game by assuming that portability will become the dominant criterion for human telecommunications and by presenting a network concept based on this assumption. Universal portability may well prove to be the norm of the future, and if so then the questions of "equivalence" with wireline telephony lose some of their force. Wireless access is seen less as an equivalent substitute for wire, and more as a superior technology displacing an inferior one.

2. Wireless access is market-driven.

   Like cellular radio, and unlike traditional telephony, the PCN model views wireless access as being driven by the consumer market, rather than a technical alternative to be deployed by the guardians of the network. If anything, the PCN model is bolder than the cellular model because it is less held back by the car phone application. Somewhat more than the cellular model, the PCN model anticipates rapid technological evolution and a proliferation of diverse applications. This is a natural outgrowth of the emphasis of the PCN model on radio's unique capabilities. It is the desire of the end-user to exploit these unique capabilities, which forms the *raison d'être* of PCN as a business.

   The strong market orientation of the PCN model also implies, I believe, a much more relaxed approach to standards. The PCN model holds that wireless access operators should be allowed to provide whatever the marketplace is willing to buy. Whereas the telephony model is almost paralyzed by its concern over standards, and even the cellular model is very conscious of a need for standards to ensure roaming compatibility, the PCN environment seems to be prepared to allow a more Darwinian process to winnow out new technologies and service offerings.

3. Wireless access should be completely unregulated.

   The market orientation also implies that it is the consumer, and not the regulator, who disciplines the wireless access operators. If the PCN operator does not offer the right service at the right price, with the right quality, the free market will siphon off his customers. Pricing, service offerings, quality standards—under this model, all should be left to the free market process.

4. Wireless access can sustain unrestricted competition.

   The PCN model implies, I believe, that entry by new operators should be relatively unrestricted. Of course, once the PCN franchises become a reality (which is not likely until the mid-1990s), those first in the game will tend to argue that further competition should be limited. However, such arguments will be hard to sustain. Digital technologies will enable such an efficient exploitation of the spectrum that I believe spectrum shortages will disappear, except perhaps in a handful of major metropolitan areas. Wireless local exchange services will become a strong competitive service (under the PCN model) similar to the highly competitive interexchange segment today.

In summary, the PCN model takes the service concept for wireless access to the antipodes. Unregulated pricing and service offerings, unrestricted entry for competitors, and heterogeneous technologies, totally dedicated to placating the market—we have come very far from the telephony model. A perhaps less attractive feature of the PCN model is that it tends to follow the YGWYPF ("you get what you pay for") principle as to service quality. This philosophy, which seems part and parcel of a full-fledged market-driven approach, has wrecked other services such as citizens band radio. There is certainly some danger that PCN services may self-extinguish in a similar manner, particularly where the *laissez-faire* principle extends even to the proposal that no interference regulation is necessary, that PCN devices might be offered on a "sufferance" basis (e.g., Part 15 of the FCC rules) [41]. We will return to this issue in Chapter 6.

These three basic service models have been drawn (perhaps slightly artificially) so as to show a continuum. The telephony model is the most conservative, the most authoritative, and carries the most historical baggage. The PCN model is the brashest, the most underspecified, the most questionable, and possibly the most exciting. The cellular model seems to occupy an intermediate position on all axes, embracing some of the network orientation of the telephony world, but also conscious of the strong market pull. Figure 5.9 represents an attempt to display this continuum of service concepts.

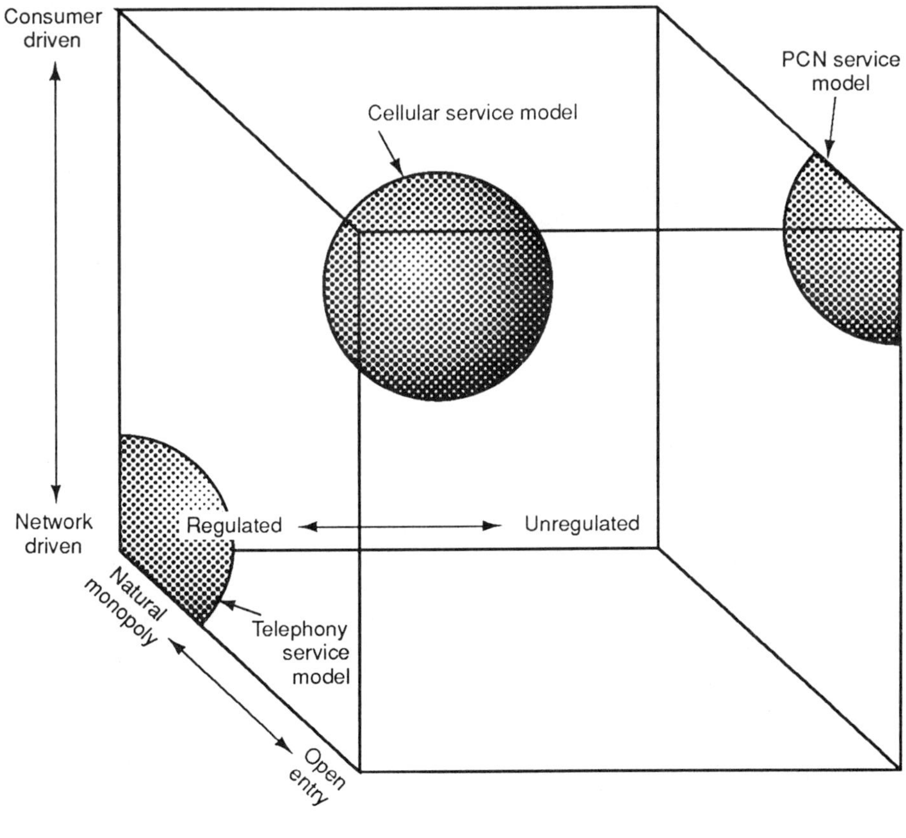

Figure 5.9 The three major service models for wireless access.

5.2.4 Wireless Access in Private Networks

The three previous models all tend to assume a common carrier status for the wireless access operator. It is also possible to imagine wireless access systems developing as privately owned systems, closely administered by a corporate or government entity for a closed set of users. Some of the important characteristics of the private carrier model follow.

1. The system administrator/operator can restrict access to the system.

    Unlike the common carrier models, the private carrier model assumes that the system administrator can decide specifically who will be allowed to transmit or receive, what equipment they must use, where and how they will use it.

    For example, the residential cordless telephone market is a form of wireless access today that essentially follows the private carrier model. The cordless phone base unit "allows" only a single remote terminal to communicate. I do not ordinarily have the expectation that I can take my cordless handset to my neighbor's house and use it on his system (notwithstanding the fact that this may sometimes be possible).

    Extend the cordless phone to a cordless or wireless PBX concept and the implications are perhaps clearer. Let us imagine an office with a PBX supporting 100 cordless extension phones. It resembles a local-area network. The system administrator will specify (and control) who receives a cordless unit, where the recipient can use it, and so forth. Visitors will not (in all probability) be able to walk in off the street with their own handset and start talking (although some visions of PCN go so far as to posit this scenario).

    We are used to these rules in the world of wireline PBXs today. The extension to wireless systems is straightforward. Indeed, there are many areas where similar restricted-access procedures will almost undoubtedly apply, such as factory-floor wireless communications systems, and many forms of dispatch networks.

2. Private wireless access systems will be loosely regulated.

    There is no "franchise" as such in a private system. It is not a public utility. There is often no "pricing" although there may be accounting of costs on a user-by-user basis. All this suggests that regulation will probably be limited to noninterference matters (and the Part 15 model may indicate that even this aspect may not be regulated).

3. Private wireless access will probably not be strongly standards-oriented.

    Since a single operator will purchase and administer the entire system, end-to-end, there will be less need for standardization of the radio interface. Again the example of wireline PBXs would indicate that systems will quite likely evolve along manufacturer-proprietary formats.

The private model is likely to be quite dynamic and vigorous. Major users have already begun to step forward into the arena of radio-access and wireless networks. Wireless PBX proposals are proliferating. Those on the common carrier side who believe that such systems should be folded into the emerging public models may be disappointed. The major advantage of the private carriers is that they need not wait for the resolution of the many troublesome regulatory problems facing the common carrier models for wireless access.

The four service models we have examined—telephony, cellular, PCN, and private carrier—form the major branches of the applications tree for wireless access (Figure 5.1). In the following sections of this chapter, we will move up each branch in turn to examine each applications set in some greater detail. The following sections of this chapter will deal with the telephony-oriented applications set (Section 5.3), the mobile applications set (Section 5.4), the PCN applications set (Section 5.5), and the private carrier applications set (Section 5.6).

5.3 "TELEPHONY-DERIVED" APPLICATIONS OF WIRELESS ACCESS

The telephone companies' interest in radio as a loop access technique goes back several decades. Analog FM-based *rural radio* has been used sporadically since the 1950s. However, rural radio remained largely a subscriber-selected communications alternative (generally a last resort), rather than a mainstream carrier-selected access method. The use of modern radio-based access systems by telephone common carriers to serve fixed-location (i.e., nonmobile) customers began in the 1970s with the introduction in Canada and elsewhere of so-called *thin-route point-to-multipoint microwave systems;* although there is no generic term for the service provided by these systems, it has sometimes been called *subscriber radio.* Due to the lack of available spectrum, however, thin-route microwave never developed in the United States to a significant degree, and its inherent economics and lack of spectrum efficiency have restricted most international uses to low-density applications in remote areas around the world. It was not until 1986 that more spectrum-efficient digital radio technology became available, which led to the creation by the FCC of a new class of local exchange radio service, known as *Basic Exchange Telecommunications Radio Service* (BETRS), sometimes shortened to *basic exchange radio* (BEXR). By 1990, more than 60 BETRS systems had been installed in the United States and it had emerged as the leading form of wireless access within the North American regulated telephone environment.

The applications for which these services have been deployed tend to follow the telephony service model; in particular, the term *basic exchange radio* implies the provision of a basic telephony service (rather than enhanced services based on radio's capabilities for mobile or portable communications). However, while both types of local wireless access have generally been sold strictly as lower cost alternatives to conventional cable plant, a number of similar systems have been used in a transportable configuration, which does recognize some of the unique characteristics of radio transmission.

5.3.1 Rural Radio

Radio has been used as an occasional solution to "loop-type" communications requirements in extremely remote areas since the 1950s and perhaps even earlier. Usually this has involved the adaptation of mobile radio equipment, patched in (often manually) to the public network (Figure 5.10). In the United States, the Rural Electrification Administration (REA)—a federal government agency responsible for financing and fostering rural telephony—has taken an especially active interest in this field, to the extent of supporting and even funding a variety of experiments in

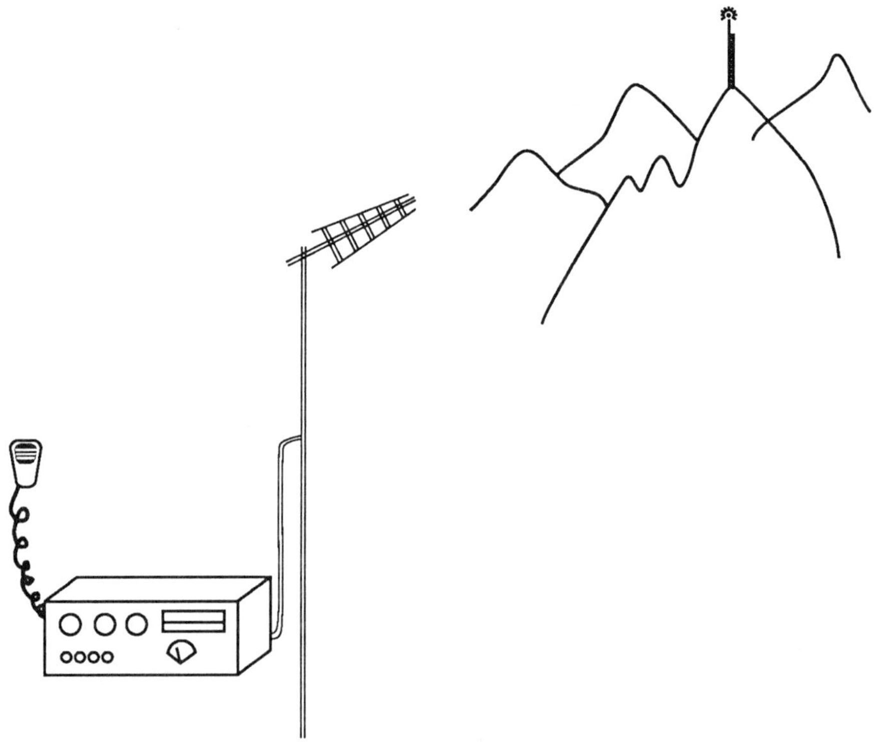

Figure 5.10 A rural radio link.

radio-based access dating back to the early 1950s. Claude Buster, the former chief of the transmission branch of the REA, has written an interesting historical summary of these early projects, which, since it has not been published and may not be readily available, has been included in Appendix 5A.

Rural radio, based on analog FM transmission, never proved satisfactory for large-scale applications, and never gained acceptance within mainstream telephony in North America. Its chief shortcomings were the high cost, power requirements, poor equipment reliability, and the poor spectrum efficiency. Indeed, traditional analog rural radio often barely qualifies as wireless access. Nevertheless, it is still sold in situations around the world where there is no alternative. It is the solution of choice for an individual user who wishes to solve his own communications problem without having to wait for the local telephone authorities to implement a more general solution. I have seen installations of this kind in operation on isolated ranches from Arizona to Argentina. The cost of a single installation ranges from \$3000 to \$10,000, in my experience—which can be a lot less than the construction charges for cable. There are undoubtedly tens of thousands of "bush radios" in operation today throughout the world providing various levels of interconnection to the public network.

5.3.2 Thin-Route Microwave Subscriber Radio

In the mid-1970s, an alternative to traditional rural radio emerged with the development of thin-route point-to-multipoint microwave subscriber radio (Figure 5.11). These systems were the first to be designed specifically for a wireless access fixed-location application, and they represented a major advance in performance.

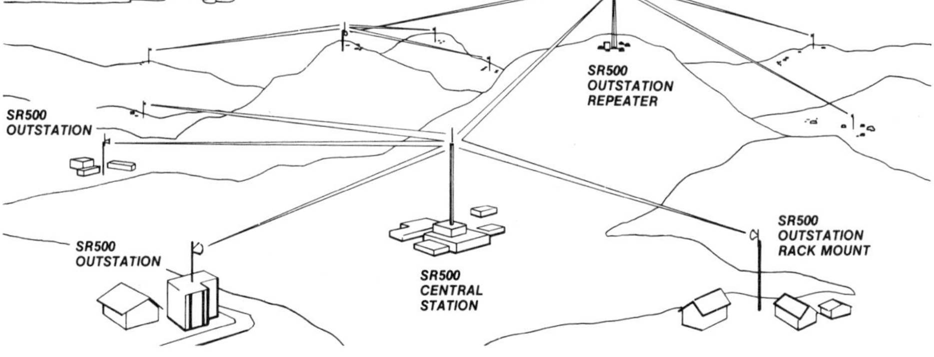

Figure 5.11 A wideband subscriber radio system. (From SR Telecom.)

To my knowledge, the pioneer in the development of this type of system was the SR Telecom company in Canada [42]. From initial prototypes in 1974 and operational tests in Saskatchewan in 1975, SR Telecom developed a successful line of wireless access systems that had provided by 1990 more than 25,000 "lines" of radio-based access worldwide. Most countries have authorized this type of service; however, the FCC has not allocated the necessary frequencies for point-to-multipoint microwave operation, so these systems are not widely used in the United States [43].

Such systems are normally deployed at frequencies between 1 and 3 GHz (depending on the available frequencies in particular markets). They employ a wideband transmission scheme (i.e., carrier bandwidths greater than 1 MHz), with a time-division architecture (see Chapter 6). The first systems developed by SR Telecom used pulse amplitude modulation (PAM) and a polling access technique; most product offerings now rely on standard 64 kb/s pulse code modulation (PCM) combined with a contention-type access protocol (e.g., Aloha). The subscriber radio system interfaces to the network by means of line cards (Figure 5.12), so that it has the appearance (from the network's perspective) of a loop carrier or "radio concentrator" [44].

The system architecture is fairly simple, incorporating a central station, subscriber stations (or "outstations" in the SR Telecom terminology), and repeaters. There is a great deal of flexibility in the configuration of the system (Figure 5.13).

Another characteristic of such systems is that they tend to utilize outstations to serve *clusters* of individual subscribers that are actually connected to the radio outstation *by wire.* The system economics tends not to favor the use of individual outstations for each subscriber; thus, subscribers are grouped into what SR Telecom calls *outstation serving areas,* which comprise groups of subscribers (up to 16 individual lines) sharing one radio outstation. The link between the outstation and the central station is via radio. This orientation toward clustered applications is quite important in understanding the economics and the range of appropriate applications for this technology (see Part IV).

Since subscriber radio systems are based on 64 kb/s clear channel transmission (the common denominator of wireline service) they can handle most network services without any special arrangements. SR Telecom is currently developing an ISDN capability for the SR 500 system [45]. The chief limitation of subscriber radio is its cluster-oriented economics, which constrains these systems to a rather narrow band of applications (this is discussed at greater length in Part IV). As well, the relatively inefficient use of the spectrum will limit the ability of these systems to compete for high-density, urban PCN-type applications.

Most applications of subscriber radio are in areas that are quite remote. The significance of the term *thin route* may be understood by reference to a published case study involving the use of an SR 100 system in northern Sweden [46]. This example involves a nearly 100-km link involving point-to-point microwave, single-channel analog radio, copper wire, and thin-route point-to-multipoint microwave—

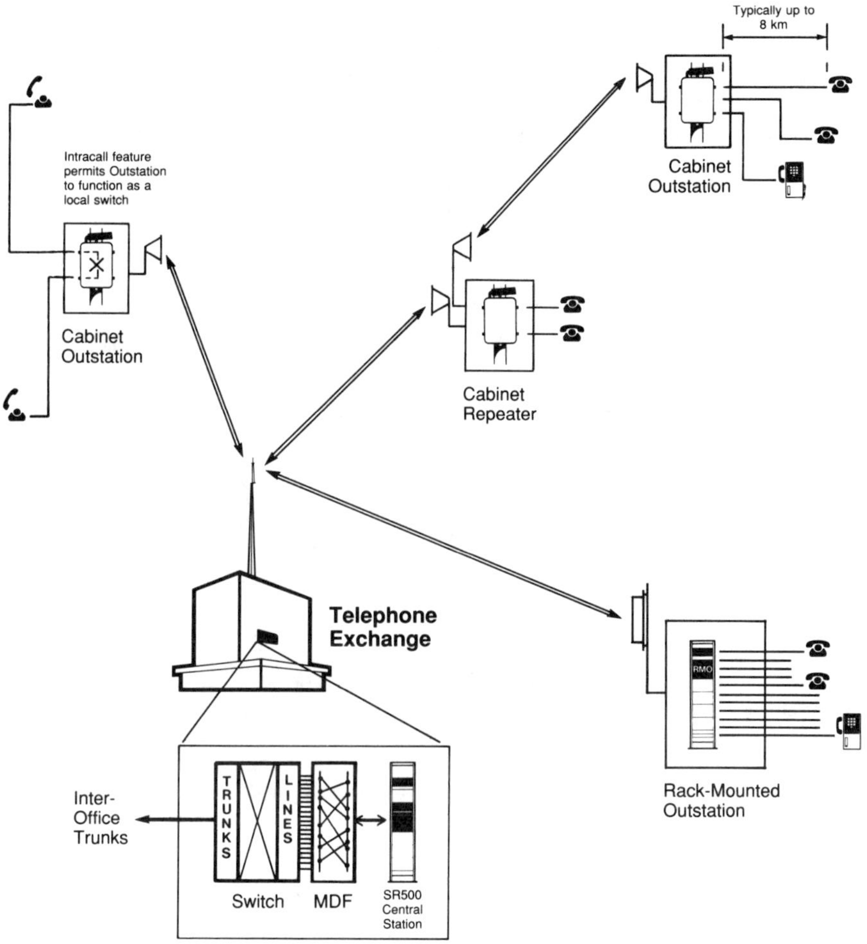

Figure 5.12 Subscriber radio interface options. (From SR Telecom.)

all to serve five subscribers. This is a good example of the diversity of techniques employed in very remote, low-density applications.

5.3.3 Basic Exchange Radio—"Wireless Local Loop"

In the 1980s, a combination of technological advances and regulatory changes led to the creation of a new class of wireless access service in the United States, based on the use of spectrum-efficient narrowband digital transmission and designed specifically for use by telephone common carriers as an alternative to copper wire. BETRS,

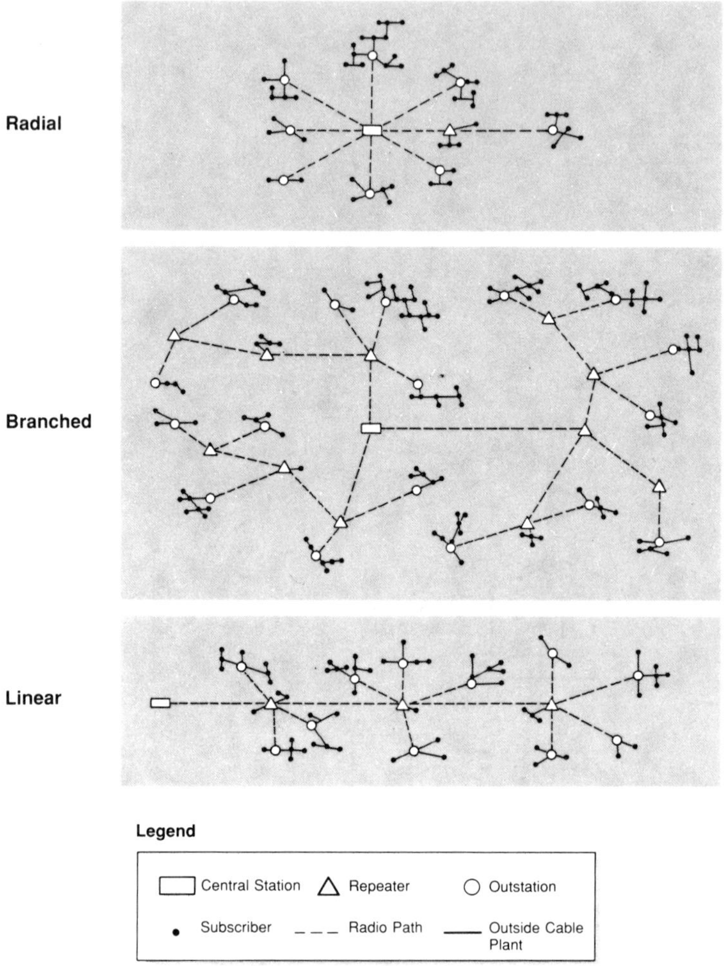

Figure 5.13 Subscriber radio configuration options. (From SR Telecom.)

or BEXR, is the first true mainstream application of wireless access within the regulated telephone industry in the United States. It now appears that this service concept will also spread to other countries.

5.3.3.1 Technical Outlines and Systems Concepts

To support a broad-based, high-capacity, high-quality telephone service, radio systems had to be developed that were much more spectrum efficient than either the analog FM of old rural radio or the wideband microwave of subscriber radio systems. The pioneering firm in this effort was International Mobile Machines Corporation (IMM), which developed in the early 1980s a frequency- and time-division multiple-access (FDMA/TDMA) technology capable of providing four telephone-quality trunked voice circuits on a standard 25-kHz mobile radio channel (in the 450-MHz band). By using digital voice coding at approximately 16 kb/s and high-level modulation (16-ary DPSK), IMM was able to achieve a spectrum efficiency of 80 (trunked) telephone circuits per MHz, compared to from 2 to about 7 circuits per MHz for contemporary wideband subscriber radios and analog FM cellular radio [47]. This unprecedented spectrum efficiency allowed for large numbers of telephone subscribers to be served using only a small number of radio-frequencies, which in turn opened up the planning horizon for telephone engineers to consider wireless access for much larger scale, more mainstream applications involving hundreds or even thousands of access lines per exchange.

Working with several telephone companies, including Mountain Bell (US West), General Telephone of the Southwest (GTE), and S&A Telephone Company of Allen, Kansas, IMM developed the specifications for a product using this narrowband TDMA digital radio technology in a wireless local-loop application. In some cases, the telephone companies were themselves developing technical specifications for a BETRS-type product [48] and assisted IMM in translating its advanced digital radio technology into a field environment. This system was subsequently marketed under the name *Ultraphone,* and became the leading BETRS-type radio system, with more than 60 system installations by 1990. The system architecture was very simple, following the wireless loop carrier model of the older microwave systems [49].

Early BETRS configurations tended to differ from the wideband systems in two respects. First, they did not (at least initially) employ repeaters, which were far less important for coverage at 450 MHz. Second, the higher spectrum efficiency and superior economics of BETRS-type equipment favored the use of a single-line subscriber station, rather than the outstation clusters that were standard with wideband microwave. Figure 5.14 shows a typical system configuration. Figure 5.15 portrays a block diagram of the basic system elements and interfaces.

5.3.3.2 BETRS Regulatory Background

Interest in the use of radio for basic exchange services was tempered initially by uncertainties over spectrum availability in the United States. Traditional rural radio had always been classified as a secondary service within the rather limited pre-cellular mobile radio allocation. Wideband microwave systems were not authorized for use except on a very limited basis; there was no general allocation. To address this issue, a Rural Radio Task Force was formed in June 1985 (partly in response to IMM's successful technical demonstrations early that year), headed by the REA and comprising members from all the major telephone trade associations, as well as sev-

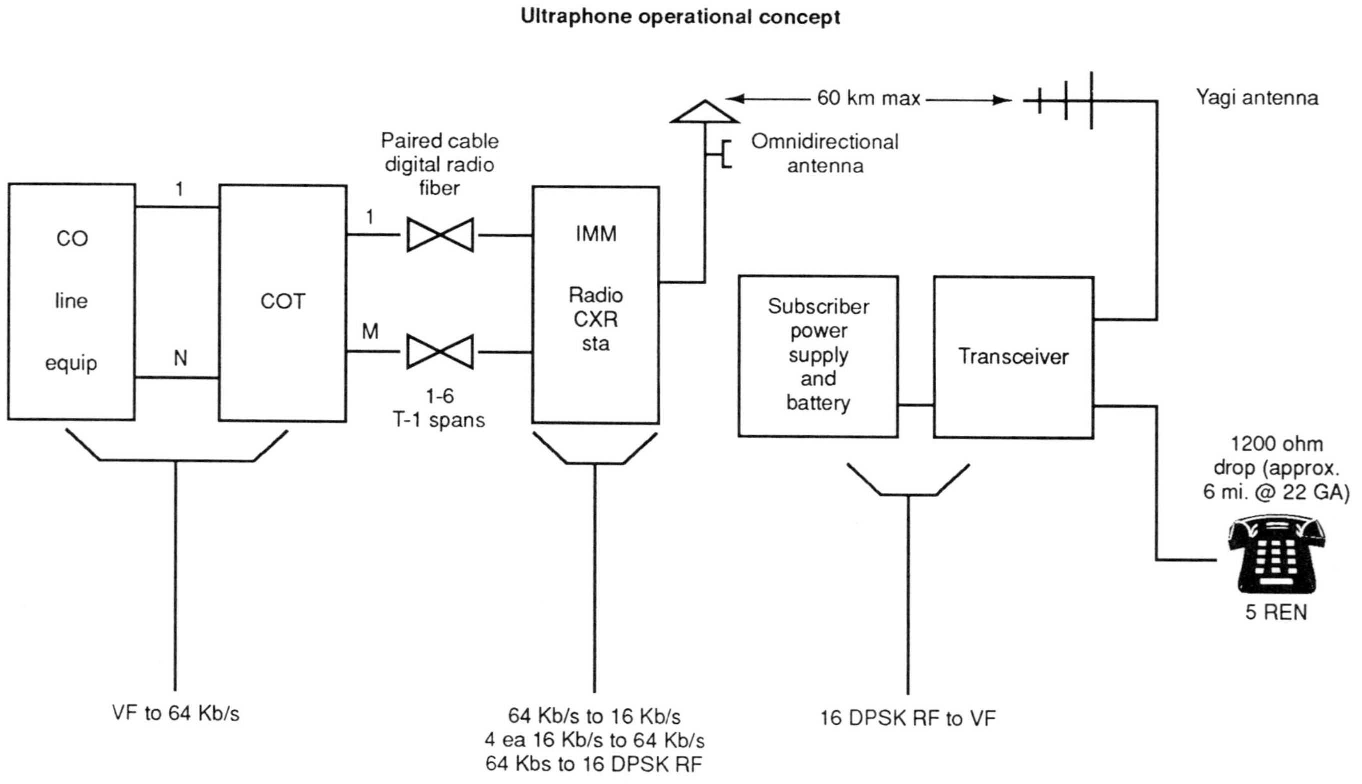

Figure 5.14 BETRS system configuration.

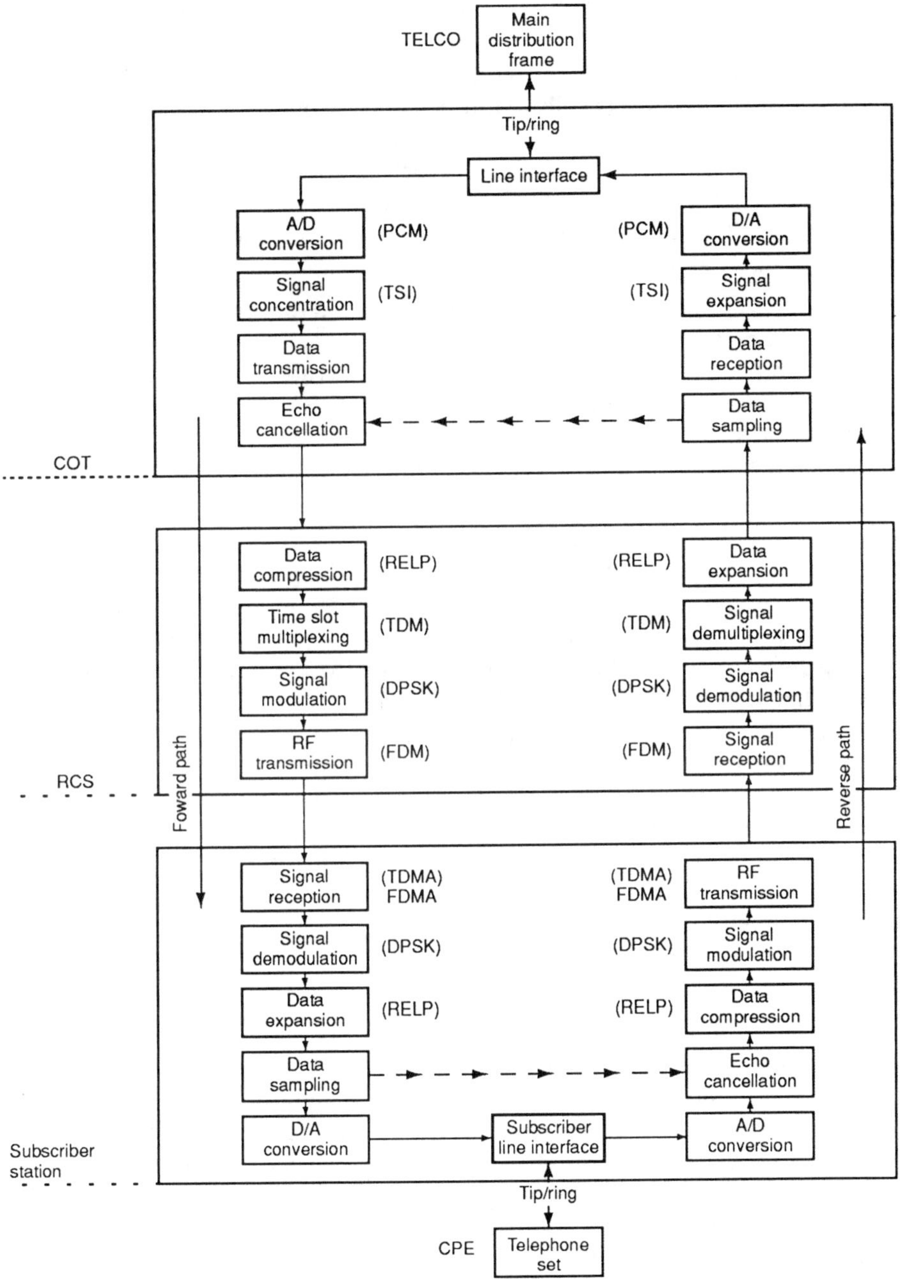

Figure 5.15 A functional diagram of a BETRS system. The BETRS subscriber station: larger box on left is the battery-back up power supply; smaller unit on right side of pole is the radio transceiver.

eral individual carriers. The task force began the process of preparing a petition to the FCC to allocate spectrum for a new type of service, *basic exchange telecommunications radio.* A number of key presentations were made to the FCC in the course of this effort, including one by Mountain Bell and the other US West telephone companies. This pioneering presentation discussed several specific case studies (see Section 5.3.3.3), and argued persuasively in favor of allowing the carriers to use radio as one of the alternatives for providing basic service. It is worth quoting the presentation on this point, since it sets the tone of the "telephony model" very clearly apart from the "cellular model":

> Our US West cellular provider—New Vector—and the other cellular providers are addressing a particular marketplace—major metropolitan areas. The service they provide is truly mobile. Their objective is attainment of market share in a new a profitable venture. We [i.e., Mountain Bell, Northwestern Bell and Pacific Northwest Bell—the regulated operating companies] on the other hand, are focusing on expense reduction. This is a key difference in perspective. We are not expanding our product line. We are asking for the option to use whatever technology is available to provide basic telephone service. [50]

Selected portions of the text of this presentation (which contains illuminating data on several potential rural radio applications) have been reproduced in Appendix 5B.

The Rural Radio Task Force submitted its final petition on May 9, 1986 [51]. The subsequent Notice of Proposed Rule-Making [52] attracted more than 40 comments [53], and led to the Report and Order [54] by which BETRS was established. In its final form, BETRS was authorized as a co-primary service (with mobile radio) in the 150- and 450-MHz mobile radio common carrier frequencies, and in 50 specific frequencies in the 800-MHz private radio allocation. Since then the FCC has also extended BETRS to the cellular frequencies, although, in practice, only the 450-MHz band has been used. The REA has developed a number of technical applications guides for BETRS, which constitute some of the best available literature on wireless access applications [55].

5.3.3.3 *Early Applications of BETRS*

Several of the early applications of wireless access under the BETRS allocation have been described in the literature. The first BETRS system was installed by Mountain Bell in September 1986 to serve eight subscribers near the town of Glendo, Wyoming [56]. It was a classical small rural radio application, serving a number of ranches separated from the town by a canyon and a reservoir that effectively rendered cabling impossible. The radio base station was located on a bluff about 12 miles from the Glendo switch (a step-by-step electromechanical exchange), connected by an 18-GHz digital microwave back to the central office terminal. Individual subscriber installations were as shown in Figure 5.16.

Figure 5.16 A BETRS subscriber unit.

The second BETRS system was installed in November 1986, by the S&A Telephone Company of Allen, Kansas [57]. S&A was a small independent—a self-described "mom and pop" operation serving about 800 access lines. The owner, Art Biggs, had obtained an REA loan for a project to upgrade 142 party-line customers to single-party service. Mr. Biggs has given a colorful description of the program and its high costs, and his exploration of radio alternatives, which is reproduced in Appendix 5C. In the end, S&A installed a radio system for about 50 customers.

One of the next sites was a less remote application in Ripley, Mississippi, where South Central Bell trialed its first system beginning in 1987 [58]. Although only 11 subscribers were served initially, the Ripley system was engineered to provide complete coverage for an entire wire-center serving area. The trial served as the field site for subsequent Bellcore evaluations of wireless access [59]. Subsequently BellSouth became the first major telephone company to standardize on BETRS.

BETRS applications and trials multiplied from 1987 onward. One of the largest was a system installed on the Navajo Indian reservation in Arizona in 1988 [60], serving more than 150 customers. Another system was installed in Silver Lake, Oregon, by Pacific Telecom, for more than 200 customers. Such applications pointed the way toward a wireless exchange concept. By 1990, there were more than 60 systems in operation, serving several thousand customers—a small but growing beachhead for wireless access within the world of regulated telephony [61].

5.3.3.4 Extensions of the BETRS Concept

With the success of initial BETRS trials in the United States, interest in the BETRS or *distributed radio* concept began to grow, although the natural conservatism of the telephony model has slowed actual deployments in many cases.

Canada. Canadian telephone companies were familiar with subscriber radio in its wideband format through the activities of SR Telecom. However, the application of wideband microwave systems was often limited to relatively remote northern villages and camps. The first Canadian operator to consider the application of radio-based access on a large scale was Alberta Government Telephones (AGT), which in 1986 issued a formal procurement specification for what they called *digital loop replacement radio* (DLRR) [62]. This 150-page specification was targeted at a province-wide rural multiparty-line upgrade program, known as *individual line service* (ILS). The total ILS program encompassed 108,000 access lines, of which 20 to 80% was forecast for radio—which would have amounted to a very significant program. Spurred in part by AGT's initiative, the Canadian government authorized a BETRS-type spectrum allocation [63].

Eventually, AGT proceeded rather tentatively with wireless access, installing several systems (including both BETRS-type and wideband systems from IMM and SR Telecom, respectively); however, due to scheduling requirements for completing the program, radio was actually little used in the AGT ILS program. Nevertheless, the regulatory precedent was set and it seems likely that Canadian interest in wireless access will be strong in the future [64].

Japan. The Japanese telephone authorities have evinced a strong interest in radio-based access, primarily as a measure for rapid restoration of service in times of disaster and for temporary uses. Figure 5.17 shows the basic systems concept embodied in a procurement document from Nippon Telegraph & Telephone (NTT) in 1986 [65]. To my knowledge, however, NTT did not actually deploy wireless access systems of this kind during the 1980s.

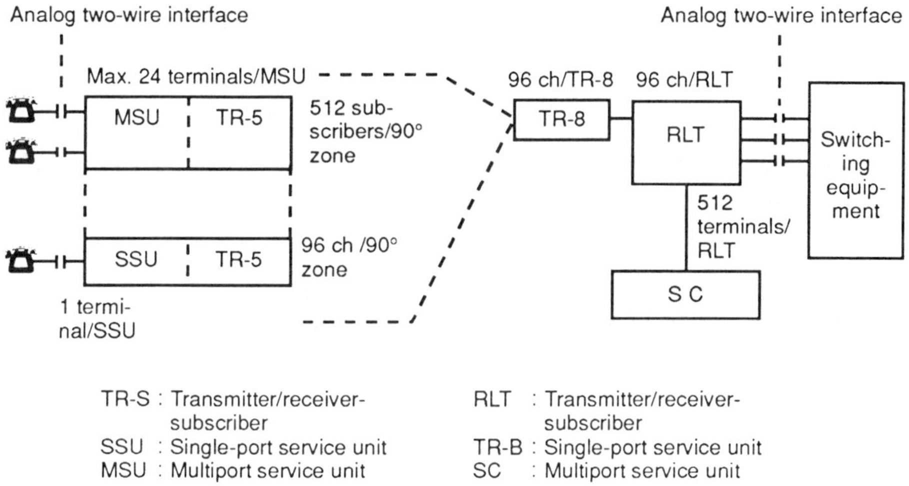

Figure. 5.17 Wireless access system described in NTT procurement. (From NTT procurement document.)

CCIR. The international radio and spectrum management organizations have also begun to study the concept of high-capacity BETRS-type radio applications [66]. A more generic approach is the concept for *future public land mobile telecommunications systems* (FPLMTS) of the CCIR Interim Working Party 8/13. FPLMTS encompasses fixed applications that are functionally equivalent to BETRS [67].

Bellcore and BEXR. Bellcore (i.e., Bell Communications Research) is one of the technical arms of the post-divestiture Bell System and in many ways the closest thing in the United States to a central standards organization for telecommunications equipment. Bellcore is developing its own approach for basic exchange radio, partly in response to requests by several of the Bell operating companies for evaluative assistance in deploying IMM's BETRS equipment [68]. Using the term BEXR (for basic exchange radio), Bellcore has prepared what is perhaps the most comprehensive generic specification currently available [69]. This document draws heavily on the previous Bellcore technical reference for wireline digital loop carrier systems [70].

The Bellcore system concept is shown in Figure 5.18. It is indeed a wireless loop carrier architecture, very similar to IMM's original system, except for the incorporation of a repeater. As with most Bellcore documentation, the BEXR advisory is quite thorough and is applications-oriented (rather than technology-oriented).

5.3.3.5 Advantages of BETRS

The major advantages of BETRS systems over previous approaches to wireless access in the telephony environment are:

1. Spectrum efficiency, which implies high capacity and a lower cost per subscriber
2. Suitability for distributed subscriber populations (as opposed to clustered populations).

BETRS was a breakthrough in terms of spectrum efficiency. It represents the first commercial use of such highly spectrum-efficient digital radio, achieving 80 circuits per MHz of spectrum (versus only 5 or so for single-cell analog cellular, and even less in some wideband microwave systems). This allowed planning of systems comprising hundreds or thousands of users per exchange area (or potentially even more), without imposing unreasonable spectrum requirements.

The economics of BETRS equipment also permits the use of single-line subscriber radios as the standard subscriber terminal configuration. This broadens the spectrum of applications beyond the clustered settlements typically served by wideband microwave. It enables the development of a true radio carrier serving area (see Section 5.3.4.3), which is the most revolutionary applications framework within the telephony model.

Viewed in the larger context, and in light of the other service models, BETRS seems rather staid. It takes little advantage of what cellular and PCN proponents view as the unique attributes of radio. Nevertheless, BETRS equipment has now been standardized at many of the largest U.S. telephone companies, including a majority of the Bell operating companies. In the course of gaining acceptance as both a service concept and as a specific hardware configuration within the conservative telephony community, BETRS has gone through an exhaustive process to develop the detailed methods and procedures for system analysis, deployment, and maintenance.

BETRS is thus perhaps the "sleeper" in the wireless access picture. While BETRS is still limited by the assumptions of the telephony service model (i.e., it is viewed strictly as a wireline substitute), and is moreover shaped by its rural radio

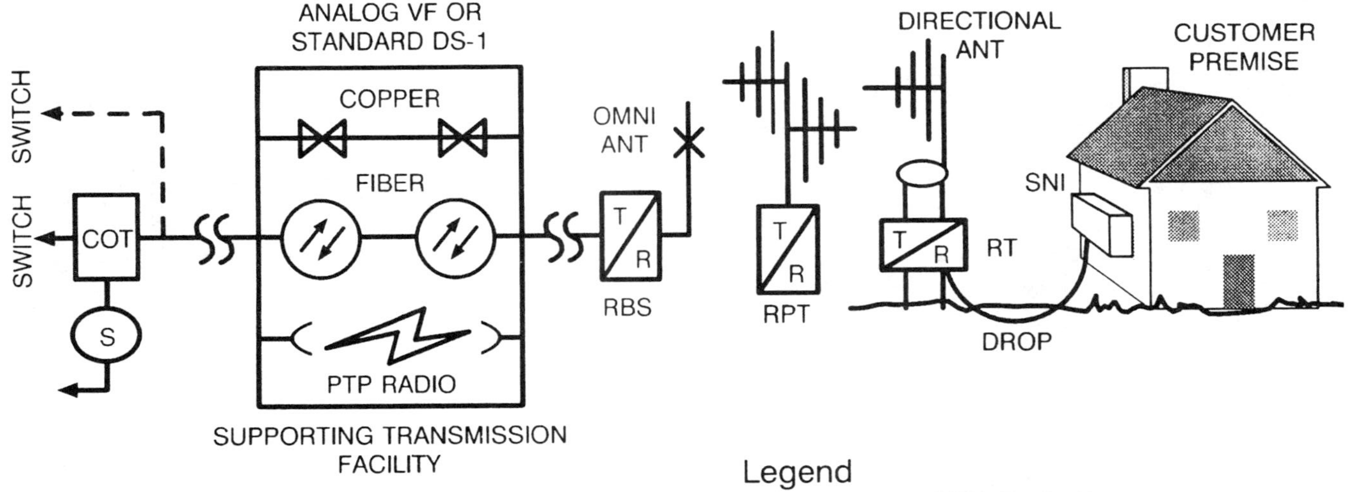

Legend

Switch - any standard local exchange office
COT - Central Office Terminal
S - Standard operations system interface
PTP Radio - any point-to-point radio external to the system
RBS - Radio Base Station
RPT - Radio Repeater
RT - Remote Terminal
Drop - Telco standard copper facility
SNI - Standard Network Interface
ANT - Antenna (type varies)
T/R - Transmit/Receive
Customer premise - mounting arrangement and connections vary

Figure 5.18 BEXR system concept from Bellcore.

heritage (which tends to view wireless access as suited primarily to remote areas), it nevertheless represents a milestone in the development of true wireless access technologies and services. Ultimately it may be absorbed into other emerging wireless access applications concepts (such as PCN). However, for the time being, BETRS possesses a higher "reality quotient" than any other true wireless access application.

5.3.4 A Hierarchy of BETRS Applications

What are the specific applications of BETRS technology?

The BETRS application set shares several basic premises. First, as noted previously, the hallmark of the telephony model of wireless access is its spartan emphasis on basic telephone service (sometimes called Plain Old Telephone Service, or POTS). The following statement is drawn from the GTE specification for a rural subscriber radio system, developed in 1983 (which may be the earliest formal description of this kind). It succinctly captures the meat-and-potatoes perspective of the regulated common carriers toward wireless access:

> *Basic Requirements*
>
> The rural subscriber radio system would be used, instead of land lines, to serve rural subscribers in certain remote locations. The system must be able to perform the following:
>
> Be able to be used as primary telephone service and be completely transparent to the subscriber.
>
> Have good transmission qualities and meet normal Telephone Company transmission standards.
>
> Have adequate capacity to serve several hundred rural subscribers and offer a P.01 grade of access to the network.
>
> Have the capability to serve single-party subscribers, four-party subscribers, private-line special service circuits, and pay stations.
>
> Be more economical than land lines to serve rural subscribers in remote areas.
>
> Be highly reliable and easily maintained. [71]

The Bellcore BEXR document elaborates on the basic access theme, stressing the rural character of wireless access which, by assumption, has become the BETRS stereotype:

> Radio-in-the-loop applications are most prevalent in the less densely populated rural areas . . . Radio-in-the-loop is designed to eliminate all cable feeder facilities and the majority of cable distribution facilities while providing basic exchange service on a point-to-multipoint basis directly to the end subscriber. [72]

The other index for most BETRS applications is the requirement to build or replace a *long* local-loop circuit. Long loops are often terribly expensive (as we shall see in Chapter 7), and radio will often win on a first-cost basis against copper where new construction is required. Based on this principle, more specific catalogs of potential applications have been offered, including the original BETRS petition by the Rural Radio Task Force [73]. The Bellcore BEXR requirements document identifies six categories of potential applications for wireless access:

1. To serve new growth along existing [cable] routes where copper facilities have exhausted. This application will include demand for party-line upgrades. Reinforcement costs along existing routes are often very expensive and radio facilities may be appropriate. . . .
2. To serve areas where no facilities currently exist. For example, radio would be used to avoid extending existing facilities that currently do not reach new housing developments.
3. To serve areas where existing service is provided on a wire facility that has reached the end of its service life. Plant requirements are expected in areas where open wire currently provides a non-acceptable quality of service due to age, lightning problems, ac induction, etc.
4. To provide emergency restoration facilities, or to provide temporary service for one-time events, for construction contractors or for the purpose of postponing long-term cable projects until it is certain that development will actually occur. . . .
5. To replace Community Dial Offices (CDOs) where it is more economical to serve an entire area with radio rather than a CDO plus the normal feeder and distribution facilities.
6. To bring telephone services to offshore islands. [74]

In one sense, the telephony model does simplify the analysis of wireless access applications: as one of a set of completely transparent and mutually substitutable transmission alternatives (including dedicated copper pairs, loop electronics, and possibly optical fiber), *wireless access is preferred wherever it costs less than other methods.* However, as we discuss in Chapter 7, the comparison of true and complete costs of such different alternatives is anything but straightforward. In practice, all access systems—wireless and wireline alike—incur certain *common costs,* which are shared among many users, and *per-line costs,* which are completely assignable to individual subscribers. As well, in any given specific case, there are what we may call *sunk costs* (the investment in the undepreciated access facilities already in place) and the *marginal costs* of installing new equipment. The advantages of wireless access over wireline access depend very much on the specific context of the application: how many subscribers, the manner in which they are grouped spatially (e.g., clustered or dispersed), the status of the existing feeder cable facilities, the size of the exchange and the status of nearby exchanges, and so forth.

Generalizing from the guidelines provided by Bellcore and others, there are a number of typical loop situations that can indicate a likely wireless access application.

1. *Rural unserved subscribers.* These are the traditional rural radio applications, typically involving subscribers living in remote areas well beyond the existing exchange boundaries or feeder facilities.
2. *Rural party lines.* Upgrading multiparty facilities calls for new cable facilities, which is often sufficiently expensive in rural areas to justify a wireless access solution.
3. *Growth in rural areas.* The simple phenomenon of growth in the subscriber base can create demand for long loops in rural areas, which, depending on the status of the feeder facilities, may be very expensive to serve with traditional wireline means.
4. *Replacement of rural feeder facilities.* The field service life of conventional outside plant varies greatly from region to region, climate to climate. However, even under the very optimistic assumption of a 50-year service life, more than 2,000,000 lines would need to be replaced annually in the United States. The loss of a rural feeder is a particularly costly and difficult process. There is often a lengthy period of deteriorating service and stop-gap maintenance. Everyone familiar with rural telephony has seen cases where the outside plant (i.e., the feeder plant) is being milked for its last years of service life, during which time the customers using those facilities are actually receiving substandard service.

The four preceding typical applications all carry the assumption that the loops involved are relatively long (i.e., rural), often with a large percentage of the total loop circuit comprising the single-line "drop" facility. How long is long? Of course, there is no absolute answer because loop economics is very dependent on the unique characteristics of each particular exchange. As we will see in the following subsection, a number of studies have been made using a variety of loop-length cutoff criteria. The average loop in the United States is about 10 to 12 kilofeet, and most telephony-oriented wireless access applications lie beyond twice that distance given today's radio equipment economics.

However, there are other "typical" wireless access applications that are not as distance-dependent, including the following.

1. *Temporary service.* The need for a temporary access line can arise in a variety of situations. For example, resort areas often have large seasonal populations that make it difficult to justify the fixed investment associated with copper facilities. Radio can serve as a "redeployable" access facility. When growth outstrips the available facilities in a given area, radio can be used as a temporary solution to provide service until the wireline facilities can be extended. Wireless access has been used in this manner by several U.S. telephone companies, sometimes in areas that are not at all rural in the conventional sense, but suburban (especially at the fringe of growth in a metropolitan area).

2. *Emergency service.* Another distance-insensitive application of wireless access involves the use of radio to replace, back up, or supplement existing wireline facilities in response to a natural disaster, or even a man-made "emergency" such as a Fourth of July weekend at the shore.
3. *Data services.* Finally, there are some applications in which radio may play a role in supporting advanced data services that cannot be easily or cheaply provided over existing wireline facilities.

The preceding categories represent typical uses of BETRS-type wireless access. However, whether or not wireless access equipment will actually be deployed depends a great deal on the analytical framework that is applied to process the facts about plant status and service requirements of the target exchange. That framework may be incomplete (for example, they often neglect the time dimension, i.e., the life cycle costs of the alternatives). It may be constrained by inappropriate assumptions. To take a very simple, but exceedingly common example: Consider two adjacent exchanges, each with certain service requirements that could be served by radio. There are not enough customers in either exchange to achieve an acceptable per-line cost (i.e., too few customers to reach the "critical mass"). A single wireless access system, however, could serve all the target customers in both exchanges, with acceptable costs (since more users are sharing the common cost elements of the base station). However, the telephone company in question makes the assumption that exchange boundaries must be respected, and customers from Exchange A cannot be served by a wireless access system that interfaces to the switch in Exchange B. Of course, in time, the use of wireless access will tend to eliminate the exchange boundaries, or at least render them much less important for planning purposes. Nevertheless, such wire-biased planning frameworks can inhibit the use of wireless access in particular situations.

The development of an adequate applications methodology for planning wireless access applications is a significant undertaking. On the one hand, it involves stripping away the wireline-oriented assumptions that are so deeply ingrained in the industry thought patterns. On the other hand, there are a number of respects in which the conventional frameworks must be amplified to take account of radio's unique capabilities. We devote most of Chapter 7 to detailing a more complete (albeit still preliminary) applications framework.

Conceptually, I have found that BETRS applications can be usefully organized in a hierarchy of applications frameworks, beginning with the most obvious and traditional situations where radio can be used and expanding in several stages to a much broader approach to wireless access within the regulated telephony environment. The six basic levels of analysis of BETRS applications are:

1. Traditional rural radio applications
2. Wireless digital loop carrier
3. Radio carrier serving area

4. Wireless digital access
5. Transportable and emergency applications
6. Integrated digital radio applications.

Each of these applications frameworks is addressed briefly in the following sections. It should be emphasized that the equipment in all cases is the same; what is different is the way in which it is used.

To discuss these different applications concepts, let us envision a typical wire center or exchange area as we would find it prior to the application of wireless technology (Figure 5.19). The exchange consists of a subscriber base served by a local central office (Class 5 telephone switch), which may be digital or analog, electronic or electromechanical. Running out from the central office are the major feeder lines, each containing dozens or hundreds of individual wire-pairs. These in turn branch off into smaller distribution cables, and finally down to individual lines (wire-pairs) connecting individual subscribers.

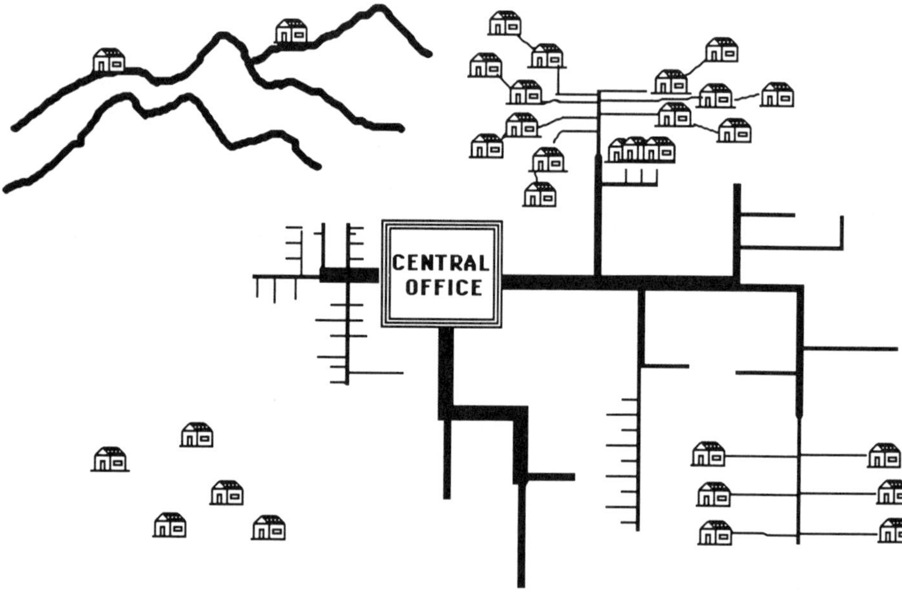

Figure 5.19 Idealized wire center.

We now look at the different ways in which basic exchange radio may be deployed in this traditional wire center.

5.3.4.1 Traditional "Rural Radio"

Radio is often considered as an alternative where there are a small number of subscribers located beyond the reach of the existing feeder and distribution network, often outside the formal boundaries of the serving area. Typically, these unserved homes lack service because of some geographical barrier, such as a river, bay, or mountain; in some cases, it is simply the great distance from the central office that constitutes the barrier to service.

Radio may be perceived in such situations as the only practical alternative (Figure 5.20). Most *first application* sites for BETRS (i.e., the sites used by the major telephone companies to evaluate and prove-in BETRS equipment) fall into this category [75]. The costs of using wire to reach such customers are so prohibitive that no detailed economic comparison is needed. This simplifies the operator's decision to experiment with new technology.

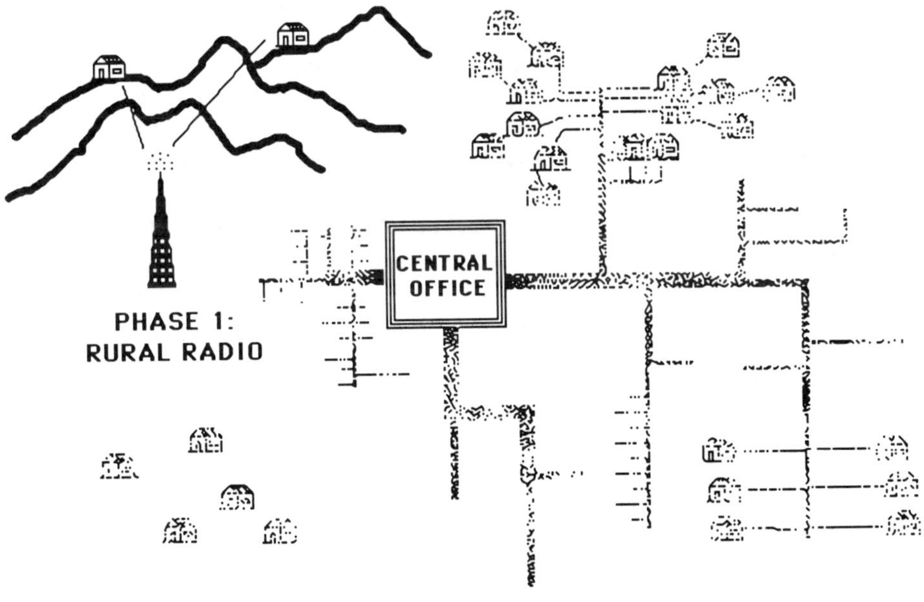

Figure 5.20 Rural radio application.

However, rural radio applications based on remote unserved populations are individually small (typically no more than a couple dozen customers, and often only one or two), and taken in the aggregate they probably do not constitute more than a few thousand access lines throughout the United States (although the world market is much larger). Moreover, rural radio lies well outside the mainstream of telephone engineering. Most telephony engineers, even those who work their entire careers in rural outside plant environments, will never have occasion to specify rural radio.

5.3.4.2 *Wireless Digital Loop Carrier*

The next step in thinking about basic exchange radio usually is stimulated by a feeder exhaustion problem: that is, a situation where the existing feeder does not have enough pairs to support the growth along the route (Figure 5.21). A shortage of wire-pairs may also arise from a gradual deterioration of the cable facility resulting in the loss of pairs.

This is now seen as a classic *digital loop carrier* (DLC) application. Developed in the 1970s and early 1980s, DLC is a method for gaining pairs electronically (hence, the alternative designation of *pair-gain* systems) by means of time-division multiplexing new circuits on existing pairs. The industry standard DLC is undoubtedly the SLC-96 (and its successors), manufactured by AT&T, although there are many others [76]. The emerging new generation of multiplexed feeder systems based on fiber optics is known as *SONET* (for synchronous optical networks [77]). Of course, there are also analog loop carrier systems employing frequency-division multiplexing, which date back to the 1920s [78]. Today, however, digital systems (electrical and optical) are the norm.

The logic of DLC is simple: By generating new circuits without having to plow new feeder cable, substantial savings are realized. The equipment configuration consists of a central office terminal, which is located at the interface between the feeder and the switch, and a remote terminal, located at the other end of the feeder cable. Both are essentially just time-division multiplexers. The cost of such electronically derived pairs varies with a number of factors, but is typically quoted in the range of $200 to $300 per new line.

However, DLC is still essentially a band-aid on existing copper cable in most situations. If the cable has deteriorated too much, DLC often cannot be supported without replacing the feeder itself. Moreover, the cost of the new drop line or distribution plant from the remote terminal to new subscribers cannot be avoided by conventional DLC. Thus, in situations of low population density, even the use of DLC on the feeder portion of the route may still leave long and costly copper loops at the end of the feeder.

The wireless approach recognizes the inherently DLC-like architecture of basic exchange radio, with the additional advantage of avoiding long copper drop lines by locating the subscriber stations on or near the customers' premises. This wireless DLC application has proven-in for a number of rural applications [79].

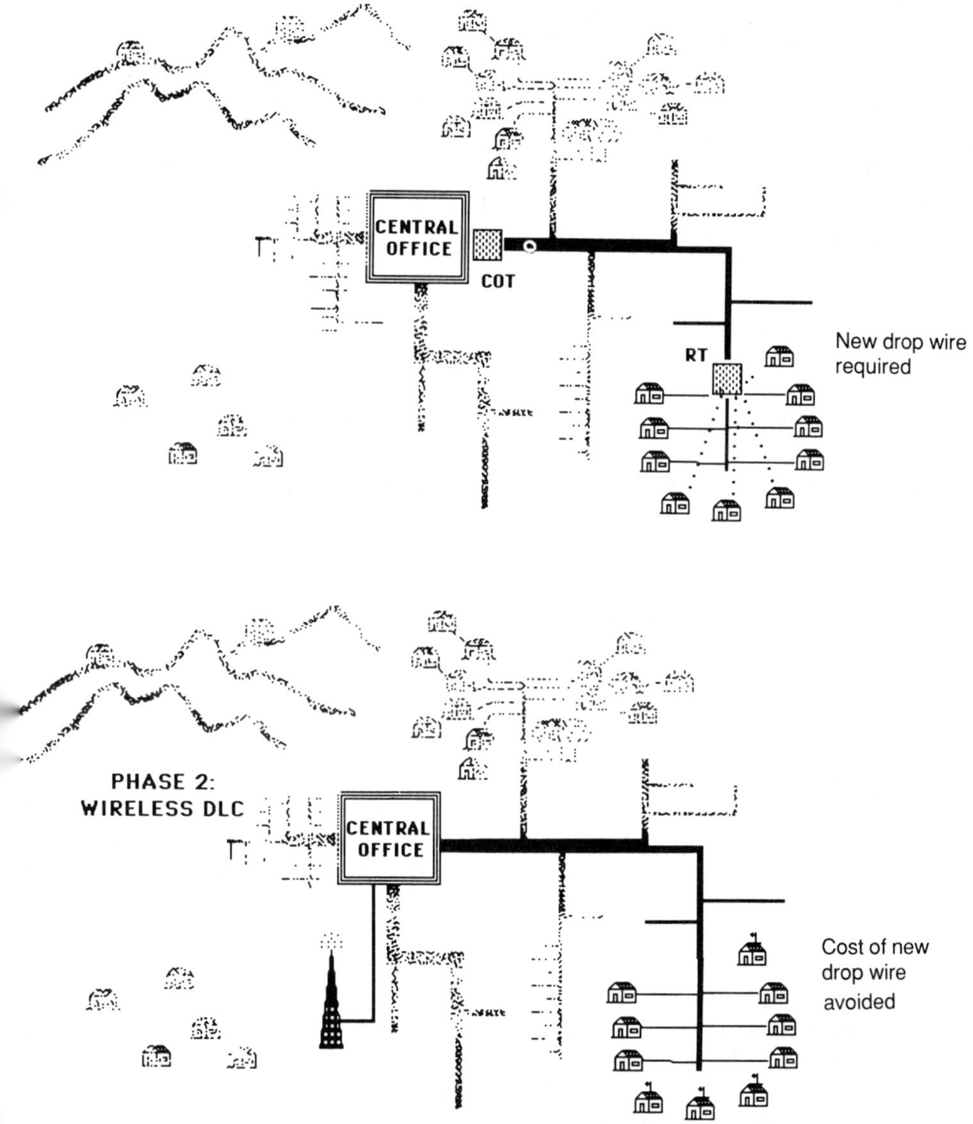

Figure 5.21 DLC applications.

In fact, the wireless DLC model appears to have gained the widest acceptance among those who follow the conventional telephony service model. The DLC paradigm provides a rich set of benchmarks and guidelines for the integration of wireless access within the local exchange environment. Bellcore's TR-57 reference on standard DLC is a strong model for the basic exchange radio requirements document [80]. Other similar documents define interfaces between the DLC and a digital switch [81], operations systems and network management [82], alarms and diagnostics [83], and, looking toward the future, digital trunk-level interfaces, or so-called *integrated DLC* [84]. Much of this material can be translated directly to wireless DLC applications, providing a powerful applications framework.

5.3.4.3 Radio Carrier Serving Area

Wireline *cable routes* are planned one dimensionally, so to speak, with the objective of minimizing the lengths of the cable runs and thus the cost of the facility (Figure 5.22). Each route is like a tributary stream, draining telephone traffic from a particular watershed. When analyzed for the application of digital loop carrier, these regions are often called *carrier serving areas* (see Chapters 7 and 8 for discussion and references). Nevertheless, the word *area* should not mislead us. The essential variable in determining how to provide service is the length of the loop, and minimizing the lengths of all customers at the same time is a difficult constraint (not dissimilar conceptually to the famous "traveling salesman" problem).

Point-to-multipoint radio systems provide wide-area coverage. It frees the planner from the "traveling salesman" framework, and essentially allows for service to be provided to any one subscriber without worrying about how that subscriber is situated with respect to other subscribers or with respect to existing feeder routes. Radio allows the telephone planner to address service requirements not just along a one-dimensional cable route (as in the DLC approach), but in two dimensions over the entire cell. Thus, several additional degrees of freedom are introduced including:

1. A larger subscriber population drawn from a wide area can more easily support the cost of upgrades and lower per-line costs.
2. The specific geographical grouping of individual service requirements is of little concern, as long as they fall within the wide area coverage (see Chapter 7).
3. Radio allows planners to implement, through overlays, new layers of service offerings that may not be economically feasible on a cable route basis.
4. Radio can provide a sort of service safety net for a large area, to handle service problems that may come up without regard to cable route plans and schedules.

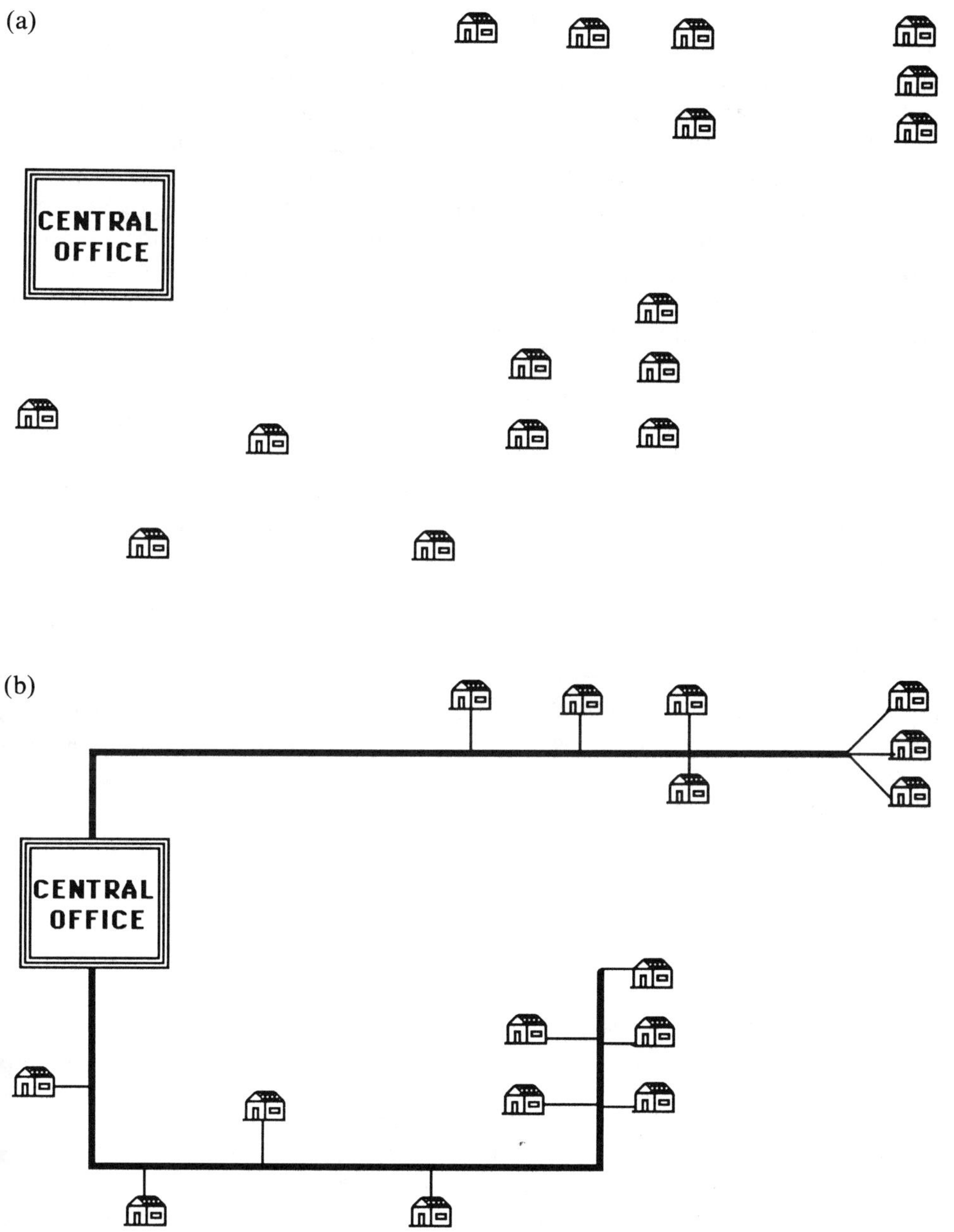

Figure 5.22 Cable-route coverage concepts; (a) before cable, (b) after cable.

At IMM this concept is called the *radio carrier serving area* (RCSA) [85]. In my experience, this is often a key conceptual breakthrough: When outside plant planners realize that they do not have to deploy radio as though it were wire, they begin to truly appreciate the revolutionary impact of wireless access on their industry.

To those coming from the cellular service model—where "cells" are the standard unit of analysis—this discussion may seem to belabor the obvious. Yet the wireline psychology of the telephony service model is so strong that most planners and engineers in this environment continue to conceptualize radio in a linear manner. Indeed, the Bellcore BEXR document reflects this thinking. In discussions with telephony engineers, I have sometimes compared the wireless DLC framework to the idea of using an airplane as a ground-transportation vehicle. It is certainly possible to drive an airplane around as though it were a bus, taxiing around the runways. But an airplane is meant to fly. Radio is uniquely suited to area coverage.

The simplest approach to area coverage is to build up multiple wireless DLC applications within a single system (Figure 5.23). As we will see in Chapter 7, the economic advantage of wireless DLC over conventional DLC *for a single cable route* is often difficult to establish, because the cost of the feeder is considered a sunk cost, and DLC electronics are a relatively mature, inexpensive technology. But whereas wireline DLC requires electronic multiplexers on each cable, a single wireless DLC system can serve all the feeder routes in the exchange—indeed all the routes in several exchanges.

The RCSA is a generic radio applications concept, similar to the cell in the mobile environment (see Figure 5.24). It is the building block for larger networks, which we call *radio overlay networks* (see Chapter 7). The economics of RCSAs forms the core of wireless access planning in today's telecommunications network where voice (and voiceband data) is still the predominant service.

5.3.4.4 Wireless Digital Access

The three preceding applications frameworks all assume that wireless access is being used to deliver voice services (including voiceband fax and data) more economically than conventional cable. The fourth applications framework addresses the vertical dimension of digital services (Figure 5.25). I refer to this as *wireless digital access.*

Consider the development of ISDN basic rate services as an example of an advanced digital service that cannot be handled by today's voice network. As we saw in Chapter 4, the emergence of such services will exacerbate the inequalities between urban and rural service (and between business and residential, or large business and small business). Assume for the moment that basic rate ISDN develops into a highly desirable communications service, eagerly sought by the market. The delivery of

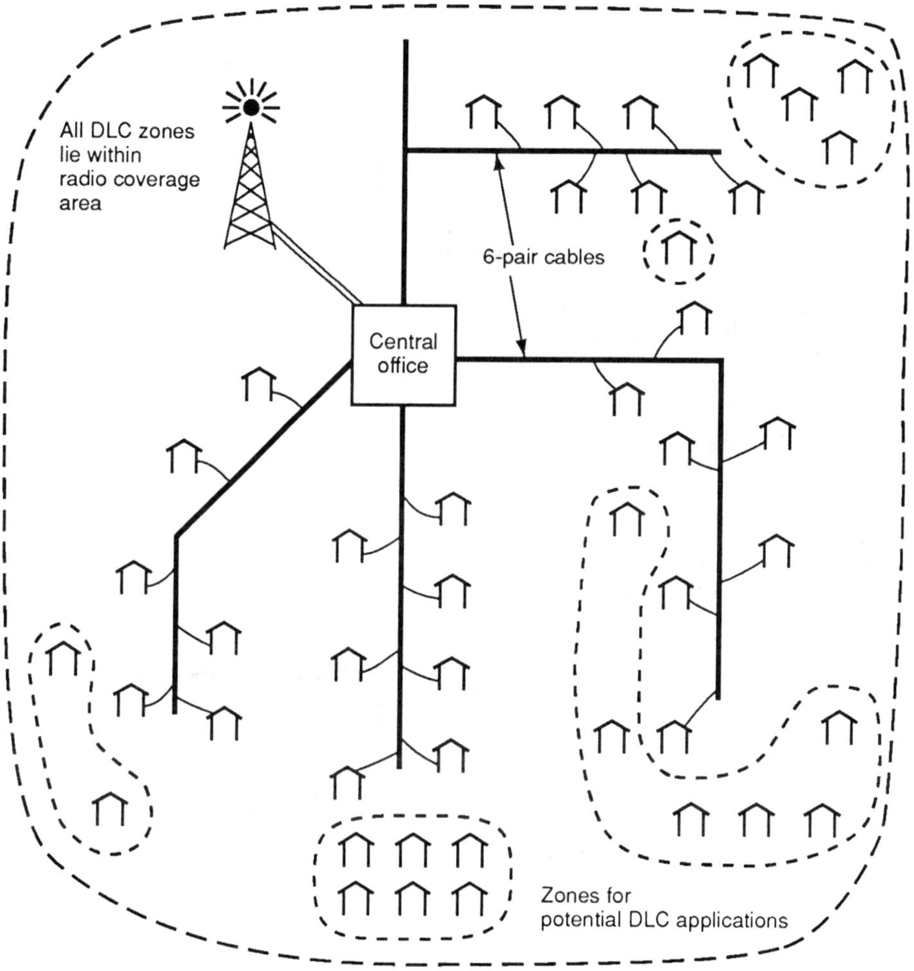

Figure 5.23 Multiple wireless DLC applications.

ISDN in many areas will require the overhaul, if not the total replacement, of existing access plant. In the wireline environment (which includes fiber optics), as a practical matter, a "critical mass" logic will rule; there must be enough customers for the new service, located in the right places, to offset the large costs of regrading the feeder and access facilities (by plowing fiber, for example). Hot spots of ISDN demand in high-density urban centers and industrial parks will draw the investments and receive the

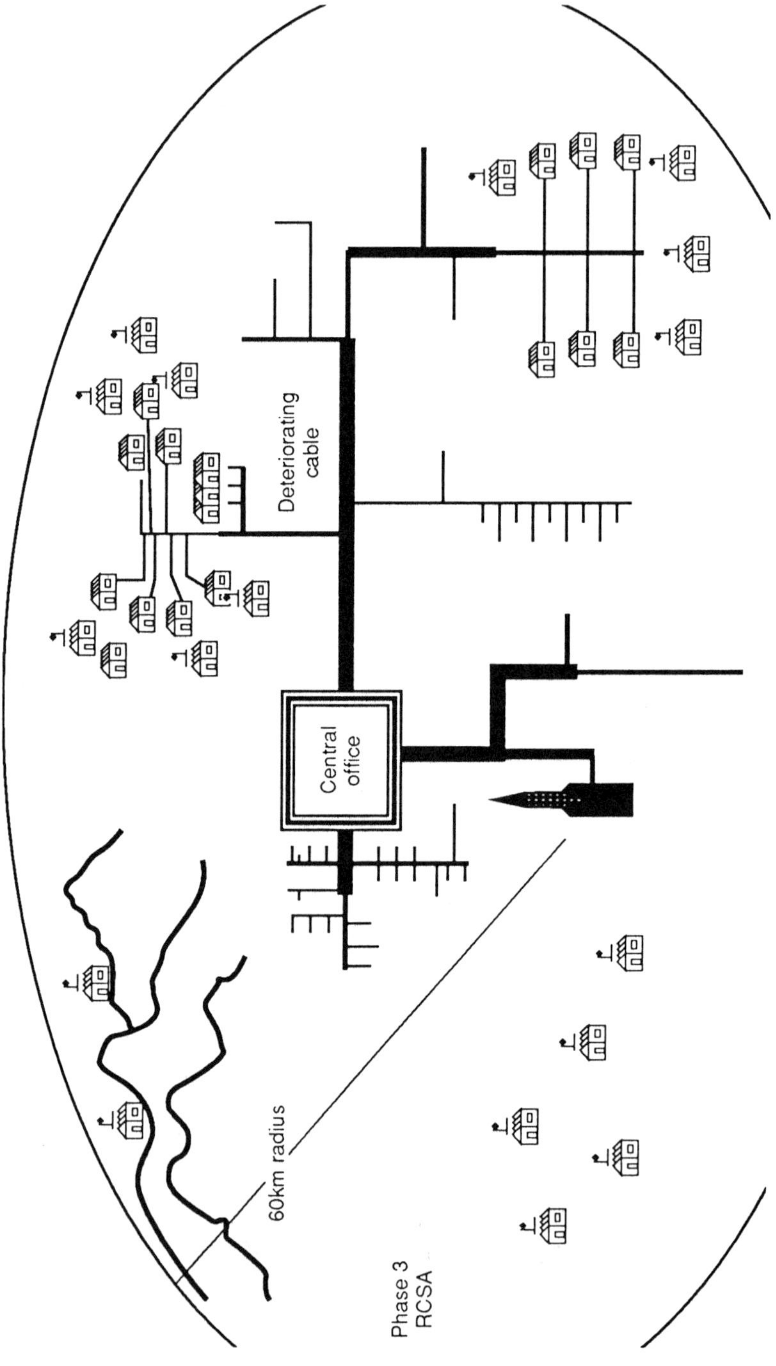

Figure 5.24 Radio carrier serving area.

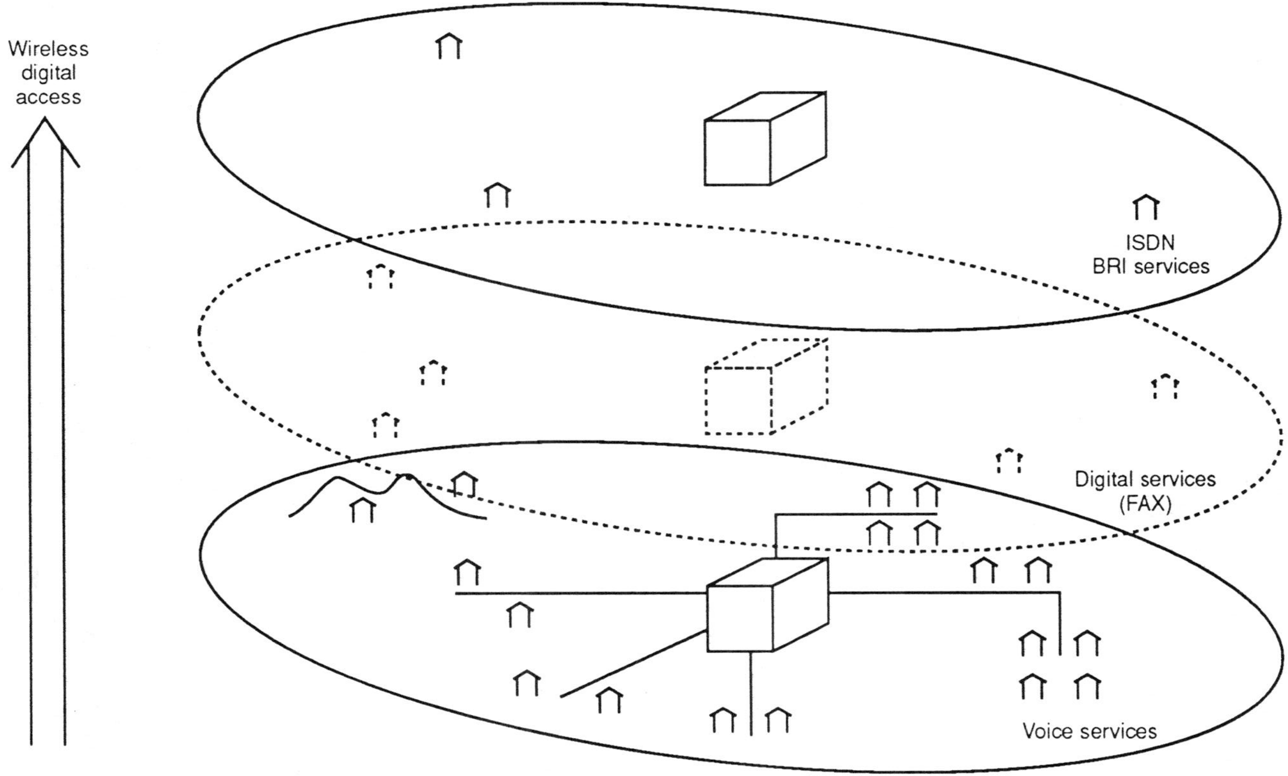

Figure 5.25 Wireless digital access—enhanced services by radio.

new services. Customers in more dispersed rural areas will languish below the threshold for critical mass (Figure 5.26).

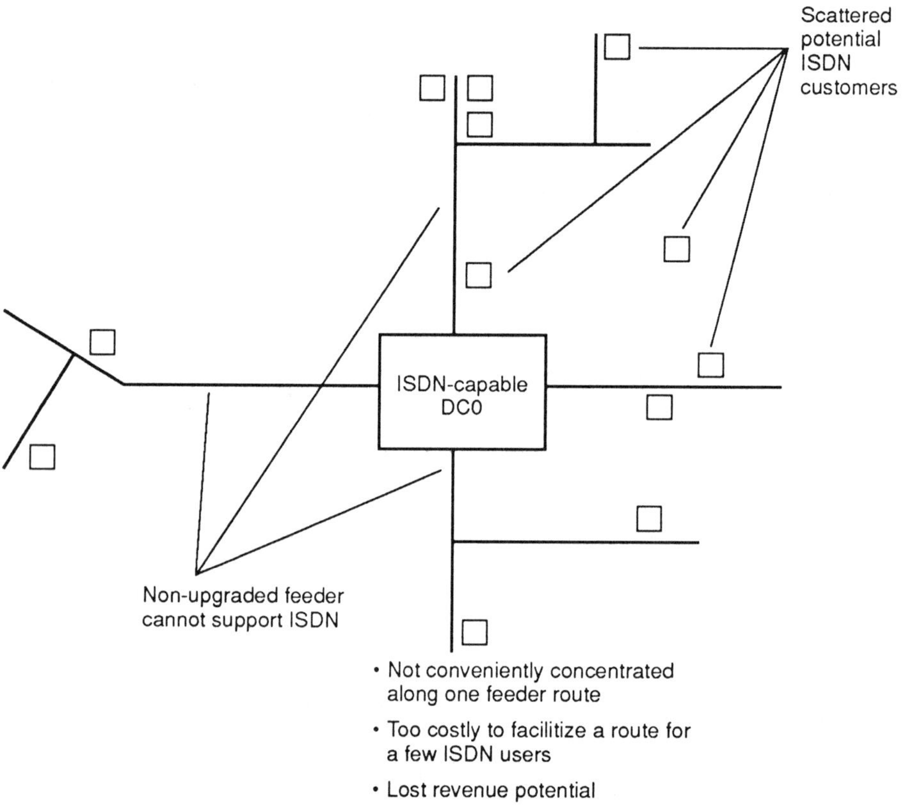

Figure 5.26 Potential ISDN demand in a sparsely settled area.

Radio can address these customers, using the powerful area-oriented economics of the RCSA, regardless of how they are individually distributed with respect to the existing feeder network (Figure 5.27). Digital radio can readily address digital services up to and including basic rate ISDN (and potentially even beyond). Individual customers can be served within a *digital services RCSA,* using ISDN radio terminals. Later, when a "critical mass" of ISDN demand has developed along a given cable route, the entire feeder facility may be upgraded with, for example, fiber optics. At that point, the ISDN radio terminals can be redeployed elsewhere, without stranded investment. Moreover, to the extent that demand for digital services may tend to exhibit high churn rates (with customers frequently adding and dropping the services), radio may remain a cost-effective method of delivering these services for a considerable period of time.

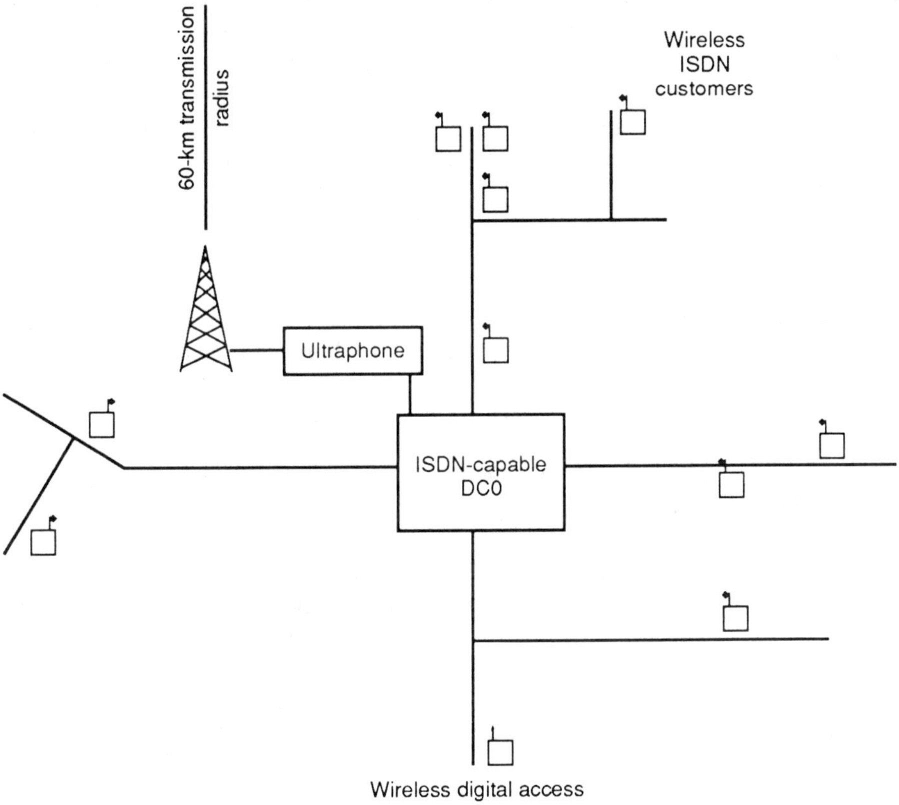

Figure 5.27 Wireless delivery of ISDN access.

Most planners in the telephone industry are still voice-service-oriented, at least when it comes to loop plant in the local exchange. Wireless ISDN or other wireless digital services attract less immediate interest than voice capabilities. This is even more true of the cellular service model, and apparently the PCN model also. Nevertheless, it is my opinion that *digital services will be one of the strongest drivers of wireless access* over the next 20 years, perhaps second in importance only to mobility.

5.3.4.5 Transportable Applications

As BETRS gained acceptance within the telephone industry, it was not surprising that some companies began to find applications that transcended basic exchange service, taking advantage of radio's unique attributes as a transmission medium that does not depend on physical plant. Many of these applications arose opportunistically. For example, in the summer of 1988, shortly after the installation of a BETRS system for the Dubois telephone company of Dubois, Wyoming, a large forest fire broke out in and around the Yellowstone National Park. Part of the fire zone lay within the radio coverage area of the Dubois system, and a number of BETRS subscriber units were mounted on vehicles to provide a rapidly relocatable or *transportable* communications capability [86]. In 1990, a similar fire near Telluride, Colorado (also the site of a BETRS system), became another occasion for this colorful application as a US West release noted:

> The new [system] is playing a crucial role in the Forest Service's battle against an out-of-control fire . . . in a remote area where there are no conventional telephone lines or cellular service. [BETRS access] is provided through a mountain top radio that emits signals so no underground cable is necessary. . . . The [BETRS subscriber station] was installed on a tree which has been burned by the fire three times—and it's still working! The fire fighters were thankful for having real telephone service. [87]

Other uses of a similar nature have emerged, including the use of BETRS-type equipment to respond to natural disasters such as 1989's Hurricane Hugo in Puerto Rico [88], the crash of an Avianca airliner on Long Island in January 1990 [89], and in support of the military operations in Operation Just Cause in Panama [90]. Less dramatic circumstances can also generate a need for a rapidly relocatable, reconfigurable access network. For example, one urban telephone company has used a BETRS-type system to provide expanded holiday service to the resort area for the Fourth of July holiday. Another interesting application involves the use of BETRS-type equipment inside the containment building of a nuclear power plant in southern California, where occasional temporary communications needs (e.g., those associated with the refueling operation) are often complicated by the fact that the existing cable conduits in the concrete structure are clogged with copper [91]. In this case, the radio transmission is partly distributed by means of a leaky coaxial cable.

Applications such as these point toward an expanded concept of the role of radio in the access network. Indeed, these BETRS-type applications are beginning to reach out toward true mobile subscriber applications [92]. This trend will probably continue and, the cellular model notwithstanding, full mobility will probably become a part of the BETRS applications set.

5.3.4.6 *Integrated Digital Radio (IDR)*

This leads to a more comprehensive formulation of the role of wireless access in the telephony industry: the concept of a fully integrated digital radio platform, capable of supporting both fixed and mobile applications in the local exchange environment. Alan Sacuta at IMM has developed this architecture as the framework for the next generation of BETRS-type systems [93].

IDR recognizes the many similarities between a fixed and a transportable or mobile application. The nature of the network interfaces and much of the common equipment can be the same. This helps realize savings in administration, maintenance, and network management; it also increases dramatically the flexibility of the operator (Figure 5.28).

5.3.5 Scope of the BETRS Applications

How large is the potential for wireless access in the regulated local exchange environment? Are we talking about scattering 10,000 wireless "lines" over a continent (the rural radio paradigm)? Or tens of millions (the PCN paradigm)?

Quantifying the scope of the BETRS application set is not easy. The data needed for a proper economic comparison between wireline and wireless alternatives is not readily available. In general, the analysis requires information about the marginal or incremental costs of individual lines; unfortunately, the data are usually available only as averages or aggregates. The outcome of the analysis also depends on site-specific characteristics of the exchange and the existing access plant (especially the feeder cable). While these factors are often well understood by the local engineers, the information is not collected in a form suitable for horizontal market studies.

Nevertheless, a number of attempts have been made to assess the potential demand for BETRS-type service for the United States as a whole. These have followed three methodologies:

1. Analysis of wireless access "indices," which tend to correlate with increased requirements for cable construction, especially in rural areas (e.g., households without phone service, party lines, etc.)
2. Top-down models, based on aggregate data (e.g., census data)
3. Bottom-up models, based on data from individual wire centers.

5.3.5.1 *Wireless Access Indices*

Unserved Telephone Demand. The most obvious application for wireless access—rural radio—is the provision of service to households that are not served by tele-

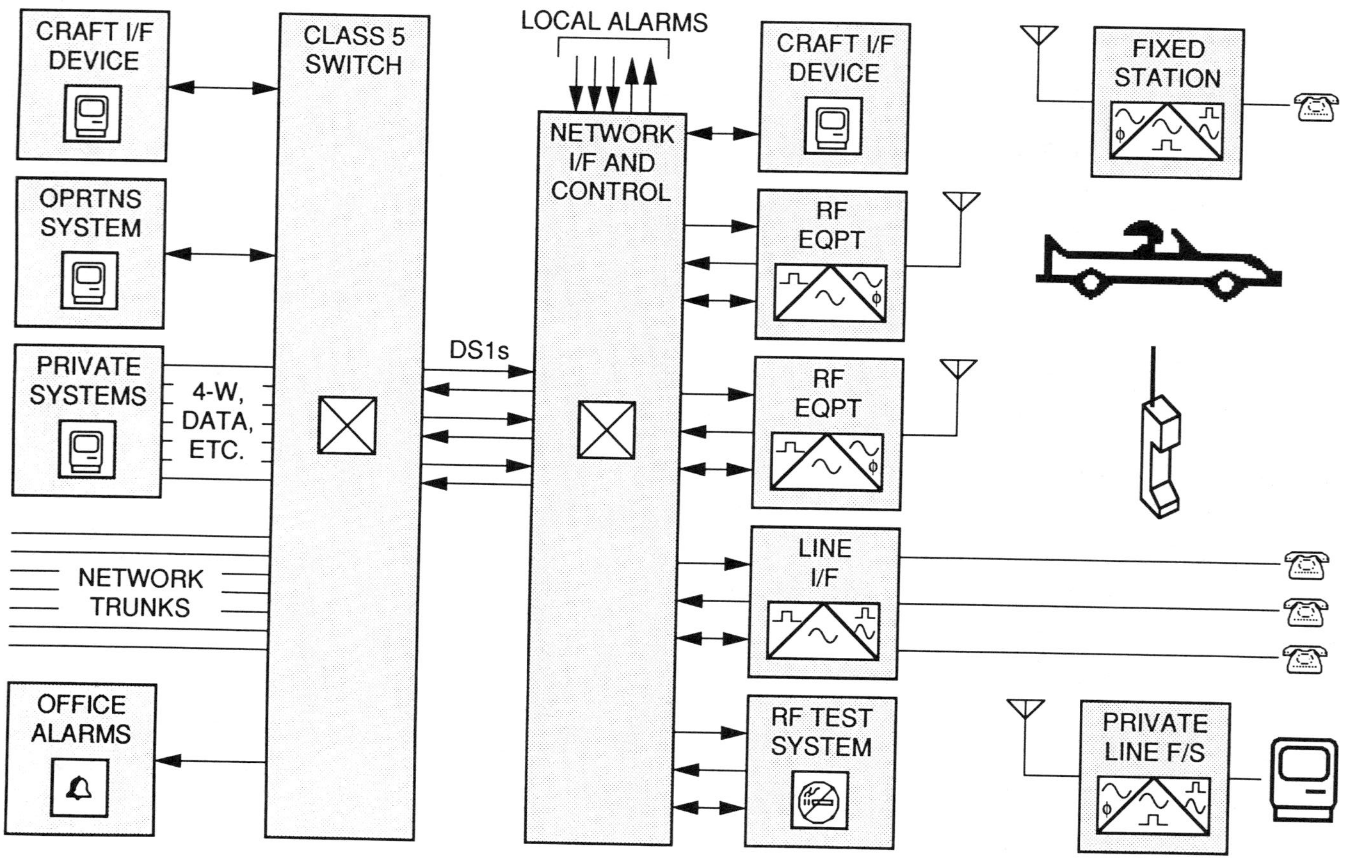

Figure 5.28 Integrated digital radio.

phone service today. Several million households are without telephone service in the United States; however, the majority are located in urban areas, often people living below the poverty line. The unserved that are relevant for BETRS are those who (1) want telephone service, (2) can afford it, and (3) do not have it because they live in remote areas where the cost of conventional wire and cable is too high.

The original petition by the Rural Radio Task Force to establish BETRS made estimates of two groups of relevant unserved subscribers. First, analyzing the geographical information on telephone company franchise boundaries, it was estimated that there were approximately 150,000 potential subscriber households located in uncertificated areas (i.e., areas of the United States where there is no certified telephone company) [94]. These prospective customers were projected to be heavily concentrated in a few states (Alaska, California, Idaho, Maine, Nevada, Oregon, Texas, and Washington State accounted for 87% of the total).

The petition also estimated the number of potential customers living in certificated areas who lack service because facilities were unavailable. Citing a 1984 REA survey, the task force projected about 330,000 households in this category [95]. A subsequent commercial market study, also focusing on unserved population segments, forecast similar numbers [96].

In my experience with IMM, however, the unserved population proves elusive (at least in the United States). Generally, it occurs as small pockets of unserved customers (usually less than 10 in any given area). Overall numbers are usually not large. One estimate for Colorado showed from 500 to 2000 unserved customers for the entire state, mostly scattered in extremely remote locations in mountainous regions [97].

In a few instances in the United States, substantial unserved populations exist within a well-defined geographical area, and wireless access may prove to be an effective solution. For example, the Navajo Indian Reservation located in the Four Corners region of Arizona and New Mexico has a substantial unserved population, representing a demand for perhaps 20,000 lines or so [98]. Certainly, outside the United States, many regions of the world literally have no telephone plant in place at all, and the relevant unserved populations are very large.

Party-Line Upgrade. The BETRS petitioners estimated that among REA borrowers (mainly small, rural telephone companies), the number of wireless access lines in demand would be 400,000 to provide party-line upgrades [99]. Somewhat larger numbers are derived from the annual survey published by *Telephony* magazine, which indicates that several million party lines still remained in the United States at the end of the 1980s [100]. There were also still more than half a million party lines in Canada [101].

These party lines will probably be eliminated in North America over the next 10 to 20 years. Most of these party lines are the residual "hard-core" of high-cost, high-maintenance, hard-to-upgrade rural lines, left over after many decades of investment in single-party service. Most are located in very rural areas. While recognizing that radio applications are site-dependent, I believe that at least 20% and perhaps as many as 50% of these party-line subscribers could be most economically served by radio. (The 20 to 50% range is drawn in part from telephone company estimates, such as that contained in the Alberta Government Telephone DLRR program specification previously referenced [102].) Thus, this upgrade effort may generate a significant demand for wireless access in rural upgrade programs—if the telephone companies can accelerate the standardization processes in time to catch this window of opportunity.

Growth in the Subscriber Base. Rapid growth in the subscriber base is another source of potential demand for wireless access. As we will discuss further in Chapter 7, wireline systems are not well suited to handling rapid and unpredictable growth, especially in rural areas where the economic penalty for incorrect forecasting can be high.

The average growth in the subscriber base of access lines is about 2 to 3% per year in the United States [103]. Overall, this means there are about three million new access lines every year, with perhaps 20% of these in rural areas (allowing for the looseness of definitions of "rural"). A study by a major U.S. independent telephone company during the mid-1980s disclosed that this rural growth was skewed toward the longer loops, with at least 35% of new rural subscribers being added beyond 18 kilofeet from the central office [104]. As we have seen in Chapter 2, the long-term trend throughout the nation as a whole is toward longer loops. Based on our analyses at IMM, we believe that at least 10% of the growth in the subscriber base every year (half the "rural" growth)—some 300,000 lines or so annually—occurs in areas where radio is a candidate for the access facility.

Replacement of Deteriorating Outside Plant. Earlier we used an extreme figure of 50 years for the service life of existing copper plant. In fact, the normal life of outside plant (especially feeder cable) is probably no more than 25 to 30 years, and much less in some areas. This implies that 3 to 4% of the existing base—some five million or so access lines—are subject annually to either replacement or substantial rehabilitation. Since about 20% of these lines (more or less) are located in rural areas, that means that a million rural lines in the existing base are likely to be rebuilt every year. I believe that there are several hundred thousand lines of facilities upgrade or replacement every year that are candidates for a radio solution.

5.3.5.2 Top-Down Studies

A number of studies have been conducted using aggregate census and telephone company data to assess the demand for BETRS-type service in the United States. These have shown a total addressable market in the range of several million lines, with annual potential for several hundred thousand lines of BETRS-type installations.

One of the most comprehensive analyses was performed by Bell Northern Research, as part of a joint project with IMM in 1988 [105]. The study drew on a database containing the location of every wire center in the United States, as well as the location (averaged) of every census block group/enumeration district (a census unit comprising approximately 400 households each). The distance from the center of each block group/enumeration district to the nearest exchange switch was calculated. A cutoff distance was applied to exclude close-in subscribers, and the residual was further manipulated for factors such as subscriber income, growth rates, etc., to estimate the "churn" that could result in radio applications.

The results of this top-down study showed that for cutoff distances (representing the ostensible point at which radio becomes less expensive than wire for solving a service problem) ranging from 15 to 30 kilofeet, a total addressable market of between about 1.2 and 5.5 million loops existed in the United States (Figure 5.29).

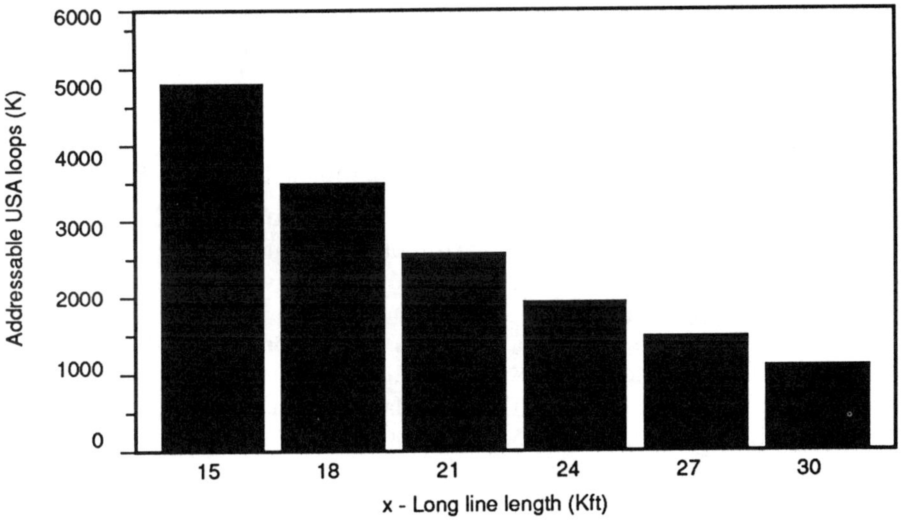

- Subscribers with block group to wire center airline distance greater than x Kft are addressable.
- Growth of subscriber base according to growth rates of number of households.
- 4% long line replacement rate.
- Negligible effect of penetration by other technologies and ISDN deployment.
- Wireless digital access
- Validity of assumptions questionable at shorter loop lengths.

Figure 5.29 Relationship between looplength and size of the BETRS market. (BNR study.)

A similar top-down study was carried out by IMM in 1985, working with data supplied by United Telephone Company. The details of this analysis were reported in IMM's comments in the BETRS Notice of Proposed Rule-Making [106]. The study concluded that potential BETRS applications of all types in the United States numbered several hundred thousand lines per year.

5.3.5.3 *Bottom-Up Studies*

A complementary approach has been adopted by a number of telephone companies to analyze specific wire centers within their serving area for the applicability of radio and thus build up a total estimate of the wireless access requirement.

One example of a Bell company study of radio in the loop analyzed individual wire centers and loop length data for five states (Texas, Oklahoma, Missouri, Kansas, and Arkansas) [107]. Various scenarios were analyzed. The base case was defined as follows:

1. 30-kilofoot crossover (i.e., radio is assumed to be more economical than a 30-kilofoot wire loop)
2. 50% of new growth beyond 30 kilofeet is served on radio
3. 3% of existing lines beyond 30 kilofeet are rehabilitated on radio
4. 50% of possible single-party regrades are completed on radio. [108]

Based on these assumptions, the study found a requirement for 125,000 "lines" of wireless access over a five-year period. Extrapolating to the entire United States, this would approximately agree with the 1.2 million lines of wireless access in the BNR study at the 30-kilofoot cutoff level.

As an example of another study that was even more practically oriented, US West analyzed its planned 1990 Rural Facility Improvement Program (targeted mainly at single-party upgrades) [109]. The data are illustrative of rural loop costs and configuration variables (Table 5.1). It is interesting that the average cost of wire-

Table 5.1
Cable Costs for Typical Rural Upgrade Program

Site	*Customers*	*Estimated Cable Cost*	*Per Line Cost*
A	1	$50,000	$50,000
B	1	$50,000	$50,000
C	1	$50,000	$50,000
D	15	$214,000	$14,267
E	11	$99,000	$9,000
F	30	$270,000	$9,000
G	39	$246,000	$6,308

Table 5.1
continued

Site	*Customers*	*Estimated Cable Cost*	*Per Line Cost*
H	357	$2,215,000	$6,204
I	216	$1,300,000	$6,019
J	237	$1,302,000	$5,494
K	214	$1,035,000	$4,836
L	280	$1,300,000	$4,643
M	126	$556,000	$4,413
N	598	$2,172,000	$3,632
O	184	$665,000	$3,614
P	277	$963,000	$3,477
Q	708	$2,428,000	$3,429
R	55	$181,000	$3,291
S	131	$403,000	$3,076
T	190	$549,000	$2,889
U	1030	$2,961,000	$2,875
V	311	$866,000	$2,785
W	285	$700,000	$2,456
X	807	$1,807,000	$2,239
Y	74	$154,000	$2,081
Z	450	$907,000	$2,016
AA	508	$858,000	$1,689
BB	175	$293,000	$1,674
CC	568	$759,000	$1,336
DD	403	$336,000	$834
Total	8282	$25,689,900	$3,102

line access in this sample was about $3100 per line, and the cost of wireless access was estimated at about $3300 per line. On a straight least-cost basis, 40% of this program would go to radio. Even after other factors were considered, 23% of the lines were still best served by radio. Obviously, as the costs of wireless access comes down these applications will grow more attractive. This point should be stressed. Overall, the potential for wireless access applications on the BETRS or BEXR model appears to be on the order of several million lines at today's costs (i.e., about $3000 per line for BETRS equipment). As costs come down toward $2000 per line, which is a reasonable near-term target, BETRS applications should expand.

5.4 MOBILE-ORIENTED APPLICATIONS OF WIRELESS ACCESS

Radio is uniquely suited for communicating with a moving vehicle, and it has been the application to mobile communications that has dominated the development of the two-way radio industry since Marconi. Ships were the first mobile application. The earliest commercially successful radio-based service was the Marconi company's ship-to-shore network in the early decades of this century [110], and for a time the U.S. Navy became the custodian of wireless technology, even to the extent of imposing a government monopoly of the use of the airwaves (during World War I) [111]. The development of the mass automobile culture during the 1920s led (inevitably, one is tempted to say) to the first experiments with mobile radio in terrestrial vehicles [112]. Radio was also adapted for airplanes [113]. Such applications clearly admitted no other communications solution. Following the introduction of superior FM technology in the 1930s [114], two-way radio received an enormous impetus during World War II, when radio communications networks were deployed on a mass scale for the first time, leading to designs and manufacturing economies truly suitable for a large mobile communications industry. After the war, the mobile communications industry began its spasmodic progress toward the mass marketplace. In my previous book, I detailed the technical and political intrigues of mobile radio over the past 50 years, culminating in the creation of "cellular" service in the 1980s [115].

The idea of using mobile radio as a form of wireless access, however, has taken longer to coalesce. For a long time, mobile radio systems tended to be designed and operated as closed systems, with extremely limited options for accessing the external network. For example, an airplane pilot could talk to the control tower, and possibly to other airplanes, but there was generally no way to gain direct access to the public telephone network. Despite the occasional patched-in call, most radio communications systems still do not interconnect to the public network. Of course, cellular radio represents an effort to broaden the scope of interconnected mobile communications; nevertheless, in 1990 there were still at least 10 times as many noninterconnected radio communications devices in the United States as there were cellular phones. These included millions of units in dispatch systems used by police, fire fighters, and other security forces, by businesses to manage fleets of service or delivery vehicles, by taxicab operators, two-way radios used on construction sites and in certain work environments, marine and aeronautical radio, and citizens band radio. Mobile radio still lay primarily outside the compass of the public telecommunications network.

Thus, while mobile radio has enjoyed a much longer history than the exchange radio configurations discussed in the previous section, its role as an access vehicle is far less developed. Whereas exchange radio applications inherit the conceptual superstructure of telephony, embodied in the library of technical references and advisories published by Bellcore and others, and in operational form by the methods and procedures of the telephone common carriers, the mobile radio tradition is relatively impoverished when it comes to the technical underpinnings of network access and service requirements.

In this section, we briefly survey some of the more important applications models for mobile-oriented wireless access, starting of course with cellular radio.

5.4.1 Cellular Mobile Radio

The basic application of cellular radio—voice telephony for users in moving vehicles, with frequency reuse and hand-off—is now well understood [116]. The history of the cellular idea and its tortured regulatory odyssey are described in my previous book [117].

Cellular radio is not yet up to the implicit standards of true wireless access. The quality of the communications channel and the quality of the network access (*interconnection* in the industry's terms) lag substantially behind the wireline network. In many respects, it is simply what it set out to be: an improved car phone for businessmen in large cities. Nevertheless, it seems clear that the transition to a full-fledged wireless access service is under way.

Viewing cellular as a wireless access mode raises several challenging issues, including:

1. Improvements in the quality of the basic communications channel
2. Improvements in the system architecture, especially the evolution of an "open architecture," in an environment of rapid technological change
3. Expansion of cellular service offerings beyond voice telephony
4. Integration of the cellular network with landline networks.

5.4.1.1 Improved Radio Transmission

Upgrading the cellular voice channel to network quality will involve the transition from current analog radio transmission technology to digital air interface standards now being developed in the United States, Europe, and Japan.

Digital radio technology will improve the level of cellular access in two important ways. First, digital techniques will improve the transmission quality of the typical cellular call. The mobile radio channel is probably the most severe of all communications environments [118]. The signal is subject to unpredictable and uncontrollable degradations that can overwhelm an analog receiver. The "smart" digital radios of the next generation will be able to compensate for the difficulties of the channel by applying intensive mathematical techniques, known generically as *digital signal processing* (see Chapter 6). This should enable cellular (and other mobile applications) to perform up to wireline standards.

Second, digital technology enhances communications *privacy.* The lack of privacy in analog cellular is now regarded as a significant shortcoming by most cellular operators, notwithstanding recent cosmetic legislation prohibiting eavesdropping [119]. Digital systems will bring an inherent level of privacy to wireless networks that will be superior (in many cases) to the privacy of the existing wireline network. They also lend themselves readily to encryption for applications that require *secure* communications.

5.4.1.2 Cellular System Architecture

A simplified block diagram of an analog North American cellular system is shown in Figure 5.30. Note that while the interface to the public network is fairly standard, and the air interface is standard, the internal system link between the switch and the cell-site is not standardized. The emerging networks at the beginning of the 1990s were still based largely on proprietary technical interfaces controlled exclusively by individual manufacturers. Cell-sites and mobile switches are manufacturer-specific subsystems that are not interoperable. System interworking, for hand-off or call initiation, is still rudimentary [120]. To a considerable degree, individual cellular systems are still isolated from one another both technically and operationally. As of 1990, the public network features of cellular service were derived from the interfaces to the landline telephone system (which were controlled and defined by the landline carriers), rather than being inherent to the cellular systems themselves.

This state of affairs exists because many manufacturers of cellular infrastructure in North America have pursued a conscious strategy of trying to "lock up" their customers. It is similar in many ways to the "IBM era" in the computer industry, when a semi-monopoly power was exercised by the dominant manufacturer through proprietary interfaces that kept customers tied to the original vendor for system expansions and peripherals. It is now recognized in most quarters that this strategy is an impediment to the evolution of modern networks. In an industry characterized by a high rate of technological change, the problem with proprietary interfaces is that they often lock operators and large subscriber populations out of the newest innovations as they emerge.

Undoubtedly, cellular systems will develop over the next decade toward a more mature network configuration, with standard interfaces and detailed performance and interface specifications. Operators will insist on the ability to build up their cellular networks from elements of equipment supplied by different manufacturers. This is how the wireline network is built today; operators have considerable freedom to combine switches from one vendor with transmission equipment from another vendor.

In addition, as we noted above, the cellular industry is transitioning to a new air interface. The customer base will be largely analog for many years, however. This presages the development of hybrid, or so-called *dual-mode systems,* which will support both analog and digital mobile units. Indeed, third-generation digital systems are already being discussed, and it seems likely that in the future a far greater diversity will exist at the air interface. The FCC has recognized this in their so-called "Flexible Cellular" Rule-Making, in which they specifically allow for the implementation of different radio technologies at the air interface [121].

Indeed, the most likely scenario is one in which the cellular industry will move toward a system architecture that is the polar opposite of today's configuration. For reasons we examine in greater detail in the next chapter, there will be technological

(a)

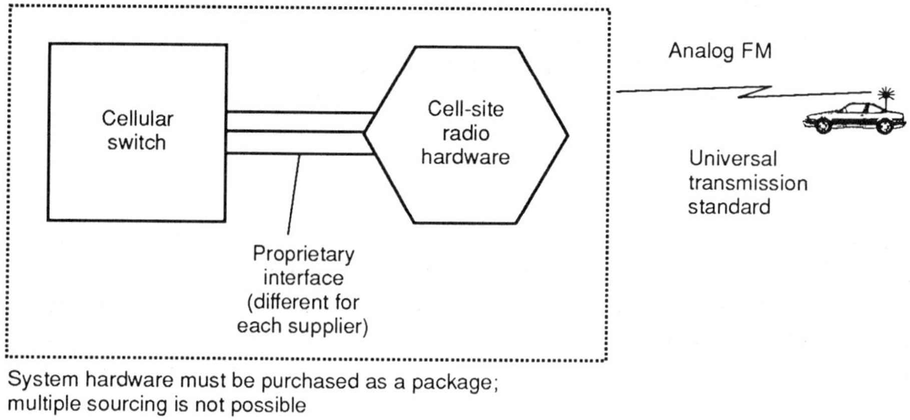

(b)

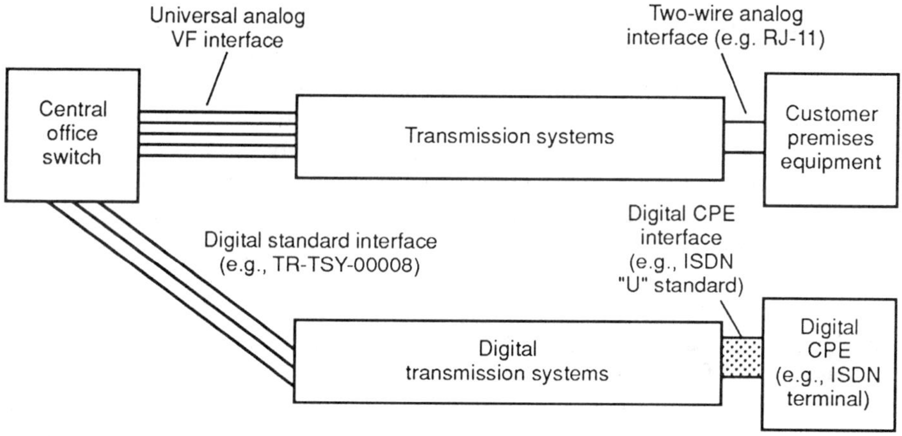

Figure 5.30 Functional diagram of cellular network; (a) today's cellular architecture, (b) today's wireline architecture.

diversity, *including proprietary standards,* at the air interface, where evolving subscriber requirements will drive a continuing ferment of innovation. On the other hand, the nonfunctional diversity of interfaces between system elements, especially between switch and cell-site hardware, will be rendered into a series of standard, even plug-compatible interfaces.

There is precedent for this in the architecture that has been developed for digital cellular radio in Europe, under the banner of the *Groupe Speciale Mobile* (GSM).

The GSM specification is elaborate and forward-looking [122]. It calls for an open architecture, with standardized interfaces between system elements—in particular between the cellular switch and the cell-site RF hardware [123]. In addition, Signaling System No. 7 is specified for all important intrasystem and intersystem control links. Figure 5.31 shows a basic block diagram of the GSM system.

5.4.1.3 Cellular Service Offerings

An expanded range of services is crucial to the acceptance of cellular as a full wireless access vehicle. Today's analog systems, and even those of the first digital generation, are optimized (in many respects) for voice communications. They handle data, if at all, in a catch-as-catch-can manner. They employ primitive signaling methods [124]. They are out of step with the march toward a more intelligent network. Even the basic diagnostic interfaces are often lacking.

A digital transmission technology at the air interface will improve this situation. For example, the North American digital standard allows for packetized data at slow and fast speeds [125]. This is clearly a step in the right direction, looking toward both telemetry applications (slow) and user data transmission (fast). The initial digital standard still may not support such services as fax (group 3), analog voiceband data of the kind generated by most computer modems, or even over-dialed DTMF tones [126]. The implementation of low-bit-rate digital voice coders does indeed complicate the transmission of some of these nonvoice signals; nevertheless, it seems clear that a circuit which in today's world will not support fax transmission, not allow the use of computer modems, and not allow access via touch-tone pads to voice mail or bank-by-phone systems, does not really qualify as full access.

It is my impression that most cellular operators are still running hard to stay abreast of the demand for basic voice telephony from their mobile subscribers. There is a tendency to give short shrift to other services in the elaboration of standards and the development of equipment. This will undoubtedly constitute a source of competitive vulnerability for the cellular industry with respect to the emerging PCN services, which are conceptually much more oriented toward intelligent networks and enhanced service offerings.

5.4.1.4 Cellular Networks

Unlike basic exchange radio, the cellular network incorporates a switching hierarchy. Therefore, the quality of the access provided by a cellular system is in part determined by the capabilities of the cellular switch, and in part by the nature of the trunk interface to the larger network.

Most of the effort toward a second generation of cellular systems has been focused on the air interface and the transition to higher capacity digital radio tech-

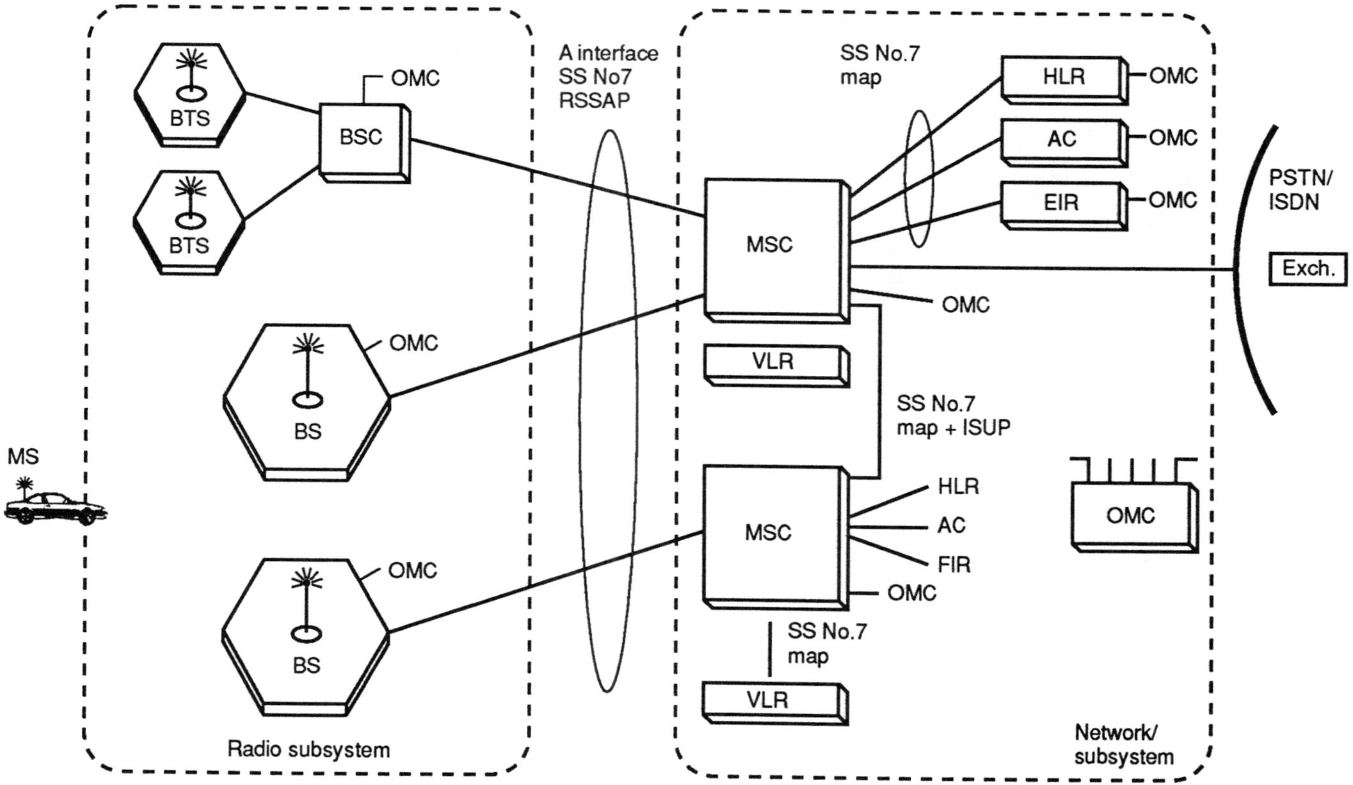

Figure 5.31 A functional diagram of a GSM infrastructure.

nology. Yet because the cellular model encompasses its own switching hierarchy, the cellular second-generation architecture must also address the switch/network side of the system. In the first-generation systems, cellular switches have often been rather underpowered for the application, particularly as it has become clear that the control and processing load for mobile subscribers is several times higher per Erlang than it is for wireline telephony [127]. In other words, because of the need for hand-off and other continuing call-control information, a cellular system serving 250,000 mobile users may generate a processing requirement equivalent to a million or more wireline access customers. The deficiencies on the switch/network side have contributed to higher blocking rates, dropped calls and improperly executed hand-offs [128].

Remedying these problems with a more robust switching architecture is the other side of the transition to a second generation of cellular systems (often overlooked in the concern about the air interface). In particular, it appears that more distributed architectures, where processing is pushed out to the cell-sites or offloaded to intermediate processors (known as *base-station controllers*), will be preferred to the highly centralized architectures that characterized the first-generation systems [129]. This is the direction in which the GSM network appears to be heading (Figure 5.31). Another important advance will be the use of standardized, high-speed signaling networks (e.g., Signaling System No. 7) to link many of the elements of the cellular network.

Finally, the integration with the landline network will require that cellular systems step up to the requirements of ISDN and other "intelligent network" applications. As new capabilities emerge from the wireline network (which is still clearly the conceptual and technological vanguard for service evolution), there will be pressure on cellular to become fully transparent to these innovations. So far, this appears to be an area that has received too little attention from cellular network planners. Once again, if this neglect of anything but voice service continues, it will undoubtedly constitute a competitive vulnerability for cellular operators as they face PCN-type competitors armed with more advanced technology and more advanced thinking about the requirements of full-featured access.

5.4.2 Noncellular Mobile Radio

Noncellular mobile applications are very diverse, and they collectively tend to lag even behind cellular radio in progress toward true wireless access. The vast majority of noncellular mobile radio systems do not provide automatic user-controlled interconnection to the public network, and are used instead for private networks, dispatch, and *net* operations (where all users communicate with all other users, or where a central base station communicates with all mobiles simultaneously).

In the early 1980s, the FCC created a new class of private dispatch carriers that were also authorized (with some restrictions) to provide interconnected service to the public network—i.e., to allow mobile users to make a telephone call. This service is known as *special mobile radio* (SMR), and it has grown strongly over recent years as a superior dispatch-oriented service in major metropolitan areas. The restrictions on interconnection have been progressively lifted, and there is reason to anticipate that SMR will become more like a cellular service in the future, focusing on interconnected traffic more and more.

SMR is hamstrung, however, by the lack of open architecture; even the air interface is in most cases a proprietary format exclusive to a single equipment manufacturer. This means that unlike cellular, where the standard air interface allows vigorous competition in the terminal market, many SMR operators and subscribers are captives of the dominant manufacturer in the industry. As long as the leading manufacturers of SMR equipment (who are also major participants in the cellular market) can control the evolution of this industry so closely, it seems likely that progress toward a broader wireless access mode will be limited.

A recent proposal before the FCC has called for the modification of some of the SMR rules to allow for a more flexible, higher capacity, more transparent SMR architecture that could compete effectively with cellular [130]. If this initiative succeeds, then SMR might be able to develop into a true wireless access system positioned midway between cellular (with its full-fledged switching infrastructure) and basic exchange radio (with its simplified interface and limited mobile capabilities). Indeed, one BETRS manufacturer has begun to develop a mobile subscriber capability. If implemented within a BETRS framework, then at least from a system architecture perspective something very close to a digital SMR would result.

5.5 THE PCN SET OF APPLICATIONS

By the late 1980s, it was widely accepted that radio-based access would develop under the aegis of the cellular industry. In part, this was a reaction to the 15 years it had taken to establish the cellular framework. It was felt by many that setting up a new framework would certainly take longer than finding a way to expand the cellular allocation. Moreover, after a rambunctious beginning the cellular industry had settled down nicely. Most of the franchises had ended up in the hands of subsidiaries of the major telephone operating companies, which may have seemed a rather fortunate outcome, since these were precisely the companies best equipped (in the minds of many) to do the job well. It was a comfortable scenario: a dozen or so major operators competing nationwide (two to a franchise), drawing on substantial capital and technical resources to see them through the birthing pangs. It seemed a reasonable middle course between the all-encompassing monopoly franchise concept under which the cellular idea had been hatched in the early 1970s (with an undivested AT&T as the franchisee) and the turbulent vision of wide-open competition among speculators and undercapitalized, unsophisticated entrepreneurs who had flooded the FCC with so many license applications in the early and mid-1980s that the licensing procedure (and even the Commission's physical facilities) had been overwhelmed [131]. Both undesirable situations had been avoided, an apparently stable industry framework had emerged that seemed to guarantee sufficient competition to benefit the consumer, but not so much that it might debilitate the players. Unlike most other segments of the telecommunications world, there was no "dominant carrier" problem. It seemed as though that rarest of regulatory ambitions had been realized—a level playing field.

Yet there were flaws in the cellular service model from the beginning, which grew more significant as rapid growth strained the cellular architecture to its limits. We may view the emergence of PCN concepts as a response to the prospective unraveling of the cellular approach. These flaws included:

1. The actual lack of effective competition in many markets under the duopoly
2. An obsolescent and inefficient radio technology, reinforced by a an FCC-mandated air interface, which inhibited technical innovation
3. Overemphasis on mobile (i.e., vehicular) communications, to the exclusion of other applications.

The fleeting illusion of cellular sufficiency was ended by two seminal regulatory initiatives: the FCC's Notice of Proposed Rule-Making on "flexible cellular" in October 1987 [132], which at one stroke rewrote most of the rules for cellular service in the United States, and, even more important, the publication by the British Department of Trade and Industry's report on "Phones on the Move" in January 1989 [133], which ignited the PCN furor.

The FCC's "Flexible Cellular" Rule-Making

In a Notice of Proposed Rule-Making (NPRM) issued on October 15, 1987, the FCC punctured the cellular stasis by indicating its intention to abdicate its role as the promulgator of technical standards for the cellular industry. Its stated purpose was to clear the way for new technology, which the FCC felt would be hampered by government-mandated standards:

> We believe it would be inappropriate at this time to embark on a proceeding to select technical standard for future cellular systems. Such a course would be premature given the early stage of development of new cellular technologies and is likely to discourage technical innovations. . . . Instead we believe [we should] provide technical freedom in the rules in order to enable the introduction of new technologies. This would foster the development of competing technologies which could then be evaluated in the market. [134]

But the NPRM went much further. The FCC indicated that it would no longer require that different cellular systems necessarily adopt the same standard or be fully interoperable. A system in New York might use a different radio technology than one in Philadelphia. Even the channelization of the spectrum might be changed. Many of the service restrictions were also lifted.

In one bold step, the technical lodgepole of the framework was knocked out. The mandated air interface standard was eliminated, in favor of a free-market approach which, the FCC hoped, would lead to the more rapid emergence of new digital radio technology, to solve the increasingly severe problems associated with the antiquated analog FM air interface originally chosen for cellular. The cellular idea, which was premised from its inception on a managed market and the availability of a nonproprietary radio technology that could support a huge, commoditized manufacturing base, received a conceptual transplant from the world of consumer electronics and computers.

In essence, the FCC removed much of the regulatory scaffolding that had been erected during the launching of cellular service. Many implications of this decision were unrecognized initially. One result was to clear the path for new radio technologies (plural) that will greatly increase the capacity of cellular systems, and undermine the rationale for limiting the number of cellular licensees. The original arguments over the structure of the cellular industry focused on the difficult question of how many operators might share the cellular spectrum, to ensure adequate competition without unduly harming the operating efficiencies of the system. The duopoly—allowing only two operators per area—was a compromise between the desire to avoid a backward step toward monopoly (since mobile radio had been a competitive market for decades prior to cellular) and the technical arguments against splitting up the spectrum among too many competitors, which, given the available analog technology, would have substantially reduced the overall system capacity (due to loss of trunking efficiency). These debates are summarized in my previous book [135]. The highly spectrum-efficient technologies of digital cellular will recast these conclusions. The traffic capacity per MHz of spectrum (or per square mile of area covered) will increase by at least a factor of 10. So much added capacity will become available (even without additional spectrum) that additional operators could be authorized without significant losses in trunking efficiency. This would eliminate the technical rationale for the duopoly.

Even more unsettling is the implicit vision of a technologically competitive marketplace. Unlike the traditional mobile radio industry where antiquated standards have meant technological stagnation and the domination of the industry by low-cost manufacturers, the new digital cellular industry will resemble segments of the computer business, where sophisticated, R&D-intensive suppliers compete on the basis of technological differentiation, and the overall pace of change is very rapid. This will enable technology to follow market requirements more aggressively, as we are already seeing in the portable segment of the cellular business.

Surprisingly, despite the revolution that it wrought, the NPRM was implemented essentially without any retrenchments in December 1988. Cellular radio operators were free to seek whatever markets they might covet, with whatever technology seemed attractive. But looming over them was the innuendo of expanded competition. And PCN is the ambiguous term under which the proposals for this competition have been marshalled.

"Phones on the Move" and PCN

The other key event in breaking up the existing cellular framework was the publication in the United Kingdom of regulations aimed at creating a new class of communications services, under the heading of personal communications networks (PCNs) [136]. The concept of PCN was a heady mixture of competition for the cellular duopoly (which was perhaps perceived as more onerous in England, with just two operators nationwide, than in the United States), an emphasis on so-called *personal* applications (as opposed to vehicular applications), and an even more far-reaching vision of true competition with the existing wireline local exchange monopoly, based on ubiquitous wireless access. Its lofty goals notwithstanding, the PCN proposal was also motivated explicitly by a desire to overcome the lack of competition in the British cellular industry. David Hendon, a well-known DTI spokesman and one of the authors of the PCN proposal has written:

> The UK cellular radio market has often been quoted as an example of how competition can stimulate a market. . . . At the same time, the limitations of a competition where new entrants are excluded have shown up fairly clearly. The cost of subscribing to a cellular service remains high and the quality of service has been variable. The operators say that this is due to the difficulty of installing enough equipment to cope with the demand for the service. Although both operators have been making significant efforts to keep up with the demand . . . the Government remained of the view that the consumer would benefit by additional competition. [137]

In short order, three operating consortia were licensed to offer PCN service in England, beginning in the early 1990s [138]. Technical standards are nearing completion and will apparently follow the European digital cellular standard known as GSM. It remains difficult at this time, however, to describe PCN in a definitive manner, since by most accounts it will be some time before operations are significantly under way.

Nevertheless, the U.K. initiative generated much excitement in the United States, including a number of ambitious filings by operators interested in creating a Yankee version of PCN [139]. Yet since the term PCN was undefined, it became a vessel in which different and sometimes conflicting visions might be conjured. In an effort to bring greater focus to the PCN question, the FCC undertook its own Notice of Inquiry in June 1990 [140].

We will return to the exquisitely topical question of the many meanings of PCN in Part V. However, in our enumeration of the various applications of wireless access, PCN may be understood today as a motley covey of wireless applications, flushed out by the regulatory gestures (on both sides of the Atlantic) toward a much more open regime for spectrum regulation, which trades on the charisma of portable handsets and in general transcends "cellular" as a model. If indeed we accept the tenets of the PCN model—namely, that neither technology nor service opportunities will be excessively fettered by centrally administered regulations—then we may expect a much more free-market, more technology-friendly scenario, with a proliferation of applications large and small, and a flowering of consumer alternatives.

Within the PCN applications set, it is often unclear where the technical and market boundaries should be drawn [141]. While there is some rhetoric in favor of a universal communications device that would subsume and replace *all* these applications with a single piece of hardware that would function equally well in the car, in the home, at the office, and on the street, more sober analyses tend to identify a number of distinct market segments that will probably develop along parallel but independent lines. With some limited exceptions, none of these forms of wireless access had reached the stage of commercial deployment as of this writing (mid-1990).

In this section we review these ideas very briefly, as applications concepts. The lack of commercial standards and of real hardware and solid operating experience comparable to either BETRS or cellular makes it dangerous to be too specific. The size of the various market niches for these applications is uncertain, as are the detailed customer requirements. The architectural details are often still to be determined. The following sections, therefore, contain a strong dose of personal opinion. The reader is forewarned.

5.5.1 Wireless Pay Phone (Telepoint)

The so-called *telepoint* application has evolved as an ambitious extension of conventional cordless telephony (see Section 5.5.2) to provide a short-range access capability via portable radio terminals operating in public areas such as train stations, airports, shopping centers, and generally in any urban public area where telephone access is desired. As of mid-1990, this concept had been implemented in the United Kingdom using CT-2 technology (for cordless telephones, second generation) [142]. Other countries have announced plans for similar experiments [143].

The coverage provided by telepoint is limited to relatively small zones (typically a transmission radius of a few hundred meters or so), without hand-off or the ability to receive incoming calls from the network. The lack of hand-off and subscriber location capabilities (needed for incoming calls) has been justified on the basis of avoiding the costs of the "cellular" network overhead. In effect, telepoint can be seen as a wireless alternative to the conventional pay phone. The subscriber carries the portable CT-2 unit with him, and when he enters a telepoint zone, indicated by signs, he can initiate calls in the ordinary manner. A number of slightly different versions of the telepoint air interface have been put forward, although as of this writing there seems to be a convergence in the United Kingdom toward the so-called *common air interface* (CAI).

5.5.2 Residential Cordless Telephony

Another application that has been swept into the web of PCN is the popular low-end cordless extension phone used by consumers in their homes and backyards [144]. This is potentially an enormous market, at least in terms of units. Various estimates place the number of cordless extension phones in use today in the United States between 20 and 50 million, all based on analog radio technology, which is widely considered to be inadequate. Donald Cox of Bellcore has referred to cordless telephony, I believe appropriately, as an extensive market survey and "sales experiment" for tetherless (i.e., wireless) access [145]. In other words, it has proven the existence of the market, despite technical insufficiencies. The application of digital radio technology could result in a superior cordless extension phone, with much greater privacy and higher transmission quality [146].

In today's configuration, the customer purchases a two-part device, consisting of a base unit, which installs in the customer's home and connects to in-house wiring, and a portable unit (Figure 5.32). The PCN-oriented versions of the cordless phone often propose a neighborhood base unit, mounted outside on a pole, with a range similar to telepoint (e.g., 1000 ft) and sufficient to reach a number of residences within its small cell. One example of this approach is the PCS proposal made by Bellcore and described in Cox's article [147] (Figure 5.33).

Indeed, CT-2 and telepoint concepts evolved from contemplations of solutions to the "problem" of cordless phones. I am not certain, however, that it makes sense to translate the poolside extension phone from the configuration show in Figure 5.32, where the individual subscriber is in control, to the configuration of Figure 5.33, where a telephone common carrier (or some similar large entity) is now the service provider to a large group of customers. For one thing, the cordless service in today's configuration is effectively free to the consumer, in the sense that once he buys the cordless unit (for generally less than $100), he incurs no further service charge from the phone company. Moreover, a base unit located inside someone's home is inherently cheaper to build and maintain than a device that must be ruggedized to sit on a pole out of doors. It is not clear to me what the advantages of the Bellcore proposal for this application would be over today's cheap, but effective K-Mart or Sears cordless phone (especially once the Sears phones go over to digital technology, which will probably happen at about the same time that PCN becomes a reality).

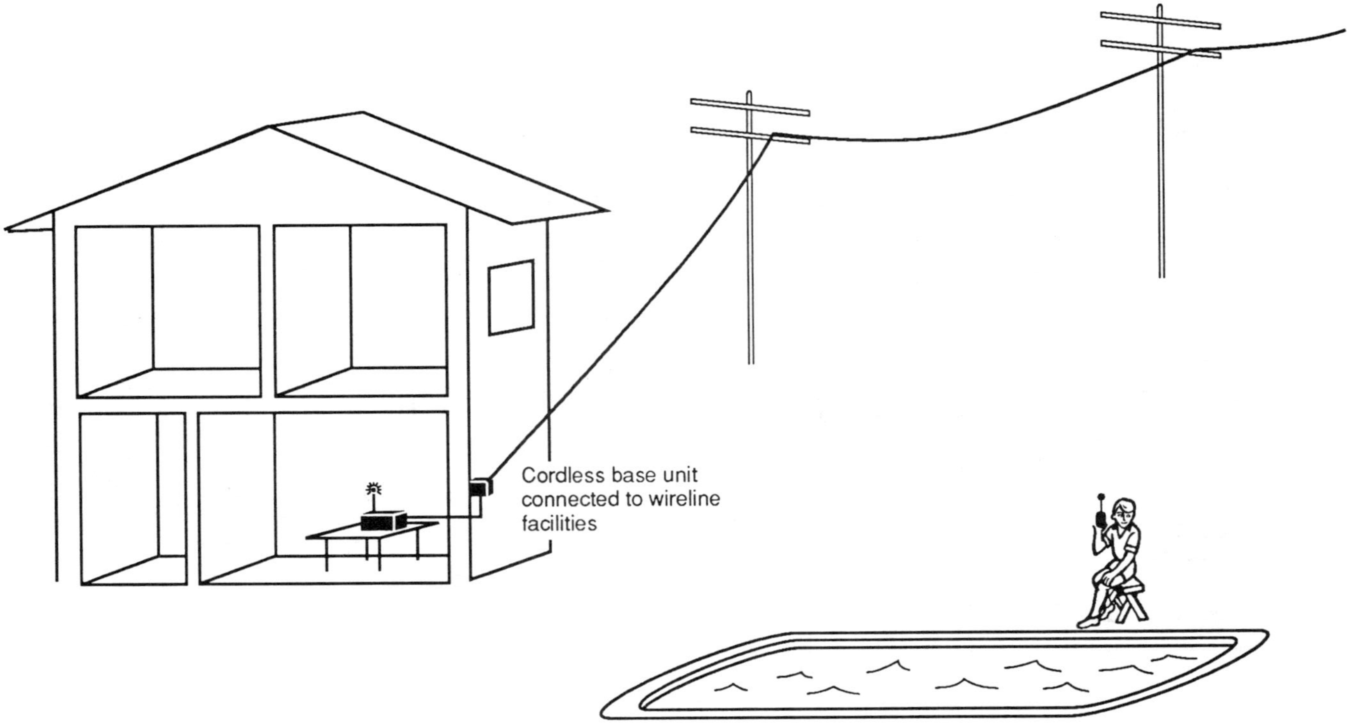

Figure 5.32 Today's cordless extension phone configuration.

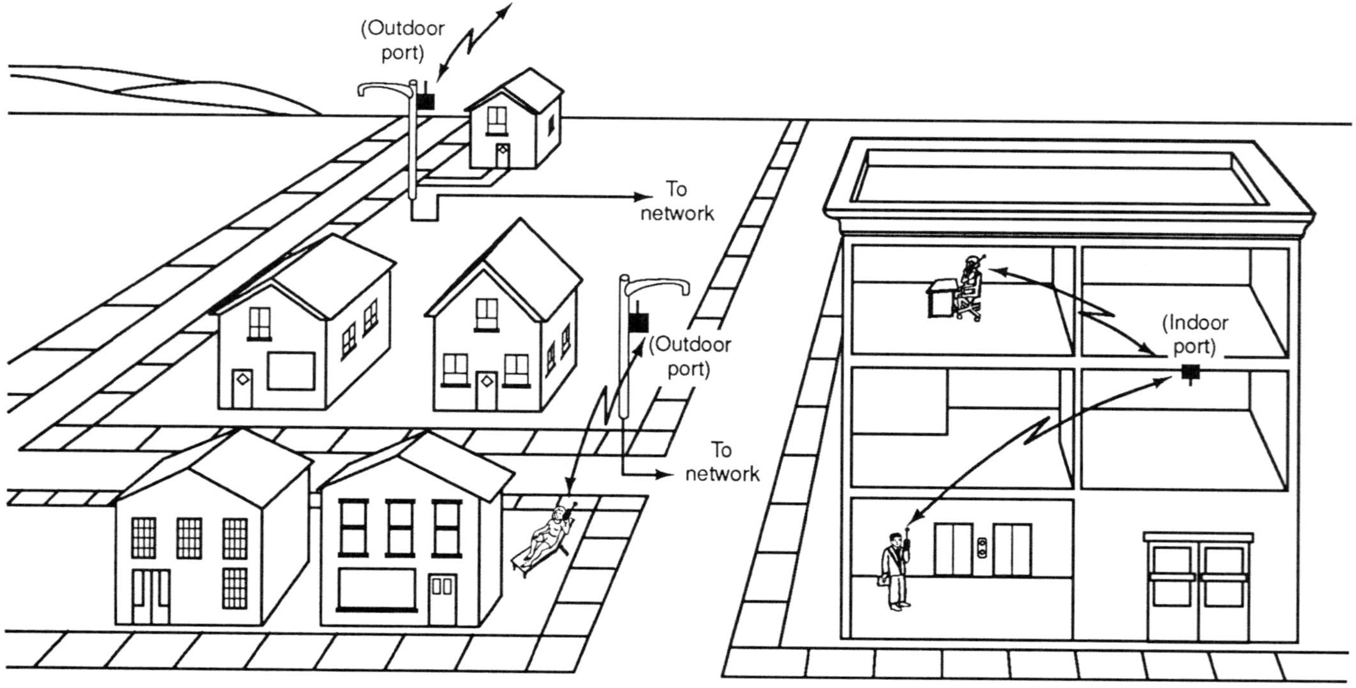

Figure 5.33 A neighborhood PCS as described by Bellcore. (From Bellcore 1013 Document.)

5.5.3 Wireless PBX and Other Applications

Another set of applications involves the use of cordless or wireless telephone sets in office buildings or campuses. This has been the traditional domain of the private branch exchange (PBX), with significant competition from CENTREX service (where a telephone company essentially provides a central-office-based service that emulates the functions of a PBX). In either case, the application and the architecture is straightforward. The Bellcore configuration in Figure 5.33 encompasses wireless in-building uses. More architectural detail is shown in Figure 5.34, which is derived from a proposal made by a major U.S. telephone company for a *wireless access business system* [148]. More than two dozen proposals for such systems have been put forth, many originating from today's PBX suppliers, although most have yet to reach the marketplace [149].

Personally, I believe that this will prove to be a very powerful application of short-range wireless technology. It could even become the dominant wireless access application, in terms of customers and Erlangs of traffic. Far more phone calls originate today from office and business environments than from moving vehicles, and even if cellular phones become a standard fixture in cars over the next 10 years, most people will still spend much more time in a work or factory environment than in an automobile.

I also believe, however, that wireless PBX and similar applications will tend to evolve along manufacturer-proprietary formats. Development of a common air interface will be difficult and there is actually little rationale for doing so. PBX services tend to be restricted to a single company and a single location. The location is quite often an "office"—meaning a physically integral workspace (although it may cover many floors in a building). Some very large applications would involve a campus that could cover many acres and a number of different buildings. However, the requirement that a single device be capable of being used in any building, on anyone's PBX, would impose undesirable constraints on the evolution of the technology. It is doubtful whether in-building applications will play a large part in common-carrier-oriented applications of PCN and, indeed, it may be that "wireless PBX" as a category belongs more appropriately in Section 5.6 on "Private Network Applications."

5.5.4 Residential Neighborhood Access

The "Phones on the Move" report stirred much interest because of its proposal that PCN might become a competitive alternative to the conventional wireline telephone network.

> The UK has more than 20 million telephone subscribers who access the public network through a copper wire link. There is no inherent technical reason why

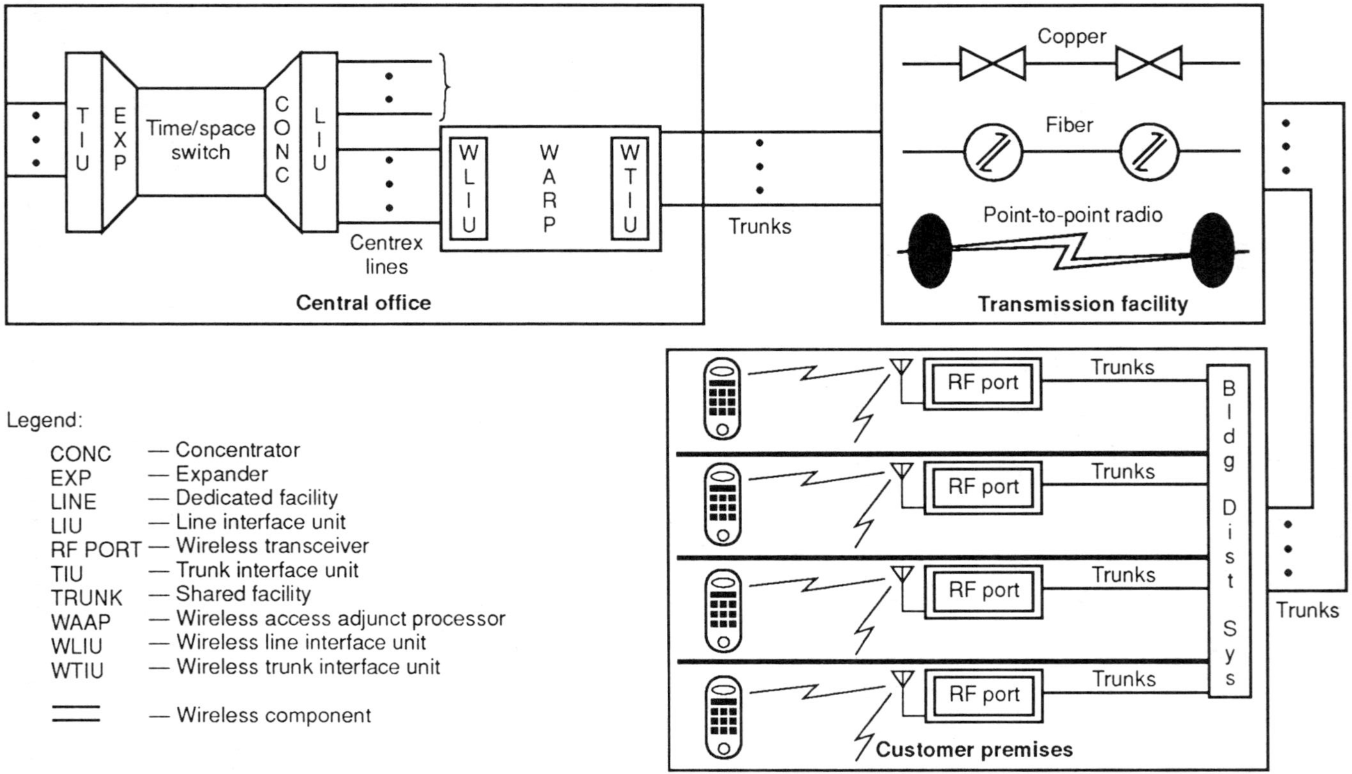

Figure 5.34 Architecture for a wireless access business system. (From BellSouth.)

> these subscribers should not access public telephone networks through mobile radio means. This possibility is sometimes referred to as "radio tails." For many purposes it would be a lot more convenient for subscribers. [150]

The concept of "radio tails" has been discussed as a complement to fiber-based access:

> Both fiber and radio will be used to implement the local loops of tomorrow. Fiber in the local loop is only cost effective, however, for clusters of lines . . . [and] for large and medium sized businesses. For residential users, however, it only makes sense to take fiber to the street end and connect a number of customers to a box at the end of the street. . . . For this last 100 meters to 200 meters [one] could use copper or . . . radio tails. Well over half the costs of a typical local telephone operation are associated with the maintenance of copper cables. . . . Radio tails would avoid these costs. [151]

The use of radio as a "last quarter mile" distribution vehicle (with either fiber or copper feeder plant) is beginning to generate interest at several major telephone companies. For example, a recent NYNEX initiative specifies a wireless loop application for suburban and urban areas, to be implemented preferably at the frequencies discussed for PCN (1.7 to 2.3 GHz) [152].

The use of PCN to provide basic access for residential customers at fixed locations crowds into the BETRS application set discussed in Section 5.3. We will look specifically at the similarities of radio and fiber in Chapter 9. I again have doubts as to whether this application can be seamlessly integrated with others in the PCN set. Basic telephone service, as we have seen, is held to a higher standard than most portable or cordless applications. As well, there is the regulatory boundary between basic service providers (monopoly local exchange common carriers) and the competitive, unregulated operating model implied by PCN, which militates against mixing these applications in a single framework.

.5.5 Personal Communications

Finally, at the core of the PCN idea is the constellation of still rather poorly defined "personal" communications concepts. The application is usually described in terms of science fiction—an ultrasmall device, as personal as a wristwatch, capable of functioning in any environment, for voice, data, and even video, an "office in a pocket," a phone number assigned at birth, and so forth. "Personal communications" thus amalgamates an ambitious wish list, including:

1. Complete portability
2. A single device capable of accessing the network in a variety of environments (fixed, residential, office, mobile)

3. A person-oriented numbering plan and the network intelligence to implement the plan
4. A minimum package of personalized services, such as voice telephony, paging, fax, and voice mail [153].

Since the "Phones on the Move" report was published, there has been a hurricane of conferences, papers, and articles extolling the personal communications concept. Most focus on the potential consumer appeal of such services. A few have begun to examine the challenges of building a network capable of supporting these services [154].

Time will tell whether such a seamless, all-encompassing architecture is feasible, but personally I rather doubt it—not because it might not be technically possible, but because it is in some ways an outmoded vision that today's consumer market really does not seek. It is a bit like the idea of standardizing a single personal computer that would be capable of running any kind of software and plugging into any sort of network. I tend to the view that heterogeneity will be the hallmark of the communications future. Smart network interfaces with inherent format translation capabilities will allow for considerable diversity at the user's end (see Chapter 11).

5.6 PRIVATE NETWORK APPLICATIONS

God, an entomologist has said, must have been inordinately fond of beetles, for She made so many different kinds. As we cross the line from "true" common carrier wireless access applications into the world of private radio networks, we have entered the radio equivalent of the kingdom of the beetles. A few moments of reflection will bring to mind how pervasive and diverse such radio devices have become, from medical alarms to pagers to baby monitors to CB radios in cars to the ubiquitous Motorola "bricks" found in almost every variety of work site. Keep in mind that even after the explosive growth of cellular radio over the past six or seven years, there are still between 5 and 10 times as many radio devices in private networks as there are in the common carrier environment. It is not unlikely that this ratio many remain unchanged for some time to come. Even if we limit the survey to voice communications devices, we must consider not only the fairly visible police and taxi cab and other dispatch radio systems, but also a wide variety of industrial radio networks, from security guards and construction workers to air traffic control and (not to be forgotten) a vast array of military and defense applications of radio.

A complete catalog of the various private network applications involving wireless technologies would be beyond the scope of this book. Nevertheless, such applications will almost undoubtedly constitute one of the most vigorous areas of radio systems development in the next decade. As digital technologies upgrade the common carrier applications, they will also transform and expand the noncommon carrier applications. While on the one hand I believe that many of these private networks will develop network access (*interconnection*) as an ancillary capability—it's always nice to be able to make a phone call in a pinch—most of these applications are not *per se* access-oriented. The primary function of air traffic communications is the control of airplane activities. The primary function of police radio is to respond to emergencies. Nevertheless, a quick glance at the drawers full of these "beetles" is perhaps tonic against the creeping (and sometimes galloping) ethnocentrism of the cellular tribe in thinking themselves the Adam of the radio universe.

5.6.1 Defining Characteristics of Private Carrier Applications

It is perhaps foolhardy to attempt to generalize about such a colorful gamut of applications. There are some characteristics, however, which are shared by many of the "beetles," and a few distinctions that may prove taxonomically useful.

First, we may distinguish between applications that are essentially local-area networks—wireless LANs, also called radio LANs (RLANs) and even local-area wireless networks (LAWNs)—and those that are metropolitan in scope (wireless MANs). There are also networks that are even larger in scope, covering nations or continents, which some have labeled wireless wide-area networks (WANs). Of these, the now well-established very small aperture satellite terminal (VSAT) networks are probably the most important at present.

5.6.2 Wireless LANs

Wireless LAN applications typically incorporate the following characteristics:

1. They are *closed systems,* or, to put it another way, *closely administered systems.* By definition, private networks are not common carriers. They are not required to serve all comers. This means that the system architecture can be optimized around an assumption that only a defined number of authorized users will be served. Moreover, the equipment used by those individuals is likely to be supplied from the central system administrator. The areas of coverage are strictly defined, as are perhaps many other parameters.

   This makes such systems very different from, say, a cellular operator serving a large city. The cellular common carrier does not know who may try to access his network at any time, what equipment they may use, how rigorously that equipment conforms to the air interface specification, etc. He is bound, at least broadly, by the nature of his service commitment as a common carrier to offer a standard service to all comers, and he must anticipate and plan for traffic peaks and valleys and coverage gaps that he cannot really control.

   A wireless PBX or LAN, by contrast, can be run like a wireline LAN, with strict access controls and tight technical requirements. Users may be assigned various priorities, and the system operator has considerable latitude to change the rules to control user behavior.

2. Wireless LANs must support *voice and data,* with a growing and perhaps even primary emphasis on modern LAN-like (as opposed to public network-like) data requirements. Some wireless LANs will emerge as exclusively data-oriented; others, toward the wireless PBX side of the scale, will be predominantly voice-oriented. But the LAN concept implies a growing convergence of voice and medium- to high-speed data over a seamless private network, and wireless LANs will follow this trend.
3. Wireless LANs will probably stress *flexible bandwidth* assignments. This is a corollary of mixing voice and data (and perhaps video). Unlike public networks that will tend to force users into Procrustean bandwidth categories (at least until the emergence of true asynchronous mode or fast packet-switching engines some time in the next century), LANs will allow for bandwidth-on-demand much earlier.
4. Wireless LANs will emphasize *security* to a greater extent than public networks. The applications environment—chiefly corporate and governmental—will drive a strong requirement for high levels of privacy and security.

With respect to specific applications in the wireless LAN category, we may certainly expect to see *wireless office* networks proliferate in the near future [155].

Another wireless LAN application involves the *factory of the future,* some feeling for which may be gleaned from the following description, which builds up from one subapplication, namely, work-in-process inventory management, to encompass a whole microcosm of in-factory communications, even to its own "vehicular" communications system:

> Methods for transporting parts-in-process in a futuristic, but realistic JIT [i.e., "Just-in-Time"] manufacturing environment is one of the thrust research areas . . . for Intelligent Manufacturing Systems. Analysis has shown that an inexpensive, agile, mobile robot fleet, capable of navigating without any type of track, could easily accommodate a majority of the material flow requirements for a JIT manufacturing system. . . . A truly autonomous guided vehicle (AGV) that does not use a tether requires a radio system for control. . . . Furthermore, radio systems are useful for providing fast and inexpensive connections for often-moved manufacturing equipment and computer terminals. Radio will also accommodate reconfigurable voice/data communications for other facets of factory operation. For example, RF tags, which can be easily attached to parts-in-process, provide a paperless data storage medium and are becoming an increasingly popular replacement for paper work order cards. . . . [156]

The variety of short-range closed-system wireless applications (not all may be properly termed "wireless" LANs) is undoubtedly very great. A recent paper on "Premises Radio" identified four other wireless applications (four more beetles to catalog!), which suggest the opportunities for creative design [157].

1. *Telemeter/telecontrol systems:* This is, in effect, an extension (with two-way capabilities) of the remote control device that we use to change channels on a TV set. Within a factory environment, there are many situations where remote control or remote reading of meters or gauges may be safer, quicker, cheaper.
2. *Mobile data entry:* This application is similar to the concept of inventory management by means of a portable radio terminal. (Note that some rental car companies are now using this kind of capability for checkout.)
3. *Premises paging:* Again, this is a straightforward extension of the wide-area paging concept to a smaller scale application, for paging in a factory or office (now often done over loudspeakers).
4. *Automatic vehicle identification:* This application uses radio to identify moving vehicles in response to queries from a stationary radio checkpoint device. Possible uses could include automatic charging at tollbooths on the highway; as radio-equipped trucks roll through, they are registered by means of a radio-handshake with the electronic tollbooth and their accounts charged for the appropriate toll without having to stop.

The possibilities seem endless indeed.

5.6.3 Wireless MANs

The use of radio to create private or closed networks that are metropolitan in scale is the province of many traditional mobile radio applications, including police and emergency services, metropolitan transit systems, and many dispatch services. Whereas wireless LANs will tend to be limited in range (to perhaps one mile at most), wireless MANs will require coverage out to and somewhat beyond the horizon (typically transmission radii of 20 to 100 miles). Today's wireless MANs are also overwhelmingly voice oriented, often at levels of transmission quality that are appalling (at least by comparison with public networks).

The coming of modern digital radio technology to these markets will undoubtedly help expand these applications. Already, mobile data systems are being used by some couriers and a few taxicab fleets. However, it is still too early to see clearly how wireless MANs will evolve beyond today's framework.

5.6.4 Wireless WANs: VSAT Networks

The use of small satellite terminals to create wide-area networks linking hundreds or thousands of nodes, each fed by its own local-area network, has become a standard communications option for large organizations seeking to network among widely separated facilities. A number of good general references are available that describe the applications [158], technology [159], and economics [160]. Three brief examples will suffice to give some feeling for the flexibility of VSAT networks:

1. *Wal-Mart:* In 1984, the Wal-Mart department stores became the first very large VSAT customer, with a network that came to encompass more than 1400 Wal-Mart stores and distribution centers over 25 states. The network was installed to provide two-way data and voice communications, as well as "fully addressable and encrypted broadcast video, training personnel, videoconferencing, and high-impact advertising" [161]. The before-and-after description of the applications shows how close it comes to a true wireless access implementation:

   > Prior to the installation of the VSAT network, headquarters had to initiate a telephone call each day to every Wal-Mart store so that data could be exchanged. At most, a store would be in direct contact with headquarters for only 20 minutes. At the present time, Wal-Mart stores have a 24-hour-a-day contact with the hub for their data network. This has allowed them to implement new applications, such as electronically approving bank card transactions. With VSAT voice, Wal-Mart is able to provide high-speed, consistent communications in place of the slower, less certain services offered by 400 different phone companies, many in rural areas and some without touch-tone capabilities. [162]

2. *Holiday Inn:* The Holiday Inn VSAT network links some 2000 hotels with centralized databases and other facilities via standard packet networks, to pass information on room availability, guest reservations, and so forth. Again, the description of the application shows the similarity to wireless access:

   > The network comprises a central communications hub at Holiday's Memphis, TN headquarters and remote VSAT terminals at each of its hotel properties. . . . The locations having a VSAT terminal have dedicated digital connections directly to terminal equipment. No telephone company facilities or modems are required. . . . [163]

3. *Chrysler:* The automaker is installing what will be (at least briefly) the world's largest VSAT network, linking more than 6000 dealerships and corporate facilities in North America.

   > Chrysler will use the network to route all asynchronous data traffic from dealer sites to its corporate data processing facilities via X.25 links, while its manufacturing facilities and corporate offices will be connected via SNA/SDLC networks. Chrysler plans to use the VSAT network to support several new strategic applications, including a way to use headquarters' mainframes to assist mechanics at local dealerships with car repairs. In addition, Chrysler plans

> to use the network to let salespeople order and confirm delivery dates for cars from a showroom computer. . . . [164]

One of the attributes of most VSAT networks is that they provide interconnection or access to the public telephone network for users at all the remote nodes. Thus, in principle, they are capable of being used as a wireless access vehicle. There are two chief problems with VSAT as a mainstream access vehicle. First, the delay introduced by the satellite hop results in an inherent deficiency in VSAT voice circuits compared to terrestrial access circuits. (There is much debate over whether the delay of a single satellite hop is really important to the customer.) Second, the high cost of today's satellite terminals requires a fairly concentrated and high-value stream of traffic to be cost-justified [165]. While satellite terminal costs will undoubtedly come down, so will the costs of terrestrial wireless terminals, and the space segment (i.e, the satellite itself) will continue to be expensive. Nevertheless, in its niche—that of linking geographically dispersed LANs into a wide-area private network, VSATs have proven to be highly successful. And it may well be that VSAT networks will nibble at the edges of terrestrial wireless access applications, or they may be combined to produce completely wireless end-to-end private networks with interconnection capabilities.

This concludes our around-the-world tour of wireless access applications. As with any such excursion, there are undoubtedly some interesting niches that we skipped or did not do justice to. I have perhaps given undue emphasis (in the minds of some readers) to the fixed, noncellular applications of wireless access, such as BETRS and VSAT, but I feel this is justified by the higher reality quotient of these digital wireless systems compared to the as-yet unborn (as of this writing) digital incarnations of cellular radio and PCN. I feel strongly that true wireless access must mean wireless *digital* access, designed to network standards, and as such, the mobile services do not yet truly qualify. Within a few short years, however, digital radio will have become the standard for all the mobile and portable services and I am sure the emphasis in the second edition of this book some years hence (not necessarily involving the same author) will be quite different.

NOTES

1. For example, Federal Communications Commission (Office of Science and Technology, Technical Standards Branch), "Cellular System Mobile Station—Land Station Compatibility Specification," OST Bulletin No. 53, July 1983; and, subsequently, EIA Interim Standard IS-3-D, "Cellular System Mobile Station—Land Station Compatibility Specification," March 1987. See also IS-19 and IS-20, both from the EIA; see also the special issue of the *Bell System Technical Journal,* January 1979, on the AMPS architecture.
2. For example, "Total Access Communications System: Mobile Station—Land Station Compatibility Specification," Issue 4, March 1989 (Draft), reprinted by Racal Vodafone.

3. For example, "Nordic Mobile Telephone: System Description," NMT Doc. 1.1980, February 1980, and associated specifications for the NMT mobile switch, base stations, and subscriber terminals; see also "Nordic NMT-900: Automatic Cellular Mobile Telephone System," NMT Doc. 900-1, January 1985, and its associated specifications.
4. Electronic Industries Association, EIA/TIA/IS-54-A, "Cellular System Dual-Mode Mobile Station—Base Station Compatibility Standard," March 1991, available from the EIA, 2001 Pennsylvania Avenue, Washington, D.C. 20006; see also IS-55 and IS-56, forthcoming, covering "minimum performance standards" for dual-mode mobile and base stations respectively.
5. For example, Keiji Tachikawa, "Development of Digital Cellular Systems in Japan," presented at the 1991 Pan-European Digital Cellular Radio Conference, Nice, France, February 5–6, 1991.
6. For example, Groupe Speciale Mobile, "General Description of a GSM Plan," Version 3.0.0, release date November 15, 1988, as well as numerous subsidiary and supporting standards documents. See also Kari Ahola, "Europe's GSM: Passage to Digital," *Telephone Engineer & Management,* September 15, 1990, pp. 51–60; Mike Pinches, "GSM Service Infrastructure Takes Shape," *Telecommunications,* December 1989, pp. 21–24; also, numerous papers associated with recent cellular and wireless conferences, such as The 1991 Pan-European Digital Cellular Radio Conference, Nice, France, February 5–6, 1991, from which note especially Thomas Haug, "Developing GSM Standards," and Nigel Cawthorne, "GSM Set to Enter the International Cellular Market."
7. British Department of Trade and Industry, "MPT 1375, Common Air Interface Specification," May 1989.
8. See, for example, William J. Franklin, "Personal Communications Services: The Next Cellular?" *Cellular Business,* February 1991, pp. 6–7. In many discussions, CT-3 is synonymous with DECT (see note 9).
9. See C. Buckingham, G.K. Wolterink, and D. Åkerberg, "A Business Cordless PABX Telephone System on 800 Mhz Based on the DECT Technology," *IEEE Communications Magazine,* January 1991, pp. 105–110; also A. Bud, "System & Network Aspects of DECT," Fourth Nordic Seminar on Digital Mobile Radio Communications, June 26–28, 1990, Oslo, pp. 6.3.1–6.3.12; H. Oeschner, "Radio Aspects of DECT," Fourth Nordic Seminar on Digital Mobile Radio Communications, June 26–28, 1990, Oslo, pp. 6.4.1–6.4.6; R. S. Swain, "Services and Facilities of DECT," Fourth Nordic Seminar on Digital Mobile Radio Communications, June 26–28, 1990, Oslo.
10. See British Department of Trade & Industry, "Phones on the Move," January 1989.
11. Bellcore, *Generic Framework Criteria for Universal Digital Personal Communications Systems (PCS),* FA-NWT-001013, Issue 2, December 1990; also Federal Communications Commission, "In the Matter of Amendment of the Commission's Rules to Establish New Personal Communications Services," Docket 90-314, adopted June 14, 1990, released June 28, 1990.
12. D.C. Cox, "Portable Digital Radio Communications—An Approach to Tetherless Access," *IEEE Communications Magazine,* July 1989, pp. 30–40.
13. "Future Public Land Mobile Telecommunications Systems," any of a number of drafts and reports of meetings of the CCITT Interim Working Party 8/13.
14. Federal Communications Commission, "Notice of Proposed Rulemaking, In the Matter of Basic Exchange Telecommunications Radio Service," CC Docket No. 86-495, Adopted December 18, 1986. For specifications, see U.S. Department of Agriculture, Rural Electrification Administration, "BETRS Radio Systems," Bulletin 1751H-703, June 20, 1991, Draft. This is a 40-page, single-spaced "manual" for BETRS applications that constitutes one of the most comprehensive practical discussions of wireless access to date. The chief author of this draft is Lamar Moore, head of the Transmission Branch of the Telecommunications Division of the REA.
15. Bellcore, *Generic Requirements for Basic Exchange Radio Systems,* TA-TSY-000911, Issue 2, July 1989.
16. See, for example, Raymond Steele, "The Cellular Environment of Lightweight Handheld Portables," *IEEE Communications Magazine,* July 1989, pp. 20–29.

17. Not that this has not been attempted; the best short summary of wireless access applications is contained in Donald C. Cox, "Portable Digital Radio Communications—An Approach to Tetherless Access," *IEEE Communications Magazine,* July 1989, pp. 380–40.
18. For example, the PLRS/JTIDS battlefield communications system, fielded by the U.S. Army in the 1980s. See W.F. Lynch, "TDMA–JTIDS Overview Description," MITRE Technical Report 8413, MITRE Corporation, Bedford, Massachusetts; also "Army Firms Up JTIDS Planning," *Defense Electronics,* August 1982; "JTIDS Coordinates Tactical Elements," *Defense Electronics,* January 1982; "Joint Tactical Information Distribution System: A View from the Joint Program Office" and "JTIDS: A View from Industry," *Signal Magazine,* March 1982.
19. R.I. Betteridge and Jack Pulford, "The Aurora (Automatic Roaming Radio) Cellular A.M.T.S. System," *Communications,* September 1981, pp. 51–57.
20. Almost all of these were using digital TDMA BETRS equipment manufactured by IMM Corporation; the only other supplier in this time frame was Rockwell, which was to my knowledge in the trial phase with several customers; see "CXR-424 Collins Exchange Radio," product description available from Rockwell.
21. Outside the United States, all the wireless digital access lines were based on subscriber radio equipment, available from a number of suppliers.
22. See G. Calhoun, *Digital Cellular Radio,* Boston: Artech House, 1988, pp. 89 ff. for a discussion of the factors requiring a "de-rating" of many published cellular subscriber numbers.
23. For example, from a presentation by Jim Caile of Motorola at the Day Group seminar on Digital Cellular, June 21, 1990, New York.
24. Bellcore, *Generic Requirements for Basic Exchange Radio Systems,* TA-TSY-000911, Issue 2, July 1989, Sec. 6.1.
25. The list of relevant Bellcore references is extensive; see the list of references at the end of Bellcore's BEXR Advisory, *Generic Requirements for Basic Exchange Radio Systems,* TA-TSY-000911, Issue 2, July 1989.
26. *Ibid.,* p. 10.
27. T.M. Taylor, "Power and Energy in the Local Loop," *IEEE Communications Magazine,* March 1991, pp. 76–82; this article deals with power in the loop mainly from the standpoint of the deployment of fiber optics.
28. See R.M. Pepper, "Through the Looking Glass: Integrated Broadband Networks, Regulatory Policy and Institutional Change," Federal Communications Commission OPP Working Paper Series, No. 24, November 1988, p. 55.
29. If a BETRS-type transceiver uses 5 W of power in the idle or stand-by mode (power in the transmit mode, which is higher, is not generally relevant for cost calculations since the duty cycle is so low), this amounts to 3.6 kWh/month; with electric rates typically between $0.03 and $0.10 per kilowatt-hour in rural areas, the cost of subscriber power is less than half a dollar per month. For shorter range wireless loops, the cost will be only a penny or so per month at today's rates.
30. Talon, Cheesman & Associates, Request for Declaratory Ruling, submitted to the FCC, #88-1831.
31. When PCN was introduced in the United Kingdom, it caused an immediate drop in the value of all cellular stocks, including those in the United States. This reflected investor concern over the impairment of the monopoly aspect of the cellular license.
32. Federal Communications Commission, Notice of Proposed Rule Making, "In the Matter of Amendment of Parts 2 and 22 of the Commission's Rules to Permit Liberalization of Technology and Auxiliary Service Offerings in the Domestic Public Cellular Radio Telecommunications Service," Docket 87-390, released October 15, 1987.
33. See G. Calhoun, *Digital Cellular Radio,* Boston: Artech House, 1988, p. 114, for a discussion of this issue.
34. *Ibid.*; also see "Dawson Pushes for 'Flexible' Spectrum Allocation & Strikes Blow to Cellular's Push for 12 MHz More," *Industrial Communications,* May 31, 1985, p. 9.

35. Federal Communications Commission, "In the Matter of Amendment of the Commission's Rules to Establish New Personal Communications Services," Docket 90-314, adopted June 14, 1990, released June 28, 1990.
36. G. Calhoun, *Digital Cellular Radio,* Boston: Artech House, 1988, pp. 77 ff.
37. Obviously, most cellular operators are reluctant to release actual blocking figures; however, it is widely understood that blocking rates are very bad in some cells during peak calling hours. See G. Calhoun, *Digital Cellular Radio,* Boston: Artech House, 1988, pp. 113 ff.
38. California Public Utilities Commission, "Investigation on the Commission's own motion into the regulation of cellular radiotelephone utilities," I.88-11-040, November 23, 1988. See also "Comments on International Mobile Machines Corporation," in response to this investigation, filed February 28, 1989.
39. G. Calhoun, *Digital Cellular Radio,* Boston: Artech House, 1988, pp. 57–58.
40. See note 38.
41. See Part 15 of the FCC Rules, sections 15.247 and 15.249, revised and republished annually.
42. SR Telecom was founded and is headed by Don Beaupré; he and his team deserve recognition as early pioneers of the wireless access industry. See Donald M. Beaupré, "Five Years of TDMA Subscriber Radio," ICC '83, June 19–22, 1983, Boston; Donald M. Beaupré, T. Le-Ngoc, and H.J. Zavitz, "A New Point-to-Multipoint Microwave Radio System," 6th International Symposium on Subscriber Loops and Services (ISSLS 84), Nice, October 1–5, 1984, pp. 47–51 (reprints available from SR Telecom); "Point to Multipoint Microwave: Spanning the Globe," Global Communications, Second Quarter 1988; Tho Le-Ngoc, "New Protocol for Multipoint Radio Systems," *Telephony,* December 23, 1985; W.S. Roy, S.W. March, and H.J. Zavitz, "Subscriber Radio Brings Telecom to the Islands," *Telephone Engineer & Management,* July 1, 1984; Tho Le-Ngoc and H.J. Zavitz, "SR500—A Point to Multipoint Microwave Radio System with a Random-Request Demand-Assignment Time-Division-Multiple-Access (RR-DA-TDMA) Scheme," paper available from SR Telecom; for other "non-SR Telecom" references, see V. Sargeant and J. Steele, "Planning Telephone Networks for Rural and Remote Areas using Digital Radio Concentrator Systems," *Proceedings of the I. E. Australian Engineering Conference,* Paper 314, 1981, pp. 138–145; N. de Couesnongle and C.R. Garnier, "IRT 1500 Integrated Rural Telephony System," *Philips Telecom Review,* Vol. 41, No. 2, June 1983; Telecom Australia, "Guidelines for the Application of Digital Radio Concentrator Systems," Planning Applications Bulletin, No. 46, September 1981; K. Wu and M.J. Miller, "Point-to-Multipoint Digital Subscriber Radio Systems in China," *Proceedings of the Pacific Telecommunications Council,* Honolulu, January 1985, pp. 269–274; M.J. Miller and S. V. Ahamed, *Digital Transmission Systems and Networks, Volume 2—Applications,* Computer Science Press, Rockville, Maryland, 1987, pp. 61–63.
43. FCC, Memorandum Opinion and Order. File Nos. 2476-CF-P-88 through 2535-CF-P-88, regarding Applications of Leaco Rural Telephone Cooperative, Inc. for Construction Authorizations in Point-to-Point Microwave Radio Service and Request for Waiver of the Rules, adopted June 22, 1989.
44. Henry Hall, Lennart Westerlund, Karl-Axel Åhl, "Rural Telephony with Radio Concentrators," *Tele,* February 1983, pp. 37–49.
45. Alain Forestier, "From Today to Tomorrow: A Pathway for ISDN Implementation with the SR500," presented at IEEE Montec Conference, Montreal, November 9–11, 1987; also, Tho Le-Ngoc, Mike Stashin, and Ali Elkateeb, "ISDN Services over Microwave Systems," *Communications Engineering International,* August 1989.
46. Henry Hall, Lennart Westerlund & Karl-Axel Åhl, "Rural Telephony with Radio Concentrators," *Tele,* February 1983.
47. See G. Calhoun, *Digital Cellular Radio,* Boston: Artech House, 1988, Chap. 15, for a discussion of the various approaches to quantification of the concept of spectrum efficiency.

48. For example, Network Planning Department, General Telephone Company of the Southwest, Product Specification for Rural Subscriber Radio, April 20, 1983.
49. Representative articles include: Paul Clements, "BETR: Digital Radio in the Loop," *Telephone Engineer & Management,* August 15, 1988; Richard G. Saunders, "Ultraphone—Wireless Digital Loop Carrier System," *Proceedings of the National Communications Forum,* Vol. 42, No. 2, pp. 1860–1866, 1988; James F. Mullen, "Digital Radio Systems for the Local Loop Offer Rural Areas a BETRS Way," *Outside Plant,* July 1989.
50. Kris Shelton and J.H. Kleymann, "A Report by the Mountain States Telephone and Telegraph Company, Northwestern Bell Telephone Company, and Pacific Northwest Bell Telephone Company in the Matter of the Use of Radio Systems to Provide Fixed Rural Telephone Service," submitted to the Federal Communications Commission, CC Docket RM-4882, August 1, 1985, p. 1.
51. "Petition for Rulemaking to Establish Basic Exchange Telecommunications Radio Service", submitted by the REA et al. to the FCC, CC Docket No. 86-495, RM-5442, May 9, 1986.
52. Federal Communications Commission, "Notice of Proposed Rulemaking, In the Matter of Basic Exchange Telecommunications Radio Service," CC Docket No. 86-495, Adopted December 18, 1986.
53. See especially the "Comments" and "Reply Comments" of International Mobile Machines Corporation, dated March 30, 1987, and April 27, 1987, respectively.
54. Report and Order, FCC 87-387, January 19, 1988.
55. U.S. Department of Agriculture, Rural Electrification Administration, "BETRS Radio Systems," Bulletin 1751H-703, June 20, 1991, Draft.
56. George G. Goldstein, "Wyoming Town Steps into the 20th Century with Digital Local Loop," *Telephony,* June 15, 1987.
57. Art Biggs, "Telco Tries Wireless System for Local Loop," *Telephone Engineer & Management,* October 1, 1985, pp. 123–125.
58. C.E. Jones, "Digital Radio Passes Test in Local Loop," *Telephone Engineer & Management,* May 15, 1989.
59. See *Technical Analysis Report—IMM Ultraphone 100 Wireless Digital Loop Carrier System,* AR-TSY-001273, Issue 1, November 2, 1989. The chief reference for the field portion of the evaluation was F.P. Tuhy, "Bellcore Test Plan for Digital Loop Carrier Systems," October 5, 1987; Bellcore did not have a formal test plan for the radio portion of the system at the time of these field studies; therefore, the radio link was evaluated later, with tests carried out mainly at IMM's Philadelphia facility by Bellcore personnel. *Note: The distribution of this document may be limited to Bellcore and authorized clients.*
60. Scott Chesbro, "A New Era for Navajo Phone Service," *Telephony,* February 6, 1989.
61. Other BETRS-related references include the following papers presented at the 1988 National Radio Engineers' Conference, Orlando, September 20–22, 1988: W.R. Hampton, "Economic and Deployment Considerations for Exchange Radio"; Robert J. Maguire, "Basic Exchange Telecommunications Radio (BETR) Technology"; M.J. Morris, "The Application of Microwave Time Division Multiple Access Techniques to the Provision of BETR Service."
62. Alberta Government Telephones, Technical Specification for AGT Rural Individual Line Service (ILS) Digital Loop Replacement Radio (DLRR) System, June 13, 1986.
63. Canadian Department of Communications, Telecommunications Policy Branch, Radio Systems Policy, "Spectrum Utilization Policy on the Use of Certain Public Correspondence Bands in Canada," Draft issued February 17, 1988, RP 013, Notice Number DGTP-003-88, enacted October 1988. See also Comments on IMM Canada Ltd. on this proceeding, May 31, 1988.
64. Manitoba also issued a much briefer RFI for wireless local loop systems at about the same time; see Phil Rubin, Manitoba Telephone System, "Request for Information on Rural Radio Systems," September 29, 1986, 10 pp.

65. Nippon Telegraph & Telephone Corporation, Procurement Documentation: Equipment for a High Capacity Subscriber Radio System, Public Notice No. 61-303, October 22, 1986.
66. CCIR Study Group, U.S.A., "Narrowband Digital Basic Exchange Radio System," Proposed Amendments to Reports 380-2 (Mod I) and 1057 (Mod I), Document 9-10, April 11, 1989.
67. CCIR Interim Working Party 8/13, Adaptation of Mobile Radiocommunication Technology to the Needs of Developing Countries: Draft Report, Document IWP8/13-47 Rev.1, June 6, 1989.
68. See *Technical Analysis Report—IMM Ultraphone 100 Wireless Digital Loop Carrier System,* AR-TSY-001273, Issue 1, November 2, 1989. *Note: The distribution of this document may be limited to Bellcore and authorized clients.*
69. Bellcore, Technical Advisory TA-TSY-000911, *Generic Requirements for Basic Exchange Radio Systems,* Issue 2, July 1989. See also, as a source document: D. Ridgely Bolgiano, IMM Corporation, "Functional Criteria for Wireless Digital Loop Carrier Systems," Draft 13, October 3, 1988 (95 pp., available from IMM Corporation).
70. Bellcore, TR-TSY-000057, *Functional Criteria for Digital Loop Carrier Systems,* Issue 1, April 1987, and later supplements.
71. Network Planning Department, General Telephone Company of the Southwest, *Product Specification for Rural Subscriber Radio,* April 20, 1983, p. 1.
72. Bellcore, Technical Advisory TA-TSY-000911, *Generic Requirements for Basic Exchange Radio Systems,* Issue 2, July 1989, p. 7.
73. "Petition for Rulemaking to Establish Basic Exchange Telecommunications Radio Service," submitted by the REA et al. to the FCC, CC Docket No. 86-495, RM-5442, May 9, 1986, p. 4; see also IMM Corporation, "Comments in the Matter of Basic Exchange Telecommunications Radio Service," CC Docket No. 86-495, March 30, 1987.
74. Bellcore, Technical Advisory TA-TSY-000911, *Generic Requirements for Basic Exchange Radio Systems,* Issue 2, July 1989, pp. 7–8.
75. George G. Goldstein, "Wyoming Town Steps into the 20th Century with Digital Local Loop," *Telephony,* June 15, 1987; other similarly remote first applications have including Fort Davis, Texas (Southwestern Bell), Etzikom, Alberta (AGT), Espanola, New Mexico (Contel), Kingman, Arizona (Citizens Utilities)—some of the most beautiful but also the most desolate spots in North America.
76. For a short survey of the development of loop carrier, see F.T. Andrews, "The Evolution of Digital Loop Carrier," *IEEE Communications Magazine,* March 1991, pp. 31–35; the standard Bellcore reference on DLC is TR-TSY-000057, *Functional Criteria for Digital Loop Carrier Systems,* Issue 1, April 1987, Revision 1, with several supplements. For a representative survey of equipment and manufacturers, see Stuart M. Dambrot, "Digital Loop Carrier Systems: The Machines and the Manufacturers," *Central Office,* December 1986, pp. 62–65.
77. For a brief, current introduction to SONET concepts and history, see R. Ballart and Y.-C. Ching, "SONET: Now it's the Standard Optical Network," *IEEE Communications Magazine,* March 1989, pp. 8–15; also the special supplement to *Telephony,* June 1990, entitled "SONET Special: Metamorphosis of the Network." See also the references in Chapter 9 of this book on fiber optics in the loop, many of which discuss SONET.
78. E.F. O'Neill, ed., *A History of Engineering and Science in the Bell System: Transmission Technology (1925–1975),* AT&T Bell Laboratories, 1985, Chap. 5.
79. For example, Paul Clements, "Wireless Rural Telephony: A Case Study of the Application of Wireless Digital Loop Carrier in a Rural Telephone Exchange," published by IMM Corporation, Philadelphia, 1987, 15 pp.; also Paul Clements, "BETR: Digital Radio in the Loop," *Telephone Engineer & Management,* August 15, 1988.
80. Bellcore, TR-TSY-000057, *Functional Criteria for Digital Loop Carrier Systems,* Issue 1, April 1987, and later supplements; also Bellcore, Technical Advisory TA-TSY-000911, *Generic Requirements for Basic Exchange Radio Systems,* Issue 2, July 1989.

81. Bellcore, TR-TSY-000008, *Digital Interface Between the SLC-96 Digital Loop Carrier System and a Local Digital Switch,* Issue 2, August 1987, and supplements.
82. See list of Bellcore references attached to "TR-57."
83. Bellcore, TR-TSY-000191, *Alarm Indication Signal Requirements and Objectives,* Issue 1, May 1986; also TR-TSY-000053, *Centralized Fault Location for Digital Transmission Systems,* April 1985.
84. Bellcore, TR-TSY-000303, "Integrated Digital Loop Carrier System Generic Requirements, Objectives and Interface," Issue 1, September 1986, plus numerous revisions and supplements.
85. By analogy to the established notion of a carrier serving area; see G.T. Hawley, "Historical Perspectives on the U.S. Telephone Loop," *IEEE Communications Magazine,* March 1991, pp. 24–28, for the evolution of these concepts; see also Chapter 7 of this book for a more in-depth discussion of carrier serving area and radio carrier serving area.
86. This application was reported in *Telephone Engineer & Management* in 1989.
87. *US West Today,* Colorado, company publication, July 2, 1990.
88. National Communications System Plans and Program Office, "Air-Lifted Equipment Restores Naval Phone Service after Hugo," press release, October 13, 1989. The release includes the following: "In its first air-lift of commercial telecommunications satellite equipment to an actual disaster site, the National Communications System (NCS) established commercial and defense (Autovon) telephone service to the Roosevelt Roads Naval Base in Puerto Rico following the destruction caused by Hurricane Hugo. The Naval air base suffered a total electrical and telephone blackout after the hurricane. The successful installation of the satellite earth station and digital wireless telephones on September 28, 1989, brought phone service to naval personnel almost two weeks before regular service was restored on the base. The satellite earth station from Alascom, consisting of a C-band satellite terminal, 4.5 meter folding dish, RedCom 32-line digital PBX switch, and 8kW on-board generator, was flown in from Anchorage where its capabilities had recently been demonstrated in an NCS-sponsorsed exercise. Eight IMM radio-based Ultraphones were flown in from Washington State. . . . "
89. "Ultraphone Comes Through for Avianca Rescuers," *Telephone Engineer & Management,* March 15, 1990.
90. IMM Press Release, "Operation Just Cause: Ultraphone 200 Responds to Crisis," March 8, 1990.
91. The application was developed in conjunction with Paul Obradavitch, Director of Communications Systems for the San Onofre Nuclear Power Generating Station.
92. "Phone on the Range Helps Rancher Call Home: New System Puts World at Dell City's Feet," *El Paso Times,* December 17, 1990, pp. 1A–2A.
93. "Ultraphone 2000," product plan and related documents, internal publications of IMM Corporation.
94. Rural Radio Task Force, "Petition for Rulemaking to Establish Basic Exchange Telecommunications Radio Service," CC Docket No. 86-495, RM-5442, May 9, 1986, p. 14.
95. *Ibid.,* p. 16. See also the study prepared by the Public Utility Commission of the State of Oregon, "A Report on Oregon Households without Telephones," attached to the Comments on the Oregon PUC on the BETRS FCC Rulemaking, March 29, 1985.
96. Philip Parker and Andrew Roscoe, *The Market for BETR/Cellular/Fixed Rural Radio Telephone Service,* Economic & Management Consultants International, Arlington, Virginia, 1988.
97. From discussions with the staff of the Colorado State Public Utilities Commission, circa 1989.
98. See Scott Chesbro, "A New Era for Navajo Phone Service," *Telephony,* February 6, 1989; the estimate of total requirements is drawn from conversations with the serving carrier.
99. Rural Radio Task Force, "Petition for Rulemaking to Establish Basic Exchange Telecommunications Radio Service," CC Docket No. 86-495, RM-5442, May 9, 1986, p. 17.
100. The annual surveys plant and other telphone company statistics published in *Telephony* are probably the most readily available source.

101. Alberta Government Telephones, Technical Specification for AGT Rural Individual Line Service (ILS) Digital Loop Replacement Radio (DLRR) System, June 13, 1986; this specification was developed in connection with a 108,000-line, five-year single-party service upgrade program, in which radio was originally contemplated as the solution for 20 to 50% of the lines. Most of the Canadian provincial telephone authorities and telephone companies have developed similar ILS programs, and most have plans for radio as part of these programs.
102. *Ibid.*
103. See annual *Telephony* plant surveys.
104. United Telephone, "Technological Substitution Study for Rural Copper Loops," 1985.
105. Dan Allen, "Fixed Wireless Access," BNR Report 3N13, Network Applications, July 1988.
106. IMM Corporation, "Comments in the Matter of Basic Exchange Telecommunications Radio Service," CC Docket No. 86-495, March 30, 1987, pp. 14–18.
107. Southwestern Bell, "Forecast Model for Radio-in-the-Loop," 1988.
108. *Ibid.*, p. 6.
109. US West, "1990 Colorado Rural Facility Improvement Program," material provided by company.
110. See Hugh G. J. Aitken, *Syntony and Spark: The Origins of Radio,* Princeton, New Jersey: Princeton University Press, 1985 (1976), especially Chap. 5.
111. Susan J. Douglas, "Technological Innovation and Organizational Change: The Navy's Adoption of Radio, 1899–1919," in Merritt Roe Smith, ed., *Military Enterprise and Technological Change,* Cambridge, Massachusetts: The MIT Press, 1987, pp. 117–173. See also Hugh G. J. Aitken, *The Continuous Wave: Technology and American Radio, 1900–1932,* Princeton, New Jersey: Princeton University Press, 1985, especially Chap. 5.
112. G. Calhoun, *Digital Cellular Radio,* Boston: Artech House, 1988, Chap. 2.
113. *Ibid.*
114. *Ibid.*, pp. 27 ff.
115. *Ibid.*, Chaps. 2 and 3.
116. For comprehensive references on cellular radio, see G. Calhoun, *Digital Cellular Radio,* Boston: Artech House, 1988.
117. *Ibid.*, Chap. 3.
118. *Ibid.*, Chaps. 8, 9, and 10.
119. *Ibid.*, Chap. 4.
120. See the IS-41 standards on cellular intersystem operations: IS-41.1-A (Functional Overview), IS-41.2-A (Intersystem Handoff), IS-41.3-A (Automatic Roaming), IS-41.4-A (Operations, Administration and Maintenance), IS-41.5-A (Data Communications), all of which are available from the Electronic Industries Association, 2001 Pennsylvania Avenue N.W., Washington, D.C. 20006.
121. Federal Communications Commission, *Notice of Proposed Rule Making,* "In the Matter of Amendment of Parts 2 and 22 of the Commission's Rules to Permit Liberalization of Technology and Auxiliary Service Offerings in the Domestic Public Cellular Radio Telecommunications Service," Docket 87-390, Released October 15, 1987.
122. See note 6.
123. The GSM standard actually defines two intrasystem interfaces: the "A" interface, between the cellular switch and the base station controller, and the "A-bis" interface between the base station controller and the radio subsystems of the base station (referred to as the *base station transmitters*).
124. See the special issue on cellular radio of the *Bell System Technical Journal,* Vol. 58, No. 1, January 1979.
125. IS-54 Draft Standard, *op. cit.*
126. *Ibid.;* there is some informal assurance that the IS-54 coder will handle overdialed DTMF tones, but it is not specified in the standard; similarly, fax transmission is "handled" by the awkward step of handing off to an analog channel.

127. G. Calhoun, *Digital Cellular Radio,* Boston: Artech House, 1988, pp. 98–103 and associated references.
128. *Ibid.*
129. The architecture for cellular systems varies considerably from manufacturer to manufacturer. One of the most interesting decentralized designs in the Alcatel product based on the revolutionary distributed System 12 switch architecture.
130. Fleet Call Inc., "In the Matter of Fleet Call, Inc. for authority to assign SMR Licenses and Waiver of certain private radio service rules," petition filed with the Federal Communications Commission, April 5, 1990.
131. G. Calhoun, *Digital Cellular Radio,* Boston: Artech House, 1988, pp. 120 ff.
132. Federal Communications Commission, Notice of Proposed Rule Making, "In the Matter of Amendment of Parts 2 and 22 of the Commission's Rules to Permit Liberalization of Technology and Auxiliary Service Offerings in the Domestic Public Cellular Radio Telecommunications Service," Docket 87-390, Released October 15, 1987; also see the resulting Report and Order, adopted October 13, 1988 and released December 12, 1988.
133. British Department of Trade & Industry, "Phones on the Move," January 1989.
134. FCC, Docket 87-390, *op. cit.,* p. 2, paragraph 11.
135. George Calhoun, *Digital Cellular Radio,* Boston: Artech House, 1988, pp. 54 ff.
136. British Department of Trade & Industry, "Phones on the Move," January 1989.
137. David Hendon, "Phones on the Move—PCN's for the 1990s," paper presented at the Personal Communications Networks Conference, London, March 1990.
138. "Eric Forth Announces Operators Licensed to Run Personal Communications Networks (PCNs)," DTI Press Notice, December 11, 1989. Also reference to IBC Conference notes
139. Millicom, "PCN America" petition to the FCC, Rulemaking 7175, incorporated in Docket 90-314.
140. Federal Communications Commission, "In the Matter of Amendment of the Commission's Rules to Establish New Personal Communications Services," Docket 90-314, adopted June 14, 1990, released June 28, 1990.
141. Once again, the best short summary of this applications set is to be found in Donald C. Cox, "Portable Digital Radio Communications—An Approach to Tetherless Access," *IEEE Communications Magazine,* July 1989, pp. 30–40.
142. CT-2 deployment has been fairly widely reported in the trade press. There is also an extensive specification for the common air interface available.
143. Both France and Germany are planning the deployment of similar systems, as is Canada.
144. See Donald C. Cox, "Portable Digital Radio Communications—An Approach to Tetherless Access," *IEEE Communications Magazine,* July 1989, pp. 30–32, for a short but ample discussion of the cordless phone market.
145. *Ibid.,* p. 31.
146. See, for example, T. Hattori, A. Sasaki, and K. Momma, "Emerging Technology and Service Enhancement for Cordless Telephone Systems," *IEEE Communications Magazine,* January 1988, pp. 53–58.
147. See Bellcore, *Generic Framework Criteria for Universal Digital Personal Communications Systems (PCS),* FA-NWT-001013, Issue 2, December 1990; also Donald C. Cox, "Portable Digital Radio Communications—An Approach to Tetherless Access," *IEEE Communications Magazine,* July 1989, pp. 30–40.
148. BellSouth Services, Inc., "Generic Criteria for Wireless Access Business Systems," Issue 1, July 1990.
149. The most comprehensive program, in my opinion, has been advanced by Ericsson. See C. Buckingham, G.K. Wolterink, and D. Åkerberg, "A Business Cordless PABX Telephone System on 800 Mhz Based on the DECT Technology," *IEEE Communications Magazine,* January 1991, pp.

105–110. I am also indebted to Kas Kalba for providing information on the many wireless PBX and cordless office systems. For further information, see Kas Kalba, *Opportunities in Microcell Communications: International Status and Business Prospects,* 4th ed., Lincoln, Massachusetts: Kalba International, Inc., 1990.

150. British Department of Trade & Industry, "Phones on the Move," January 1989, pp. 3–4.
151. Peter Purton, "Radio over Fiber?" *Telephony,* November 1989.
152. NYNEX Enterprises, "Request for Information on Radio Loop Access," RFI No. 546-NSC-10B, October 12, 1990; see also NYNEX press release, "NYNEX to Test Wireless Loop Technologies," August 2, 1990, and associated filing with the FCC.
153. Richard M.K. Tam, "Personal Communications Architecture," paper presented at ComForum Conference on Worldwide Personal Communications, June 27–28, 1990, Dallas; see also the February 1991 special issue of *IEEE Communications Magazine* devoted to personal communications.
154. For example, J.J. McCarthy, "Originations and Terminations within the Personal Communications Environment," AT&T Bell Laboratories, 1990.
155. Paul R. Strauss, "On the Radio: Will LANs Cut the Cord?" *Data Communications,* September 1987, pp. 57–62; I am also indebted to Ray Simpson of O'Neill Communications for the following references: Frank J. Derfler, "LANs without Wires," *PC Magazine,* May 29, 1990, pp. 295–318; Tom Smith, "Analysts See Limited Use of Wireless LANs," *Network World,* November 5, 1990, pp. 23–25; Ira Brodsky, "Wireless Communications," *Business Communications Review,* May 1990, pp. 112–119; Sorin Davidovici, "On the Radio," *Byte,* June 1990, pp. 224a–228; and the table of spread spectrum applications included in Donald L. Schilling, Raymond L. Pickholtz and Laurence B. Milstein, "Spread Spectrum Goes Commercial," *IEEE Spectrum,* August 1990, pp. 40–45. See also Frank Owen, "Cordless Data Communications," in W.H.W. Tuttlebee, ed., *Cordless Telecommunications in Europe: The Evolution of Personal Communications,* Berlin: Springer-Verlag, 1990; and T.A. Freeburg, "Enabling Technologies for Wireless In-Building Network Communications—Four Technical Challenges, Four Solutions," *IEEE Communications Magazine,* April 1991, pp. 58–64.
156. Theodore S. Rappaport, "Indoor Radio Communications for Factories of the Future," *IEEE Communications Magazine,* May 1989, p.15.
157. Hisakuni Nogami & Masayoshi Wakao, "Premises Radio Systems in Japan," *IEEE Communications Magazine,* October 1987, pp. 49–55.
158. K.M. Sundara Murthy and Kenneth G. Gordon, "VSAT Networking Concepts and New Applications Development," *IEEE Communications Magazine,* May 1989, pp. 43–49; K. M. Sundara Murthy et al., "VSAT User Network Examples," *IEEE Communications Magazine,* May 1989, pp. 50–57; Leonard S. Golding et al., "VSATs: Expert Views on Future Trends," *IEEE Communications Magazine,* May 1989, pp. 458–64.
159. D. Chakraborty, "VSAT Communications Networks—An Overview," *IEEE Communications Magazine,* May 1988, pp. 10–24; also K. M. Sundara Murthy and Kenneth G. Gordon, "VSAT Networking Concepts and New Applications Development," *IEEE Communications Magazine,* May 1989, pp. 43–49.
160. Ranjana Sharma, "VSAT Network Economics: A Comparative Analysis," *IEEE Communications Magazine,* February 1989, pp. 31–35.
161. K.M. Sundara Murthy et al., "VSAT User Network Examples," *IEEE Communications Magazine,* May 1989, p. 52.
162. *Ibid.,* p. 52.
163. *Ibid.,* p. 53.
164. *Ibid.,* p. 53.
165. Ranjana Sharma, "VSAT Network Economics: A Comparative Analysis," *IEEE Communications Magazine,* February 1989, pp. 31–35.

Appendix 5A
A HISTORY OF THE USE OF RADIO IN THE REA TELEPHONE PROGRAM (1985)

Claude Buster
(Former Chief of Transmission Branch of the REA)

Since the earliest days of the Rural Electrification Administration (REA) telephone program, which has as its mandate to extend and improve telephone service to Americans residing in rural areas, efforts have been made to employ radio systems of various types to serve rural subscribers. The following situations are particularly well suited to radio applications: (1) where subscribers are located at considerable distances from the nearest end offices switching center, (2) where terrain considerations make conventional wireline construction very expensive or difficult to accomplish, or (3) where the subscriber density in an area is extremely low.

Currently, REA borrowers serve only 200 or 300 subscribers by fixed radio, whereas they have more than 10,000 mobile telephone subscribers. This information indicates that REA has been successful in encouraging the provision of mobile telephone service and not successful in encouraging the use of radio in fixed subscriber loop applications. One very basic cause of the lack of success in using radio in fixed subscriber loop applications has been the lack of available frequencies on a primary basis. This lack of success is in marked contrast to the extensive use of point-to-point microwave radio for trunking purposes in rural areas. The history set forth in this appendix describes past attempts to use radio and radio-based technology to provide fixed subscriber loops.

5A.1 EXISTING SERVICES AND FREQUENCIES

Some existing subscriber loop radio systems operate in the Rural Radio Service in the 150- and 450-MHz bands on a secondary basis. This means that a showing must be made that rural subscriber radio service will not adversely affect the availability or adequacy of service to mobile subscribers and that no harmful interference will be caused to stations in the Public Land Mobile Service. A rural exchange carrier applying for a license from the FCC must show the availability and adequacy of existing public communication facilities along the route, the type of each communication facility, and must provide a statement as to why it is impractical to use wireline facilities.

Other systems operate in the point-to-point microwave radio service (47 CFR, Part 21, Subpart I) in the 2- and 6-GHz bands. At one time, systems operated in the 890- to 940-MHz band for point-to-point service, but this band is no longer available to new systems for exchange carrier subscriber loop applications.

5A.2 SUMMARY OF USES OF RADIO SYSTEMS TO DATE

1. Early 1950s

REA purchased single channel VHF subscriber radio equipment from the General Electric Company and Motorola, Inc., and installed them in two rural telephone systems for evaluation purposes. The radio systems provided full duplex services to clusters of party-line subscribers. Although these systems performed satisfactorily from a technical standpoint, their maintenance costs were high due to the vacuum tube technology used at the time together with the high cost of electric power consumption at the remote subscriber terminals. Following the evaluation periods, the subscribers on these systems were connected to wire facilities, which were extended to their areas.

2. Mid-1950s

REA purchased a multi channel subscriber radio system operating in the 890- to 940-MHz band from the Budelman Radio Corporation for evaluation purposes. This radio system, also using vacuum tube technology, had high maintenance costs and high power consumption resulting in very limited use by exchange carriers.

Also in the mid-1950s REA provided some seed money as part of a development contract with the Raytheon Manufacturing Company for a two-channel subscriber microwave link in the 6-GHz band. Although the developers were requested to keep power consumption to a minimum at the remote terminals, the equipment still required a large number of vacuum tubes with relatively high power consumption. This factor, along with the high maintenance costs associated with vacuum tube technology, resulted in high operating costs. This system had limited use in rural subscriber applications.

3. 1957

REA provided some seed money and wrote performance specifications for the development of dial mobile radio systems and fixed subscriber versions of these systems to serve isolated subscriber locations. It was hoped that these efforts, intended to make mobile radio services available in rural areas served by unattended central offices, could be applied to fixed service as well as mobile uses. The Automatic Electric Company and Motorola, Inc., each provided equipment of the push-to-talk type.

The dial mobile systems proved highly successful for mobile service; the subscribers were found to be willing to tolerate the required high monthly cost and a grade of service lower than that normally associated with wire-based telecommunications services. However, fixed subscriber loop applications were not satisfactory because the channel was often shared with 20 or more mobile users and because the high monthly cost was not affordable.

4. Late 1950s

The General Electric Company entered the dial mobile radio market with its own radio equipment, together with signaling equipment made by the Secode Corporation. This system was also compatible with manual mobile radio systems such as those of Bell System companies, which used a mobile service operator. ITT-Kellogg and the AC Spark Plug Division of General Motors produced similar equipment in the same time frame. These systems were successfully applied to mobile use but were not satisfactory for fixed subscriber loop use for the reasons discussed above.

5. Early 1960s

The General Electric Company introduced mobile radios that permitted full duplex operation. This eliminated the push-to-talk method of operation and made usage similar to that of conventional wireline telephones. The mobile systems developed from 1957 through the early 1960s were used by many exchange carriers for mobile service in rural areas and a few fixed subscriber applications. The use of these systems to provide fixed service was very limited for the reasons discussed above.

6. 1964

The Secode Corporation introduced an improved mobile telephone system (IMTS) for use in the 150-MHz band. This system used Secode's own signaling equipment in conjunction with Westinghouse Air Brake Company (Wabco) mobile radios and

General Electric base station equipment. In the same general time frame, Motorola, GE, and others also introduced IMTS in the same frequency band for use by the Bell System and other companies. REA-financed telephone systems then began to use this type of equipment in lieu of the dial types mentioned above because IMTS had advantages such as full duplex capability, better transmission quality, capability of expanding to multichannel operation, and a unique numbering system that accommodated roaming traffic well. Since these systems could use several radio channels on a trunking basis, a higher grade of service could be offered than was possible with the earlier systems. A few fixed subscriber installations of this equipment were made in sparsely settled locations. However, the general lack of channels, the high monthly cost, and the secondary nature of fixed subscriber use generally made, and continues to make, the use of this equipment unattractive to both exchange carriers and their subscribers.

7. Early 1970s

Systcom, a Canadian manufacturer, introduced a relatively low cost, partially transistorized, single-channel subscriber radio system operating in the 150-MHz band. A number of these systems were satisfactorily installed in the northern part of the United States. The system serves fixed subscribers on an island in Lake Superior. However, the lack of available channels and the secondary basis of frequency use precluded the widespread application of this equipment.

8. 1973

The Hallicrafters Corporation provided a single-channel radio link to serve fixed subscribers in an isolated area in Georgia. The system, operating in the 150-MHz band and partially transistorized, was not very successful due to high maintenance costs.

Also in 1973, the Lynch Corporation's D960 subscriber line concentrator coupled with an Avantek microwave system was used to connect a cluster of fixed subscribers to a serving central office in the Hilton Head, South Carolina, area. This system, operating in the point-to-point service at 2 GHz, is one of the rare examples of a fixed subscriber radio loop system in operation.

9. Late 1970s

The radio equipment of several U.S. manufacturers was linked together to provide pay station services from isolated Puerto Rican villages to a serving central office. This service is provided where these villages do not as yet have any other form of telephone service. This service was made possible only because Puerto Rico's geographic location allowed sufficient channel capacity to be made available.

10. Late 1970s

Motorola provided a multichannel rural radio system in the 150-MHz band based on IMTS-type trunking principles. Although its design was all solid state, maintenance costs were high. In at least one installation in a sparsely settled area, a four-channel system was subject to interference from IMTS base stations located at considerable distances from this particular installation. Due to problems of this type, high maintenance costs, and a lack of available channels to handle the traffic (two channels were occupied on a call between any two radio subscribers), the exchange carrier discontinued this service in 1984 and connected the subscribers to the central office by means of open wire plant. In another installation, similar interference problems were solved at considerable expense to the exchange carrier by shifting to tertiary frequencies.

11. 1983–84

REA and the Rural Telephone Coalition explored the use of communication satellites for providing fixed station subscriber services and interoffice trunking. At the time of the study, the fixed station application was not attractive due to very high cost projections for satellite earth stations at rural residences. After considerable study, the trunking application was also rejected in early 1985 due to high costs.

12. 1984

Two REA-financed exchange carriers, serving very sparsely populated areas of Texas, have applied to the FCC for authority to operate point-to-multipoint multichannel subscriber loop microwave systems manufactured by SR Telecom of Canada. This system is successfully operating in many other parts of the world. The applications included a waiver request to operate on frequencies allocated to mass media uses since suitable common carrier frequencies were not available in that geographic area. To date, this petition has not been granted, but the matter is being pursued with the FCC. This radio equipment and the frequencies are ideal for subscriber loop applications in this geographic area because the costs of providing service would be reduced by about 30% from the costs of using conventional wire and cable facilities.

Also in 1984, Westcom, Inc., petitioned the FCC to amend its rules to authorize wireline cellular carriers in areas other than metropolitan statistical areas (MSAs) to provide cellular service to fixed and mobile customers. In May 1985, the FCC ruled that wireline and nonwireline cellular licensees may already provide fixed service incidental to their mobile services. Damage to, interference with, and other harmful effects on licensed cellular service may not occur. According to the FCC, incidental service is permitted within all areas where cellular mobile service is currently authorized.

13. 1985

A laboratory demonstration of the Ultraphone point-to-multipoint subscriber radio system developed by the International Mobile Machines Corporation was successful. By means of advanced digital modulation techniques, four voice channels are derived within the 25-kHz spectrum space allocated to one radio channel in the 450-MHz band allocated to the Rural Radio Service. This system preserves speech privacy, in contrast to cellular types, since its modulation system makes monitoring or eavesdropping extremely difficult to achieve. An REA field trial of this spectrum-efficient system is planned for early 1986 in a rural area in Kansas. In this trial, the system will operate in the 450-MHz band. If the field trial is successful and the costs are acceptable, this equipment may become part of the solution toward providing high-quality, lower cost telecommunications services where conventional wire and cable facilities are not feasible.

Appendix 5B
EXCERPTS FROM A REPORT BY THE MOUNTAIN STATES TELEPHONE AND TELEGRAPH COMPANY, NORTHWESTERN BELL TELEPHONE COMPANY, AND PACIFIC NORTHWEST BELL TELEPHONE COMPANY

Presented to the Federal Communications Commission on August 1, 1985

5B.1 INTRODUCTION

We represent Mountain Bell, Northwestern Bell and Pacific Northwest Bell—the three operating companies of US West.

We are seeking your support of a concept—that we may use any technology available to provide service.

Your response to the Westcom petition affirmed that cellular technology may be used to provide fixed service on an incidental basis. Both wireline and non-wireline companies may provide this service within established rules, i.e., the service must not be degraded, applications must go through regular channels.

Our US West cellular provider—New Vector—and the other cellular providers are addressing a particular marketplace—major metropolitan areas. The service they provide is truly mobile. Their objective is attainment of market share in a new and profitable venture.

We, on the other hand, are focusing on expense reduction. This is a key difference in perspective. We are not expanding our product line. We are asking for the option to use whatever technology is available to provide basic telephone service.

US West territory covers 14 states and 42% of the contiguous U.S. Our name is more than an image. We are still pioneering. We are bringing basic telephone service to places where there is none. And that is an expensive proposition.

Today, some cross-subsidy of marketplace and customer group exist. We are moving to an environment where the cost causer pays. We are moving toward local measured service.

All of us are familiar with the image of the lone rural customer who is left to pick up the tab after other customers depart. We don't want anything resembling that scenario to occur. We are planning to keep costs down, to keep prices down for all our customers. We want the freedom to select from the array of technology available to achieve that.

Today we will present you with some facts. We will talk about large areas with few customers. We will talk about central office and distribution plant that is in a hostile environment. We will talk about customers who must pay large amounts for construction to their premise. We will discuss customers who have decided not to install service because construction costs were too high. We will share our thoughts on substitute technology and on regulation.

We ask that you consider this information as technology discussions continue. We want to be able to provision with any technology available.

We are willing to compete.

We support the growth and development of cellular technology, of all technology in our industry.

We look positively toward the future, as new technologies lead to lower costs and new applications.

5B.2 CURRENT SITUATION

Providing plain old telephone service (POTS) to remote rural customers has always been a problem for operating companies. Rural customers are difficult to serve with conventional technology either because of distance from the CO or because geographic barriers such as water or mountains come between the customer and the CO. Rural customers are also often distant from each other. Wire can be brought to these customers, but installation and maintenance costs are high.

In the next few minutes, we will give you an overall profile of the US West rural market. We will then focus on the 10 county area known as the "Sand Hills" of Nebraska, a single central office in Harney County Oregon, and 4 individual customer requests for service in Northern Minnesota—and costs to serve will be shown in each case.

This map shows the territory served by US West. US West has 42% of the U.S. geography—but only 5 cities in the top 30 MSAs, 41 in the top 300.

How rural are we?

	Total Central Offices	Offices Under 1000 Lines	Offices Under 500 Lines
PNB	250	80	43
NWB	656	268	135
MTN	795	319	185

Let's narrow the view to the Sand Hills of Nebraska. Here 10 counties cover 12,346 square miles. This is comparable to New Jersey plus Connecticut in size. 62% of the households—3,274—are described as rural in census information. An average CO covers 686 square miles. An average customer is 7.4 miles from the CO. A 7.4 mile loop costs Northwestern Bell $86 per month. Contrast this with the average outstate loop cost of $18 per month. We must explore ways to bring the costs down in the Sand Hills.

Distance and nature both play a part in the case of Harney County. This county is located in Southeastern Oregon, and covers a 5,300 square mile area about the size of Connecticut.

PNB provides service for some 500 telephone lines from its central office in Crane, Oregon, located in the north central area of Harney County. Although, most of the users live in and around Crane, 165 customers are spread throughout the remainder of the county.

Local residents, and the PNB outside plant network, are exposed to a number of severe environments/conditions during the year. Winters bring freezing fog and high winds which cause cable and carrier failures, and radio outages. The extreme heat of Eastern Oregon summers create failures to an analog wire carrier system that is rapidly deteriorating and will soon have to be replaced. Water levels in the lakes of this region have been rising during recent years and threaten to cover existing cables in some areas. The central office in Crane is also being threatened and may be moved to another site that is on higher ground if flooding continues.

The present outside plant facility arrangement includes a mixture of analog carrier (mostly Kellogg K31), radio (450 MHz), and open wire. The radios extend up to 77 miles from the central office and the open wire carrier systems extend up to 62 miles from the radio terminals. The longest loop makes use of a total of 127 miles of radio and wire. A total of six carrier systems handle between 11 and 39 subscribers each.

As demonstrated by the description of this network, these customers are spread extremely thin over a vast area. These 165 customers are almost exclusively ranchers. It is not uncommon for these ranchers to travel as much as 10 to 15 miles from their telephones and still be on their ranch properties. In addition, there are areas in Harney County that offer no telephone service at all.

In 1984, PNB spent more than $65,000 on outside plant maintenance and repair for the 165 customers located outside of Crane. This cost is nearly $400 per line for maintenance alone. Outside plant personnel have estimated an expenditure of approximately $2,300,000 to replace 225 miles of cable and carrier. This replacement should essentially bring this network up to PNB standards for transmission quality and service reliability. This expenditure, when spread over 165 customers, equates to $14,000 per line.

At an expenditure of $14,000 per line, it makes economic sense for PNB to shop around for an alternate means of serving these customers. Certainly if customers were asked to spend $14,000 for access lines, they would seek some other means to gain network access, i.e., bypass. But with access line charges below $20 per month, there is no competition for phone service in Harney County.

Because of the unique problems in Harney County, PNB should be allowed to seek alternate technologies for serving these customers. These alternatives should include all types of radio applications where PNB can demonstrate economic benefits to PNB or service improvements to the customer.

We have presented the problem from the operating company perspective. In many cases, the customer also faces high costs. From recent requests for service in Northern Minnesota, we have selected four typical situations.

Two Grand Rapids customers required construction to gain services. The first construction was 2,000 feet, and the cost to the customer was $556. The second job was 1,085 feet, and the customer paid $200. The customers were willing to pay these charges in addition to normal installation and monthly rates.

Another Grand Rapids customer was quoted $2,540 for construction. That customer is still weighing the cost against the benefits of being on the network.

And then there is the customer who is located 163 KF from our Coleraine CO. The customer is in a nobody's-land. An Independent Company in the area is not interested in providing service to this customer. We quoted charges of $6,980. The customer decided against service.

Is there a way to keep rates down for those already on the network, can we find new ways to serve those customers who find construction too costly?

We believe the solution may be jeopardized if we are precluded from using a technology because it is perceived to be a product.

5B.3 TECHNOLOGICAL SOLUTIONS

We will now review radio technology briefly, and then share the results of some exploration we have done on rural applications.

Radio systems can be divided into two types:

1. Lumped Radio System: A lumped radio system in which each radio link provides a number of voice frequency (VF) channels. The number of channels can, of course, be significantly smaller than the number of customers to be

served if switching is provided. A conventional wire distribution system is required from the location of the lumped radio remote terminal to the customer main station. A lumped radio system is usually employed when a number of customers are close to one another but are remote from a CO. Figure 1 illustrates a typical scenario in which a lumped radio system might be employed.

2. Distributed Radio System: A distributed radio system is one in which each radio link provides a single VF channel. Each customer is provided with a radio transceiver, and the need for any wire distribution plant is obviated. A distributed radio system appears to be an obvious choice when customers are both remote from one another and from the CO. Figure 2 illustrates a typical scenario in which a distributed radio system might be employed. Note that a radio channel need not be dedicated to a single customer but can be shared by as many customers as it will support, provided that appropriate logic is provided for each customer.

Although the primary focus of this paper will be on distributed radio technology, it is important to point out that radio (in the form of lumped radio systems) is already playing a major role in information distribution for the Bell Operating Companies. Distributed radio systems (including cellular systems), on the other hand, have only recently been investigated as a serious alternative for basic exchange service (in more than just a few isolated examples).

Distributed radio technology has progressed rapidly, making radio more attractive economically than in the past. Much of this progress is directly attributable to the development of cellular radio.

It is our position that the telephone companies should be able to utilize any of the new distributed radio technologies (including cellular, narrowband, amplitude companded single side band, and digital radio technologies) as the circumstances dictate. In this way the FCC can maintain the benefits of current regulatory policy and also avoid significant degradation of the quality, availability and affordability of service in high cost areas.

Furthermore, the resulting fixed service should not be bound by the restrictions, both technical and procedural, currently on cellular technology. The host of previously mentioned new technologies offer the opportunity to apply the most efficient system individual circumstances dictate; thus maximizing the benefits of spectrum efficiency and the resulting economies that will be realized.

The need to provide landline telephone equivalent service should not be hampered by the technical considerations of limited power, antenna height, and cell handoff. The advantages to be gained by technical flexibility in providing low cost, spectrum efficient radiotelephone systems for non-urban subscribers should not be sacrificed to maintain compatibility with a mobile system targeted primarily to the urban business community.

The initial allocation of cellular frequencies was made to insure sufficient capacity for mobile telephone service in the major metropolitan areas. Most experts in the industry agree that the demand for mobile telephone service in the geographic area of the United States outside the major metropolitan areas will never require the use of more than a fraction of the frequencies currently allocated to cellular.

The additional allocation to cellular discussed in the Notice of Proposed Rulemaking in Gen Docket No. 84-1231, adopted November 21, 1984, provides an opportunity for the FCC to set aside a discrete block of frequencies, for use outside the high capacity urban areas, for fixed basic telephone service without impacting legitimate mobile telecommunications requirements. Such an allocation to fixed radio telephone service would allow market forces to dictate the pace and scope of use of the radio spectrum to replace costly outside plant.

Those same forces would determine the appropriate use of radio channels based on local demand for mobile service and the degree of cost-effectiveness in using radio to provide basic fixed location-telephone service.

We would like to share the results of two studies which illustrate the cost effectiveness of radio.

A study of Harney County conducted by Northern Telecom estimated the cost of a fixed cellular system at $1.3 million plus installation costs. The system would require four cell sites and a switch controller located in the local DMS-10 switch.

Although the cost of this system is high (approximately $10,000 per user) it is lower than estimates to rebuild PNB's existing network with copper and wire. (Approximately $14,000 per user.) One of the cell sites was located in an area currently not receiving telephone service. Other sites have the potential of reaching locations outside PNB's service area.

As a second example, Mountain Bell has outlined the cost to serve customers in Glendo, Wyoming under three scenarios.

1. Network Distribution Services estimate the cost to be $190,000 to $230,000 using a combination of microwave and buried cable.
2. Industry experts estimate the cost to be $160,000 to $210,000 using cellular technology. However, cellular's limited range would not allow all customers to be served.
3. Our estimates for the cost of a digital radio system is $80,000-$90,000.

Based upon our cost projections, savings of $110,000-$140,000 could be realized by using digital radio rather than traditional technology.

5B.4 CONCLUSION

We view cellular radio or other radio systems as a technology, not a product. Therefore, we can see no reason why this technology need be provided under a separate subsidiary. A separate subsidiary was not required when we began substituting glass fiber for copper cable. Radio technology is, in fact, already being used to reduce the costs to serve in US West today.

We consider the use of such new technology as part of, and indeed, a responsibility of our current franchises to provide local service. However, if this market is to be opened up to competition in the form of unregulated non-wireline companies, we feel there should be some changes in the "ground rules." Should the regulated wireline company be required to be the Last Resort Provider to customers in remote locations while an unregulated competitor could select its customers based on profitability or on the potential for future enhanced radio services?

We appreciate the time you have given us today, the opportunity to talk about the Sand Hills, Harney County, and Grand Rapids. We ask that you factor this market into your decisions. We ask that we may continue to use whatever technology is best operationally and financially.

Thank you.

Appendix 5C
A PIONEERING WIRELESS TELEPHONE PROJECT

Presentation to the United States Telephone Association by Art Biggs, President of S & A Telephone Co. (Kansas), July 5, 1985

Thank you for that introduction. I deeply appreciate this opportunity to talk to you. Part of my appreciation for this opportunity is derived from the fact that speaking to national organizations hasn't been my thing in the past (and probably won't be in the future). This is my opportunity to have the undivided attention of a very distinguished nationwide audience of grass roots telephone people.

Speaking of grass roots telephone people, just for my own curiosity I'd like to see a show of hands of anyone in this audience who got their start by living in a house with a central office where it was a family project to provide the operator services. Are there any survivors here of that era other than me? (The rest of you evidently started from a higher plateau than we did.)

The other part of my appreciation is that I very much enjoy talking about the fact that I will very soon be installing the world's first wireline quality wireless local loop system for the purpose of upgrading 4-party service to private service in Allen, Kansas. My involvement in dreaming about the wireless local loop concept has been almost a lifetime thing. It has too frequently seemed that the need for service was where the cables were not.

Each time I have had to build a high cost infeasible wireline extension, I have wished for wireless telephone.

Each time I have had to dig in rock, I have wished for wireless telephone.

Each time I have experienced endless and senseless cable damage inflicted by hordes of disoriented and uncaring rural water contractors, I have wished for wireless telephone.

And each time I have struggled with the need to install a buried cable on a right-of-way that is already too crowded with other buried utilities, I have wished for wireless telephone.

Our company is very small. It's a mom and pop type telephone company with only about 770 stations. My dad had it before me. He started buying it from a bankrupt Chicago-based company in 1938. I was a preschooler then, and little did I then suspect all the postholes that I was to dig before I reached adulthood. (In fact, in hindsight it sometimes appears a wonder that I even reached adulthood.)

We have two exchanges: Scranton, Kansas, and Allen, Kansas, with central offices separated by about 30 miles. The overall density is about 3.6 subscribers/mile of cable. And we serve an area of about 170 square miles with about 215 miles of cables. The plant includes two automatic electric step offices and outside cables that are about 98% buried. Total investment is about $1.5 million ($1,900/customer) and the depreciated value is about $1,000,000 ($1,200/customer).

We have about 80% 1-party service and 20% 4-party service. With $1.5 million invested at this time, it will take on the order of another $900,000 to bury cables to upgrade the remaining 20%. This would raise our investment to the level of $3,200/customer. The Scranton exchange has good subscriber density and was upgraded to private lines in 1977 by burying grease-filled cables. At that same time we tried to upgrade the Allen exchange but were stopped by our 4-party subscribers via the Kansas Corporation Commission because they didn't want to see the rates increase from $5.75 to $8.75. In 1979, we tried again but with the same result.

In 1980, we got our first experience with wireless telephony when we installed a digital microwave for toll service at Allen. That wireless experience has been such a cost-effective and reliable one that it increased my longing to apply similar principles to the local loop. But that opportunity never seemed to be available until I began to read about cellular mobile telephone and its wireline-like quality of service. About the same time that cellular began to gleam in the eyes of telcos nationwide, I was making a third application to the REA for a third try to bury the cables for upgrading the Allen exchange to 1-party service. While awaiting REA's response to the loan application, I was also pestering every cellular supplier in the business to see if there was any practical possibility of using cellular mobile radio as a sort of wireless station carrier. What I determined was that it could be done, and that some of the companies already had it on the drawing boards, and considered it a practical way to go (in high cost areas).

In studying for cost comparison, I discovered that the Allen exchange had 49 rural 4-party loops and 142 4-party customers. This meant that I was about to spend something on the order of $890,000 for cables alone, to pick 93 customers off of the existing buried loops to make all private lines. Thought of in these very rough terms, the cost comes to about $9,600/house. Since this figure includes in-town burying of all aerial plant at a cost of approximately $158,000, a closer estimate appears to be on the order of $8,500/house for the rural.

Frequently, at the mention of this cost, telephone people assume that I must have enormously long lines out there, but remember, when I say $8,500/house, I'm only talking about those 93 targeted customers that we are going to lift from those

46 4-party lines. We are going to leave our buried air core cables in service and continue to use them in conjunction with the wireless station carrier. By leaving this cable in, we thereby cancel the abandonment of $130,000 worth of cables. Our maximum air line distance to the furthest customer is less than ten miles.

Granted, putting the cables in would buy some things that wireless won't. Like a 25% overbuild in pairs, but those pairs aren't worth anything until you use them. Ultimate use of such overbuild is a pretty iffy thing in many rural areas (and especially in my rural area).

At this stage, I thought cellular radio at about $5,000/station looked pretty hopeful. But then two significant things happened: 1) the FCC changed the time of cellular frequency availability delaying it in excess of two years. 2) Claude Buster, Chief of the REA Transmissions Engineer, in Washington, D.C., tuned his home TV slowly through UHF channels 80 through 83, and found that he could hear cellular conversations with poor but decipherable quality. This pretty much eliminated any likelihood of my using cellular for this purpose.

At the same time, however, another even better technology has been developed. It is called "Ultraphone," and it is soon to be manufactured by International Mobile Machines in Philadelphia, Pennsylvania. I understand you received a preview of this system about two years ago. Well, here is a brief update. Ultraphone is an omnidirectional, Time Division Multiple Access, total digital wireless station carrier system, which in its initial thrust, will interface step equipment by simply connecting the outgoing line pairs to it. Ultraphone will then radiate typically a 20-mile radius, if needed, to a receiver at each customer's location which would in turn connect to the existing telephone housewiring and existing station apparatus. Wireline quality service would be provided by this means. The designers believe that up to 40-miles radius coverage can be achieved with special provisions.

The wireless local loop has the following characteristics and possibilities:

1. Total Digital Operation—This eliminates noise problems in transmission and also eliminates the requirements for duplexers. . . .
2. Wireless Privacy—The digital aspect of Ultraphone enables it to provide privacy of communications with a quality that rivals the privacy of digital dish-to-dish microwave that many of you are confidently using in your toll networks today.
3. Spectrum Efficiency—the system makes four talking paths out of each of today's IMTS 25 kHz channels. Each talking path works similar to the linefinders you are now acquainted with, in that when the customer goes off hook, a random talking path is seized.
4. Installation—At the central office, installation will consist of the rolling of two refrigerator-sized consoles into available floor space and connection of dial-tone pairs to it. A simple antenna will be installed on my existing 100-foot microwave tower, which is immediately adjacent to the central office, and coax will connect the antenna to the wireless loop system. The customer installation

will consist of the mounting of a box, about the size of two loaves of bread, which will be connected to an outside demarcation point, the existing telephone housewiring, a remote small 120 VAC operated 12 VDC source, a 12 VDC emergency backup battery, and a coax to a small antenna. Additionally, a backup emergency 12 VDC jack will be included for connection to the car cigarette lighter in case of very long power outages. Surge protection will be included at all Ultraphone exposures and also at the house power entrance.

5. Cost—We believe that the first fifty units (both ends) can be installed for under \$3,000/station. As compared to the cable cost of \$8,500/station. Overall, if we buy 100 Ultraphones, the cost would be about \$300,000 as opposed to about \$900,000 for cables to do the same job. A "no frills" central office end only could be installed for about \$50,000.
6. Mobility—Though it's not our immediate intention, the initial configuration of Ultraphone could be mounted in vehicles and used for mobile phones. It is anticipated that, after very large-scale integration produces smaller and cheaper Ultraphones, it could become an awesome competitor to cellular both in the 450 and 800 MHz bands if the Powers That Be permit it to happen.

We have asked the REA to permit us to do an REA field trial of 30 units for a year to determine the practicality of Ultraphone in this environment. Permission has been granted, and the installations will take place in the Spring of 1986. The initial 30-unit system will be installed in addition to the 4-party service. The 4-party service will not be disconnected until the wireless loop system has proven itself for several months. It has been interesting to experience the customer reaction to the talk about the coming of the wireless loop system. There is very high interest, and as near as I can detect, they are all for it. There is certainly no shortage of guinea pigs willing to have holes drilled in their houses for the initial test.

I believe I am about to demonstrate that the initial design of the wireless loop system is a very good and cost-effective alternative to burying cables where distances greater than four miles are involved. As the system grows cheaper and better, its practical distance as a cable alternative will grow shorter than four miles.

I have tried to leave some time for questions, but I have to warn you. This wireless local loop has been sort of a family thing with us. My 16 year old daughter Marcia gave a talk on it at school which ended as follows:

> If you have any questions don't ask me, because this is all I know about it, but if you ask my dad, be prepared to listen to two or three hours of information—.

So, being duly warned, we're open for questions.

Chapter 6
TECHNOLOGY FOR WIRELESS ACCESS

6.0 THE END OF FM

Wireless access is based on a series of breakthroughs in radio transmission technology. Until recently the only radio transmission methods in common use (at least in applications relevant to local exchange access, including cellular mobile radio) were based on analog modulation of the amplitude and frequency of the radio signal (AM and FM). These technologies have a number of merits, including their relative simplicity and maturity, especially for equipment manufacturers. But the field of analog radio is and has been a technological backwater for a long time. Indeed, FM radio (the ubiquitous transmission technique employed in mobile radio systems even today) has not changed fundamentally in several decades [1]. "There is very little incentive for changing the present FM pattern of systems design," wrote Daniel Noble (then executive vice-president for Motorola) in 1962—when FM was already 25 years old [2]. The same words could have been written in 1972, or even 1982. For at least 40 years, FM has been the alpha and omega of mobile radio.

This long FM epoch has been in some respects a comfortable period for the mobile radio industry, marked by:

1. *Stable technology.* The dominance of a single technical architecture allowed equipment suppliers to concentrate on improving manufacturing efficiencies and product packaging in what became a commodity-like market, much like the television industry.
2. *Straightforward standards.* Where industry-wide standards were seen as necessary (e.g., for analog cellular), it was easy for technical committees and government agencies to agree on the air interface since there were really no alternatives to FM.
3. *Few problems with IPR (i.e., intellectual property rights, principally patents).* Since FM was based on nonproprietary technology (the inventor of FM, Edwin Armstrong, having donated his invention to the government in connection with the war effort during World War II [3]), there were no significant

patent barriers (or other intellectual property rights issues) to the development of a competitive industry.

Nevertheless, analog FM is not suited to true wireless access applications, for a number of reasons including:

1. Its relatively inefficient use of the radio spectrum
2. Its high infrastructure costs, especially for telephony-type traffic
3. Its inherent lack of privacy
4. Its unsuitability for many of the new digital services.

I discussed the advantages and deficiencies of FM systems at great length in my previous book [4]. Suffice it to say here that as long as FM was essentially the only practical option for the radio link, it was difficult to implement true wireless access on a large scale. Analog cellular radio represented a brute-force attempt to build the first really large systems using FM—and the cellular industry has encountered such serious technical constraints in the very early stages of operation that it is being forced to abandon the FM standard after only a few years of operation, and to undertake a painful transition to an entirely new air interface [5].

Fortunately, today we are in the midst of a real technological revolution in radio, the first in decades. The simple, inefficient analog techniques of yesterday are being superseded by complex and powerful digital radio technologies, in which individual circuits are built up out of layers of extremely sophisticated software (indeed, some of the highest octane code in any commercial application anywhere) and then combined into ingenious multiple-access systems to achieve great gains in cost, quality, and capacity. Moreover, the technological ferment breeds a new kind of industry. The new techniques are bewilderingly diverse. This gives rise to more difficult choices, calling for a different management style. Companies find themselves facing a constant threat of obsolescence and competitive technological pressure more typical in, say, the computer industry than the tradition-bound world of mobile radio.

In short, the strategic framework has shifted:

1. Technology is no longer stable. The FM generation may have lasted 40 years, but some of the new signal processing techniques now emerging may have technological lifetimes of as little as 3 to 5 years (for a voice-coding algorithm, for example) before they are superseded by something better.
2. No single, all-encompassing standard will be achievable. There will be competing technologies in different applications niches, and even within a niche there may be several formats in uneasy coexistence; the era of monolithic standards is over.
3. IPR will play an increasingly important role. As competitive strategies shift from the commodity virtues of low-cost manufacturing and innovative packaging to the high R&D environment where proprietary technology will be

much more important, patents and control of intellectual property will loom larger for the industry than at any time since the creation of the RCA patent pool at the end of World War I.

Pandora's tool box is open again, and it is becoming possible to envision true wireless access systems, on a large scale, which can equal or surpass the performance of traditional wireline access networks.

In this chapter we review some of the basic concepts underlying this radio revolution. It is largely a revolution in digital radio—a field that is still in many respects an awkward hybrid. Indeed, the shotgun marriage of digital engineering and radio engineering is one of the more interesting liaisons of the contemporary technological scene. It involves an intriguing clash of cultures. The Digital Tribe, now swarming in from the computer industry (broadly construed to include telephone switching systems), consists of the (mostly young) software hotshots, who are aggressively exploring the implications of viewing the world in terms of black and white, one and zero. They are adept at reducing a typically messy reality, full of impossible nuances, to a clean numerical simulacrum—and then manipulating the simulacrum with the power of mathematical and logical precision. They stand clearly on the side of the Artificial, and may smooth away many "details" of the Real World, but they are willing to stand on the results of their methods.

The Radio Tribe, on quite the other hand, is a coven of wizards, steeped in the black art of teasing the Ether. Stereotypically, they are usually older, crustier individuals, addicted to bizarre nocturnal pleasures (e.g., ham radio), a little out of step with the computer revolution. But they know from experience that in the real universe everything radiates, and everything interacts, and that nothing real is ever, at bottom, truly "digital." I once sat through an animated and logger-headed discussion between a group of radio engineers and a group of software engineers, who were trying earnestly to talk about the same subject—in this case it was a design problem that appeared to be producing unfavorable performance in a radio system—in two mutually unintelligible languages. For the software people, the problem was to be stated in terms of "bit-errors" and diagnosed as a matter of the relationships between various "entities" (i.e., abstract software objects), coding "levels" (e.g., "Are we talking about the data link level or the signal transfer level?"), and protocols (i.e., the arbitrary formats to be agreed on by the all-powerful systems architects). The software people were convinced it had to be a logical trap, somewhere in the code, that was not functioning properly. The radio people were equally convinced that they were seeing some subtle, but real, interaction between the system and the external environment. At a certain point, a frustrated radio engineer got up and walked over to the radio transmitter rack and passed his hand through the space in the shelf between two of the circuit boards. The bit-error rates went gently up and down, as though the system were purring in response to his invisible electromagnetic stroking. As it turned out, someone had—in the interests of neatness (often a fatal flaw)—laced together the coaxial cable leading to the antenna and the power cable for a short stretch. Score one for the old guys. And although I myself owe greater allegiance to the digital tribe, I had to admit that there was a certain mute eloquence in the demonstration of the ultimate authority of the real world.

In this chapter I shall not attempt a thorough technical primer. There are many books available that go into more than adequate depth on radio systems design [6]. There are also excellent texts on the principles of digital communications in general, many of which cover radio quite well [7]. Instead, in the applications-oriented spirit of this book, I shall limit this chapter to an overview of what I believe are some of the most important concepts underlying the digital radio revolution.

This chapter will not address the technology embodied in the switching and network infrastructure, which stands on the other side of the radio interface to the subscriber. This is not because the technology on the network side is any less impressive or important; however, once again I believe there are adequate treatments of the evolution of switching and intelligent network technology [8]. Also, as important as these developments are, and as significant as they are in determining the requirements that any access technology must fulfill, they are not the principal enablers of *wireless* access *per se*.

Section 6.1 attempts to arm the nontechnical reader with a handful of very basic concepts of digital radio. Section 6.2 addresses the broad technical requirements for wireless access (especially as these transcend the objectives normally set for today's radio systems, such as analog cellular, that fall somewhat short of true wireless access). In Section 6.3, we survey the technologies that are brought together to create a digital radio-telephone circuit, while in Section 6.4, we examine some of the ingenious systems architectures that are used, or may be used, in wireless access applications.

6.1 A PRIMER ON DIGITAL RADIO

6.1.1 The Vocabulary of Radio

Radio is a form of electromagnetic energy, propagating through space at the speed of light, which may be visualized as a sine wave (see Figure 6.1). For our purposes, the radio wave has three important characteristics:

1. *Amplitude:* The magnitude or height of the crests and troughs of the sine wave.
2. *Frequency:* The number of crests that occur in one second; the basic measure of frequency is the *hertz,* which is defined as one full cycle, one crest-to-crest event. Most radio frequencies are measured in thousands or millions of cycles per second, kilohertz (kHz), or megahertz (MHz), respectively. (*Wavelength* is the inverse of frequency—the distance between two crests; it is normally measured in terms of meters, millimeters, and so forth.)

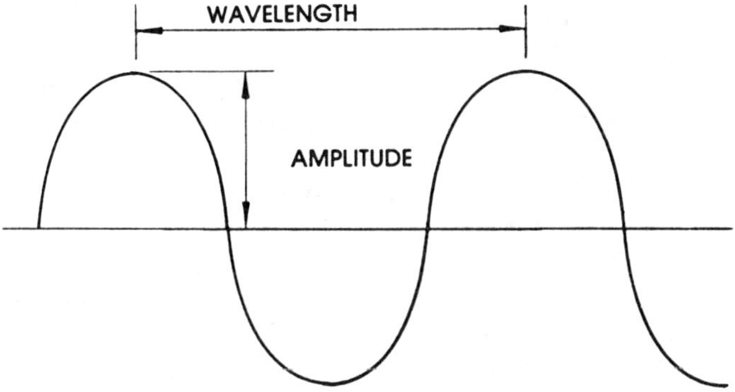

Figure 6.1 Idealized radio wave. (From G. Calhoun, *Digital Cellular Radio*, Artech House, Boston, p. 203.)

3. *Phase:* The angle of inflection of the wave at a precise moment in time. It is normally measured in terms of degrees of arc.

As we shall see, all three of these parameters may be used to carry information, such as the speech signal of a telephone call.

Most radio transmissions begin with a continuous wave of a fixed frequency and amplitude, called the *carrier*. The signal is then superimposed on this carrier by one of several methods described below. The frequency of this carrier is usually stated as so many MHz (in the range of interest here). For example, the carrier frequencies for North American cellular radio are in the 800-MHz region.

The carrier frequency is very important because different carrier frequencies interact in different ways with the physical environment; for example, the propagation characteristics of the radio transmission are determined chiefly by the carrier frequency. Some frequencies provide only line-of-sight propagation; others can bounce or reflect off of the ionosphere to propagate thousands of miles over the horizon. Some frequencies tend to be absorbed by foliage or atmospheric moisture while others will penetrate without much loss (see Section 6.1.3).

Information, such as a voice signal, is impressed on the carrier wave by varying either the amplitude, frequency, or phase (or some combination of these). This process is called *modulation* (this is standard radio terminology, although the term *modulation* is used in several slightly different ways by communications engineers). The variations in amplitude, frequency, and so forth may be continuous, in which case they are called *analog* modulation, or the variations may take the form of discrete changes, in steps, levels, or pulses, which is known as *digital* modulation. The two most familiar forms of modulation are (analog) amplitude modulation (AM) and (analog) frequency modulation (FM). But today several dozen different forms of modulation (mostly digital) exist that are under consideration for various wireless access applications. We will discuss the different forms of modulation in Section 6.3.

For most, but not all, of the applications discussed here, two carrier frequencies are actually used for each radiotelephone circuit: one for transmitting from the subscriber terminal to the base station, and one for transmitting from the base station to the subscriber terminal. For example, the BETRS allocation in the United States (see Section 5.3) uses carrier frequencies near 454 MHz for transmissions by the base station and 459 MHz for transmissions by the subscriber terminals. It is important to remember that in many of these applications a single circuit consists of transmissions on two actual frequencies (see Figure 6.2). The use of two frequencies is one of the easiest ways to implement *duplex* operation, which means that both parties can speak at the same time, in a natural manner. Although many of the older radio communications are *simplex,* or "push-to-talk," where only one party can transmit at any one moment, all of the systems discussed here are duplex because network access

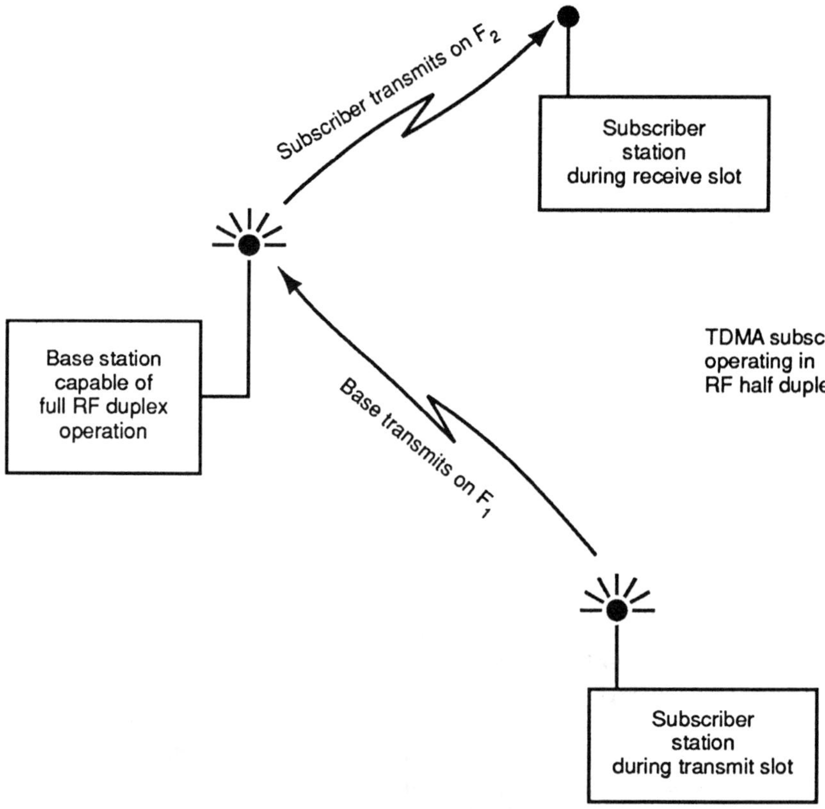

Figure 6.2 A two-frequency duplex radio circuit.

is by its nature a duplex process. The separation between these paired frequencies is called the *carrier separation.* Carrier separation can have a significant impact on cost and performance; in general, narrow separations (5 MHz is relatively narrow) are more expensive to implement than wide separations. This is one reason the cellular separation was established at 45 MHz, much wider than pre-cellular mobile radio. Finally, in some applications, there is only one carrier frequency, which is time-shared by the subscriber and base station transmitters. This is known as a "ping-pong" or time-division duplex architecture (see Section 6.4.1.2).

The modulated carrier actually occupies a narrow region of the spectrum around the nominal or unmodulated carrier frequency, sometimes called the *center frequency* in this context (Figure 6.3). The width of this region—the *occupied bandwidth*—is also measured in kHz or MHz. This is what is commonly referred to as a radio *channel* (although the term *channel* is also applied conceptually to the duplex pair of frequencies). Some systems use a fairly narrow occupied bandwidth, typically 25 or 30 kHz per transmission: these are often called *narrowband.* Others use transmission bandwidths as wide as several MHz; these are (not surprisingly) called *wideband.* (A similar term, *broadband,* often refers to the information-theoretic concept of a circuit with much greater information-bearing capacity than what is normally needed to carry a single voice circuit. *Bandwidth* may be construed logically or physically, and although the two constructions are correlated in practice, they are not identical. For example, in so-called spread spectrum systems, a channel with a narrow logical bandwidth is carried on a wide physical bandwidth, which is precisely what gives spread spectrum systems some of their unusual theoretical characteristics. The reader is cautioned that the term *bandwidth* is a particularly slippery one.)

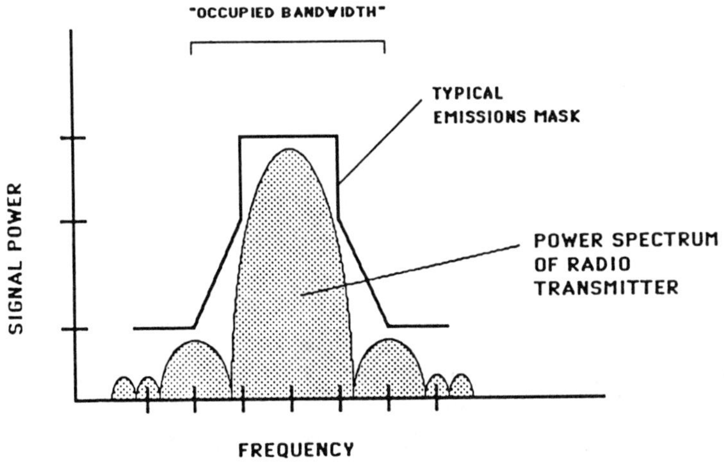

Figure 6.3 Emissions characteristics and channel definition. (From G. Calhoun, *Digital Cellular Radio,* Artech House, Boston, p. 203.)

The radio signal is affected by noise and interference. Noise is considered to be the result of "random" environmental processes that produce radio energy, such as lightning or automobile ignition sparks, or of the thermal noise of the atomic particles of the receiver itself. The ratio of the signal strength to the strength of the noise is the signal-to-noise ratio (SNR), or carrier-to-noise (C/N). It is the most basic measure of signal quality, although it must be used with caution in discussing the quality of digital signals [9].

Interference is a form of signal degradation resulting from other man-made radio emissions. There are two types of interference that are normally considered relevant at the applications level: *adjacent channel interference,* which occurs when energy from the modulated carrier "spills over" or "spreads out" into the adjacent channels (Figure 6.4), and *cochannel interference,* which occurs when two transmissions on the same carrier frequency reach a single receiver (Figure 6.5). The ratio of the carrier to interference is called the *carrier-to-interference ratio* (C/I). (Most of the time, when radio engineers talk about "C/I" without specifying which type of interference they mean, the reference is to cochannel interference, which is the most important in decisions concerning base station placement, frequency plans, and reuse opportunities in the deployment of actual systems.)

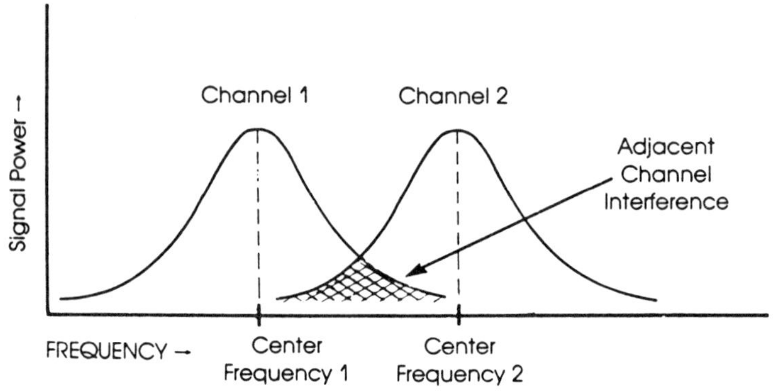

Figure 6.4 Adjacent channel interference. (From G. Calhoun, *Digital Cellular Radio,* Artech House, Boston, p. 204.)

All of these terms reflect the analog origins of radio, where noise and interference produced quite different qualitative effects. The trend in digital radio is to simplify and reduce all sources of signal degradation to their one and only important effect on the digital signal: the occurrence of bit-errors (see Section 6.1.3). The *bit-error rate* (BER) is now commonly used in place of many other measures, except for the retention of the C/I concept, usually pegged to a reference BER level.

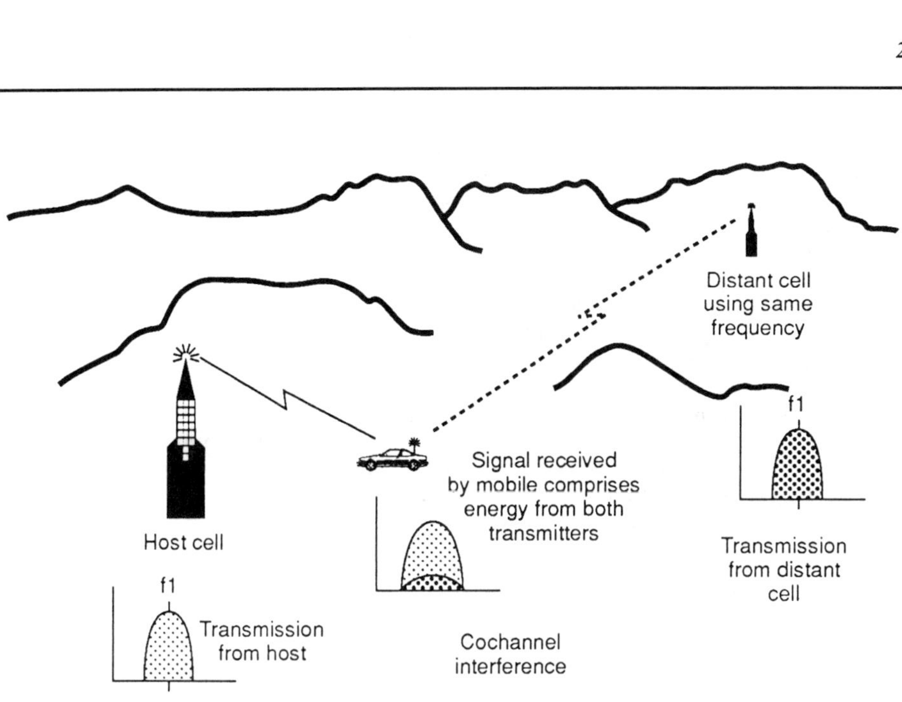

Figure 6.5 Cochannel interference. (From G. Calhoun, *Digital Cellular Radio,* Artech House, Boston, p. 205.)

Another basic propagation concept is the *link budget,* which determines the *expected* quality of a radio link [10]. Just as a financial budget is used to ensure that available funds are sufficient to cover necessary expenses, a link budget is designed to make sure that enough signal will survive the transmission process to achieve the required SNR or BER for acceptable system operation. A link budget is measured in terms of *decibels* (dB) of signal power (or more precisely dBm, where the "m" stands for milliwatts). In a simple link budget analysis, certain factors "add decibels" to the link, such as (obviously) the transmitter or (less obviously) the gain on the receiver antenna. Other factors "subtract decibels" from the link, such as blocking terrain or fades (see Section 6.1.2). The receiver will normally require a certain number of decibels of signal above its thermal noise floor, plus an appropriate margin for fades, to achieve a given BER; if the link budget arithmetic results in enough decibels at the receiver, the signal should get through with acceptable quality. A typical link budget is given in Table 6.1. Note that this is an extremely simplified propagation analysis.

6.1.2 The Fate of the Radio Signal

The happy result of a successful radio transmission is that the message is received and understood by the proper receiver. And yet the radio signal is extremely subtle, almost ethereal indeed. Consider the fact that the amount of energy contained in all

Table 6.1
Simplified link budget

$$G_s = L_p + F + L_t + L_m + L_i - G_t t - G_r$$

where G_s = system gain in decibels
L_p = free space path loss, in dB
F = fade margin, in dB
L_t = transmission line loss from waveguide or coaxials used to connect radio to antenna, in dB
L_m = miscellaneous losses such as minor antenna misalignment, waveguide corrosion, and increase in receiver noise figure due to aging, in dB
L_b = branching loss due to filter and circulator used to combine or split transmitter and receiver signals in a single antenna
G_t = gain of transmitting antenna
G_r = gain of receiving antenna

Source: David Smith, *Digital Transmissions Systems.*

the radio signals actually impinging on all the active receivers in a fair-sized cellular system (say, 25,000 subscribers) during an *entire year* of operation is approximately equal to the energy of the linear momentum generated by a grain of rice falling from the fork to the plate. Well, maybe two grains of rice. This electromagnetic wisp must be infinitesimally divided, subdivided, and modulated to carry tens of thousands of hours of human speech under some of the most challenging transmission conditions in any communications application. Along its way, the signal is subject to any number of threats or obstacles that may prevent it from reaching its intended recipient. The failure of a radio link for even a few milliseconds will produce a noticeable degradation of the communications channel. A handful of such failures and the system breaks down. From this perspective, the mobile radio link is an almost inconceivably fragile thing.

We will briefly survey these obstacles to successful radio transmission, since they define some of the most important design constraints for wireless access systems architects.

6.1.2.1 *Free Space Loss*

Wireless access applications are all *point-to-multipoint* architectures, which means that the subscriber terminals may be located anywhere within a fairly large coverage area extending out from the base station (either a cell or a sector of a cell). In the simplest case—an omnidirectional transmitter—the received power of the signal diminishes steadily as we travel away from the transmitter. In a vacuum, in "free space," the signal strength would diminish as the inverse of the square of the distance—the famous *inverse square* propagation law. In other words, if the received signal is 1 W at 1 mile from the transmitter, it will be one-quarter of a watt at 2 miles, and one one-hundredth of a watt at 10 miles.

In practice, since wireless access systems are not set up in outer space, the path loss is more severe that the inverse square law would predict. Path losses for mobile radio systems may be better modeled as the inverse of the cube of the distance, or of some higher exponent up to the fifth or sixth power. Some models call for inverse square out to a certain distance (usually associated with the horizon) and then inverse third, fourth, or higher powers beyond that. Such statements reflect the averaging of terrain, atmospheric, and other real-world effects. Such estimates are useful only in a very general way to indicate feasible transmission limits. The actual path loss is highly dependent on frequency, as well.

It should be emphasized that radio propagation analysis is still largely an empirical field, especially in new applications at new frequencies. We will look at this issue in greater depth in Chapter 8.

6.1.2.2 Attenuation Effects

The radio signal may be partially blocked and its energy absorbed or attenuated by physical obstacles in the environment. The absorbing body may be almost anything, from a rain drop to a mountain range. The specific cause and the severity of the attenuation depend chiefly on frequency. For example, transmissions at frequencies below 1 GHz are essentially unaffected by rain or atmospheric moisture. On the other hand, transmissions at frequencies above 10 GHz are often severely affected, and frequencies above 30 GHz are largely unsuitable for lengthy outdoor paths (Figure 6.6). As a general rule, lower frequencies have greater penetrating power and will propagate further. (Keep in mind that this is not always desirable; in some applications a limited range is desired to facilitate reuse.) The higher the frequency, the greater the attenuation, and the more transmitter power required for a given distance. For example, for TV transmission at the lower VHF band (50 to 90 MHz), the FCC authorizes a maximum transmitter power of 100 kW. For UHF stations, operating at 500 to 800 MHz, the FCC authorizes station power of up to 5000 kW to achieve approximately the same quality over the same range.

Historically, the development of radio technology has proceeded from low frequencies to higher and higher frequencies, as technology has continually pushed out the "spectrum frontier." Many of the applications now emerging will need large blocks of new spectrum, and this may only be available at higher frequencies. Almost no more spectrum is available below 1 GHz, without displacing current users. But the use of frequencies above 1 GHz entails a propagation penalty due to higher attenuation. At some point, this penalty may become so great that point-to-multipoint

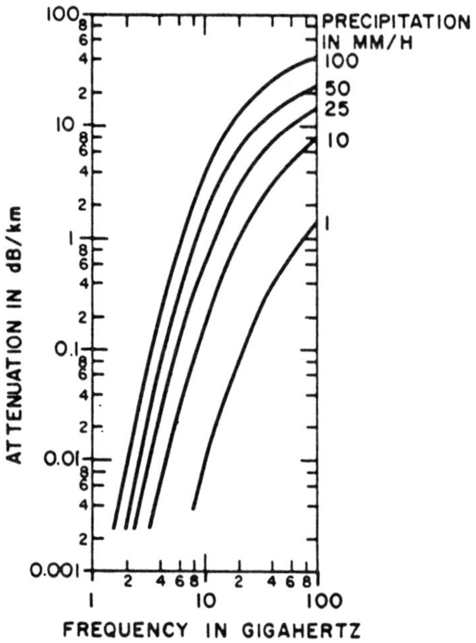

Figure 6.6 Rainfall attenuation as a function of frequency. (From William C. Jakes, *Microwave Mobile Communications,* p. 89.)

transmission is no longer feasible, and highly directional antennas must be used for *point-to-point* transmission. The "microwave" systems used by the telephone companies for long-distance point-to-point communications exemplify this.

Another important effect of attenuation is *shadowing,* where buildings or hills create radio shadows. Shadowing can be severe in heavily developed urban environments and at higher frequencies. Early propagation studies for mobile radio found shadows as deep as 20 dB (i.e., reductions in signal strength by a factor of 100) over very short distances, often literally from one street to the next, depending on the orientation of the transmitter and local building patterns [11].

Another significant source of attenuation is foliage in wooded suburban or rural areas. For example, one study measured an attenuation of 2.4 dB per 100 ft at 210 MHz through a grove of live oaks [12]. Now, 2.4 dB may not sound like much—but it means that nearly half the signal power is lost in penetrating just 100 ft of greenery. Another text reports that for a suburban New Jersey setting, "the average received signal strength [at 836 MHz] in the summer when the trees were in full leaf was roughly 10 dB lower than for the corresponding locations in later winter" [13].

The signal fading producing by shadowing is often referred to as *slow fading,* because from the perspective of a receiver in a moving vehicle, entering and exiting from such a shadow takes a perceptible amount of time (seconds) since the area of the fade is large. In some urban applications, however, the emergence from a shadow behind an obstacle may be quite sudden, for instance, if a mobile user drives out from behind a large building and onto a cross street with a clear path to the transmitter.

6.1.2.3 *Reflection and Multipath Effects*

A radio wave may also be *reflected*—from a hill, a building, a truck, an airplane, or a discontinuity in the atmosphere. Reflections can produce a multitude of different transmission paths between the transmitter and the receiver. This is known as *multipath propagation* (Figure 6.7). Reflection and multipath propagation are a two-

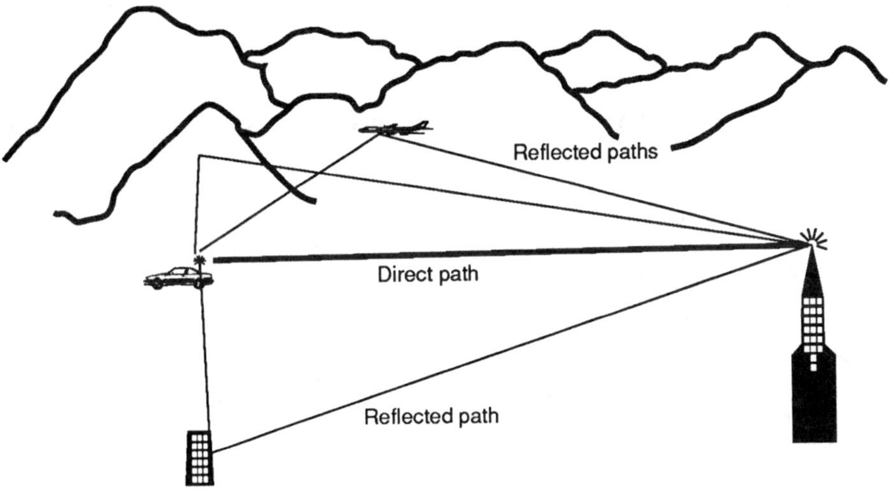

Figure 6.7 Multipath propagation. (From G. Calhoun, *Digital Cellular Radio,* Artech House, Boston, p. 213.)

edged sword for the radio engineer. On the one hand, they allow radio waves to "bend around corners" and to reach behind hills and buildings and into parking garages and tunnels (Figure 6.8). Practicing radio applications engineers will often cite situations where they were able to use multipath constructively to improve coverage in a difficult area [14]. On the other hand, multipath propagation creates some

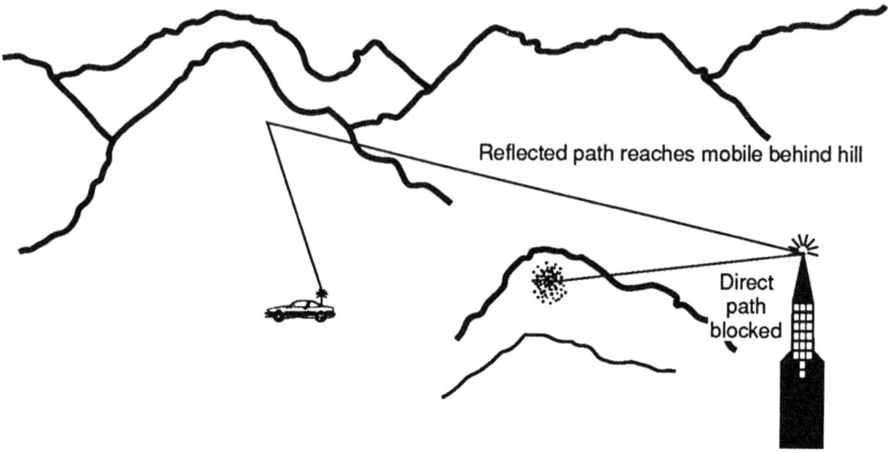

Figure 6.8 Multipath allows propagation around obstacles. (From G. Calhoun, *Digital Cellular Radio*, Artech House, Boston, p. 213.)

of the most difficult radio transmission problems. The three most important multipath issues for the radio engineer are:

1. The *delay* spread of the received signal
2. Random phase shift, which creates rapid fluctuations in signal strength, known as *Rayleigh fading*
3. Random frequency modulation due to different *Doppler shifts* on different paths.

Delay Spread. Because the signal follows several paths, and because the reflected paths are longer than the direct path (if there is one), the multiple signals arrive with a slight additional delay (Figure 6.9). Thus, a digital symbol (such as an amplitude pulse) will travel several different paths, each resulting in a slightly different time of arrival at the receiver; the effect is to smear or spread out the signal in time. Figure 6.10 shows the actual appearance at a receiver of a single transmitted pulse [15]. In a digital system, particularly one operating at a high bit rate, the delay spread causes each symbol to overlap the preceding and following symbols in time, producing *intersymbol interference* (ISI). In Section 6.1.3, we discuss ISI in greater depth.

The delay spread is essentially a fixed feature of the environment for a particular application at a particular frequency; the engineer must design around it. The longer the average delay spread, the more of an ISI problem it creates. In some environments (e.g., most indoor applications) the delay spreads will be negligible, because all paths are very short. In other environments, the delay spread may be sub-

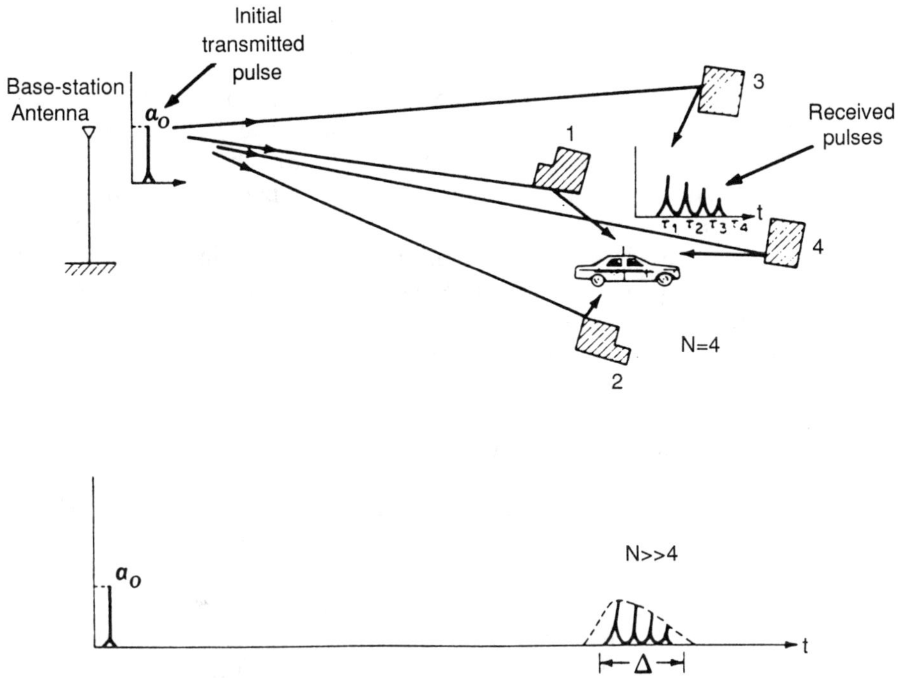

Figure 6.9 An illustration of delay spread. (From G. Calhoun, *Digital Cellular Radio,* Artech House, Boston, p. 214.)

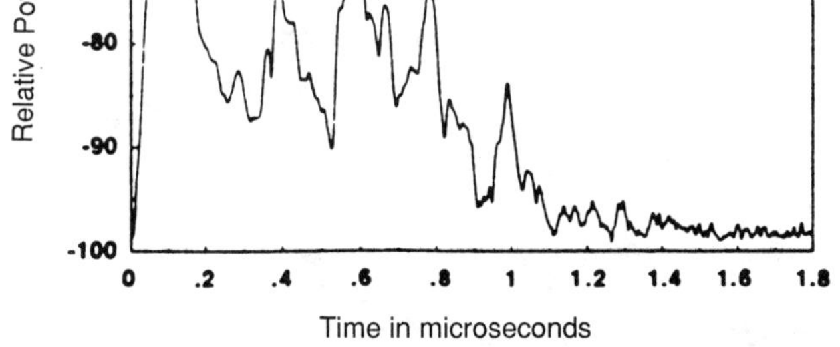

Figure 6.10 A measured profile of average received power *versus* time delay at 850 MHz. (From G. Calhoun, *Digital Cellular Radio,* Artech House, Boston, p. 215.)

stantial. For example, the mountain ranges around Salt Lake City have been informally reported to produce delay spreads of as much as 50 to 100 μs.

Rayleigh Fading. The second important effect of multipath propagation is that the reflected radio wave may undergo drastic alteration of some of its fundamental characteristics, particularly phase and amplitude. The phase of the reflected signal may be rotated such that it arrives out of phase with the direct-path signal. If the reflected signal and the direct-path signal are exactly 180° out of phase, they will cancel each other at the receiver. In effect, the signal disappears. Any partial out-of-phase relationship between the various reflected paths and the direct path may therefore reduce the strength of the received signal to some degree.

These fades are best conceived as a spatial phenomenon. A fade is a small region in space where the precise confluence of signals traveling over multiple paths from a given transmitter combine to produce a measurable reduction in signal amplitude. The location of a fade at a particular frequency is therefore a function of the specific location of the transmitter and various reflectors (most of which are stationary). For a given transmitter at a given frequency, the signal field can be visualized as a kind of Swiss-cheese full of radio "holes" of varying depths. The physical size of these holes is related to the frequency and wavelength. At the frequencies of interest to us for wireless access applications, the "holes" are quite small, only inches across at most. But they may be very "deep"; in the deepest fades the signal strength is reduced by 10,000 to 100,000 times from its expected value (i.e., the received signal level predicted based on the direct path alone). A receiving antenna that is located in one of these holes may get virtually no signal at all, whereas moving just a few inches will take it out of the fade and restore the signal to its desired levels.

The signal field is peppered with radio fades of varying depths. At 800 MHz, maximum fades occur every few inches. In between there are thousands of shallower fades. In fact, the distribution of these "holes" can be statistically described by what is known as a *Rayleigh distribution* (after Lord Rayleigh, the great turn-of-the-century English physicist), and for this reason the phenomenon is often referred to as Rayleigh fading. The environment is sometimes called, from this perspective, the Rayleigh environment (Figure 6.10).

Rayleigh fades can only be described statistically; it is impossible to predict the exact location of a fade. For a fixed-location subscriber antenna, Rayleigh fading is important mainly in terms of positioning the antenna when it is initially installed (see Chapter 8). If a poor signal level indicates the existence of a multipath fade, moving the antenna a few inches or feet will usually improve the situation (much like manipulating the "rabbit ears" antenna on a television set).

Now, however, imagine an automobile cruising through this radio Swiss-cheese universe at 60 mph, or 88 ft/s. The antenna whips through thousands of "holes" of varying depths every second, causing the received signal strength to fluctuate rapidly between the expected levels and fades of 40 dB or more. An amplitude monitor on a mobile receiver will produce a graph like that of Figure 6.11. Viewed thus in the temporal dimension (the horizontal axis here is time), Rayleigh fading appears to be a very dynamic phenomenon. Since the fades are very rapid, this is sometimes called *fast fading.* These signal amplitude fluctuations constitute a very difficult challenge for any mobile radio application, and even more so for cellular architectures where there are multiple transmitters operating on a given frequency [16]. Countering Rayleigh fading is the dominant design objective for a digital radio technology that must operate in a mobile or portable manner. Indeed, paradoxically, the slow-moving portable applications are the most severely affected because the receiver antenna may spend longer times in the deepest fades.

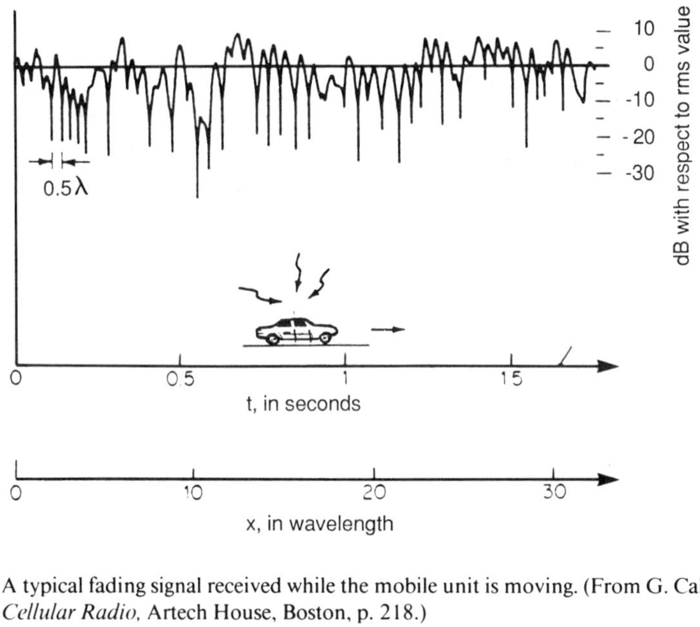

Figure 6.11 A typical fading signal received while the mobile unit is moving. (From G. Calhoun, *Digital Cellular Radio*, Artech House, Boston, p. 218.)

Doppler Shifts (Mobile Applications Only). The movement of the receiver relative to the transmitter produces a shift in frequency of the received signal, known as the Doppler shift (after Christian Johann Doppler, a nineteenth-century Austrian physicist who first called attention to frequency shifts caused by relative motion). Much as the sound of a horn on a moving car appears to the stationary observer to be slightly higher in pitch when the car is approaching rapidly and slightly lower when the car is receding, so radio transmissions are frequency-shifted due to the relative motion of the vehicle. Moreover, the Doppler shift affects all the multiple propagation paths, some of which may exhibit an upward shift, and some a downward shift, all at the same instant (Figure 6.12). The net effect is to produce small, random frequency changes (random FM), similar to the Rayleigh amplitude modulation described previously. Doppler-induced random FM is correlated with vehicle speed for a given frequency. At 22 mph, transmitting at 900 MHz, the median Doppler offset is about 30 Hz, which with certain radio technologies can "introduce distortion objectionable to the ear" [17].

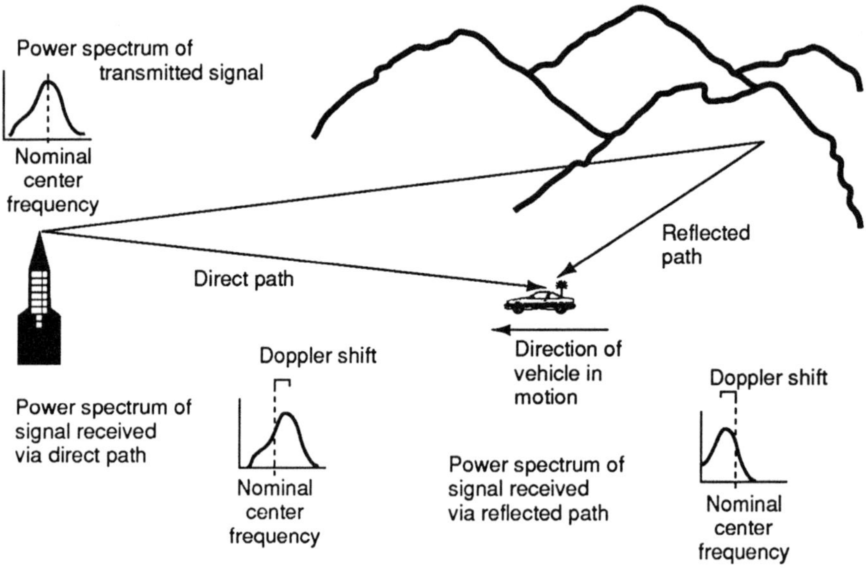

Figure 6.12 Illustration of the Doppler effect on received signals in a mobile multipath environment. (From G. Calhoun, *Digital Cellular Radio,* Artech House, Boston, p. 220.)

Importance of Multipath Effects. Multipath propagation is the most challenging aspect of point-to-multipoint radio transmission systems used in all wireless access applications. In general, multipath is usually not very important for stationary applications, such as basic exchange radio or subscriber radio described in Section 5.3 [18]. For *mobile* applications, however, multipath creates an extremely challenging transmission environment. Figure 6.13 shows the combined effects of slow and fast fading on the received signal amplitude for a moving receiver. Coping with this fluctuation in signal level determines most of the important decisions about coding and modulation technique, as we shall see. The multipath-induced delay spread drives the need for intensive equalization and other countermeasures to be discussed below.

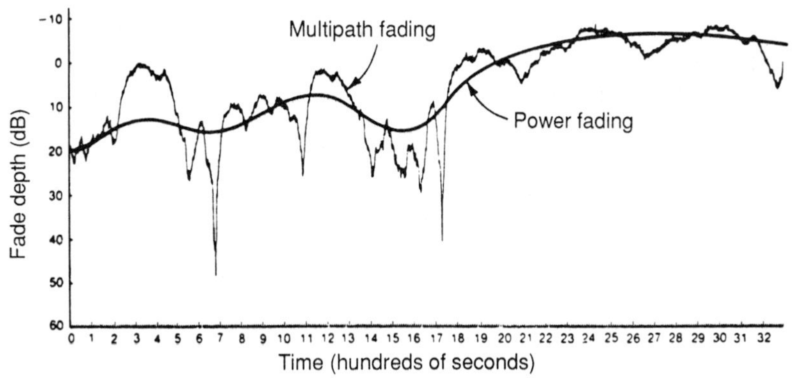

Figure 6.13 Signal levels for a radio receiver in motion, reflecting multipath and fading. (From G. Calhoun, *Digital Cellular Radio,* Artech House, Boston, p. 224.)

6.1.3 The Vocabulary of Digital Communications

About 30 years ago, a quiet revolution in telecommunications began with the introduction of the first commercial digital transmission systems, known (somewhat enigmatically) as *T-carrier* [19]. It was quiet in part because it involved esoteric military techniques and communications concepts that often defied the common-sense intuitions of laymen and engineers alike. As well, digital technology was initially placed into the network in the form of "black boxes," manipulating conventional analog voice inputs and providing conventional analog voice outputs, by means of who-knew-and-who-cared-what alchemy. Its far-reaching implications were poorly understood; indeed, T-carrier was applied for years simply as a cheaper method of handling trunk connections between central offices [20]. There was no palpable impact on customer services, no aggressive exploitation of the unique characteristics of digital (as compared to analog) transmission. As recently as the late 1970s and early 1980s, knowledge of digital concepts and techniques was quite limited, even within the technical branches of the telephone industry.

Yet digital communications is a true revolution and, like all successful revolutions, it has become an orthodoxy. Over the past three decades, digital techniques have been applied to almost every sector of the telecommunications network, from switching to microwave radio, to fiber optics, to PBXs and packet networks (local area and metropolitan area). Today, the North American network (excluding the access plant) is approximately half analog and half digital. But virtually all of the new network equipment being installed uses digital technology, and within another 10 to 15 years the digital transformation of the core network will be complete. The application of digital communications technologies and concepts has become pervasive.

There is no career in telephony in the 1990s that is not thoroughly grounded in digital communications principles.

Having said this, it would seem that another resumé of digital communications might be unnecessary. Indeed, we have developed a powerful and well-organized language for talking about communications problems in digital terms, and there are many excellent texts [21]. My goal here is provide just enough of a framework for the nontechnical reader to get through the material in the remaining sections of this chapter without getting lost.

The Three Stages of the Digital Communications Process

The central concept of digital communications is the conversion of an analog speech input, which is a complex signal composed of many superimposed frequencies with rapidly and continuously varying amplitudes, into a stream of numbers ("digits"). In most cases, the numbers are simply ones and zeros. Indeed, once the principles are understood, any signal—even Beethoven's Ninth Symphony—can be *digitized,* or converted into a vast stream of ones and zeros. Later these numbers, or *bits,* can be used to reconstruct the symphony in a form indistinguishable from the original. Photographs, x-rays, seismographic readings, and television broadcasts—in fact, almost any representation of any real-world phenomenon—can be digitized according to more or less the same principles.

Why bother? Translating Beethoven's Ninth into ones and zeros may seem an odd operation to perform, since the music is both produced and consumed (heard) in analog form. We will discuss the advantages of digital communications below at greater length, but the basic reason for digitizing is that it allows us to apply very powerful mathematical tools to manipulate the digital signals in a number of useful ways. This is known as *digital signal processing* (DSP).

How is it done? The digital communications process is often described as a three-stage process as shown in Figure 6.14. First, the analog input (e.g., speech) is converted into digital form. This is known as *A/D conversion,* and is sometimes sim-

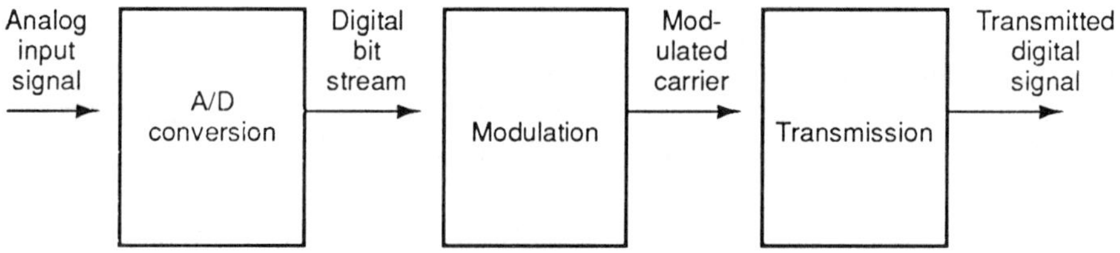

Figure 6.14 A block diagram of the digital communications process. (From G. Calhoun, *Digital Cellular Radio,* Artech House, Boston, p. 151.)

ply (and misleadingly) referred to as *coding*. Next, the digital information is converted into a form suitable for transmission, by *modulating* a radio carrier wave, for example. Finally, the signal is *transmitted*. During the transmission stage, a number of important signal processing steps may be performed. For example, a very common form of signal processing is *multiplexing*, which involves the combination of several individual circuits in a single communications channel (Figure 6.15). In fact, the historical reason for the development of T-carrier (the first commercial digital communications system) was precisely to facilitate the use of time-division multiplexing, to reduce facilities costs by combining 24 digital voice circuits on a single wire-pair. (Note that this simplified view of the digital communications process is in several ways conceptually inadequate, I believe. Terms such as *coding* and *modulation* are often ambiguous and may be used in different ways by different people. In my previous book, I have discussed a reformulation of this model in terms of two-level codes, i.e., logical codes and physical codes, and their interrelationships in transmission and reception of digital signals [22]. We shall return to this issue in Section 6.3.)

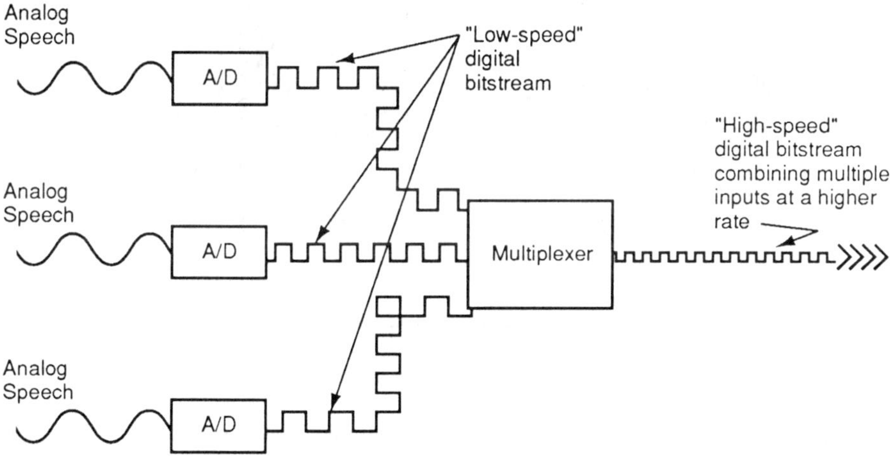

Figure 6.15 Multiplexing. (From G. Calhoun, *Digital Cellular Radio*, Artech House, Boston, p. 152.)

6.1.3.1 A/D Conversion

The digitization process involves three basic steps: (1) sampling, (2) quantizing, and (3) coding.

Sampling. The analog speech input is a waveform of astonishing complexity (Figure 6.16). The most common method to reduce this complex input to a manageable numerical form is to measure the amplitude of the waveform at regularly timed intervals. The resulting amplitude samples will become the basis of the digital signal to be transmitted, in lieu of the actual waveform itself. According to the famous sampling theorem of modern communications theory, if we take enough samples, we can reconstruct the original waveform to any desired level of accuracy (Figure 6.17). (Actually, the sampling theorem states that a sampled signal contains all the information in the original signal at frequencies below one-half of the sampling rate. For example, the standard wireline digital transmission systems use sampling rates on 8000 samples per second, which means that they capture all the voice information at frequencies below 4000 cycles per second, which is considered adequate for telephony.)

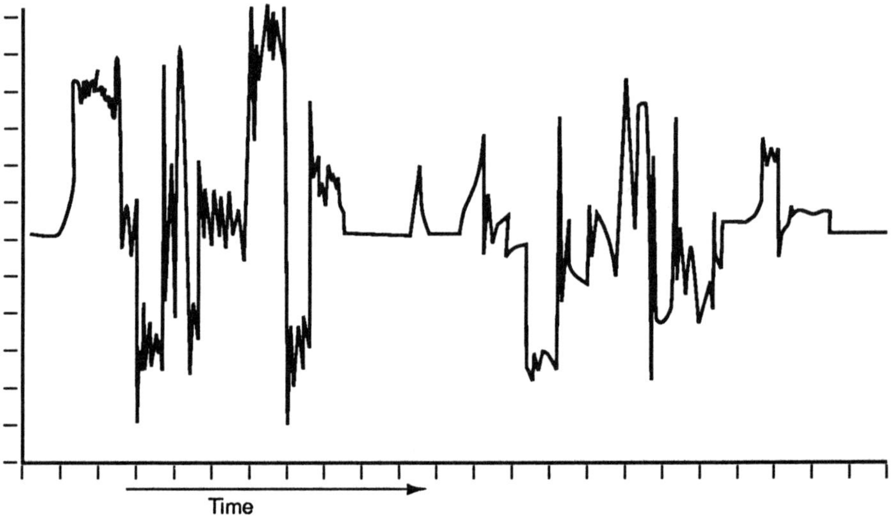

Figure 6.16 Analog input waveform (e.g., speech). (Adapted from G. Calhoun, *Digital Cellular Radio,* Artech House, Boston, p. 155.)

Quantization. Once the amplitude of the waveform has been sampled and a value obtained, the next step in reducing it to digital form is to *quantize* that value, which means rounding it off to the nearest unit on a specially designed scale. For example, assume that we define the allowable range between the minimum acceptable amplitude value (which would normally be zero) and the maximum acceptable amplitude, and then we divide this range into 10 equal steps. Each sampled amplitude is then rounded off to the nearest of these 10 steps.

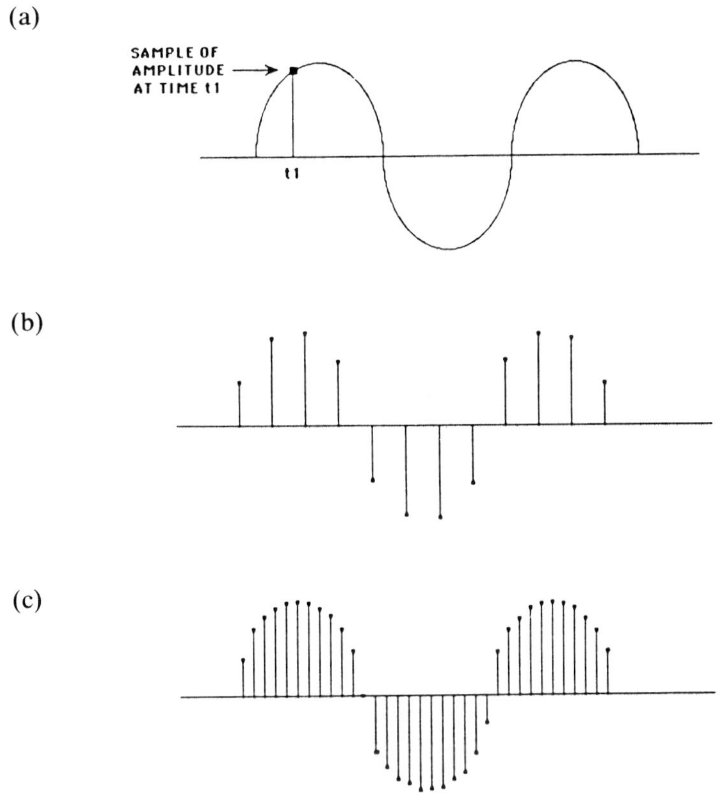

Figure 6.17 Sampling of an analog waveform; (a) analog wave form input, (b) lower sampling rate misses much information, (c) a faster sampling rate allows a more faithful reproduction of the original waveform.

One result of quantization is the loss of information due to rounding. Looked at differently, quantization introduces a measure of "noise" or "error" into the signal (Figure 6.18). Quantization noise may be reduced by using a logarithmic scale instead of a linear scale (Figure 6.19). The scale is finer at low amplitude levels and coarser at high amplitude levels, with the result that quantization noise is approximately a fixed percentage of the signal amplitude [23]. The design of the quantization scale is one of the important design parameters for voice coding algorithms.

Coding. The conversion of the quantized amplitudes into a binary code completes the A/D process. There are many alternatives. For example, consider a case in which an amplitude sample has been measured and quantized on a 128-step scale, and the

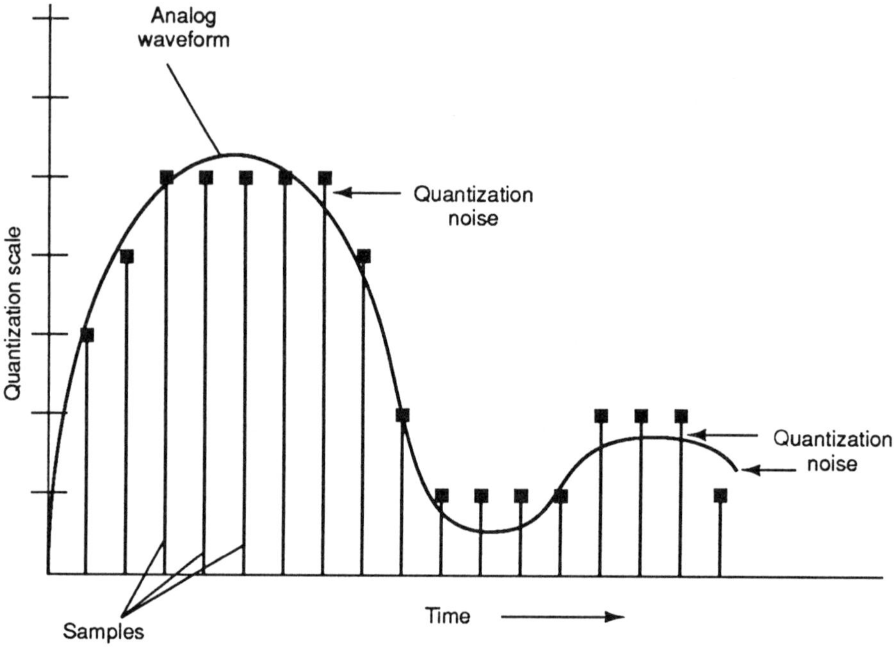

Figure 6.18 Quantization. Individual samples of signal amplitude are rounded up or down to nearest step on the quantization scale. The "rounding error" is referred to as quantization noise. (From G. Calhoun, *Digital Cellular Radio,* Artech House, Boston, p. 159.)

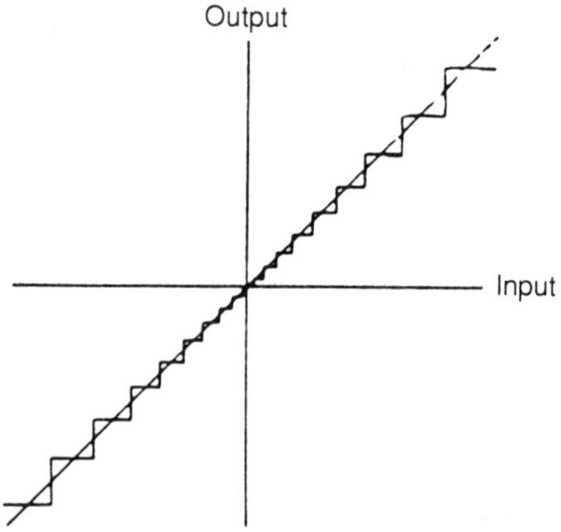

Figure 6.19 Logarithmic quantization scale. (From G. Calhoun, *Digital Cellular Radio,* Artech House, Boston, p. 161.)

result is that this particular sample has received a value of "100" on that scale. In principle, the value "100" could be transmitted as a single *symbol.* We could design the transmission system around an "alphabet" of 128 separate symbols (represented by 128 discrete electrical voltage levels, for example), each corresponding to a single level on the quantization scale. Such a system would be very efficient, since each sample would require only a single electrical pulse to be fully encoded. It would require, however, that the receiver be able to make very fine distinctions, say, between a pulse of 99 mV and a pulse of 100 mV. In a noisy channel (and almost all channels are noisy), this may not be easy and may lead to high error rates.

The opposite alternative would be to encode the value "100" with a string of seven symbols, each representing a "1" or a "0," to stand for the binary representation of "100"—namely 1100100. Transmitting seven pulses per sample would take seven times longer, of course, and reduce the capacity of the channel by a factor of 7, which may seem a severe loss of efficiency. But the detection problem for the receiver is now much simpler. It only has to decide whether each pulse is a one or a zero. Indeed, in a noisy channel, this "digital" nature of the signal at its microscopic level—that is, symbol by symbol—is what gives digital transmission its tremendous robustness. A signal made up solely of ones and zeros can tolerate a very high level of noise or interference without any ill effect; unless the noise is sufficient to cause a "1" to be altered to a "0," it will simply not be registered by the system at all. For this reason, most digital systems utilize binary encoding strategies.

6.1.3.2 Modulation

Once the analog input signal has been converted to a digital bit stream, the information is encoded a second time to put it in a form suitable for physical transmission. For example, most A/D conversions are carried out by circuits that are entirely electrical, so that the digital output of the A/D stage is a series of electrical pulses. If we intend to transmit this information over a radio channel, the electrical pulses from the A/D phase must be converted into a coded radio signal. (Although we will discuss only radio systems here, note that even when the intended transmission medium is electrical, the output of the A/D converter must generally be recoded in a specialized format intended for distance transmission over lengthy electrical circuits, a process sometimes called *line coding,* but functionally identical to modulation in the radio application.)

The conceptually simplest form of digital radio modulation is called *amplitude shift keying* (ASK). This consists of simply sending pulses, turning the transmitter on and off, with "on" signifying a "1" and "off" (at the specified sampling interval) signifying a "0." The word *keying* calls attention to the historical association with the telegrapher's key, which is a wireline form of ASK (albeit complicated by the information value assigned to symbol length differences between "dot" and "dash"). For a number of reasons, however, ASK is not widely used in radio systems.

A more useful method for radio is *frequency shift keying* (FSK), where ones and zeros are represented by two different frequencies offset slightly from the nominal carrier frequency (Figure 6.20). FSK is widely used today for relatively low data-rate radio systems. For example, the control channels in today's analog cellular systems use FSK for transmitting call setup information.

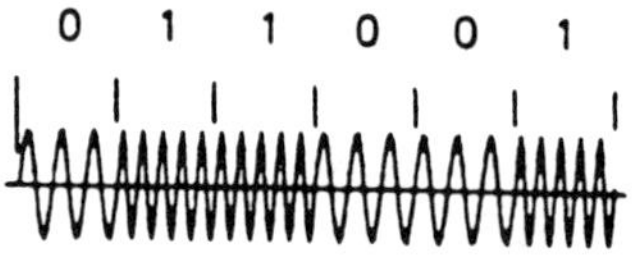

Figure 6.20 Binary frequency shift keying. (From G. Calhoun, *Digital Cellular Radio*, Artech House, Boston, p. 166.)

A third approach is *phase shift keying* (PSK). Binary PSK is the simplest form (Figure 6.21). PSK-based systems are very popular in wireless access designs, including BETRS-type radio, subscriber radio, and the next generation of digital cellular systems.

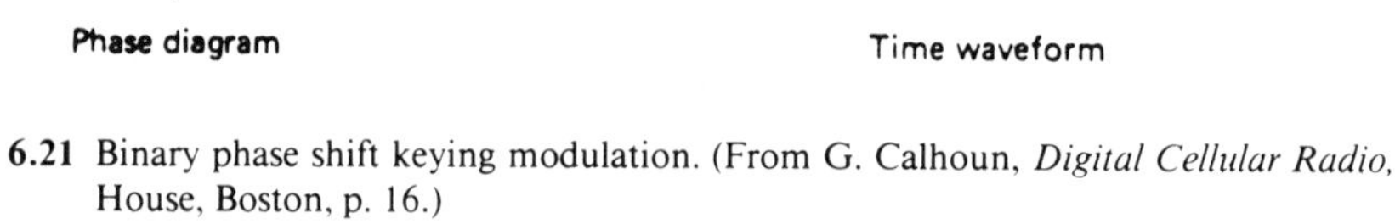

Figure 6.21 Binary phase shift keying modulation. (From G. Calhoun, *Digital Cellular Radio*, Artech House, Boston, p. 16.)

Sophisticated modulation techniques are the core of modern wireless access technology, and we examine them further in Section 6.3.

6.1.3.3 Transmission and Signal Processing

Once a signal enters the transmission channel—whether it is a copper wire, an optical fiber, or the ethereal realm of radio—the signal begins to degrade. The deterioration of the mobile radio signal is far more severe than for other types of channels, for reasons discussed in Section 6.1.2. Rain, foliage, and other features of the environment can absorb huge amounts of the transmitted signal. Hills, mountains, buildings, airplanes, and trucks can reflect and distort the signal, creating all the problems of multipath transmission. Lightning storms, electric power lines, and even the sparks in automobile ignitions generate radio noise, which can obliterate the faint signals reaching a distant receiver. All in all, ensuring that a mobile radio transmission reaches its intended destination is probably the single greatest challenge in all of communications.

Here is where digital techniques pay off—in the transmission stage. For one thing, digital signals can be manipulated to achieve much greater carrying capacity in a given channel—to use the limited radio spectrum more efficiently. The "muscular software" of today's voice-coding algorithms can be truly astonishing (Section 6.3.1). Also, modern mathematics has yielded powerful new signal processing techniques that can increasingly counteract the hazards of the transmission channel itself—*much* more effectively than is possible with analog techniques. For example, *equalization* is a technique that allows the receiver to digitally "clean up" the degraded signal it receives and recover seemingly lost information (see Section 6.3.3). *Error-correction* is still another way to attack the problem of replacing "lost" information, correcting bit-errors by taking advantage of purposeful redundancies in the encoded signal to combat poor channel conditions (see Section 6.3.4).

In the following sections, we will examine some of these techniques in greater detail as they pertain to the design of wireless access systems.

6.2 SYSTEM PERFORMANCE OBJECTIVES

The design of a wireless access system involves the application of certain technical means, in a certain environment (i.e., a certain constellation of constraints), in pursuit of a certain set of objectives. The purpose of this section is to examine some of the general design goals that shape most wireless access architectures. First, however, we shall summarize the most important transmission environments within which wireless access systems are being designed to operate.

6.2.1 Transmission Environments

There are four basic transmission environments for wireless access applications:

1. Indoor, or in-building
2. Outdoor short range (up to 1 km maximum, and typically less than 500 m)
3. Outdoor long range (up to 60 km or so)
4. Satellite links.

In each environment there are three important classes of applications—fixed, mobile, and portable—that may be mixed in some cases and separate in others.

The *indoor radio environment* is characterized by small delay spreads (nearly negligible, in many cases), and low RF power requirements. Yet the physical layouts of most buildings may preclude direct "optical" radio paths in many situations. Steel-reinforced construction may present special problems for propagation. In most cases, however, the environment is determinate or fully specifiable (especially in terms of its interference and propagation characteristics). It is also usually subject only to controllable or predictable changes (such as when a factory is reconfigured); this presents the opportunity to design for and operate a tightly administered system optimized for known propagation and interference characteristics. This environment will be characteristic of wireless PBX and related applications.

The *short-range outdoor environment* is less predictable. Weather, foliage, and unforecastable alterations of the physical setting (the neighbors add a garage, for example) become more important. On the other hand, there are usually no narrow corridors of steel and concrete with which to contend. The radio path is probably more often direct. Power is still not a major concern, nor is Doppler or delay spread. Applications in this environment include telepoint (which overlaps the indoor and the short-range outdoor environments), residential cordless telephony, and many PCN applications, as well as urban wireless loop applications.

The *long-range outdoor environment* is the classical mobile radio environment for cellular systems (as well as BETRS-type applications). It is characterized by severe multipath effects, significant Doppler (for mobile applications), potentially large delay spreads, and high power requirements. Atmospheric conditions can become important in determining system performance. This is the least "engineerable" of all radio environments. Many propagation and interference characteristics can only be estimated statistically. Moreover, the environment is undergoing unforecastable changes on a more or less continuous basis.

Finally, the *satellite link environment,* which is relevant for VSAT applications, some "air telephone" systems, mobile satellite applications, and possibly in future portable telephony applications (e.g., Motorola's "Iridium" project), is a direct optical path, essentially devoid of multipath problems, but subject to very demanding power requirements and an inherent signal delay that precludes or limits its use in certain applications.

6.2.2 Quality Objectives

The most important quality objective for wireless access systems designers is the elusive requirement for "telephone quality" voice communications. Unfortunately, in the end, "quality" is not truly "quantity"—that is, it cannot be reduced to a quantitative assessment performed mechanically. Speech is a human characteristic and human beings must be called on to evaluate speech quality. It is also clear that speech quality is multidimensional. Speech coding experts recognize many different ways in which the speech signal may be degraded, describing them with terms such as "clipping," "distortion," "aliasing," "delay," "echo," and "quantization noise." A machine may be able to measure the amounts of different types of degradation present in a signal, but only the human listener can tell which ones matter, and how much.

Nevertheless, the introduction of low-bit-rate coders (as described in subsequent sections of this chapter) has forced the industry to grapple with the problem of assessing speech quality in a reasonably objective manner. Different types of coders may produce quite different effects on speech quality and we need some methodology for discussing these effects scientifically. Two important measures of voice quality have been developed, both of which are based on *controlled subjective evaluations* (e.g., by carefully selected panels of listeners):

1. *Mean opinion score* or MOS scale, where a value of *4.0 or more* is considered to be acceptable telephone quality [24]
2. *Quantization distortion units* (QDU), a newer scale where a value of *4.0 or less* is considered desirable [25].

Essentially both scales produce a measure of the quality, or acceptability, of an input speech sample. But they do not really address the different types of degradation individually. In addition, the results are maddeningly unreliable. As one authoritative text laments, "MOS values are very difficult to duplicate in repetitions of an experiment" [26]. Moreover, the *sources* of different kinds of signal degradations lie "upstream" of these tests, so to speak; these are the specific link conditions that produce a real BER in a real channel. Some of these conditions involve randomly distributed errors, while others involve bursts of errors that are highly clustered. As well, the relationships between MOS and other important quantitative measures of radio link quality—such as cochannel interference, or multipath-induced delay spread—have not been completely calibrated. How does C/I relate to channel BER, with a given delay spread and a given vehicle speed (Doppler)? How does any of these, or all together, relate to MOS? We still lack definitive answers.

This highlights a troublesome aspect of "quality" testing. Much of the concern about voice quality focuses on the *inherent quality of the coder* itself. Coder quality is usually assessed under an assumption of perfect channel conditions—i.e., zero BER. For many applications (e.g., voice storage or synthesized message systems), this is adequate. But *coder quality under perfect conditions is almost meaningless for radio-based access systems.* As described in the preceding section, the signal degradation in the mobile radio transmission channel is so severe that system designers must plan for routine BERs of 10^{-3} or 10^{-2}, with tolerance for even worse conditions. Therefore, the more important measure is the performance of the coder under routine channel conditions.

These issues came to the fore in two recent voice quality coder evaluations associated with two different wireless access applications. In 1988, a Bellcore evaluation (using the QDU scale) of a 14.6 kb/s RELP-type coding algorithm used in a BETRS system, found (not surprisingly) that the low-bit-rate coder did not perform quite as well as 64 kb/s PCM or 32 kb/s ADPCM *under perfect channel conditions.* But when the tests were performed under simulated *actual channel conditions* (e.g., BER of 10^{-3}), the low-bit-rate RELP coder actually did better (i.e., scored a lower QDU) than 32 kb/s ADPCM and even standard 64 kb/s PCM [27].

In the second instance, a 1989 comparison of a number of low-bit-rate voice coders was conducted in connection with the selection of a coder for use in the new IS-54 North American digital cellular standard. These tests used the MOS scale to evaluate the performance of each coder, as well as today's analog FM, under various simulated channel conditions. A winner was selected. But confusion persists over the precise significance of some of the results of these tests. For example, I have listened to heated debates over the proper correlations between C/I ratios with digital modulation systems (with an assumed or specified BER) and similar ratios in analog FM systems, where the uncertainty has ranged over several decibels (a potentially highly significant range). In layman's language, there is still a good deal of uncertainty in the industry as to how to compare properly the voice quality of analog and digital modulation schemes under various channel conditions. Moreover, the interrelationship between coder performance and (for example) the performance of the equalization algorithm in the receiver, which is not part of the standard but will differ from one manufacturer to another, may result in different voice quality "outputs" for two different receivers, even assuming the same channel conditions and the same signal characteristics and, indeed, the same coding algorithm.

We should bear in mind that despite the growing interest in advanced coding techniques, experience with low-bit-rate coders in actual field applications has so far been very limited [28]. As the industry develops greater familiarity with advanced coders under adverse channel conditions, the procedures for evaluating and calibrating coder quality will become more sophisticated and reliable than they are today.

To summarize some of the important principals of voice quality:

1. For all radio applications, but especially for mobile applications, it is important also to consider normal (impaired) channel conditions as the baseline, rather than focusing on coder performance in a perfect channel.
2. The performance of the coder should be evaluated in conjunction with other signal-enhancement techniques, such as interleaving, dynamic diversity techniques, equalization, and error-correction. The emphasis found in some circles today on so-called *raw BERs* (a slippery term) or, even worse, *channel error rates* is almost worse than meaningless, at least as far as voice quality is concerned. What matters is the quality of the fully processed output, regardless of the BER at any intermediate stage.

3. Although it is true that in one sense all forms of signal impairment in a digital system are translated into a simple bit-error rate, the correspondences between various sources of signal impairment (cochannel interference, echo, delay spread, etc.) and measurable and meaningful error rates are still in need of clarification, which will only come from greater field experience with digital wireless access systems.

6.2.3 Cost Objectives

The cost of wireless access can be viewed in two different ways, depending on the application and the service model (e.g., telephony versus cellular): as a network infrastructure, comparable to wire and cable, or as a consumer product.

As *infrastructure*, for applications where wireless access is a cheaper alternative to wire (such as BETRS or in-building wireless PBX), the relevant cost target is obviously the avoided cost of conventional cable plant. Individual component costs, and especially the relative proportion of terminal and base station costs, are less important than total system costs. The total cost of wireless access per subscriber includes the cost of the wireless access terminal (transceiver, power supply, and installation), plus the allocated portion of the cost of the shared base station and network equipment. An effort will generally be made to carry out a careful economic comparison of alternatives. Sophisticated analyses will also look at the time-value of the investment, encompassing life cycle costs, depreciation schedules, timing of investment, ease of expansion and growth, and so forth. An important caveat is that the cost of wireless access as infrastructure varies with the amount of message traffic (per line) that the system is expected to carry. Telephony applications, with requirements for high availability, low blocking, and relatively high offered traffic per access line, will involve higher costs per subscriber than (for example) cellular applications. We will examine this analytical framework in considerable detail in Chapter 7.

In short, when wireless access is viewed as a form of infrastructure, the cost question is handled in much the same way as any other capital investment decision by a business enterprise.

Concerning actual cost targets for infrastructural applications of wireless access, a wide range of specific situations occurs. As noted in Chapter 2, the distribution of loop costs is generally held to be a lognormal curve (Figure 6.22). At \$5000 per subscriber, wireless access will be limited to a small number of fairly remote installations where it would compete with, or replace, very long loops over difficult terrain. At \$1000 per subscriber (total cost), wireless access would probably be cheaper than copper wire for the majority of local loop installations in the United States, even in urban areas. As a general rule of thumb, developed from analyses of hundreds of potential applications, I believe that for infrastructural applications (such as BETRS, or urban wireless loops), the total cost per subscriber should be targeted as somewhere between \$1000 and \$2000. In that range, the scope of the appli-

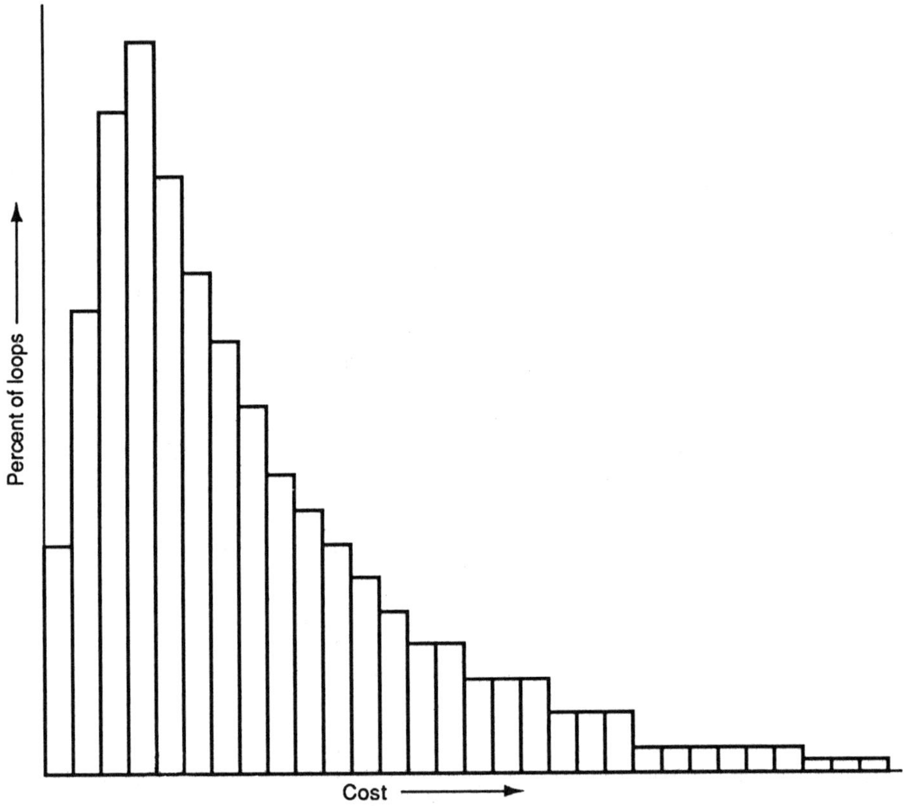

Figure 6.22 Log-normal distribution of loop costs.

cation in North America would be in the tens of millions of lines and wireless access would be a major component of the telecommunications business. For in-building applications with less stringent environmental requirements and lower power but with much higher traffic, the cost target is probably between $500 and $1000 per subscriber (again this is for total cost, including terminal, power, installation, and RF and network infrastructure).

The cellular model (and presumably PCN also) regards the wireless access terminal as a *consumer product,* purchased and owned by the end-user. The consumer's decision process is typically less formal and analytical; therefore, the design must be sensitive to the more emotional calculus of perceived value versus price, which consumers normally apply in such decisions.

Those interested in cellular and PCN-oriented applications of wireless access often feel strongly that the terminal prices are a crucial design constraint, perhaps even the dominant constraint in the choice of technology and architecture. For example, the very simplified and spectrally inefficient designs for CT-2 hardware in the United Kingdom and elsewhere were largely driven by the desire to achieve a very low cost subscriber terminal. Indeed, the architecture of analog cellular was similarly biased in favor of low-cost terminals, even at the expense of spectral inefficiency, frequency management problems, and high cellular infrastructure costs [29]. The conventional wisdom in these applications holds that a subscriber terminal price of less than $500 is vital, and terminal prices under $200 are highly desirable.

I am less convinced of the overriding importance of some of these cost objectives. The same reasoning—that terminal cost would become the dominant consumer decision criterion—was advanced during the early stages of the development of the microcomputer industry, with some manufacturers staking their business strategies on low-cost terminals (recall the ill-fated Timex Sinclair $100 computer of the early 1980s). Experience has shown, however, that consumers in the microcomputer market are more responsive to features and capabilities, at least below a threshold of $2000 to $3000 or so. The best selling systems today still carry hefty price tags; rather than reduce prices, the manufacturers keep adding features. I believe that similar considerations will drive the maturing wireless access market. When automobile prices have reached $12,000 to $20,000 for an ordinary car, whether the (factory-installed) mobile terminal costs the customer $400 or $600 will be of less significance, I suspect, than the quality and range of communications capabilities the unit will provide.

Cost targets must also be assessed as a function of the stage of the manufacturing cycle, especially for new technologies (such as the digital cellular systems emerging in the early 1990s). Analog hardware is quite far down the learning curve, with volumes in some markets approaching 10 million units in total. Digital systems will be more costly when they first appear. But the cost trends in digital system manufacturing augur well for a very rapid decline, and it is likely that digital terminals will reach price levels equivalent to analog terminals very rapidly (especially as today's marketing subsidies are shifted away from analog terminals to encourage high-traffic users to transition to the more spectrally efficient digital circuits to free revenue-producing capacity).

More importantly, the economics of the network infrastructure for digital wireless access systems will be far superior to today's analog systems. The multiplexing of many circuits on a single radio-chain (transmitter, receiver, etc.) at the base station results in very favorable cost per circuit and cost per subscriber (see Chapter 7).

6.2.4 Capacity Objectives

The earliest systems for wireless access—such as subscriber radio and pre-cellular mobile systems—were hampered by stringent capacity limitations, which limited the use of wireless access to low-density telephony applications and rudimentary high-blockage or dispatch-like mobile applications.

Today, there is general agreement that wireless access systems must be designed from the beginning to handle very large numbers of subscribers—tens of thousands to several million per system—if wireless access is to realize its potential. Such objectives imply:

1. The use of highly efficient radio technologies (see Section 6.2.5)
2. Signaling and control protocols capable of handling high capacities (i.e., contention algorithms like Aloha rather than polling algorithms)
3. Efficient switching and hand-off procedures, for applications that require mobile operations
4. Numbering plans capable of handling large numbers of subscribers.

In short, large-scale wireless access applications will require most of the network features of today's high-density wireline systems, as well as any additional network capabilities based on mobile or portable service requirements.

6.2.5 Spectrum Efficiency

Some forecasts predict that there will be 40 million cellular radio subscribers in the United States by the year 2000. If the cellular industry tried to serve these customers with today's radio technology (given the practical limitations on cell-splitting), it would likely require several hundred MHz of spectrum (at least), and that could double or triple if cellular users begin to use more air time. This is but one application of wireless access. Indeed, some analyses today postulate the immediate need for 500 to 800 MHz of spectrum to be allocated to mobile radio systems [30]—more than the entire spectrum allocation for all broadcast television and radio! In major urban centers, the population densities may produce an even more acute need. Another study has shown the spectrum shortfall for all mobile services to be on the order of 1500 MHz in major cities like Los Angeles within a few years [31]. More would be required for in-building and other private systems, not to speak of BETRS and the telephony set of applications, which could dwarf even the cellular requirements—let alone PCN (whatever it turns out to be).

Clearly something has to give. New technologies must use the available (and nonexpandable) spectrum more efficiently. One can draw an analogy between today's FM-based radio techniques and the wasteful slash-and-burn agriculture of the pioneer days when there was so much virgin land and so few people that there was hardly any incentive to use the soil efficiently. As the frontier closed and the supply of cheap land came to an end, farmers have learned to use the soil much more efficiently. So must we in the radio communications industry now learn to use our nonrenewable resource more competently.

Spectrum efficiency is an elusive concept related broadly to the traffic-carrying capacity of a particular radio system in terms of the amount of radio spectrum it uses. There is no agreement on the formal definition of the concept, however, and the banner of "improved spectrum efficiency" has been raised on behalf of many systems, some of which were appallingly inefficient. Calculations of spectrum efficiency and capacity gains are often mathematically specious. Models are advanced that mix different gain factors in ways that can be highly misleading, sometimes degenerating into numbers games, which aim to triumph by confusion.

6.2.5.1 Approaches to Increasing Spectrum Efficiency

Efficient RF Link Design. One obvious approach to increasing carrying capacity of a radio channel is to utilize radio technologies that are capable of transmitting more information in each circuit of a given bandwidth. In digital radio circuits this may be achieved by means of lower bit-rate coders, higher level modulation, and possibly by tighter filters to reduce emissions bandwidths and enhance adjacent channel operation. A number of techniques designed to achieve this goal are discussed in Section 6.3.

Efficient System Design. This encompasses a wide range of techniques for improving the throughput of a radio system based on the way in which individual circuits are bundled together and managed as a system. Simple trunking is one such technique [32]. Trunking efficiency is a very basic and important element of any efficient systems design (see Figure 6.23). A more complex technique is *digital speech interpolation,* or DSI, where idle capacity on assigned traffic channels (during periods where one party on a call is simply listening and thus not transmitting on his half of the wireless access circuit) is reassigned rapidly and dynamically to other active users. These techniques are discussed in Section 6.4.

Improved Frequency Reuse. There are a variety of techniques that promise better frequency reuse, from cell-splitting in conventional analog cellular to claims of better cochannel interference resistance for some technologies, which could allow denser packing of cells or more efficient reuse patterns.

Better Traffic Management. A data circuit that actually needs only a few tens of bits per second (for telemetry, for example) need not be allocated a full voice circuit running at a rate of, say, 13 kb/s. The use of slow associated data channels (as anticipated in the IS-54 digital cellular standard) is one way of improving the throughput of the system by segregating different types of traffic and matching bandwidth requirements for each traffic stream more effectively. Another example of manipulating the traffic as a way of improving throughput could be the use of voice mail as an adjunct to a wireless access system. If voice mail reduces the time needed to transmit short messages that are essentially one-way in nature and not strictly real-time sensitive (a classic example is a cellular call from a gridlocked commuter to his office to indicate he will be delayed), then effectively the system has gained capacity that would otherwise have been utilized for talking to a receptionist, waiting for a secretary to finish a previous call, exchanging pleasantries, taking down the message, etc.

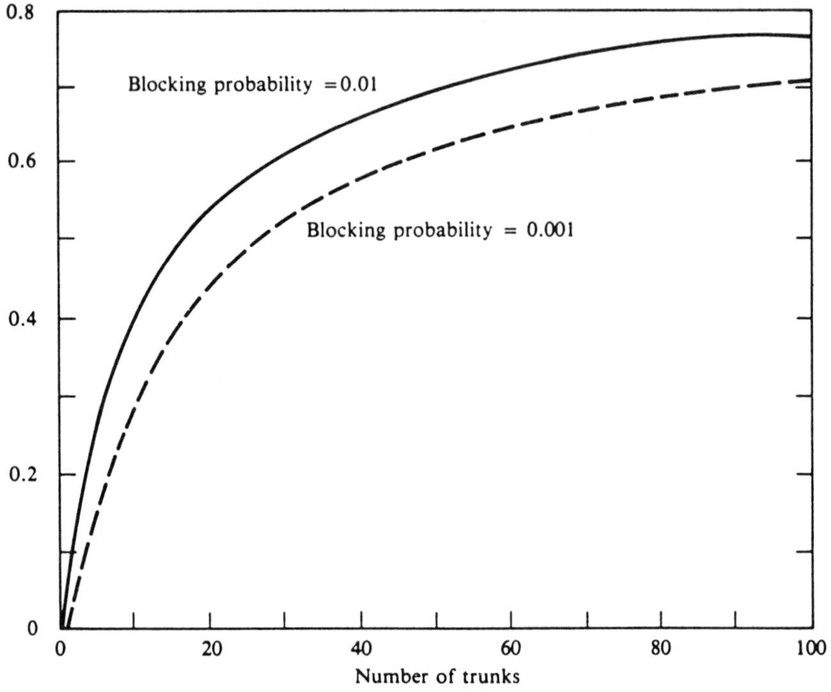

Figure 6.23 Trunking efficiency. (From Crandall and Flamm, *Changing the Rules*, p. 37.)

6.2.5.2 *Approaches to the Measurement of Spectrum Efficiency*

Several measures of spectrum efficiency can be applied fairly objectively. While no one measure captures the whole picture, they can provide at least a baseline for comparing specific alternatives.

Spectral Efficiency (b/Hz). A very common and useful measure of the efficiency of any digital radio transmission is the *spectral efficiency* or *information density*. Mathematically, it is obtained by dividing the bit rate of the transmission (in bits per second) by the bandwidth of the digital signal in hertz [33]. For example, the North American digital cellular standard (IS-54) involves the transmission of a little more than 48 kb/s on a 30-kHz channel, or about 1.6 b/Hz. The GSM standard for European digital cellular specifies a 240 kb/s transmission on a 200-kHz channel, or 1.2 b/Hz. By comparison, a fixed wireless access system for BETRS applications achieves 64 kb/s in a 20-kHz occupied bandwidth, for a spectral efficiency of 3.2 b/Hz [34]. Essentially, information density is a measure of RF modulation efficiency (see Section 6.3). For example, in the IS-54 standard, 4-level PSK modulation is specified, whereas in the BETRS system 16-level PSK modulation is used, with twice the information density. One of its drawbacks, however, is that while higher level modulation schemes will score higher on this scale, their use will be inhibited by requirements for higher SNR and C/I ratios, so that the actual capacity gain may be much smaller that the spectral efficiency numbers would indicate.

Circuit Spectrum Efficiency (circuits/MHz). Information density is a raw measure that tells us nothing about the efficiency with which the bit stream is organized into telephone-grade voice circuits. A more general and, I believe, more useful measure is *circuit spectrum efficiency,* defined as *the number of trunked telephone voice circuits per Mhz of spectrum,* within a single cell or serving area. This measures coding efficiency as well as modulation efficiency. For example, a BETRS system employing 64 kb/s transmission in the RF channel could carry one standard 64 kb/s PCM voice circuit, or two 32 kb/s ADPCM voice circuits, or four 16 kb/s RELP voice circuits (see Section 6.3 for a clarification of these terms). Obviously, we would want to say that a system that creates four circuits in the same spectrum as another system that creates only one circuit is much more efficient (assuming quality levels are acceptable).

Guardbands and adjacent channel restrictions must also be taken into account. For example, in today's analog cellular systems it is not possible to utilize adjacent channels in a given cell; in fact, it is necessary to skip at least two frequencies between occupied channels to accommodate the loose receiver designs of today's cellular units. This reduces the number of circuits per MHz.

A simple calculation of circuit spectrum efficiency can be based on the following equation:

$$\frac{\mathrm{C}}{\mathrm{B} \cdot (\mathrm{RA} + 1)} \cdot \frac{1000\ (\mathrm{MHz})}{2} = \mathrm{Circuits/MHz}$$

where C = Number of Circuits per Carrier
B = Channel Bandwidth (including guardbands) in kHz
RA = Number of "restricted" adjacent channels

The final division by 2 is necessary to account for the use of transmit and receive frequency pairs. Table 6.2 shows the circuit spectrum efficiency for several wireless access proposals.

Geographical Spectrum Efficiency (circuits/MHzkm^{-2}). An important feature of cellular systems is frequency reuse. It should be obvious that a system that uses small cells will generate more circuits per square kilometer for the same amount of spectrum than a system using large cells (Figures 6.24 and 6.25). It should also be obvious that the reuse pattern, or as it is known in the industry, the *N* factor, is important in calculating geographical spectrum efficiency. If $N = 3$ (that is, the frequencies are

Table 6.2
Spectrum Efficiency Calculations (Circuits/MHz) for Several Wireless Access Systems.

System	*Architecture*	*Channel Bandwidth*	*Circuits per Carrier*	*Coder Bitrate (kb/s)*	*Modulation Information Density (bits/Hz)*	*Adjacent Channel Restrictions*	*Circuits per MHz***
IMTS mobile radio	Analog FDMA	25 KHz	1	—	—	No	20
AMPS cellular	Analog FDMA	30 KHz	1	—	—	Yes	5.5
Digital AMPS—full rate	TDMA	30 KHz	3	13	1.6	Yes	16.7
Digital AMPS—half rate	TDMA	30 KHz	6	6.5	1.6	Yes	33.3
Digital AMPS—E-TDMA	TDMA	30 KHz	~15	6.5	1.6	Yes	~80
Microwave subscriber radio*	Single-channel TDMA	2 MHz	32	64	1	No	2–8
BETRS	TDMA	25 KHz	4	14.7	3.2	No	80
CT-2	TDD	50 KHz	1	32	0.6	No (?)	10
DECT	TDMA	~1 MHz	?	32	?	?	?
GSM	TDMA	300 KHz	8	16	~1	Yes	~7

All numbers are approximate.
*There are many slightly different versions of this equipment. The numbers cited are typical values.
**Excludes cellular reuse; includes adjacent channel restrictions.

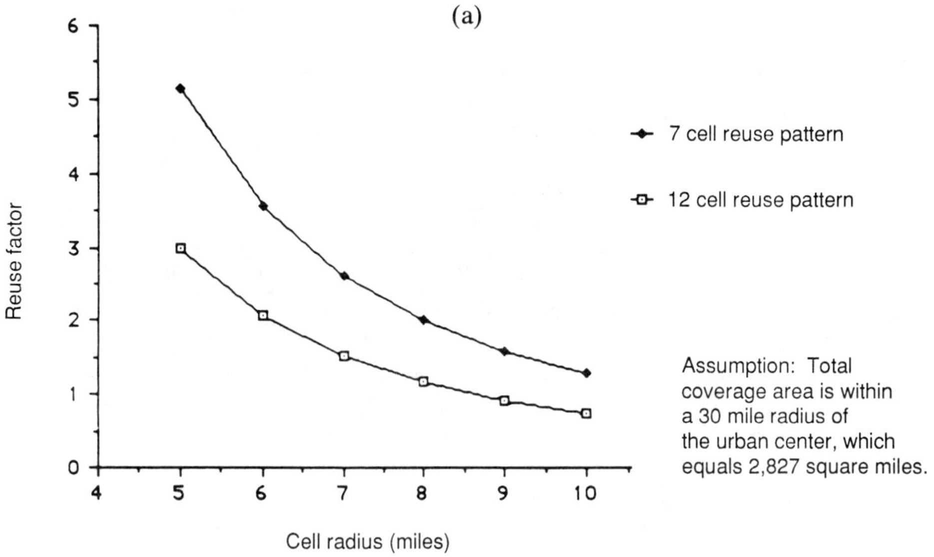

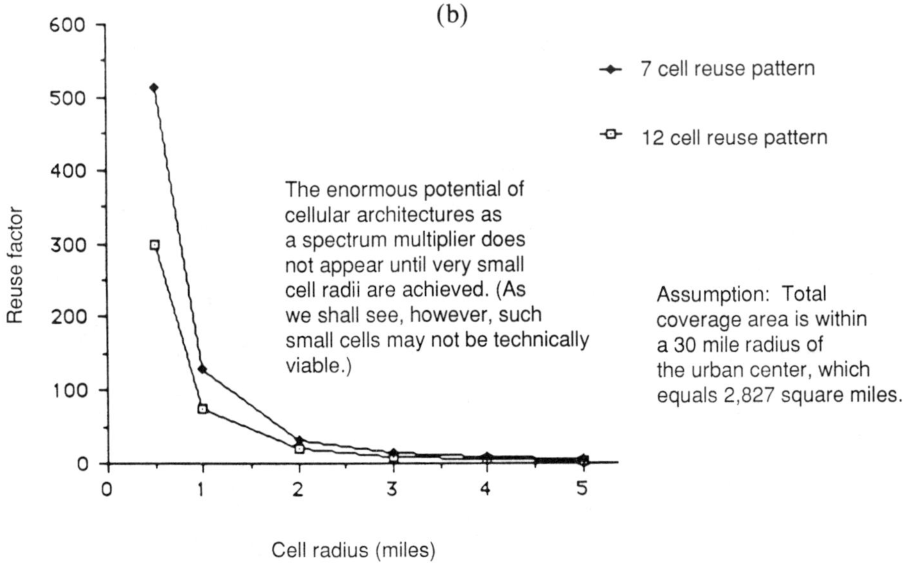

Figure 6.24 Smaller cells mean higher capacity; (a) reuse factor at moderate cell radii, (b) reuse factor at small cell radii.

(a)

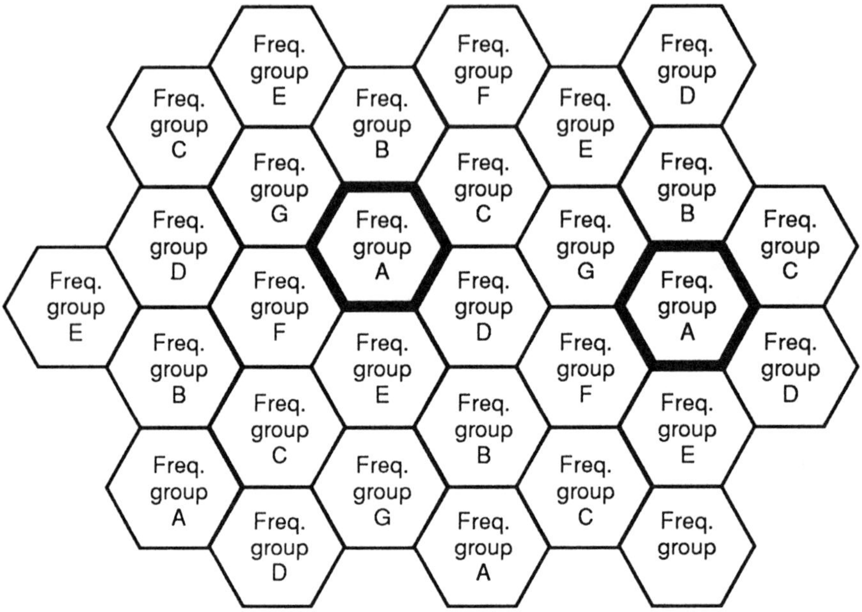

(b)

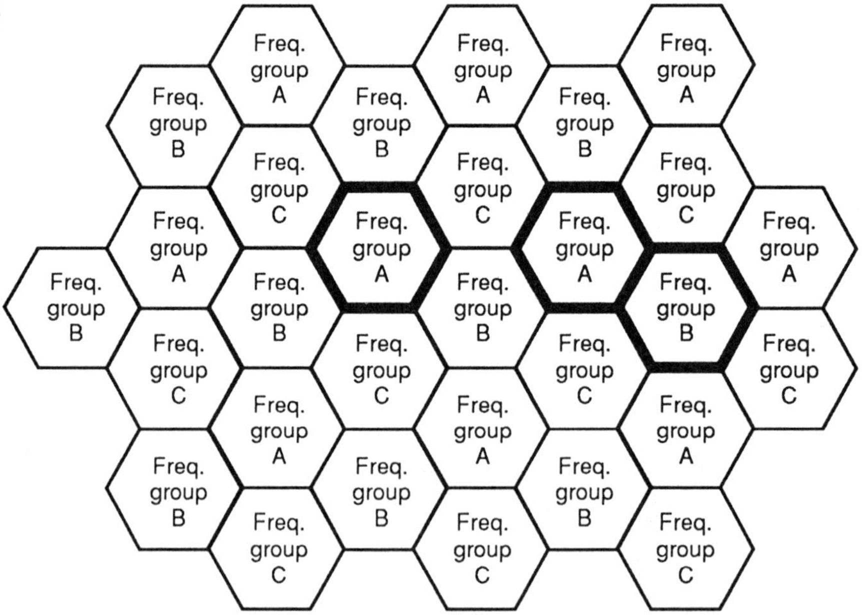

Figure 6.25 Comparison on reuse factors of 3 and 7; (a) 7-cell reuse pattern, (b) 3-cell reuse pattern.

reused in three sets, which no two adjacent cells using the same frequency set), the circuits per square kilometer will be higher than if $N = 7$ (Figures 6.24 and 6.25).

A very simple calculation yields an approximate idea of the importance of these factors. Assuming that adjacent channel restrictions do not enter into the calculation, i.e., that the number of restricted adjacent channels is less than the number of cells in the reuse pattern, the simplest equation for geographical spectrum efficiency is:

$$\frac{\text{Number of Cells} \cdot \text{Number of Circuits/Cell}}{\text{Total Area Covered}}$$

The number of circuits per cell is equal to:

$$\frac{\text{Total Spectrum Allocation}}{(B/C) \cdot N \cdot 2}$$

where B is the channel bandwidth, C is the circuits per carrier, and N is the reuse pattern. The complete calculation is therefore:

$$\frac{\text{Total Spectrum Allocation} \cdot \text{Number of Cells}}{(B/C) \cdot N \cdot 2 \cdot \text{Total Area Covered}}$$

Unfortunately, the real reuse calculations are not as simple. Sectorized cells with directional antennas and different cell sizes and even novel cell architectures (such as the dual-ring cells now being deployed in some situations) are among the complicating factors. As well, the comparison of alternative reuse schemes assumes *ceterus paribus;* for example, the cochannel interference levels are assumed to be the same for $N = 4$ and $N = 7$. Unfortunately, this is probably not the case. Obviously, the N factor or reuse pattern can only be adjusted in certain gross steps that are topographically suitable. Values for N of 12, 7, 4, 3, and 1 are possible, but $N = 5$ is not implementable in any straightforward way. The jump from $N = 7$ to $N = 4$ will almost certainly produce nonequivalent C/I readings, so we are not really measuring *equivalent* circuits per square kilometer per MHz.

6.2.6 Flexibility

Wireless access systems should be capable of supporting a standard set of communications services, starting with voice telephony, but extending (today) to other services such as:

1. DTMF tone transmission during the call (to operate voice mail services, for example)

2. Voiceband data transmissions (up to at least 2400 baud)
3. Group III facsimile transmission
4. CLASS services (e.g., call forwarding, call waiting).

The fact that some wireless access standards now emerging (such as IS-54) may not initially support some of these basic services is problematic.

In the near future, the wireline network will support more advanced digital services on a routine basis, including Group IV fax, higher data rates for voice-band transmission (4800 and 9600 kb/s), and even the ISDN basic rate of 144 kb/s. Packet data are becoming an ubiquitous feature of modern networks. I believe that wireless access technologies must be designed to support all current and future services at least up to the ISDN basic rate.

6.2.7 Forward Compatibility

The accelerating evolution of digital radio technology is a highly synergistic interplay between hardware and software. To take but one example, the frontier of "implementability" for telephone-quality voice coders advanced from about 32 kb/s in 1980 to about 8 kb/s by 1990. Part of this advance is due to innovations in voice-coding algorithms and part is due to the availability of faster DSP chips to implement more computationally intensive techniques. Strictly on its own, this increase in voice-coding efficiency would translate into a fourfold increase in circuit spectrum efficiency for a digital radio system—a clear competitive advantage in the marketplace. Each major area of digital signal processing technology—voice coding, modulation, equalization, error-correction, and access techniques (including digital speech interpolation)—has its own expanding frontier of implementability. Of course, "implementability" is a compound of many parameters, including cost, power requirements, processing delay, memory requirements, and so forth. The rapid decline in the cost of processing power is perhaps the most dynamic of these trends. Every year new algorithms become commercially feasible as the processing engines become cheaper and more powerful. In some cases, the algorithm may have sat for years on the library shelf in the form of a computationally infeasible Ph.D. thesis, awaiting the development of the enabling hardware. Indeed, the technology cycle for new generations of digital signal processing has shortened to about five years.

On the other hand, the *minimum* time required to bring an actual significant new radio system to the market as a *commercial product*—that is, to establish a new technical standard, obtain regulatory approvals, reduce the standard to a product design, set up factories, build, test, adjust, market, and reach sustainable commercial production levels—is at least 5 years for systems of this kind. In telephony-oriented applications, the product cycle is even longer, probably 8 to 10 years. Thus, there is a disjunction between the explosive forward progress of the technology and the natural inertia of the communications industry in bringing new systems into the field. Setting standards in such a technological environment becomes very challenging.

The general approach to coping with this state of affairs has already been mapped out by the successful companies in the computer industry, which have lived with this kind of technological ferment for decades. In very simplified terms, the answer is to move away from hardware-intensive systems to develop more *software-intensive* designs. Today's radio communications products (such as analog cellular radio terminals) are lumps of inflexible hardware, with all the intelligence bolted down and essentially unchangeable. If the customer wants to upgrade his communications capabilities, in most cases he has to "throw away" his current hardware and buy new (similarly inflexible) hardware. This is hardware-intensive. It might be compared to a so-called "smart" typewriter, which contains certain built-in (but unchangeable) word processing features.

The alternative evolved by the computer industry is a software-intensive system, the best example of which is the personal computer. This provides a basic hardware platform, and perhaps some built-in operating software, but the bulk of the system resides in replaceable software modules, most commonly loaded by means of a floppy disk. The functional capabilities and characteristics of the system can be altered drastically by simply loading new software. One word processor program can be replaced by another, improved version, without having to junk the hardware. The software-intensive system (and this is the model for all general-purpose computer systems today, from micro to mainframe) has a much longer commercial life cycle, and remains much more open to continuing technological advances. In time, of course, the advances in raw processing power of the hardware may render the hardware platform itself obsolescent, but this replacement cycle seems to be more on the order of 8 to 10 years and perhaps even longer in less demanding applications.

It is not necessary today to throw out a computer in order to change word processor programs. It should not be necessary to throw out the wireless access terminal (or system) to accommodate and exploit improved signal processing technologies that we already know are in the pipeline. A wireless digital access terminal is already, structurally, the equivalent of a microcomputer, or even a workstation. It is time to start moving toward more software-intensive architectures to allow these systems to grow gracefully with the flow, instead of resisting obstinately until they (and their users) are swept under.

6.2.8 Portability

Much is made of the requirement for portability, variously defined. The anthropomorphic increments of size—briefcase-size, coat-pocket-size, checkbook-size, shirt-pocket-size, cigarette-pack-size—actually tell us little about the true requirements for portability. Portability is more a functional requirement than a structural one and is thus dependent on the application. We may question somewhat the rather rigid insistence from some quarters on the value of portability *über alles.* The evolution of other consumer industries may offer valuable lessons. While no one is ready to quarrel with the principle that "smaller is better," the availability of lap-top computers has not killed the market for desk-top computers. Small, portable television sets have not hurt sales of 26-in. screens. Cordless phones have not eliminated the fixed phone (the "500 set"). Work habits (and play habits) are more important than the raw physical dimensions.

To my mind, the true need for portability is unknown and unmeasured (with the limited exception of portable cellular terminals). Whether 10 or 90% of all wireless access terminals will be shirt-pocket portable is unclear (at least to me). This raises the question as to how much importance should be assigned to this criterion in the process of making system design decisions. For example, would a shirt-pocket portable unit that provided poorer communications quality voice transmission be preferable to a larger unit (a "transportable") that offered toll-quality voice? In fact, my reading of the evolution of the wireless communications marketplace is that portability is rather far down the list of design goals. For example, it has been possible for years to buy a very small simplex radio telephone (i.e., for "one-speaker-at-a-time" conversations of the "over-and-out" variety), yet these products have never been viewed as superior to the larger duplex (analog) cellular phones.

That said, however, portability is clearly crucial to some wireless access applications, especially those in the PCN set. Some have argued that portability becomes the overriding concern for these applications, even to the point that it has been suggested that "the goal of a radio link architecture should be to minimize portable-set power consumption" [35]. It is argued that power consumption and battery limitations will dictate the selection of relatively inefficient voice coders (because efficient ones use more power), and simple radio links that eschew equalizers, error-correction, and all the other software gymnastics we discuss in the next section.

The extent to which the desire for shirt-pocket portability is allowed to drive other parameters, such as cost or spectrum efficiency, will be determined by the needs of the application and the market judgments of the manufacturers. Keep in mind that, unlike many other parameters, portability is strictly a market issue. In the next 10 years we will see a great of deal of experimentation, and in the process we will all learn a great deal about the true market requirements for portable communications.

6.3 MUSCULAR SOFTWARE: BUILDING THE DIGITAL RADIO CIRCUIT

Consider the following true case. A telephone company in the southeastern part of the United States is installing a BETRS-type wireless access system. For the most part, the installation is going smoothly. There is one particular subscriber, however, whose house is located behind a grove of tall pine trees. The trees absorb enough of the radio energy to reduce the received signal level at the house by about 6 dB—just enough to knock the transceiver below the desired threshold for 99.99% availability.

A number of solutions suggest themselves. The transmitter power could be increased, but that would involve expensive hardware modifications. Alternatively, a higher gain antenna could be installed at the subscriber station, possibly combined with a 60-ft pole (instead of the normal 25-ft pole); both, however, would increase the cost of the installation quite a bit. One of the more straightforward proposals is to get out the chain saw and cut down several of the pine trees.

But the "digital age" solution to the problem does not involve such brute force methods at all. Instead, the system engineer can rely on the "muscular software" already burned into the silicon chips in the transceiver, which can increase the effective signal level without altering the physical environment. How can this be done?

Actually, quite a few different software solutions exist. For example, it is possible to use *forward-error-correction,* to reduce the number of bit-errors in the received (and processed) signal to a level comparable to that we would expect to see if we had 6 dB higher power (or fewer trees, or a higher pole, etc.). Or we could design a more powerful *adaptive equalization* algorithm, capable of similarly enhancing the signal through the "magic" of digital signal processing. Another approach might be to substitute a different *voice-coding* algorithm program that was capable of handling a higher channel error rate (associated with lower signal strength) without a reduction in perceived voice quality. In the actual case at hand, however, it was a change in *modulation* that solved the subscriber's problem: The standard 16-level PSK modulation was reduced—by means of a software "switch"—to 4-level PSK modulation, with a resulting gain of approximately 12 dB in the signal-to-noise ratio. It was more than enough to boost the secluded subscriber over the system performance threshold, without changing a single piece of hardware or altering the installation physically in any way.

This example shows how interchangeable the real and the artificial are becoming. The fundamentals of radio transmission until now have been based on the same methods first employed by Marconi and the other pioneers—essentially, more power and less interference. With the advent of digital techniques, new concepts such as "coding gain," "information density," and "decision feedback" come to the fore.

In this section we survey briefly a number of the key areas of the muscular software that is being increasingly and effectively substituted for the chain saws. This discussion will force a reexamination of our conceptual foundation. The simple three-stage model of digital transmission used in the preceding sections (Figure 6.14) is not really adequate to describe the structure of a modern wireless access circuit in which advanced signal processing techniques are chained together in stages or layers of code that constitute the digital radio revolution. Therefore, as an introduction to this survey of signal processing techniques, let us reformulate our basic model of digital transmission in a more illuminating manner.

The Concept of Coding Revisited

The central concept of all signal processing is *coding,* in the following broad sense. Coding involves:

1. The *analysis of the information content* of an *input* signal (which may be analog or digital), followed by
2. the *representation of that information content* in the form of a numerical or *digital output,* such that
3. the original input signal can later be *reconstructed* without any loss of the original information, or, to be more precise, such that if there is any information lost, the amount of loss of information can be specified (and, in principle, controlled).

The primitive terms here—such as *information* and *representation*—are themselves fraught with conceptual, even philosophical, difficulties. But the fine points need not concern us here. The essential idea is that a digital radio circuit is built from a series of sequential coding stages. For each coding stage, the input is processed into a digital output, which has certain mathematical and physical characteristics. In some cases, a coding stage involves "undoing" a previous coding stage to produce an output that is "closer" to the original form of the original input signal; these are often called *decoding* stages.

The concatenation of coding stages in building a digital radio circuit constitutes a kind of chess game played against the forces that attack the signal during the actual transmission. Noise, interference, and multipath effects cannot be directly controlled. But by the use of powerful logico-mathematical coding methods, the signal can be "immunized" ahead of time against many of the hazards of transmission. Other coding methods allow for defects that have occurred during transmission to be "cured" after the signal has been received.

Logical-Level and Physical-Level Codes

The second important observation is that every coding process can be viewed on two levels. In one sense, coding is a logical (mathematical) process, producing an output that conforms to a defined logical format. A good example is the conversion of a decimal number to a binary number. The sequence "00111110" is the binary coded output representing the decimal input for "62" (Figure 6.26). Note, however, that the physical form of this output is not specified. It could be a series of electrical pulses, lightwave pulses, or even smoke signals. It is possible to describe and discuss the logical coding process without reference to the physical form of the code.

Nevertheless, in any real digital circuit, the coded output always possesses a real physical form. Normally, for intermediate coding stages inside the digital device, the form is electrical. During transmission, the code will be embodied in a physical form

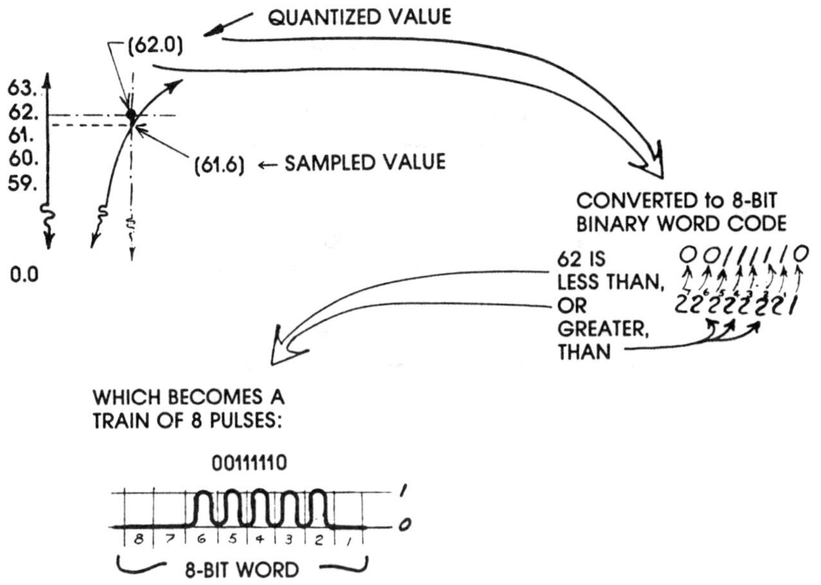

Figure 6.26 Binary coding. (From G. Calhoun, *Digital Cellular Radio,* Artech House, Boston, p. 162.)

appropriate to the physical character of the transmission channel. If it is a radio channel, the transmitted code will obviously be some type of modulated radio wave. But what may not be immediately apparent is that the logical code and the physical code are independent of one another. A given segment of logical code can be represented in many different ways at the physical level.

I am not talking only about gross differences in the nature of the physical medium. Even if we consider only electrical pulses as the physical form of the code, the same logical sequence can be represented in different ways physically. For example, the sequence "10110" might be represented in a very straightforward manner as a series of electrical pulses, where the presence of a pulse of, say, +1 V represents a logical "1" and the absence of a pulse—i.e., zero voltage—represents a logical "0" (Figure 6.27). What could be simpler? Yet early T-carrier experiments showed that simple "on-off" coding was unreliable in practice. Instead, a different and more robust scheme was developed in which zero voltage was still used to represent a logical "0" but each logical "1" was represented by a pulse of either +1 V or −1 V, alternating in sequence (Figure 6.27). This approach, known as *alternate mark inversion,* provided superior performance in the normal (i.e., imperfect) transmission environment [36].

In fact, there are dozens of different forms of what the wireline telephone engineer calls "line code," which is the physical form used to represent the logical infor-

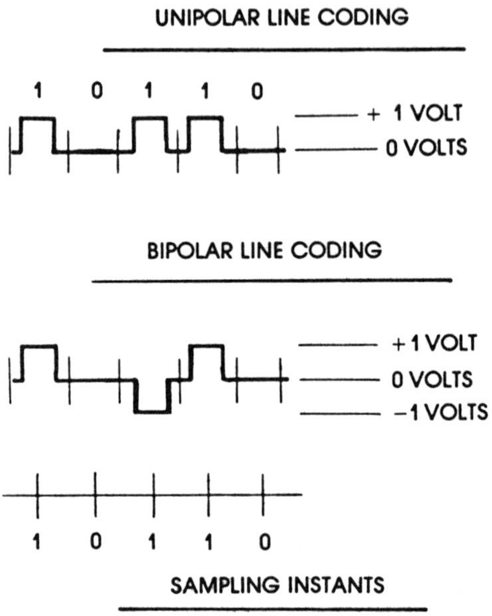

Figure 6.27 Bipolar transmission. (From G. Calhoun, *Digital Cellular Radio,* Artech House, Boston, p. 163.)

mation of a wireline digital signal. The evolution of alternative physical codes—equivalent to the use of different *modulation* methods by radio engineers—has been driven by the requirements of the physical transmission channel.

Thus, we can differentiate between those coding stages that are *primarily* concerned with the transformation of the information content of the signal into a code with a particular *logical* configuration, on the one hand, and those concerned mainly with the *physical* characteristics of the code, on the other hand. While it is true, and bears repeating, that every code "exists" on both levels, logical and physical, the distinction between those coding stages where the logical level is of primary interest and those where the physical level is of primary interest is very important. It largely defines the boundary between software engineering and hardware engineering, for example. Software engineers normally are not concerned about, and may not be very knowledgeable about, the physical transmission process. They are concerned about whether the information content has been captured in an efficient and accurate manner. They tend to be well versed in mathematics and in the properties of the input signal (e.g., acoustics, or speech physiology, in the case of voice coding). Designers of transmission hardware (including digital radio systems), on the other hand, are unconcerned about the information selection process—the "meaning" of the signals they handle. Their job is generally to transmit the input signal with no planned loss of information—i.e., no selection of information—whatsoever. For this purpose, they will quite often test their transmission systems with randomly generated (logically meaningless) sequences of bits. In fact, the techniques for generating "pseudorandom binary test signals" constitute a subdiscipline all on their own [37].

Four Basic Stages of Coding

A digital radio circuit involves at least four distinct signal processing operations, each of which is of course divided into two halves—at the transmitter side and the receiver side of the circuit. Two of these stages can be viewed as primarily encoding stages, with most of the processing "muscle" being applied *prior* to transmission, and the receiver half being relatively mechanical. The other two stages may be viewed as decoding stages, with most of the muscle being applied at the receiver side. (Admittedly, this may be a somewhat artificial construct, but I believe it helps to simplify for the layman the potentially confusing diversity of the signal processing field.)

The four primary coding stages are:

1. *Voice coding:* The conversion of the analog speech input into a digital form containing the important information in an efficient form. This stage is usually combined with channel coding, which adds information to the coded signal to allow the operation of subsequent stages (equalization and error correction) at the receive side; this is primarily a logical coding stage.
2. *Modulation:* The translation or recoding of the output of the voice (and channel) coder into a form suitable for physical transmission. The emphasis in this coding stage is primarily physical.
3. *Equalization:* On the receive side, this involves the recovery of the physical code after it has suffered the degradation of transmission; the emphasis is primarily physical
4. *Error-correction:* Also on the receive side, this stage involves the use of mathematical techniques to analyze, detect, and correct any bit-errors that may have resulted from the transmission process. It is primarily concerned with logical-level code.

A more complete description of the transmission process would replace the simple three-stage diagram used previously with a more complex coding sequence (Figure 6.28). The following sections provide brief, conceptually oriented surveys of each of these fields.

6.3.1 Voice Coding: Bandwidth Reduction

The basic challenge in developing voice-coding algorithms for digital radio-access systems is the problem of *information selection.* The bandwidth of the digital radio

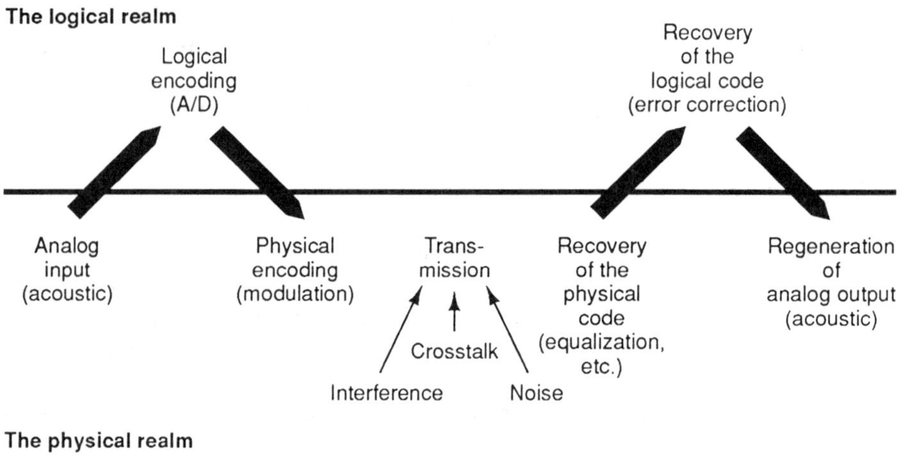

Figure 6.28 The digital communications process.

signal must be reduced as much as possible, to allow for efficient use of the finite radio spectrum. Since bandwidth for digital systems is directly correlated with the bit rate of the circuit, the first step in this process is to try to design a voice coder that captures the essential information needed to achieve a telephone-quality voice circuit with the fewest number of bits/second (the lowest bit rate).

How much bandwidth reduction is possible? We can get some feeling for the answer by looking at the structure of the information contained in the voice signal. The standard digital coding technique for wireline transmission is pulse code modulation (PCM), which uses 8000 samples per second, and encodes each sample with an 8-bit word, for a gross bit rate of 64,000 b/s. On the other hand, if we possessed a device to analyze speech automatically and convert it into written (alphabetic) form, the resulting transmission would require only 50 to 100 b/s, a figure confirmed by numerous psychological experiments to be about the semantic bandwidth (so to speak) of the human speech "channel" [38]. In other words, the semantic content of speech (the words themselves) accounts for something like 1/10 of 1% of the bandwidth of a PCM transmission. Analytically speaking, the other 99.9% of the bandwidth provides the tonal quality of the speech, the color, the emotion, the humor, the unique character of the individual voice. In fact, this "extra" information is highly redundant. Some have estimated that the true, fully nonredundant information content of the speech signal could be transmitted in 1000 to 2000 b/s [39]. The use of 64,000 b/s is clearly very inefficient.

PCM, despite being the standard voice-coding technique in use throughout the network today, is not especially adapted to voice transmission. It is a so-called *waveform coder;* it simply measures the amplitude of whatever signal waveform it finds, and generates an 8-bit word to represent that amplitude—over and over, 8000 times a second. It does not "know" or "care" whether the signal is voice, tones from a DTMF device, a computer modem, or whatever. It does not make use of any built-in intelligence to anticipate the characteristics of the input voice signal. In fact, it makes no assumption about any relationship between any one sample and the next sample. Each is processed completely independently, as though the input signal were totally random. It is, in the parlance of speech coding specialists, a "zero-memory process" [40]. Such a coder is simple, cheap (in dollars), and expensive (in bandwidth).

The digital coding process can be tailored to capture this information much more efficiently and selectively. Voice signals are highly organized, nonrandom, and incorporate huge amounts of redundancy. Each sample tends to be strongly correlated with the preceding and following samples. Beyond mere redundancy, however, we can apply our quite detailed knowledge about the nature of the speech process itself (both speech production and speech processing, or hearing) to build coders that are far more efficient in handling the voice-coding requirement for a bandwidth-limited wireless access system.

With such techniques, it is commercially feasible today (1990) to achieve toll-quality voice coding at around 16 kb/s—four times the efficiency of a 64 kb/s PCM coder—using a single DSP chip [41]. Indeed, as noted above, these 16 kb/s coders actually outperform 64 kb/s PCM in the less-than-optimum conditions found in a typical radio channel [42]. There are excellent 8 kb/s coders in the development stage that are very promising for fixed applications such as BETRS; with additional channel coding, an 8 kb/s CELP coder is the heart of the new digital cellular IS-54 standard, which will reach the market in 1992 or so. The frontier for telephone-quality coding today is around 4 to 5 kb/s, in anticipation of the so-called "half-rate" coders for digital cellular in the United States and Europe in the next few years.

Thus the capacity of a wireless access system can be boosted by a factor of 4 to 16, based on more efficient voice coding alone.

6.3.1.1 Strategies for Efficient Voice Coding

How are such dramatic advances being achieved? There are several very different strategies for reducing the bit rate.

Analysis of Inherent Redundancies in Speech Signals. As noted above, a simple waveform coder, such as the 64 kb/s PCM, assumes nothing about the nature of the analog sound input other than that it contains no important information at frequencies above one-half the sampling rate—the Nyquist theorem. Any input value is as probable at a given moment as any other; that is, the coder design assumes nothing at all about the patterns in the input signal.

Actually, of course, human speech exhibits a high degree of patterning, or redundancy, especially over short periods—up to, say, 100 ms. If the sampling rate is high, a particular sample will tend to be very similar to the samples that preceded it. The correlation coefficient between successive 8-kHz samples is generally 0.85 or higher [43]. In principle, if these redundancies can be analyzed, the coding process can be made much more efficient. Accepting the 0.85 correlation figure, we could expect to squeeze at least half of the bits out of the signal without affecting the quality—if we know which ones to eliminate.

Thus, one of the easiest, and one of the most fruitful, avenues for reducing the bit rate of the coder is to analyze the waveform for inherent redundancies. This analysis is strictly formal; it does not require any particular previous knowledge of the types of input patterning that may be experienced. In fact, as we shall see below, this operation can be extremely simple and mechanical, allowing for very economical implementations.

Analysis of Aural Performance Characteristics. The ear is a remarkable filter (i.e., information reducer) in its own right. In other words, we hear some sound better than others. Inherent biases in aural processing can be exploited to reduce the bit rate by transmitting only the information that is *perceptually relevant.* It appears that the overall perceived quality of speech suffers more from low-frequency noise than from high-frequency noise, which often seems to blend with, or to be masked by, the high-frequency hiss of fricative consonant sounds. Most of the perceptually relevant information is contained in the lower frequencies. So although most people can hear sounds up to 15 to 20 kHz, the conventional analog wireline telephone is designed to cut off frequencies above about 3 to 4 kHz. In a crude way, this represents a reduction in bandwidth that results from taking advantage of aural processing.

This approach can be extended into the digital domain with much greater precision. The study of acoustics by the Bell System began in the 1920s and 1930s [44]. Algorithms have been developed based on extensive empirical studies of human speech and hearing that allow the coder to concentrate on the most significant parts of the speech input signal and to downrate the perceptually less important elements. Significant reductions in bit rates are possible.

Parametric Modeling of Speech Production. In his childhood, Alexander Graham Bell once constructed an artificial speaking machine. It was actually an anatomical model of a human vocal tract, complete with teeth, throat, nasal passages, and a seven-section articulated tongue made of gutta-percha [45]. By carefully positioning a sound source in the "throat," Bell was able to effect a rudimentary articulation of individual English words. Others in the nineteenth and early twentieth centuries built more elaborate and apparently quite successful and intelligible speaking machines for public amusement [46].

Several generations later, armed with better techniques and materials, telephone engineers tried again. In the 1930s, Bell System researchers succeeded in constructing crude mathematical and electrical models of the vocal tract [47] (see Figure 6.29). They modeled the speech process as a *source* of sound (sound either at a given pitch or as white noise, representing voiced and unvoiced speech sounds, respec-

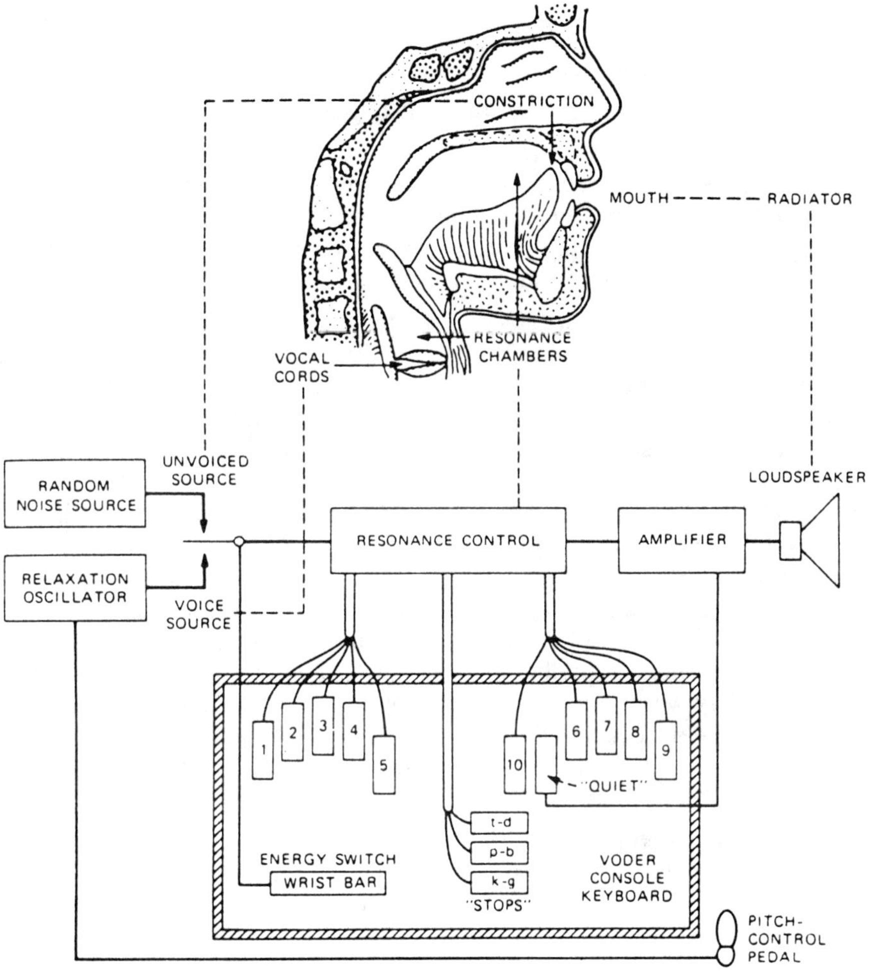

Figure 6.29 Early vocoder speech synthesizer. (From G. Calhoun, *Digital Cellular Radio,* Artech House, Boston, p. 321.)

tively) and a *filter* representing in an abstract form the passages of the mouth and nose. The "settings" of the filter could be changed, corresponding to the different positions of articulation (like the settings of Bell's artificial tongue), and the whole model could be represented by a series of mathematical equations. Instead of transmitting the actual input signal, that signal was measured, a mathematical model constructed, and the parameters or coefficients of the equations of that model were transmitted. By such means, intelligible speech was achieved at very low bit rates, although quality was poor and artificial sounding. Such techniques are often called *source coders,* because they do not encode the waveform; instead, they analyze the waveform to determine the characteristics of the source and then code the source characteristics, which can be transmitted much more economically.

Quantization of Block or Parametric Information. There is a very simple coding strategy that can be used to reduce the bandwidth of any transmission even further. Consider the following sample of text, which consists of the preceding paragraph of this book in which the word "the" (with its preceding and following spaces) has been replaced with an asterisk "*" everywhere it occurs:

> Several generations later, armed with better techniques and materials, telephone engineers tried again. In*1930s, Bell System researchers succeeded in constructing crude mathematical and electrical models of*vocal tract. They modeled*speech process as a source of sound (sound either at a given pitch or as white noise, representing voiced and unvoiced speech sounds, respectively) and a filter representing in an abstract form*passages of*mouth and nose. *"settings" of*filter could be changed, corresponding to*different positions of articulation (like*settings of Bell's artificial tongue), and*whole model could be represented by a series of mathematical equations. Instead of transmitting*actual input signal, that signal was measured, a mathematical model constructed, and*parameters or coefficients of*equations of that model were transmitted. By such means, intelligible speech was achieved at very low bit rates, although quality was poor and artificial sounding. Such techniques are often called source coders, because they do not encode*waveform; instead, they analyze*waveform to determine*characteristics of*source and then code the source characteristics, which can be transmitted much more economically.

By simply replacing a common single five-character sequence with a one-character symbol, the English text is compressed by over 5%, without any loss of information. This principle can be pursued much further. The following is a portion of the same text, with 30 common two-letter and three-letter sequences or "morphemes"—such as "ing," "and," "ed," "it," and so forth—replaced by single character symbols.

> Sev£å g±£∧§i∧¶s l∧£, arm# w@h πtt£ teçniφes % m∧§£iås, t¥eph∧¶e ±g!e£s -i# aga!. ! * 1930s, -ll Sy¡em Σsearç£s succe## ! c∧¶¡ru√& crude ma*m∧§≈å % ¥e√r≈å mΩ¥s $ * vocå -a√ *y mΩ¥# * βeç process. . . . [48]

By this very simple procedure, the text can be reduced by nearly 30% from its original length. We need only equip the receiver with a small look-up table of some 30 short sequences, and any English text message can be compressed by a significant amount.

Conceptually, this approach can be extended to digitized voice information, as we shall see. Essentially, it involves applying a second stage of coding to the output of the first voice coder. The digital bit stream may itself be analyzed for redundancies and recoded a second time quantizing the speech patterns contained in larger blocks of samples produced by waveform coders, or in the parametric information generated by source coders. The chief method of compression is to compare the output with a previously constructed dictionary of code words and to transmit the code word in the dictionary that is closest to the output of the coder. Further details are provided in the following section in the discussion of vector quantization.

6.3.1.2 Representative Coding Strategies

We shall review briefly four representative approaches that are currently being analyzed in connection with the development of digital mobile radio systems:

1. Continuously variable slope delta modulation (CVSD)
2. Adaptive subband coding (SBC)
3. Residual-excited linear predictive coding (RELP)
4. Vector quantization.

CVSD is a delta-modulation technique that was in favor several years ago and was one of the first modern coders to be implemented widely. It is now generally considered to have been superseded. Subband coding and RELP, and related coders, represent probably the two most popular techniques under investigation today for application to digital mobile telephony. Vector quantization, which is actually a general approach more than a specific technique, is one of the areas that will likely come into its own in the next five years and will strongly influence future coder strategies; the CELP coder selected for the IS-54 digital cellular standard employs elements of a VQ strategy.

Continuously Variable Slope Delta Modulation (CVSD). Delta modulation is based on the principle that, instead of encoding each sample independently of all others, a gain can be realized if we encode only changes from one sample to the next. This is a very simple way of exploiting speech redundancies. It yields a type of waveform coder that is simple, robust, and more efficient than conventional PCM.

A delta-modulation coder in its simplest form simply decides whether each sample is higher or lower in amplitude than the previous sample. If higher, it moves up one step in a preestablished quantization scale. If lower, it moves down one step. It effectively tracks the analog signal with a stepwise approximation (see Figure 6.30). Delta modulation is efficient because it only transmits the change—the "delta"—between the current sample and the previous sample, instead of retransmitting the entire amplitude scale of each sample. In effect, a "1" bit says simply: "Add one step to whatever value you had before." A "0" bit says: "Subtract one step from whatever value you had before." The coder is assumed to have a memory for at least the previous sample.

Delta-modulation coders are susceptible to two types of inherent quantization error. On the one hand, if the signal amplitude rises or falls at a rate faster than the

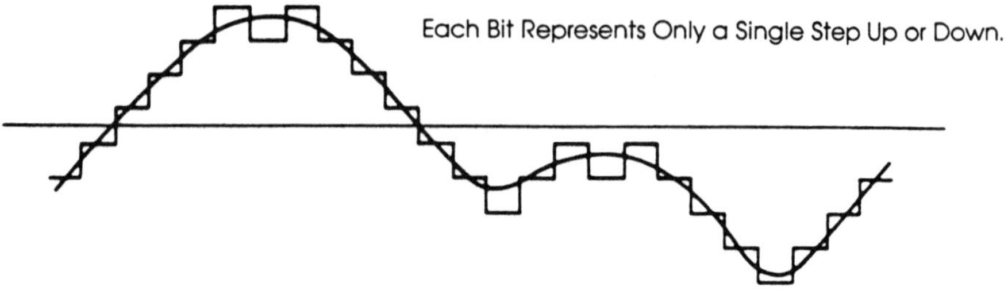

Figure 6.30 Tracking a waveform with delta modulation. Each bit represents only a single step up or down. (From G. Calhoun, *Digital Cellular Radio,* Artech House, Boston, p. 323.)

step size, the delta-modulation coder cannot keep up with it. It falls behind the analog signal, producing what is known as *slope overload.* On the other hand, when the signal amplitude is stable, the coder will oscillate between two states that straddle the actual value. This produces granular noise (see Figure 6.31). Unfortunately, these two forms of inherent error are complementary and are related to the step size of the delta-modulation coder. If a large step size is used, slope overload will be reduced, because the coder can keep up with a rapidly changing signal (see Figure 6.32). On the other hand, a large step size increases granular noise (see Figure 6.33). Perhaps surprisingly, granular noise is more objectionable than slope overload at equivalent power levels. This is because slope overload tends to occur during active speech, when the noise is masked by the speech itself and is therefore less perceptible. Granular noise occurs during periods of silence; hence it stands out.

Researchers therefore began to explore adaptive delta-modulation algorithms. The basic idea was to operate with a step size that adapted to the conditions of the signal, increasing when slope overload was detected and decreasing when the signal flattened out. For example, in a particular adaptive-delta approach that has come to be known as *continuously variable slope delta modulation* [49], the step size is

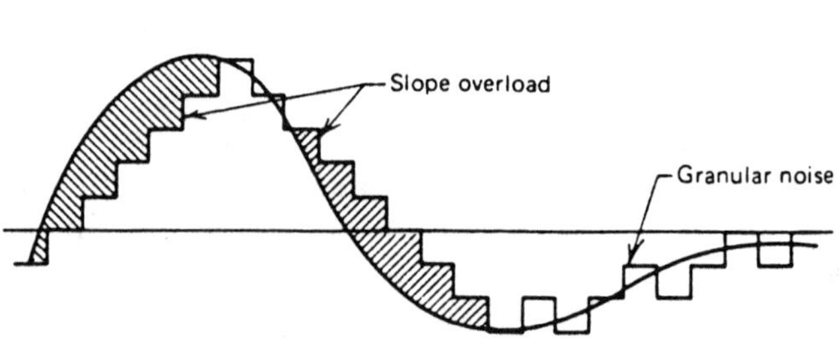

Figure 6.31 Characteristic forms of noise in delta modulation. (From G. Calhoun, *Digital Cellular Radio,* Artech House, Boston, p. 323.)

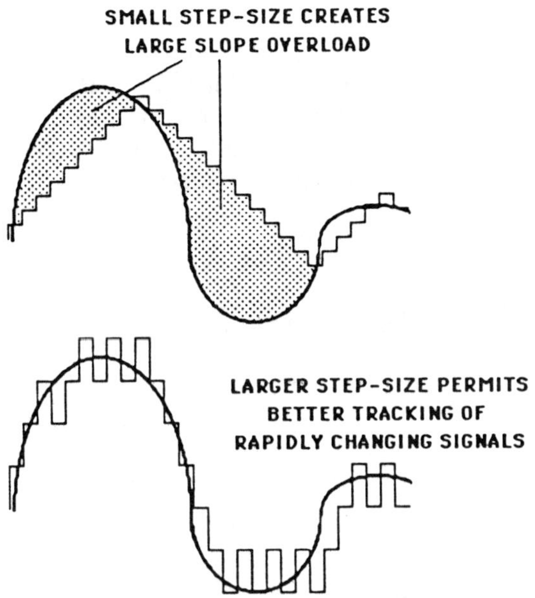

Figure 6.32 Delta modulation: the effect of large step size on slope overload. (From G. Calhoun, *Digital Cellular Radio,* Artech House, Boston, p. 324.)

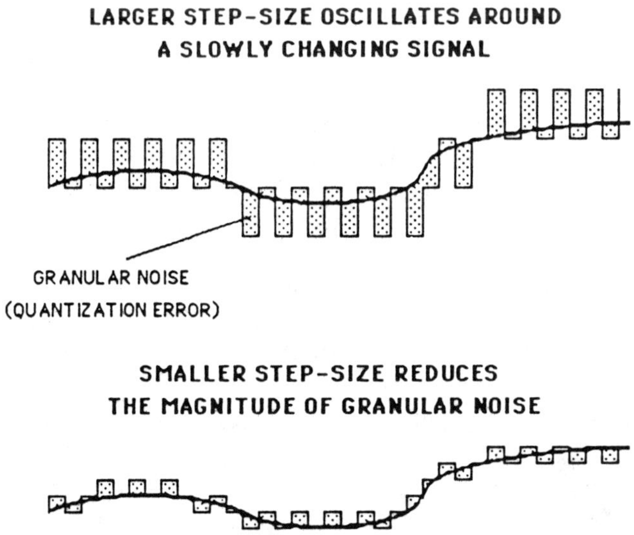

Figure 6.33 Delta modulation: the effect of small step size on granular noise. (From G. Calhoun, *Digital Cellular Radio,* Artech House, Boston, p. 324.)

increased by a predetermined amount if the coder detects four "1" bits or four "0" bits in succession, which would tend to indicate that the input signal is rising or falling faster than the coder can track it, hence slope overload is probably occurring. In between such forced increases, the step size is allowed to decrease exponentially with each sample. At some point it decreases to a very small level at which overload is detected once again, causing a new step-size increase (see Figure 6.34).

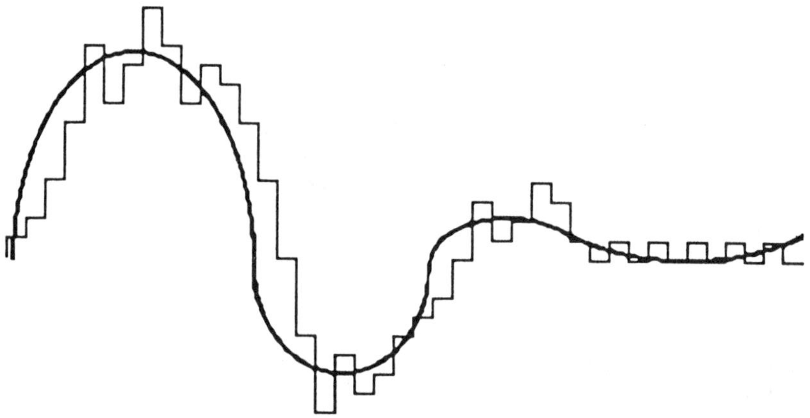

Figure 6.34 Delta modulation with adjustable step size. (From G. Calhoun, *Digital Cellular Radio*, Artech House, Boston, p. 325.)

CVSD helps to overcome the inherent problems of overload and granular noise and allows the delta coder to realize the efficiencies inherent in difference-only encoding. Telephone quality equivalent to 64 kb/s PCM has been achieved with delta coders at 32 to 40 kb/s. The technique is also relatively simple to implement.

On the other hand, CVSD coders do not deliver adequate quality at the 8 to 16 kb/s level needed for today's mobile systems [50]. Indeed, difference-encoding techniques that work on inherent speech redundancies are apparently not sufficient to reach the desired low bit rates for most wireless access applications. To achieve lower rates, researchers have turned to more powerful techniques.

Adaptive Subband Coding. Delta-modulation techniques are not inherently adapted to the transmission of perceptually important characteristics of human speech. Subband techniques, on the other hand, are based on a knowledge of the unique characteristics of human sound processing. By effectively weighting the different frequency components of the digitized signal to correspond to the perceptual weightings assigned by the ear, subband techniques are able to reduce the bit rate needed for perceived telephone quality.

The concept of subband coding is fairly simple [51]. The input signal is first filtered into several frequency subbands. Figure 6.35 shows an idealized set of subbands. In the simplest subband coders, each subband is digitized with a fairly conventional PCM-type waveform coder (see Figure 6.36).

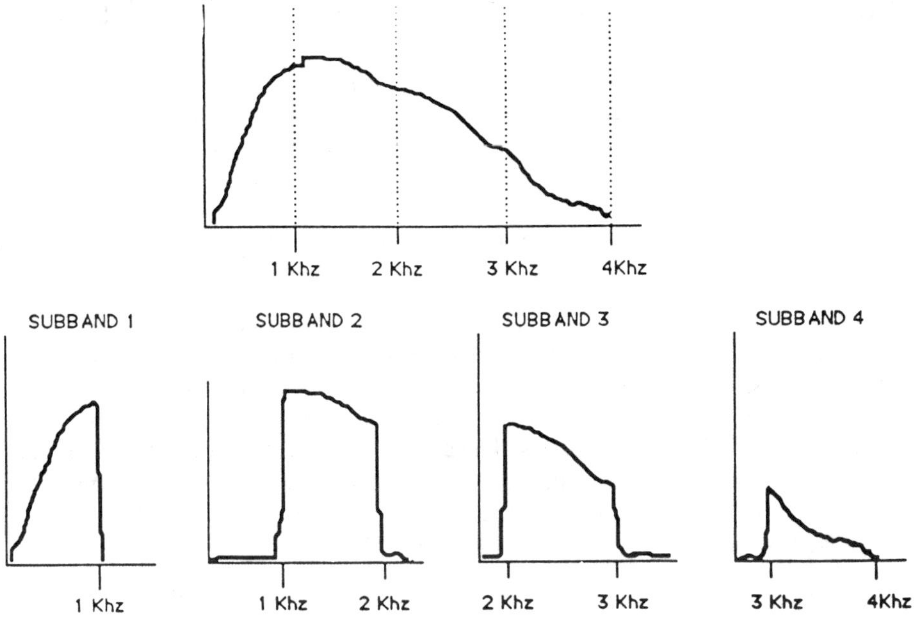

Figure 6.35 Hypothetical subbands. The original analog input is filtered into four separate frequency subbands. Each is then coded as an individual waveform. (From G. Calhoun, *Digital Cellular Radio*, Artech House, Boston, p. 326.)

Note that simply dividing the signal into subbands would not reduce the bit rate. The savings come about in several ways. First, instead of dividing the total available number of bits in the sample equally across all bands, a relatively larger number of bits can be assigned to low-frequency bands below 1 kHz, which tend to contain the most important tonal information that we associate aurally with quality, or fidelity, or speaker recognition. The higher bands tend to contain the hissing, noise-like sounds of certain consonants, such as *s* and *f*, and adequate quality can be achieved in these subbands with much lower bit rates. It is also possible to adjust the quantization step size for each band depending on the energy level in that band. This allows the system to tolerate higher levels of quantization noise by masking it under high-amplitude bands.

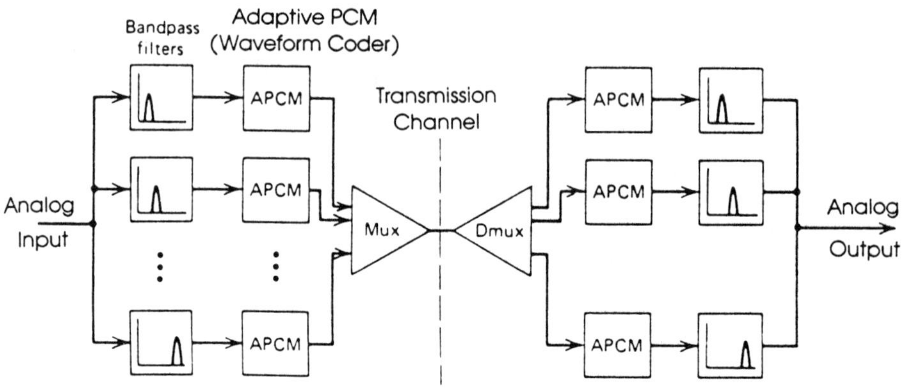

Figure 6.36 The structure of a subband coder. (From G. Calhoun, *Digital Cellular Radio,* Artech House, Boston, p. 327.)

More recently, more fully adaptive techniques have been applied. In standard subband coding, the bit allocations to each band are fixed. In adaptive subband coding, each sample is analyzed to determine in which band the greatest amount of speech energy, and information, lies. The available bits are then adaptively allocated to the various subbands. For example, in an *s* sound, which is unvoiced, most of the information will be in the higher frequency bands. If we assume an 8-bit sample and four subbands, the allocation of bits might be:

Subband A (lowest)		1 bit
Subband B	1 bit	
Subband C	2 bits	
Subband D (highest)		4 bits

On the other hand, a melodious voiced sound like the vowel *o* would generate a different bit allocation:

Subband A (lowest)		5 bits
Subband B	2 bits	
Subband C	1 bit	
Subband D (highest)		0 bits

The allocations can be changed for each sample, allowing the coder to attain good quality with fewer bits.

Unlike delta modulation, subband coding is capable of reaching bit rates of around 16 kb/s with excellent quality. Moreover, because of the underlying waveform coding in each band, subband coders tend to be fairly robust. On the other hand, subband coders are algorithmically more complex than delta-type coders, and it is only in the past few years that progress in digital signal processors has enabled the effective realization of many subband algorithms.

Residual-Excited Linear Predictive Coding. To understand the structure of a source coder, let us conceive of the vocal tract as similar to a musical horn instrument such as a saxophone. At one end there is a source of sound energy, referred to as *excitation.* In the case of the saxophone, it is a vibrating reed. The excitation source has two chief characteristics: pitch and loudness, both of which may vary, but tend to change slowly and may be conceived of as constant over short intervals, e.g., a few milliseconds. The sound produced by the excitation source is then channeled through a peculiar configuration of tubes and chambers—the body of the horn itself—which modulates the sound in a rich and characteristic way. From the perspective of the acoustical engineer, the body of the horn is a kind of complex filter that attenuates, distorts, and amplifies the excitation input to produce the throaty saxophone sound. (The same vibrating reed combined with a different "filter" will produce the clear notes of a clarinet.) The precise characteristics of this filter may vary from moment to moment as the saxophone player modifies the shape of the sound chambers by depressing different keys. With other types of horns, the player may modify the filter characteristics in other ways, using the hand in the bell for muting the French horn or a hat over the mouth of a trumpet.

It is possible to construct a mathematical model of the saxophone in which the characteristics of the complex filter are represented in a polynomial equation with a number of coefficients that determine its precise value. In principle, if we knew the values of these coefficients, along with the pitch and the gain (loudness), we could describe the sound being produced at any given moment in terms of a very few bits of information. We could then transmit these bits electrically to a receiver that incorporates an identical mathematical model of the saxophone. Plugging in the coefficients and the pitch and gain factors, the receiver model could then electrically synthesize the original sound. This could be accomplished with far fewer bits than a conventional waveform coder.

The human vocal tract can be modeled in a similar manner [52]. There is an excitation source modulated by a complex and variable filter corresponding to the vocal tract, which can be viewed as a complex of chambers and valves similar to a horn (see Figure 6.37). The major structural difference is that in human speech there are two kinds of excitation: *voiced* speech, in which the excitation is generated by the vocal chords and exhibits a detectable pitch, and *unvoiced* speech, generated by certain consonants, such as *s* and *f,* which produce noise-like, relatively pitchless sound from turbulent air passing through constricted vocal passages. Therefore, a source coder—such as a linear predictive coder (LPC)—must first determine whether the speech sample is voiced or unvoiced. If it is voiced, the pitch is extracted. The filter coefficients are determined by setting them in a feedback loop to compare with the

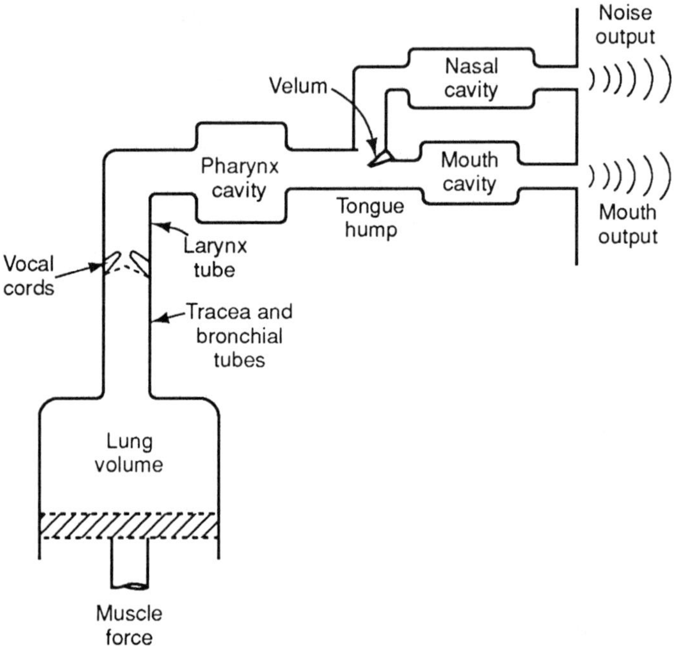

Figure 6.37 A schematic diagram of the functional components of the vocal tract. (From J.L. Flanagan, *Speech Analysis, Synthesis, and Perception,* Springer-Verlag, Berlin, 1972.)

input speech sample until the error, the difference between the modeled speech and the original sample, is minimized (see Figure 6.38). The only information that must be transmitted is:

1. The type of excitation, voiced or unvoiced
2. The pitch, for voiced excitation
3. The gain or loudness
4. The filter coefficients (parameters of the vocal-tract model) [53].

A source coder can thus be constructed to yield intelligible speech at very low bit rates. In fact, conventional LPC has largely been employed at bit rates between 1.2 and 4.8 kb/s, much lower than waveform coders. Unfortunately, the quality achieved at such low bit rates has not been acceptable for telephonic communication. LPC often generates intelligible but machine-like speech and can distort some speakers' voices severely. LPC is also far more computationally intensive than waveform coders, and until relatively recently LPC coders required large computers for implementation in real time.

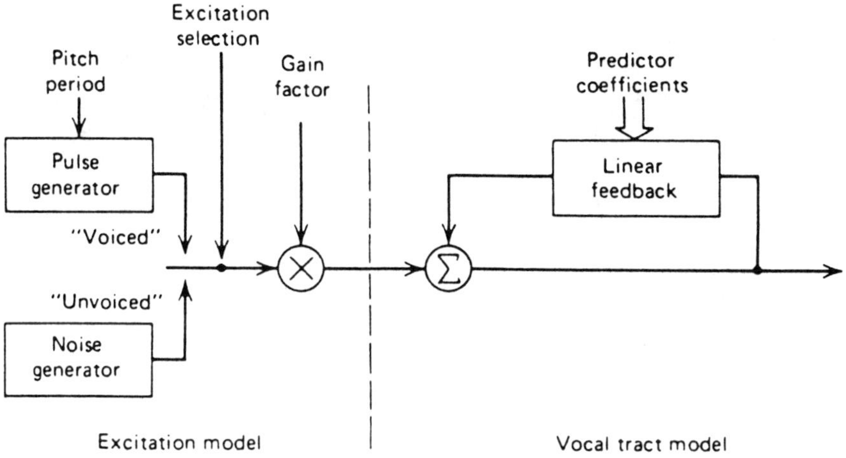

Figure 6.38 The structure of a linear predictive coder. (From G. Calhoun, *Digital Cellular Radio,* Artech House, Boston, p. 330.)

Thus, there was a crucial gap in the region of greatest interest to wireless access systems architects (see Figure 6.39). Traditionally, waveform coders, like subband coding, were unable to achieve good quality much below 16 to 20 kb/s. Source coders could reach much lower rates, but with a limit on speech quality below the telephony standards.

Recently, however, a new family of still more complex coders has emerged, starting to combine elements of both LPC and waveform coders to achieve good quality in the region of the gap. These techniques work on improving quality by better definition of the excitation source for the LPC-type coder. Instead of using a very simple measure of pitch, they invest a few more bits to create an excitation source that will more faithfully recreate the original sound. A number of techniques are currently being investigated, including regular pulse excited LPC (RPE-LPC), multipulse excited LPC (MPE-LPC), and code excited LPC (CE-LPC), among others. Such enriched excitation source coders are currently the most popular algorithms, along with subband coding, for mobile radio research. Bit rates of 8 to 16 kb/s are being tested in various implementations.

RELP illustrates the mixture of coding strategies. A speech sample of approximately 20 ms is the input to a more or less conventional LPC vocal-tract modeler, which extracts pitch, filter coefficients, and gain, and uses a feedback loop to minimize the residual or difference between the LPC output and the original. The residual, which is itself a waveform, is then encoded by means of a fairly conventional PCM-type coder, incorporating a few tricks to reduce the bandwidth of the residual, and is transmitted along with the coefficients of the LPC model. At the receiver, the

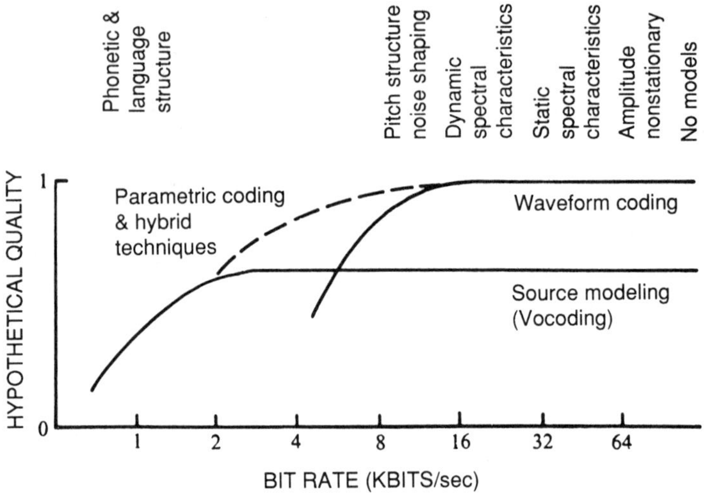

Figure 6.39 The relationship between speech quality and bit rate. (From G. Calhoun, *Digital Cellular Radio,* Artech House, Boston, p. 331.)

residual is utilized along with the pitch and gain information to define the excitation source that is put through the LPC filters.

Hybrid LPC coders are currently very popular for a number of wireless access applications. Such coders seem to be able to deliver better quality at lower bit rates than other algorithms that are currently implementable on a single microprocessor, more or less a requirement for low-cost mobile and low-power portable telephone units [54].

Vector Quantization. As suggested above, VQ is not an algorithm *per se,* but an idea that can be applied in conjunction with *any* coding algorithm to compress the data further. VQ is very powerful, but computationally very intensive. As processors become more powerful, VQ techniques will become the "turbo" factor to boost the efficiency of other techniques even further.

Vector quantization is effectively a second level of coding that is applied to the results of the primary algorithm [55]. To illustrate, let us consider a vector-quantized version of ordinary PCM.

The PCM algorithm produces, as we have seen, a stream of 8-bit words at the rate of 8000 per second. Assume that these are accumulated in a memory register in groups of six words, 48 bits per block. The second level of coding operates on these 6-word sequences. There are, in principle, an enormous number of these sequences. There are 256 possible 8-bit words. That means that the number of different possible 6-word sequences is 256^6, or 2^{48} different sequences.

Not all of these sequences, however, are equally probable. A very large number probably never occur at all. If it can be determined that a small number of these sequences can be used as a kind of codebook, every actual 6-word sequence can be compared with the codebook sequences, known as *vectors,* to find the best fit, the least difference, and then *quantized* by sending the best-fit vector instead of the actual sequence. It is similar in principle to the original quantization of the analog input signal into 8-bit words. As with all quantization processes, a measure of irreducible *quantization error* is introduced. A high degree of data compression, however, can be achieved. For example, if we construct a dictionary or codebook of 65,536 possible vectors and quantize every 6-word sequence with the best fit among these, then, instead of sending a 6-word sequence, 48 bits, we could send a 16-bit vector—a compression of the data by a factor of 3.

The principle of vector quantization can be applied to any coding technique. For example, an LPC coder that transmits 10 filter coefficients requires, say, 56 bits for each set of coefficients or vector. There are 2^{56} different possible sequences. If we could define 1024 most likely code words, however, each of which is represented by a 10-bit sequence, we could perform the search and best-fit process, and then transmit only a 10-bit vector. The data compression factor would be 5.6.

The VQ designer faces a number of complex and interesting questions. How many code vectors should be used? How large should the codebook be? The larger the codebook, the smaller the average quantization error, the more accurate the code. However, large codebooks mean that more bits must be used to transmit each vector, thus reducing the data compression gains. Also, more importantly, the larger the codebook, the longer the search to find the best-fit vector. If the search is exhaustive, i.e., the input sequence is compared with every single code vector, and a constant search rate is assumed, then it will take 256 times longer to search a codebook of 16-bit code vectors, of which there are 65,536, than a codebook of 8-bit code vectors. For a real-time process, the processing burden is a very significant constraint.

How should the code vectors be defined? The general answer is that the code vectors should be selected such that the average quantization error is minimized. This requires very intensive and skillful analysis of large numbers of speech samples.

Once the vectors are defined, how should the codebook be organized, and how should the search-and-match process be carried out so as to speed processing and allow for the use of larger and more accurate codebooks? Procedures have been developed to allow for more rapid searching. For example, tree-structured codebooks can be constructed, where, instead of searching every code word, the processor goes through a series of binary choices. Is the input sequence more similar to vector A or vector B? If A, then is it more similar to A.1 or A.2? If A.2, then is it similar to A.2.1 or A.2.2? And so forth. Another expedited process is *multistage vector quantization.* A small codebook is used, and a fast search quickly finds the best fit. The residual, or difference between the input sequence and the best-fit code vector, is measured, and then the residual is encoded with a second codebook, which is also small and amenable to rapid search. Several stages may be used. By keeping each codebook simple, the search process is rapid and better quality may be obtained more quickly (see Figure 6.40).

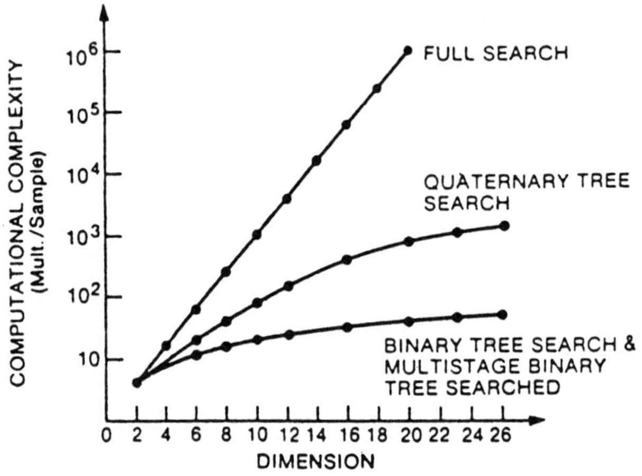

Figure 6.40 Computational complexity in VQ as a function of vector dimension and search strategy. (From G. Calhoun, *Digital Cellular Radio,* Artech House Boston, p. 333.)

In all these versions, it is assumed that the codebook has been constructed in advance and stored in read only memory (ROM) for access by the processor. (Memory limitations are another significant constraint for VQ systems.) This implies that the codebook is, practically speaking, an average across a large number of speakers. If it were possible to construct an individually tailored codebook for each speaker, the average quantization error could be further reduced for a given transmission rate. The ultimate VQ coder would be capable of retuning the codebook, optimizing it for each speaker. Who knows how many supercomputers it would take to accomplish this in real time!

The beauty of VQ is that it can be applied to the output of any algorithm. It can be applied to the degree desired or permitted by the economic and processing constraints of the system. It is true that efficient VQ algorithms will require enormous processing power, compared to the underlying techniques on which they operate. As digital signal processors continue to increase in power and decrease in cost, however, vector quantization will come into play in low-bit-rate coders appropriate for the bandwidth-limited mobile channel.

There are literally hundreds of different voice-coding algorithms, falling into many different families. It is beyond the scope of this book to attempt even a cursory survey of all these techniques. The reader is referred to a number of excellent texts and periodical publications on the subject [56].

The concepts and approaches discussed above help to point out, however, the exponential relationship between increased efficiency (i.e., reduced bit rates, holding quality constant) and computational complexity (see Figure 6.41). In one sense, this is a daunting picture. Each new step forward is won at an exponentially increasing price in terms of computing power. On the other hand, the trends in the availability and cost of computation have been declining exponentially for many years. With the advent of faster DSP chips and parallel processing architectures, it may well be possible, dollar for dollar, to apply hundreds or thousands of times more processing power to a voice-coding algorithm in a few years time. This is certainly one area where our sense of the possible has been expanded in recent years.

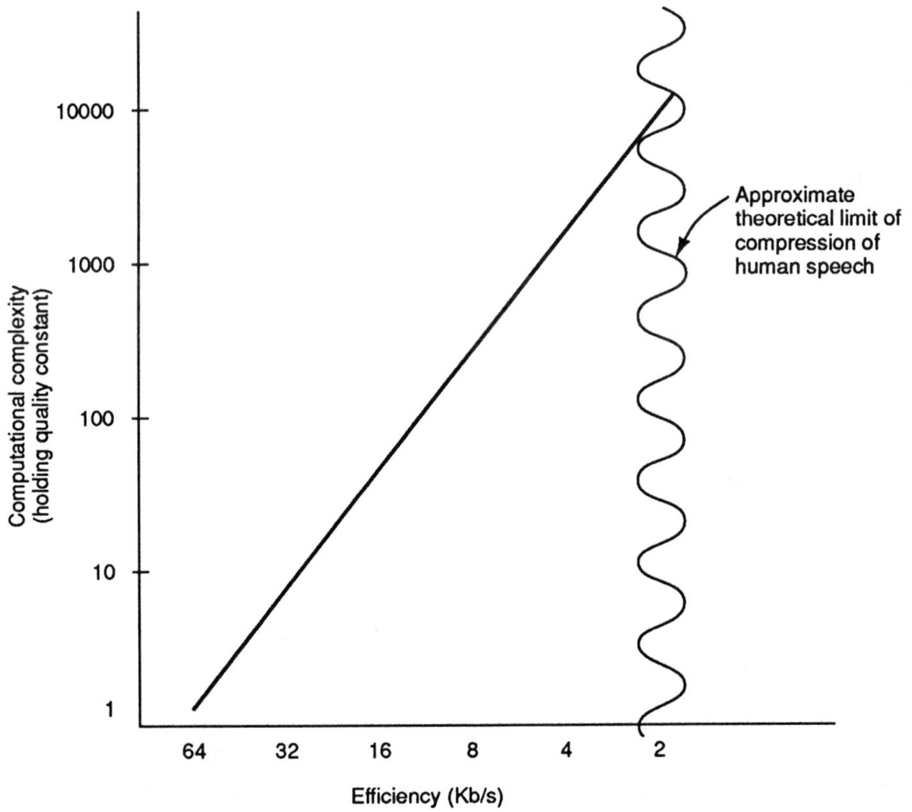

Figure 6.41 Relationship between computational complexity and efficiency (idealized).

6.3.2 Modulation: Increasing Information Density

The alternatives for RF modulation are almost as diverse as for voice coding. Where FM was once the only real option, today's wireless access designer might consider the following techniques, among others:

Proposed Modulation Technique	Illustrative Reference
Single Sideband Amplitude Modulation	[57]
Frequency Shift Keying (FSK)	[58]
Fast FSK	[59]
Tamed Frequency Modulation	[60]
Generalized Tamed FM	[61]
Double Phase Shift Keying	[62]
Minimum Shift Keying (MSK)	[63]
Multiamplitude MSK (MAMSK)	[64]
Gaussian MSK	[65]
Continuous Phase Modulation	[66]
Phase Shift Keying (PSK)	[67]
Offset PSK	[68]
$\pi/4$ Shift PSK	[69]
Raised Cosine PSK	[70]
Compact Spectrum, Constant Envelope PSK	[71]
Quadrature Amplitude Modulation (QAM)	[72]
Nonlinearly Filtered QAM	[73]

The actual choice of modulation for a particular application is determined by several considerations, including the following:

1. Information density (spectral efficiency)
2. Power spectrum characteristics
3. Amplifier characteristics.

6.3.2.1 Information Density

Designers of modulation methods for wireline voice transmission (i.e., line coding methods) tend not to be concerned with the "efficiency" of the transmission (measured in terms of signal bandwidth) since, as one text puts it, "wire pairs have no rigid bandwidth limitations" [74]. Most wireline transmissions utilize only a very small part of the true capacity of the circuit—typically far less than 1%. Advanced wireline systems such as coaxial cable and fiber optics possess so much bandwidth that there is really no incentive to use it particularly efficiently. And if need be, another cable can always be laid. For this reason, line coding is usually based on binary (two-level) logical code, to simplify the detection problem for the receiver and improve system performance. In fact, some wireline modulation systems expand the bandwidth of the signal even further—that is, they are less "efficient" (or more redundant) than even straight binary coding [75]. The commonly used Manchester coding scheme, for example, provides superior clock recovery—at the expense of doubling the bandwidth relative to alternate mark inversion coding [76]. But the wireline and LAN applications in which it is used offer so much bandwidth relative to their applications, that the bandwidth penalty is not seen as significant.

The sense of having "bandwidth to burn" has led wireline system designers to be quite willing to trade bandwidth for low-cost demodulators and robust, virtually error-free channels. This in turn allows them to implement high-quality, economical voice coders, which, however, can only survive in a virtually error-free channel. The whole philosophy of wireline digital transmission is based on robust, inefficient modulation.

For radio modulation, our thinking must be very different. Because the radio spectrum is finite, the bandwidth of the radio transmission should be as narrow as possible. Capacity shortages are chronic is almost all radio applications. As wireless access becomes a truly ubiquitous, broad-based set of services reaching tens of millions of people, increased capacity must come from more efficient use of the available spectrum. Improving spectrum efficiency and raising system capacity dictate the use of more efficient modulation techniques. But efficient modulation (as defined below) does incur a number of penalties, such as the need for more power, higher SNRs, better channel conditions or, when this is impossible (as in the mobile applications), the tolerance for enormously higher residual error rates in the channel. This in turn drives the voice-coding designers in new directions, as we touched on in the previous section.

One important way to increase the carrying capacity of a radio signal is to increase the *information density* of the transmission [77]. The simplest measure of information density is the concept of *spectral efficiency* described in Section 6.2.5: the transmission bit rate divided by the channel bandwidth, i.e., the number of bits/second/hertz, sometimes loosely referred to as bits per hertz (b/Hz). Increasing the number of bits/Hz is achieved by increasing the number of *levels* in the modulation system.

Consider the following simplified example. A two-level modulation scheme is used in which there are two predefined levels of signal amplitude, representing "1" and "0." The input bit stream is encoded one bit at a time. Each pulse, or *symbol,* contains only one bit of information (see Figure 6.42). It turns out that the best we can expect, theoretically, for a binary modulator of this kind is about 1 b/Hz with a double-sideband signal [78]. Thus, if we need 10 kb/s for our voice coder, we must use 10 kHz of spectrum for a single wireless access circuit.

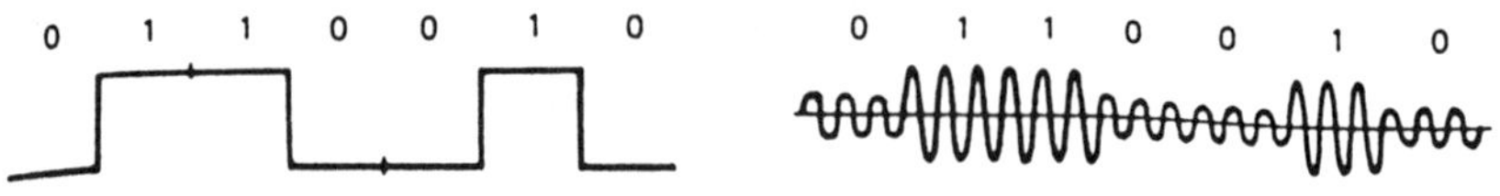

Figure 6.42 Two-level amplitude shift keying. (From G. Calhoun, *Digital Cellular Radio,* Artech House, Boston, p. 305.)

Now consider an enhanced amplitude shift keying system in which four levels are defined instead of two. The input bit stream is encoded two bits at a time. If a given sequence of two bits is "00," a pulse of level-1 amplitude is transmitted. If the bit sequence is "01," in that order, a pulse of level-2 amplitude is transmitted. A "10" sequence is transmitted with a level-3 amplitude pulse. An "11" sequence is transmitted with a level-4 pulse (see Figure 6.43). Each symbol now contains two bits of information. The bit rate is still 10 kb/s (from the voice coder, which has not changed), but the symbol rate is 5 *kilosymbols* per second. Under optimal conditions, the occupied radio bandwidth would be reduced by 50%, to 5 kHz. Spectral efficiency is now 2 b/Hz.

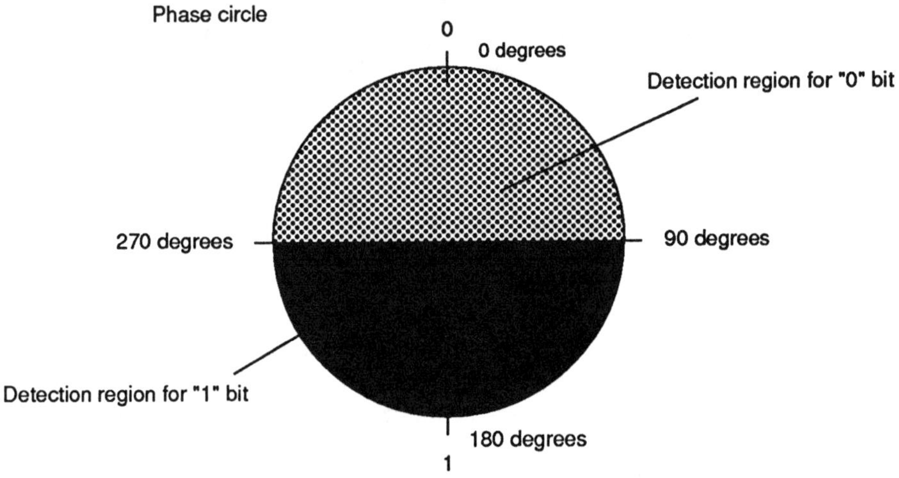

Figure 6.43 Detection sectors for binary PSK. (From G. Calhoun, *Digital Cellular Radio,* Artech House, Boston, p. 307.)

The concept of multilevel modulation is very powerful. We can extend the 4-level system to an 8-level system, with three bits encoded in each symbol, and a spectral efficiency of 3 b/Hz. A 16-level modulator, encoding four bits per symbol, would achieve 4 b/Hz. Of course, the levels can be represented by different frequencies in an FSK system, or different phase values in a PSK system. In point-to-point microwave radio, 64-level modulation schemes are becoming common enough, 6 bits per symbol, roughly 6 b/Hz. Indeed, modulation techniques are being developed that will employ 256 or even 1024 levels, encoding 8 and 10 bits per symbol, respectively [79].

It might seem that multilevel modulation is the Aladdin's lamp of radio design. Why not simply keep increasing the modulation level to achieve whatever spectral efficiency we desire? Alas, as with all such things, there is a price to pay.

Consider the problem faced by the demodulator, which must detect the incoming radio signal and decide which level in the modulation scheme the signal corresponds to at a given instant. Assume that our example is a binary phase shift keying (BPSK) system, which uses 0° to symbolize a logical "0" bit, and 180° for a logical "1" bit. We know that the radio channel—especially in mobile applications—may introduce distortions in phase. Nevertheless, the discrimination for BPSK is relatively easy. If the detected phase falls anywhere between 89° and 271°, the demodulator registers a "0" bit; otherwise it detects a "1" bit (see Figure 6.29). Such a system is robust. Distortions of the phase information can be tolerated as long as they do not exceed 90° in either direction.

A 4-level PSK demodulator must discriminate with twice the accuracy of a BPSK system (see Figure 6.44). The increase in information density is offset by the greater challenge posed to the demodulator. Obviously, if same received signal—

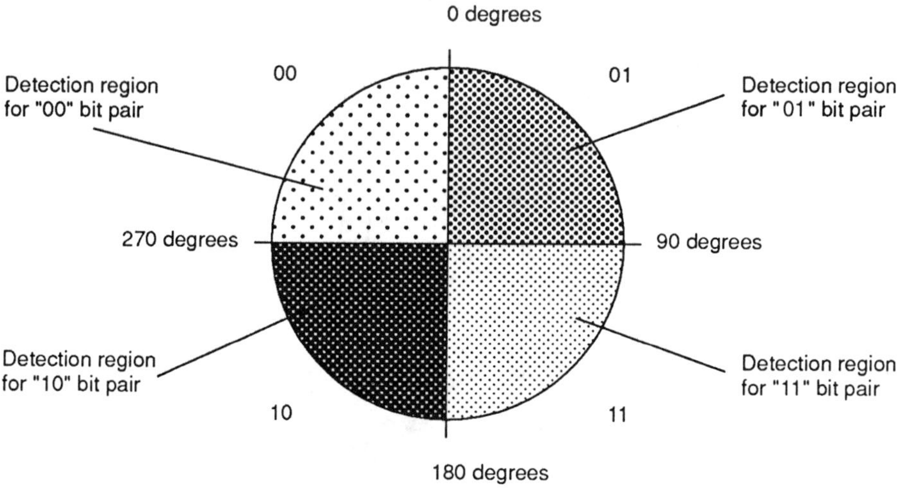

Figure 6.44 Detection sectors for QPSK modulation. (From G. Calhoun, *Digital Cellular Radio,* Artech House, Boston, p. 307.)

with a given amount of phase distortion produced by the transmission environment—is processed by both a 2-level and a 4-level PSK demodulator, the 4-level demodulator will make more erroneous decisions than the 2-level demodulator. For example, a "0" bit, encoded by the modulator as 0° of phase, and distorted by the environment to the point where it is detected as a 60° phase at the sampled instant at the demodulator, would be registered accurately by the 2-level process (i.e., 60° is still within the detection zone for a logical "0" bit), but would be registered as an (unrecognized) error by the 4-level process.

In sum, it should be intuitively clear that to achieve the same bit-error rate (BER), a 4-level PSK demodulator needs a better signal (i.e., higher signal-to-noise ratio, or SNR) than a 2-level PSK demodulator. In fact, as Figure 6.45 shows, the 4-PSK demodulator needs precisely twice the SNR (i.e., 3 dB better) that 2-PSK needs, to achieve the same BER [80] (the figures assume coherent demodulation). Or, to look at it another way, for a given received signal level (RSL), the 4-PSK system will suffer 100 to 10,000 times more bit-errors than the 2-PSK system! To achieve the same BER with a 16-level modulation scheme, we would need approximately 12 dB better SNR.

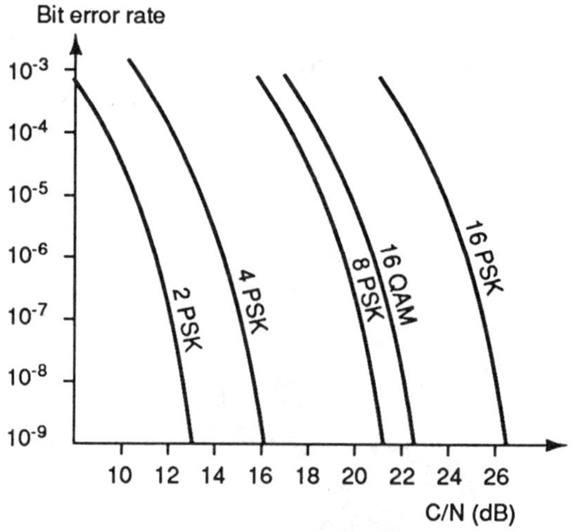

Figure 6.45 Bit error performance for various modulation formats (coherent detection). (From M.J. Miller and S.V. Ahamed, *Digital Transmission Systems and Networks, Volume 2: Applications,* Computer Science Press, Rockville, MD, p. 37.)

In other words, the higher the level of modulation employed, the less robust the system, and the higher the RSL required to achieve a desired BER. In some systems, this can be viewed as a problem of power; we can boost power to achieve a better SNR and take advantage of the spectral efficiency of high-level modulation. In BETRS applications, 16-level PSK is currently used. In the cellular-type or interference-limited systems, however, gains in capacity from higher level modulation may be canceled out, in part, by the increased vulnerability to cochannel interference.

As a practical matter today, 4-level PSK systems seem to be the popular choice today for mobile (and portable) applications. The state of the application for digital circuitry allows for cheap, robust implementations of several QPSK variants. In fixed applications, 16-level PSK is common and even 32-level systems (PSK or QAM) have been discussed. I believe that modulation schemes in this range will prove feasible in many applications (except perhaps those that are exclusively mobile), but that beyond about 32-level or perhaps 64-level modulation, the wireless access designer is faced with diminishing returns from modulation strategies.

6.3.2.2 Power Spectrum Considerations

The output of any unfiltered modulation is a transmission similar to Figure 6.46. Most of the energy is concentrated in a narrow band around the center frequency; however, there is also energy contained in the sidelobes, which may spread out into adjacent channels. This distribution of energy over a frequency band is known as the *power spectrum* [81].

Spectrum regulations normally specify an *emissions mask* to limit this sidelobe energy and reduce interference. Figure 6.47 shows a typical emissions mask laid on top of an unfiltered power spectrum. In a case of this kind, there is clearly a need for additional *filtering* to limit the out-of-band energy of the transmission. This requirement can add cost to the radio design, as well as introduce nonlinearities into the signal, which can force the use of more costly amplifiers. The inherent power spectrum of the modulated carrier is therefore another important consideration in the selection of modulation techniques.

The design of the modulation scheme itself can be used to reduce the size of the sidelobes. For example, so-called *continuous phase modulation* techniques have been developed in which the phase of the radio carrier is not subject to any sharp or discontinuous shifts, which helps to reduce the sidelobes [82]. One of the most popular of these techniques is known as *minimum shift keying* (MSK), in which two specially selected signaling frequencies are used to "key" the digital bit stream. Figure 6.48 shows the waveforms generated by MSK and 4-PSK (QPSK); note that the MSK is free of the discontinuities found in the PSK-modulated carrier, which results in much smaller sidelobes for MSK (see Figure 6.49). A form of MSK has been chosen for the European digital cellular standard (GSM—see Section 6.5) that is based partially on these considerations.

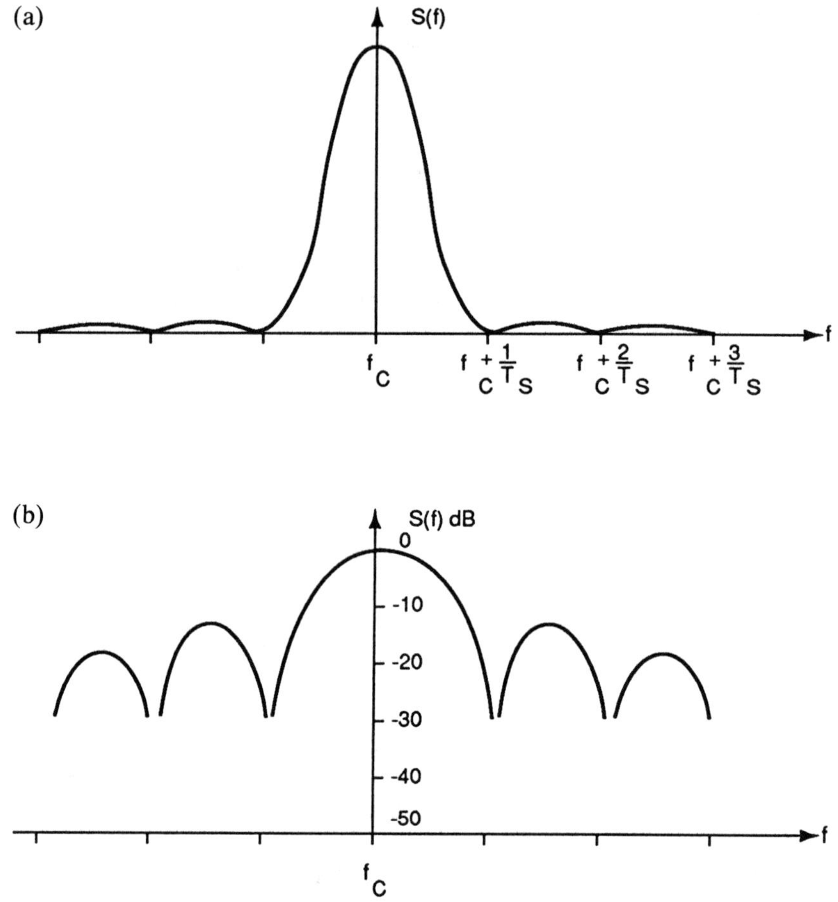

Figure 6.46 Power spectral density of PSK waveforms; (a) linear scale, (b) dB scale. (From M.J. Miller and S.V. Ahamed, *Digital Transmission Systems and Networks, Volume 2: Applications*, Computer Science Press, Rockville, MD, p. 20.)

6.3.2.3 Amplifier Considerations

As already mentioned, the choice of modulation technique also influences the selection of the type of amplifier. Modulation is normally performed at very low power, and then the modulated signal is translated upward in frequency and the power boosted (amplified) for actual transmission. Amplification can be accomplished by several well-known types of amplifier circuits (designated Class A, Class B, Class C, and so forth). Some amplifiers are designed to be highly "linear," which means essen-

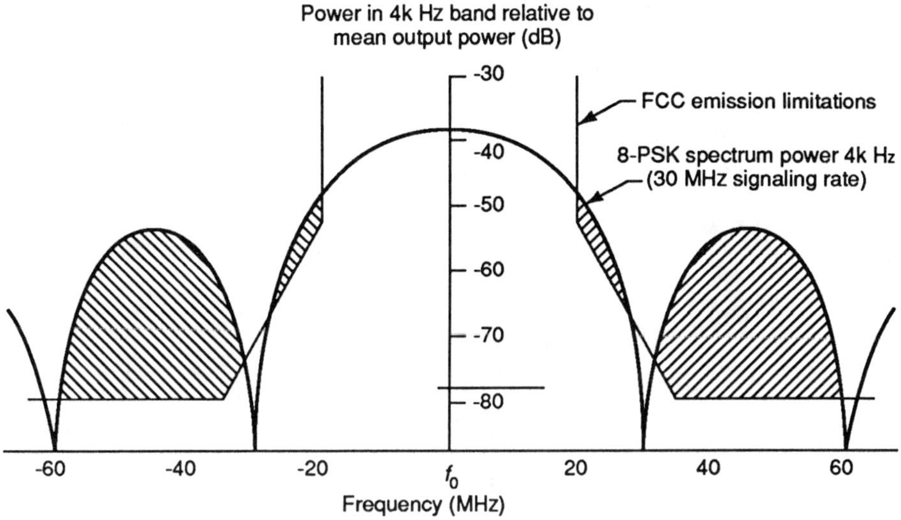

Figure 6.47 Typical FCG emission mask (11 MHz) *versus* a standard modulation power spectrum (90 Mbps, 8-PSK). (From J.C. Bellamy, *Digital Telephony,* First Edition, John Wiley & Sons, New York, p. 309.)

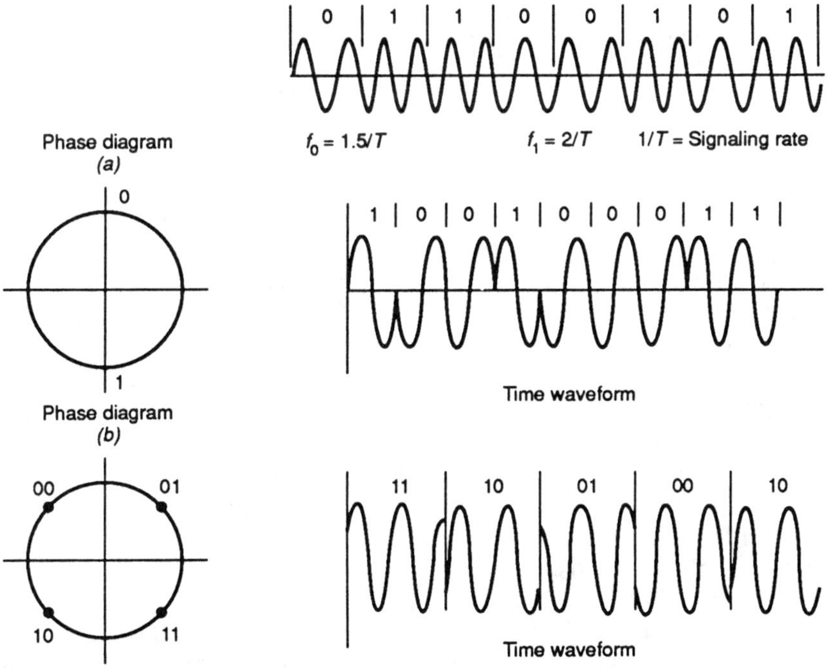

Figure 6.48 Continuous *versus* discontinuous phase modulation. (From J.C. Bellamy, *Digital Telephony,* First Edition, John Wiley & Sons, New York, p. 283.)

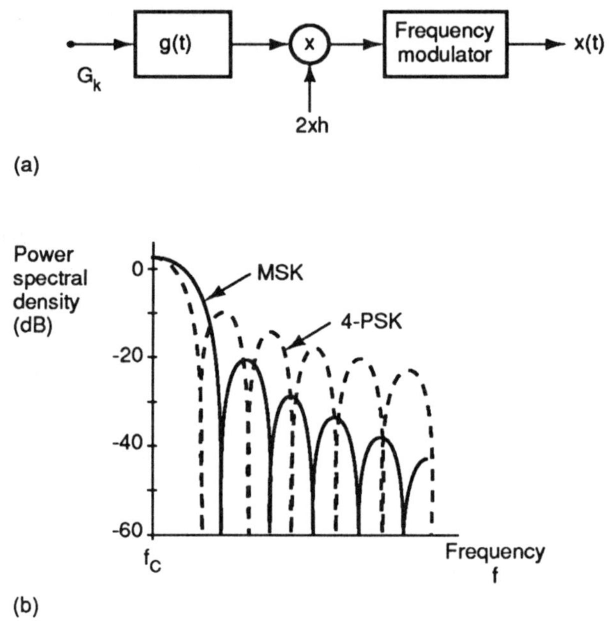

Figure 6.49 Power spectra for MSK and 4-PSK; (a) continuous phase modulation, (b) spectrum of MSK (and 4-PSK for comparison). (From M.J. Miller and S.V. Ahamed, *Digital Transmission Systems and Networks, Volume 2: Applications,* Computer Science Press, Rockville, MD, p. 29.)

tially that they do not distort the signal during the amplification process. But high linearity means higher cost. Also, the linearity is achieved by using the amplifier very inefficiently from a power perspective [83]. A modulation technique that requires a high degree of linearity, such as QAM (or any form of amplitude modulation), will force a more costly and power-hungry amplifier implementation [84].

Therefore, as noted above, some wireless access system designers have preferred so-called *constant envelope techniques* (such as MSK or unfiltered PSK, where the signal stays within a constant amplitude "envelope"). These forms of modulation allow the use of cheaper, more power-efficient amplifiers that do not need to be highly linear. The lower power may be critical if the application calls for a portable terminal (with portables, battery life is always a concern). But when a PSK modulated carrier is filtered (to reduce the sidelobes), the envelope becomes nonconstant [85]. Thus, the IS-54 North American digital cellular standard has specified a form of QPSK known as $\pi/4$-shifted PSK, one of the virtues of which is that it avoids the discontinuous zero-crossing phase shifts that generate the robust sidelobes and lead to the need for filtering, which in turn produces a nonconstant amplitude envelope and forces the use of highly linear amplifiers.

6.3.3 Equalization: Recovering the Physical Code

The digital transmitter produces carefully shaped pulses to input to the transmission channel. Indeed, pulse shaping is a specialized DSP discipline in its own right [86]. But the transmission process inevitably results in the distortion and "smearing" of the original sharp transmitted pulses, creating intersymbol interference (ISI) (see Figure 6.50). If the ISI is too great, it will cause demodulation errors by the receiver. In the radio environment, by definition, such distortions are uncontrollable; moreover the channel characteristics are constantly changing in a rapid (at least for mobile channels) and unpredictable manner, continuously altering the nature and scope of the distortion. ISI is thus a very significant challenge to a digital radio system.

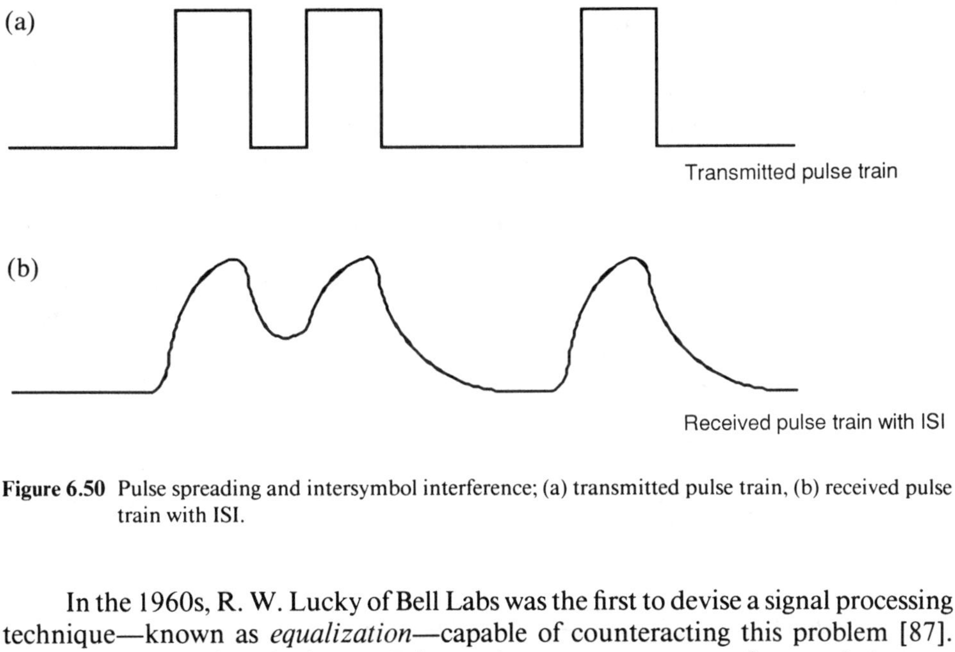

Figure 6.50 Pulse spreading and intersymbol interference; (a) transmitted pulse train, (b) received pulse train with ISI.

In the 1960s, R. W. Lucky of Bell Labs was the first to devise a signal processing technique—known as *equalization*—capable of counteracting this problem [87]. The fundamental idea is simple: Prior to the commencement of transmission, we obtain certain information about the distortion and attenuation characteristics of the channel, and then we employ this information in the receiver to correct the received signal and reshape the incoming pulses. This is done by first sending a "training sequence"—a preamble consisting of a defined sequence of pulses that is "known" to the receiver beforehand. The receiver then compares the actual received signal with what it knows the original input sequence to have been. The difference, or error, is thus known to be channel-induced. From this data, the equalizer derives sufficient information to model the channel, and it sets its processing algorithms to subtract the expected error from all subsequent samples.

If the channel characteristics can be expected to remain stable for the duration of the transmission, as may well be the case for wireline channels, the initial training may suffice. If, however, the channel characteristics are expected to continue to change significantly during the course of the transmission—which is the case with most radio applications, especially mobile and portable applications—then the initial training must be supplemented by an ongoing adaptive equalization process that is capable of monitoring the changing channel characteristics and continuously tuning the equalizer parameters to subtract the proper "error" from each succeeding received symbol. The basic idea of *decision feedback equalization* (DFE) is also quite simple. The equalizer is assumed to start out with the right setting—a correct analysis of the distortion introduced by the channel—immediately after training is completed. Now the first message sample will be analyzed and extracted from the distorted incoming signal. Once this first pulse has been decided and its value is "known," the error between that value and what was actually received can be recomputed. This actual error can now be compared with what had been the predicted error. This difference—between the expected error and the observed error—can now be used to update the equalizer's model of the channel.

The basic idea is that if the values of the symbols already detected are known (past decisions are assumed to be correct), then the ISI contributed by these symbols can be canceled out exactly, by subtracting past symbol values with appropriate weighting from the equalizer output [88].

The concept of using past decisions to aid in the making of future decisions is central to all adaptive signal processing [89]. It is one of the overarching advantages of digital transmission in noisy, imperfect channels, such as those found in most wireless access applications. As the official AT&T history of engineering in the Bell System observes:

> Feedback control information was obtained by comparing decided bits against the receiver's incoming, distorted analog pulse train in order to ascertain an error component. This technique of using prior decisions to determine an ideal upon which to base adaptation became known as decision direction; it subsequently became the basis not only for equalization but for the recovery of timing and other necessary receiver parameters. [90]

The results of dynamic equalization properly applied can be dramatic (see Figure 6.51).

6.3.4 Error Correction: Recovering the Logical Code

Just as equalization uses software to reconstruct the physical code, forward error correction (FEC) denominates another large and diverse family of algorithmic techniques used to cleanse the logical code of residual bit-errors.

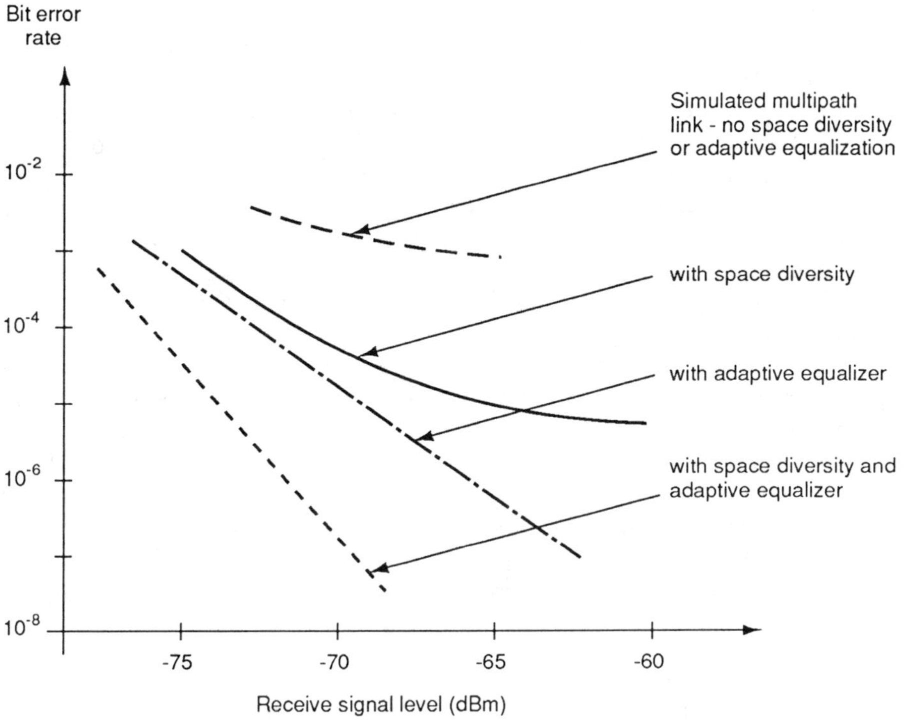

(Note: Multipath simulated notch at f_0 = 7MHz with depth of 12dB)

Figure 6.51 Effects of equalization and spatial diversity upon system performance. (From M.J. Miller and S.V. Ahamed, *Digital Transmission Systems and Networks, Volume 2: Applications,* Computer Science Press, Rockville, MD, p. 58.)

In some ways, FEC is the purest and most impressive example of the power of today's muscular software. In effect, it is possible to view error coding as another means of reducing the bit-error rate of the signal, just like increasing the transmitter power or implementing antenna diversity. The decibels that are gained from FEC—coding gain—can be added right into the link budget on the favorable side of the balance. The results can be surprising. Miller and Ahamed cite an example comparing a 4-PSK system (2 b/Hz spectral efficiency), and an 8-PSK system (3 b/Hz). Normally, the penalty for the higher spectral efficiency of the 8-PSK system would be a requirement for about 5 dB better SNR in this application. But if FEC (in this case a Viterbi algorithm with soft-decision decoding) is judiciously applied, the 8-PSK system can be beefed up to outperform the uncoded 4-PSK system by 3 dB [91]. In other words, there is a projected gain of 8 dB in effective SNR from coding alone!

This is the equivalent of boosting the transmitter power by a factor of about 6—yet, as the text comments, "the cost in terms of input data rate and transmitter spectrum width is nil" [92].

FEC is a very complex field, and it is neither possible nor necessary here to review the many different approaches [93]. All FEC techniques are based on the principle that by encoding the bit sequences prior to transmission in such a way as to introduce certain redundant information of a mathematically precise nature, the presence of transmission-induced errors can be detected by the receiver, and in many cases actually corrected. If the exact original information cannot be completely recovered, these techniques may still allow for a reasonably accurate estimation or interpolation of the missing data.

In one sense, the simplest form of FEC is to retransmit the desired message several times, with some sort of voting procedure at the receiver (best 3 out of 5, or best 6 out of 11) to determine the correct message in a very noisy channel. This is what is actually done today in some of the analog cellular control channels. Simple redundant transmission is, however, grossly inefficient in its use of the bandwidth. Modern FEC research is driven by the goal of correcting more errors with less additional information. In other terms, the maximum coding gain with the minimum additional bandwidth. The other chief constraint is the computational complexity of the more powerful techniques.

6.4 INGENIOUS ARCHITECTURES: BUILDING THE DIGITAL RADIO SYSTEM

The previous section dealt with the construction of individual digital radio circuits as multilayered coding processes implemented on powerful DSP chips. The challenge at the system-level is to combine a large number of these circuits in a finite band of the radio spectrum, within an efficient and practical access architecture capable of supporting the types of applications discussed in Chapter 5. Because of the scope of these applications, their orientation toward circuit-like communications, and the requirement for efficient spectrum utilization, we are chiefly interested in what are called *multiple-access* architectures. Multiple-access systems are designed to allow a large number of users to share a common pool of radio-telephone circuits, much like the sharing of any other common or trunked network facility. In this respect, multiple-access radio systems are unlike wireline dedicated access loop facilities where every user has a preassigned circuit. Architecturally, multiple-access radio systems bear a greater similarity to local-area networks in which the common transmission channels are available to all users. (The principle of multiple-access may not apply to certain very small scale radio applications, such as residential cordless telephony. For all the mainstream applications considered in Chapter 5, however, multiple-access is an absolute requirement.)

6.4.1 The Concept of Multiple Access

The first radio systems were designed to transmit and receive on a single frequency or "channel," and single-channel designs predominated in many mobile radio applications until quite recently (primarily due to the cost of equipping mobiles with multiple-frequency capability). Such single-channel systems usually operate like party lines. They are used in a broadcast mode, and do not provide true circuit-like (private, user-to-user) communications. A common example is a dispatch or "taxi-cab" type of system, where all users can hear each other at all times. In some applications (e.g., police radio), this may even be considered a positive feature. Such systems are normally limited to a single transmission from a single user at any one time. Just like a party line, any number of people can listen but only one can talk at any one instant. This means that single-channel systems are usually very inefficient in terms of traffic throughput.

Multiple-access radio systems encompass four main improvements over the single-channel concept:

1. The system is designed to allow multiple simultaneous transmissions by a reasonably large population of users. (Note that the transmissions may be "simultaneous" in an operational sense, when viewed on a time scale appropriate to human perceptual abilities, even when in some cases they may use time-division techniques involving sequential multiplexing of circuits—see below.)
2. The communications channels provided to users are circuit-like channels, capable of reasonably private, individual-to-individual communications.
3. The circuits are trunked—that is, there is no preassignment of any circuit to any user. Any user can gain access to any circuit, and may use different circuits for different calls.
4. The circuits are demand-assigned, i.e., assigned on demand on essentially a first-come-first-served basis. The provision of access to the radio circuit is done in an orderly, controlled fashion.

The availability of trunked, circuit-like communications channels enhances system performance in a number of ways:

1. Multiple-access allows basic communications privacy (especially important in radio).
2. It provides transparency to circuit-oriented services of the wireline network.
3. Trunking allows for much higher traffic throughput.

The improvement in capacity is especially important for a radio system. Figure 6.23 shows the importance of what is often called *trunking efficiency,* which is only possible with a multiple-access system.

There are a number of very different ways to create a multiple-access radio system. The design of a multiple-access system encompasses two major issues:

1. *Multiplexing strategy:* How are the individual circuits combined in the available radio spectrum to allow for simultaneous access by a large number of users?
2. *Channel assignment:* How are individual users assigned to specific circuits on demand?

In the contemporary debate over standards for various wireless access applications, the issue of multiplexing strategy dominates all others. We will examine several of the most important generic alternatives in Sections 6.4.2 through 6.4.6. A brief discussion of channel assignment issues is presented in Section 6.4.7. In Section 6.4.8, I offer some personal observations on some of the controversial questions surrounding the competition among alternative multiple-access architectures.

6.4.2 Frequency-Division Multiple-Access Architecture

FDMA is the perhaps the most straightforward way to build a high-capacity multiple-access system. Historically, it represented the earliest solution to the problem of sharing the finite spectrum among a large number of simultaneous users. *Frequency division,* or *channelization* as it is sometimes called, means simply that the total available spectrum allocation is divided into a large number of individual radio channels (pairs of transmit and receive carrier frequencies). Each channel is capable of carrying one full-time voice circuit (or a control circuit). Any user is able to access any one of these channels, under the direction of the system controller (see Figure 6.52).

Analog cellular radio is an example of FDMA architecture. A digital FDMA system is also possible (and was strongly advocated by some manufacturers prior to the selection of TDMA for the air interface standard for digital cellular in North America [94]). Some of the characteristics of digital FDMA are:

1. *Frequency agility:* The subscriber terminal must be able to tune to every one of the available frequencies automatically, on instructions from the system controller.
2. *Continuous transmission:* While a conversation is in process, the FDMA unit is always transmitting—a 100% duty cycle.
3. *Simultaneous transmission and reception:* The FDMA unit must transmit and receive simultaneously, which means that duplexer circuitry is required at both the mobile and the base station. This can add significant cost and volume.
4. *Narrowest transmission bandwidth:* For a given level of spectral efficiency or circuit spectrum efficiency (see Section 6.2.5), the FDMA transmission will have the narrowest bandwidth. This may be important in that narrower chan-

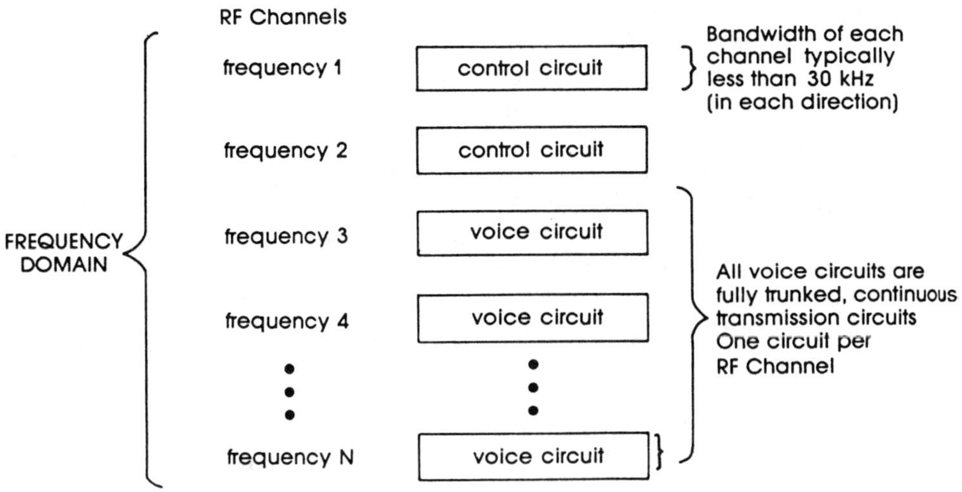

Figure 6.52 FDMA architecture. (From G. Calhoun, *Digital Cellular Radio*, Artech House, Boston, p. 284.)

nels may be relatively more susceptible to cochannel interference [95]. The narrower channel is also more susceptible to frequency-selective multipath fading.

5. *Lower transmission overhead:* In an FDMA channel, the transmission is continuous, much like T-carrier transmission on wireline systems. As in T-carrier, FDMA digital radio formats will require a few overhead bits to be inserted into the bit stream for framing and to pass certain control information. But the FDMA system does not face the challenge of reacquiring synchronization on every time-slot as burst-mode TDMA systems do.
6. *High infrastructure costs:* Because FDMA is a single-channel-per-carrier (SCPC) system, there must be equipment to support a separate transmitting chain and a separate receiving chain at the base station for every circuit. This burdens a large system with significant costs for infrastructure, compared to TDMA systems. The cost differential is felt most in applications that demand very high availability and serve high-traffic customers—in short, the telephony applications set. In BETRS applications, FDMA base station hardware is typically twice or more the cost on a per-circuit or per-customer basis when compared to TDMA [96].
7. *Hand-off complexity:* Because of the continuous transmission of an FDMA system, performing a hand-off (i.e., changing a user from one circuit to another during a call, for whatever reason) is difficult without interrupting the trans-

mission briefly. While tolerable for voice communications, the hand-off difficulties of FDMA can be fatal for data applications.

8. *Hardware-intensive, inflexible with respect to new technology:* The most serious difficulty with FDMA architecture in the digital era is that such architectures are by their nature hardware-intensive, and therefore difficult and costly to adapt for improvements in circuit-level technology (improved voice coders, for example). The only way an FDMA system can evolve is through channel-splitting. Indeed, this has been proposed as a growth path for North American cellular radio, to split the current 30-kHz channels into three 10-kHz channels in order to gain capacity [97]. The penalty for channel-splitting may be severe, however. For one thing, the RF hardware at the base stations must be replaced with three times as much new narrow-channel hardware. Second, the frequency plan for the entire system may have to be altered. And if it is true that the interference characteristics of the narrower channels are altered, even the cell boundaries and the placement of base station transmitters may be affected. In short, large amounts of hardware must be changed out in order to take advantage of new technology.
9. *Inflexible with respect to new services:* Digital FDMA systems are, in effect, optimized for a particular bit rate, which reflects an assumption about what sort of traffic will predominate on the radio network. Usually, this is voice traffic, which means that in today's environment for circuit-level technology a channel rate of 10 to 20 kb/s or so would be implemented. But as new digital services emerge that require bit rates that are significantly higher or lower than the voice-optimized channel rate, difficulties begin to mount.

   On the low-end, for example, consider a packet-data service that might have an asynchronous average data rate of 300 b/s. Or consider even a telemetry-type application in which data rates of 50 b/s or less might be required (to transmit vehicle performance data, for example). To accommodate such low rates on an FDMA channel would under normal circumstances involve assigning an entire channel to the transmission (since that is the smallest unit available). Obviously this would be grossly wasteful. (One could imagine ways to mitigate this by designing the channel structure in a way that it might effectively be shared among multiple users more or less simultaneously—but this is precisely what TDMA is all about.)

   On the other hand, how would we transmit a 64 kb/s bearer circuit (as might be required in a proto-ISDN environment, or for high-speed facsimile) over an FDMA system with 10 kb/s voice-optimized channels? One could imagine lashing together a number of FDMA channels in some manner. But it would not be easy or cheap. (Such a multi-channel device would need to create, isolate, and combine a large number of separate radio transmissions and receptions, making in effect a miniature base station.)

In summary, for a number of reasons, but especially because of the hardware-intensiveness and difficulty of accommodating circuit-level improvements in digital processing technology, FDMA is not favored by most observers for emerging wireless access architectures.

6.4.3 Time-Division Multiple-Access Architecture

TDMA is a more complex architecture. In the simplest case, a single carrier frequency is formatted as a tightly synchronized digital transmission, which is divided into a number of different time slots. Each time slot carries a burst transmission containing information pertaining to a single telephone circuit. The channel transmission rate is faster than the circuit bit rate by a factor approximately equal to the number of circuits being multiplexed on the carrier. The base-to-subscriber transmission looks a lot like the multiplex format used on a wireline T-carrier, as shown in Figure 6.53. Indeed, the single-channel TDMA architecture in subscriber radio systems developed by SR Telecom in Canada involved the transmission of essentially a T-1 circuit over the air [98].

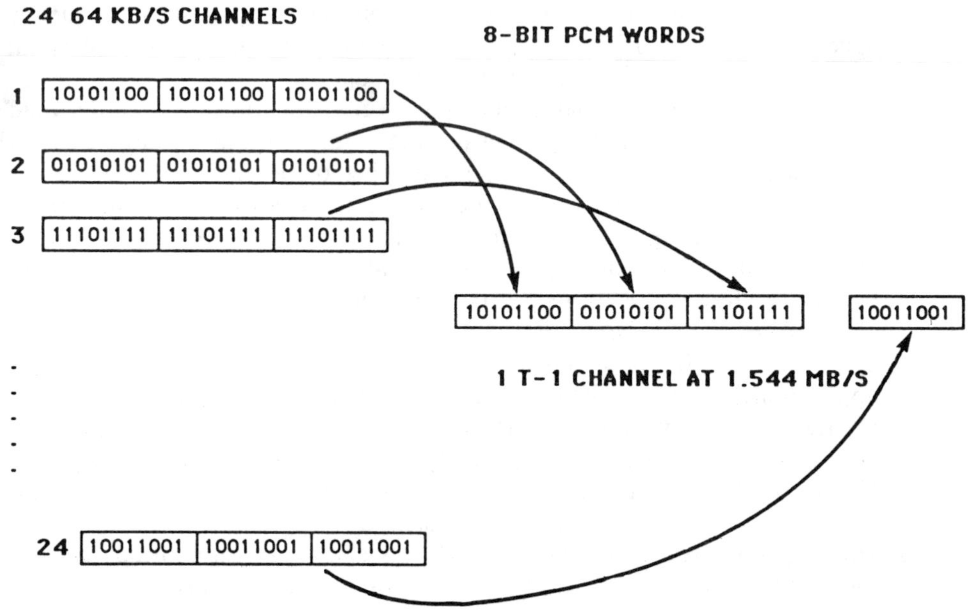

Figure 6.53 Idealized T-carrier multiplexing. (From G. Calhoun, *Digital Cellular Radio,* Artech House, Boston, p. 179.)

The tricky part comes at the subscriber's end. For each active subscriber, there is an assigned time slot (the assignment normally having been made once, at the beginning of the call). Each subscriber unit synchronizes itself with the base station transmission, and "counts" the correct number of time slots until it reaches its assigned time slot. It then activates its receiver and demodulator circuitry to decode its assigned burst. Similarly, when transmitting, the subscriber unit has to buffer the speech input and wait for its assigned transmit time slot, at which point the subscriber transmits an accelerated burst of data. It then begins accumulating data for the next burst (see Figure 6.54). In early applications, this mode of transmission was sometimes referred to as buffer-and-burst transmission.

The next step is to combine TDMA and FDMA, so that circuits are trunked in both the time and frequency domains (see Figure 6.55). In fact, this is the architecture that most people refer to when they use the term "TDMA" in connection with wireless access applications, a combined FD/TDMA system.

The significant features of TDMA include:

1. *Multiple circuits per RF carrier:* As indicated, TDMA systems multiplex anywhere from 2 to 30 or more circuits per RF carrier, depending on the application.
2. *Frequency agility:* FD/TDMA transmitters obviously must be frequency agile. Single-carrier TDMA systems, on the other hand, such as VSAT and subscriber radio, may not require frequency agility.
3. *Burst transmission:* The subscriber station transmits in bursts, rather than continuously as with FDMA. This has many implications for circuit design and system control, some of which are touched on below.
4. *Staggered transmitting and receiving slots:* It is possible and desirable to offset the transmitting and receiving slots at the subscriber station. This can allow the elimination of the duplexer circuitry required by FDMA units, to be replaced by a simple transmitting/receiving switch.
5. *Wider transmission bandwidth:* For a given level of circuit spectrum efficiency, the TDMA channel will be wider than the FDMA channel by a factor approximately equal to the number of circuits per carrier. This may improve interference and multipath performance somewhat. Actual transmission bandwidths range from 20 kHz for FD/TDMA BETRS, and 30 kHz for North American digital cellular, to 200 kHz for GSM cellular, up to several MHz for some subscriber radio systems.
6. *Higher transmission overhead:* A TDMA burst transmission format requires the receiver to reacquire synchronization on each burst. Also, since several units are sharing the channel at the same time, temporal guardbands are necessary to keep the slots separate in the event of users transmitting from different distances with different propagation delays (see Figure 6.56). Because of this, TDMA systems need more overhead than FDMA systems. The "over-

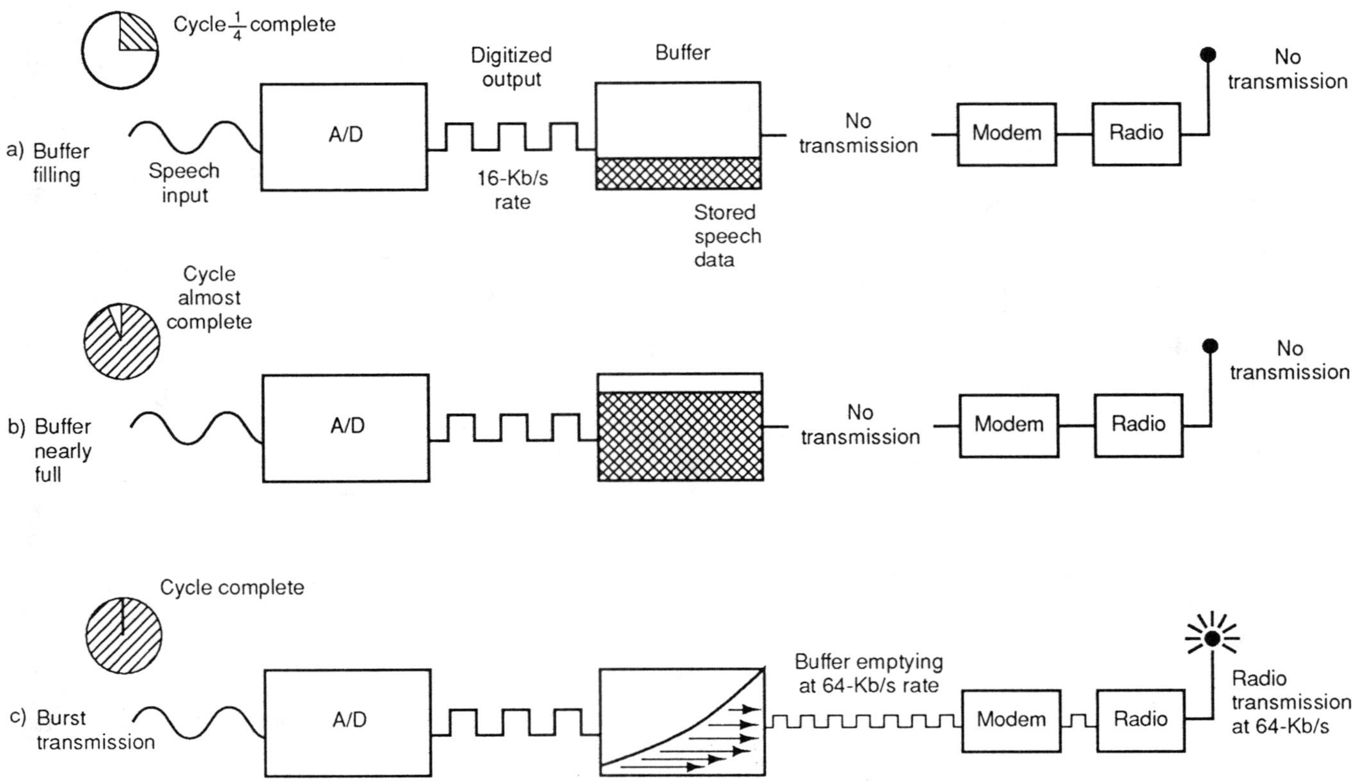

Figure 6.54 Buffer and burst transmission.

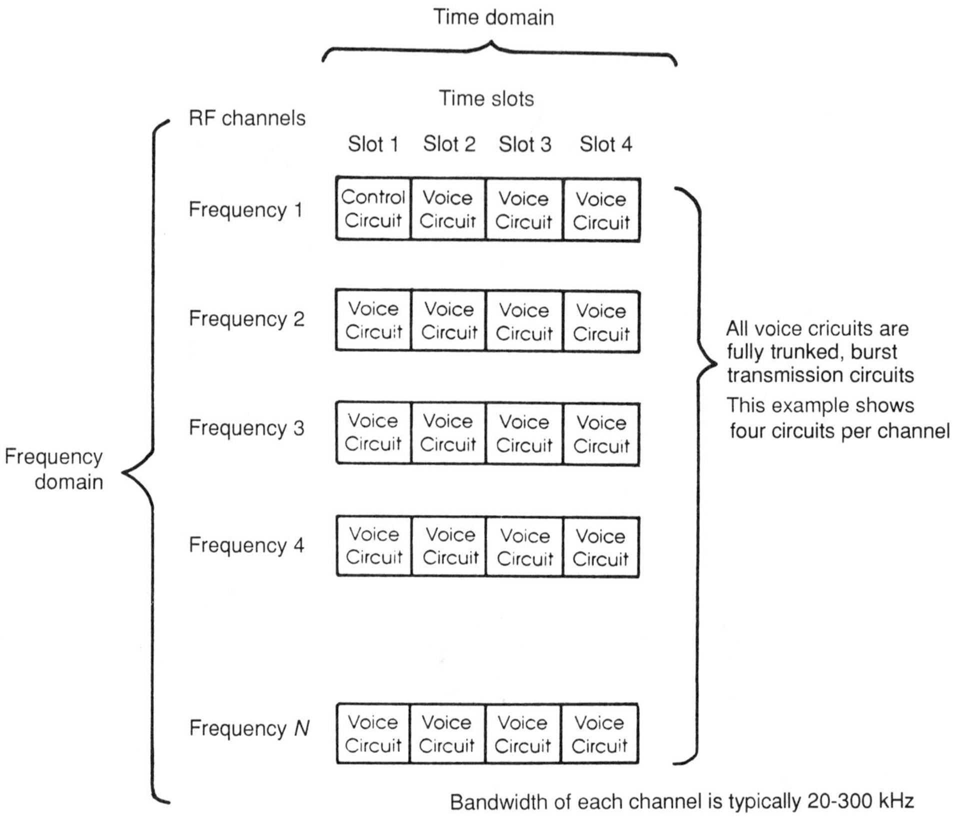

Figure 6.55 A TDMA system—the time-frequency matrix. (From G. Calhoun, *Digital Cellular Radio*, Artech House, Boston, p. 287.)

head" built into the IS-54 standard, for example, is about 19% of the channel capacity. In the BETRS system developed by IMM Corporation the frame overhead is about 9%.

7. *Infrastructure costs:* The multiple-circuit-per-carrier architecture of TDMA results in dramatic cost savings when it comes to system infrastructure. In essence, for any given application serving a given number of customers, TDMA systems will require much less RF equipment (by a factor approximately equal to the circuits per carrier of the system). Figure 6.57 gives some idea of the cost savings. In practice, this is a substantial advantage for TDMA.
8. *Hand-off:* A TDMA system can be designed to perform an essentially perfect hand-off, without loss of information, since the hand-off can occur between transmitting slots. Moreover, since the mobile unit generally has at least some

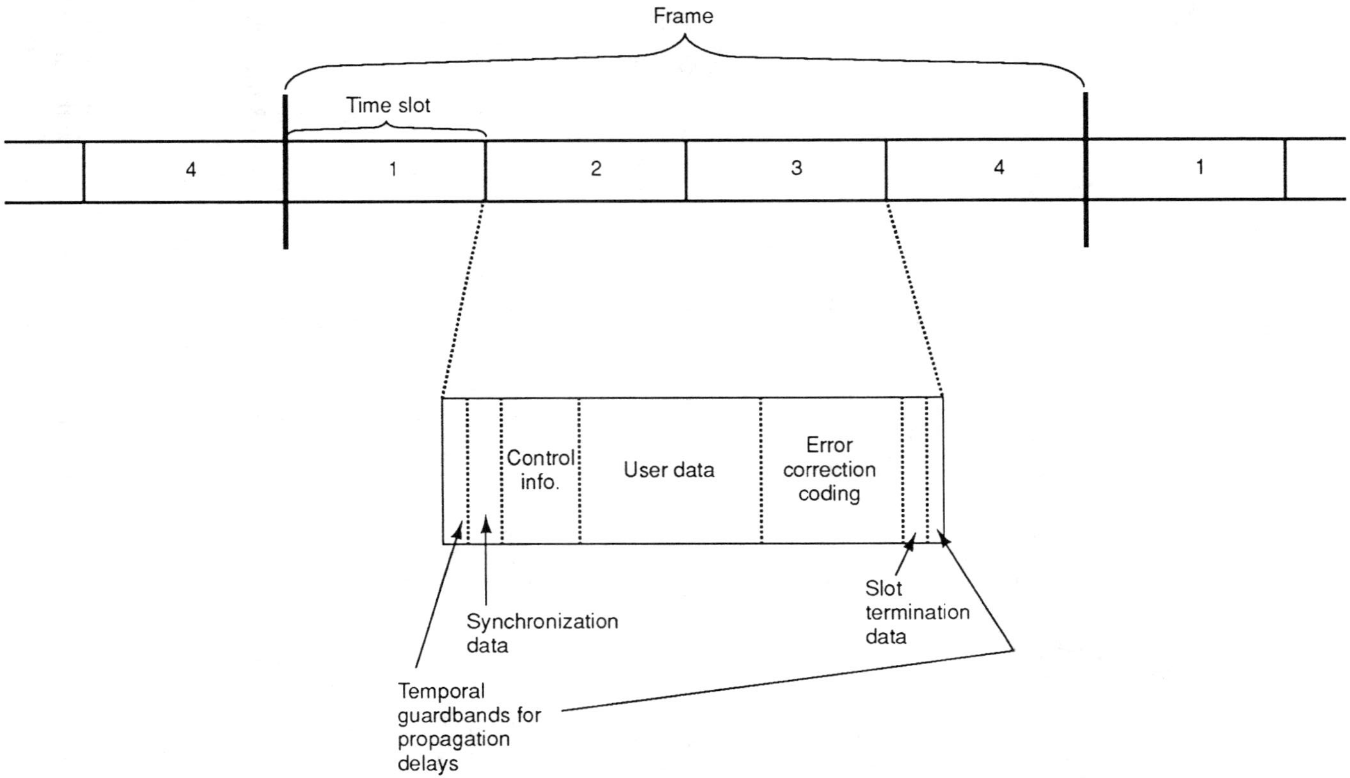

Figure 6.56 TDMA slot structure with guardbands for propagation delay.

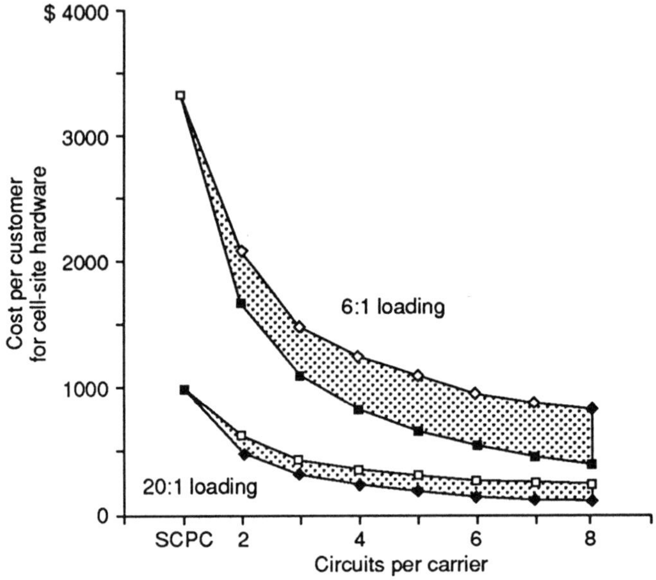

Figure 6.57 TDMA cost savings. Note: cost of RF hardware is assumed to be $20,000 for the SCPC case. Lower bounds represent assumption that no cost penalty is incurred for TDMA hardware. Upper bound reflects higher RF channel costs for TDMA hardware. Actual hardware costs variations would push these curves higher or lower. (From G. Calhoun, *Digital Cellular Radio,* Artech House, Boston, p. 413.)

"idle" time slots during which it is neither transmitting nor receiving, it can be instructed to monitor other channels and assist in the hand-off decision. Mobile-assisted hand-off (MAHO) is felt to be one of the major advantages of TDMA digital cellular architectures now coming on line [99].

9. *Software-intensive, more flexible architecture:* The TDMA channel is a software-configured channel. Changes in technology, such as an improved voice coder, can be incorporated into a TDMA format, often without changing any of the hardware. For example, if a TDMA architecture is designed to handle 4 time slots per carrier, using a 16-kb/s coder, the change to 8 time slots to accommodate a half-rate coder can be achieved by simply redefining the channel format. The transmission bandwidth and interference characteristics are not altered and the transmitter, receiver, and other RF hardware are not affected.
10. *Flexibility for new digital services:* A TDMA architecture is much easier to reconfigure, in software, to handle different types of services with different bit rates. For example, consider a fairly wideband TDMA system operating at

200 kb/s in the channel (similar to proposals made recently by Bellcore and others [100]). The voice rate (same as FDMA) may be assumed to be about 10 kb/s, which would allow, say, 15 time slots for voice telephony (if we allow an extra 33% in each slot for TDMA overhead, which is certainly generous). Sub-slots could be created for lower bit-rate services, and super-slots could be used for higher bit-rate services. A single RF channel might carry many different kinds of traffic at one time (see Figure 6.58). This type of system has already been implemented in BETRS TDMA systems [101]. Indeed, if a user wanted to transmit an ISDN basic-rate signal at 144 kb/s, he might be allocated an entire RF circuit (FDMA-fashion) for the duration of the transmission (and charged accordingly). The only "penalty" as such for this flexibility is that the slot overhead would probably remain fixed, in absolute terms, regardless of the user rate or the size of the slot. This would mean that for very small slots and low-bit-rate services there would be a proportionally larger TDMA overhead. For higher rate services, however, the overhead would diminish toward the FDMA baseline.

In summary, FD/TDMA is the current preferred access architecture for most wireless access applications. It was first developed in the United States for BETRS applications in the mid-1980s [102], and FD/TDMA architectures have been chosen as the standards for digital cellular in both North America and Europe. A number of other wireless access standards, including PCN in the United Kingdom and DECT in Europe, also rely upon FD/TDMA architecture.

In addition to its many technical advantages, TDMA is well understood, with established performance characteristics in over 20 years of commercial application. It builds on the vast base of time-division multiplexing experience incorporated into the wireline digital network. As a software-intensive system, it provides flexibility in the face of rapidly evolving digital technologies. Perhaps most important, despite its greater complexity (*vis à vis* FDMA) TDMA has proven to be a robust system in the field.

6.4.4 Time-Division Duplex—"Ping Pong"

A variant of TDMA, which uses only a single carrier frequency for both transmitting and receiving channels is known variously as *time-division duplex* (TDD), *time-compression multiplexing,* or by the more illuminating term *ping-pong* [103]. A single radio-frequency is divided into a number of time slots, and both TDD units (e.g., a base unit and a subscriber unit) synchronize on these time slots. In the simplest arrangement, there are two time slots, and the base unit and the subscriber unit transmit to each other on alternate slots (see Figure 6.59). The analogy of two players batting a ping-pong ball back and forth is not inappropriate. More elaborate TDD

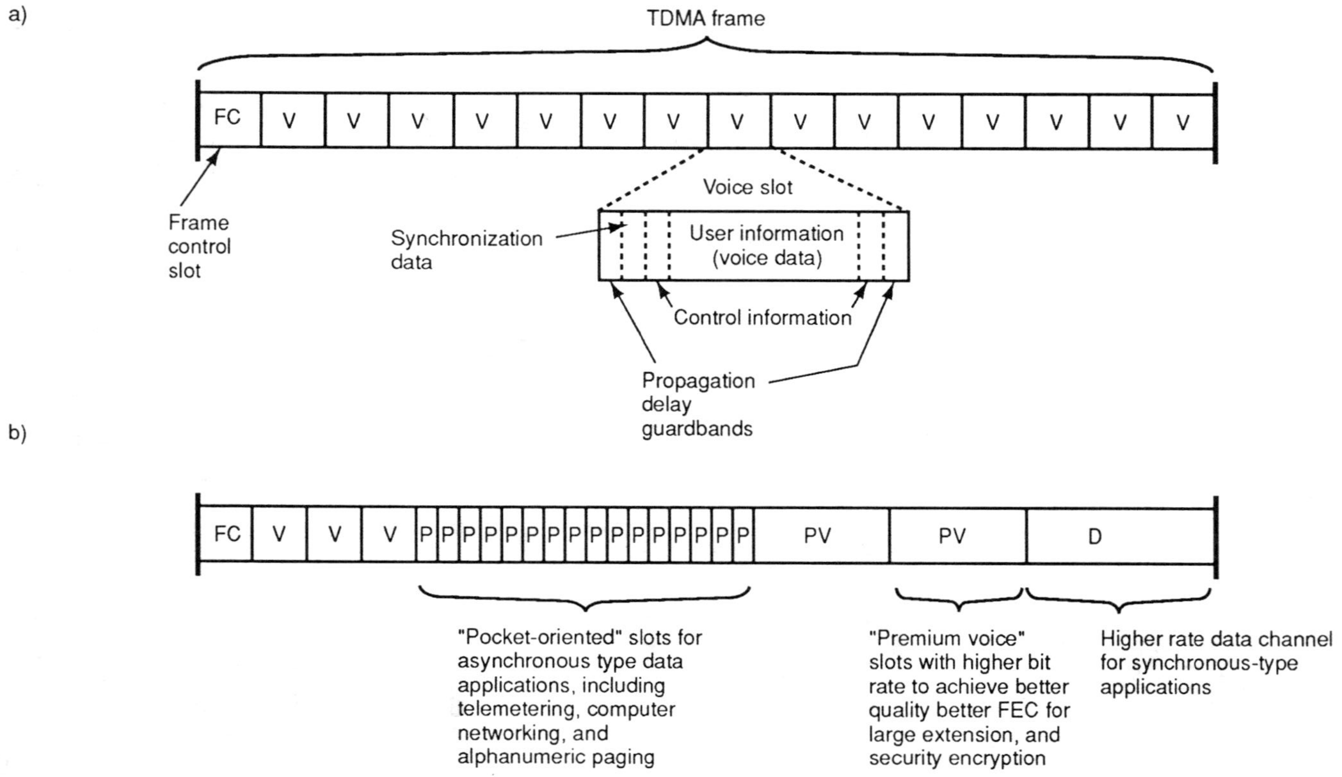

Figure 6.58 Dynamic bandwidth allocation in a TDMA architecture; (a) standard voice mode, (b) channel reconfiguration in software to handle a variety of message types.

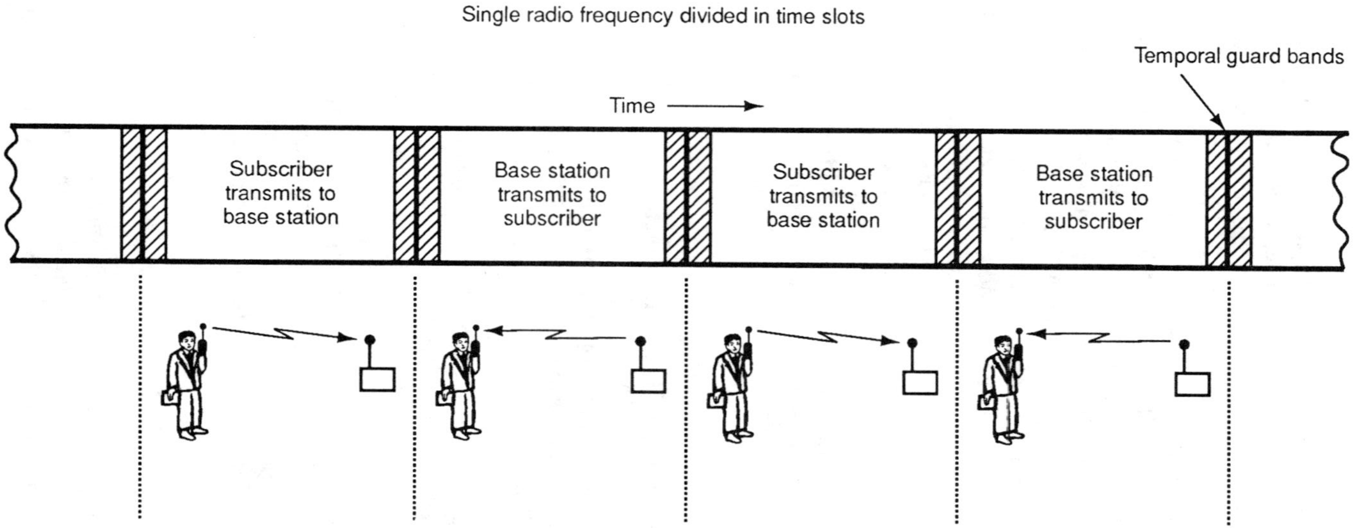

Figure 6.59 TDD architecture.

arrangements involving four, eight, or more time slots have been proposed [104]. As one colleague described such a system: "It's really ping-ping, pong-pong."

Many of the same time-division characteristics that apply to TDMA apply here as well: burst transmission, possible avoidance of duplexers (if there are more than two slots), and so forth. The chief advantage of TDD is that it can be implemented on a single frequency, rather than requiring frequency pairs as the other systems described here do. This may allow TDD to find a niche in certain gerrymandered spectrum allocations that, for historical reasons, have not been organized into paired channels. In small-scale applications where multi-frequency operation is not required, the simplicity of a single-frequency TDD unit could be the superior choice. TDD may make sense for a cordless residential extension phone, offering inexpensive implementation of a digital radio without the overhead of frequency agility.

In other respects, however, TDD is a somewhat inflexible architecture. There may be interference problems with large-scale systems that operate without a transmit/receive carrier separation. And to the extent that capacity requirements in large-scale applications may require the use of frequency synthesizers and related hardware anyway, the cost and simplicity advantage for TDD may disappear. In short, if frequency pairs are available—as they should be for any significant new application—I believe there is no real advantage to be gained from TDD in large-scale applications.

6.4.5 Code-Division Multiple-Access Architecture

CDMA represents a very different approach to building a multiple-circuit radio system. Instead of trying to keep individual transmissions separated from one another in either frequency or time the way FDMA and TDMA systems do, CDMA intentionally combines a large number of transmissions on the same channel at the same time. In other words, CDMA allows the different transmissions to interfere with each other in a controlled fashion. The "trick" is that each transmission is subjected to a special additional level of encoding before it is sent. This extra layer of coding imparts a unique "signature" to each transmission that survives even when many transmissions are mixed together. Every transmitter has its own unique code. The receiver, in turn, "knows" the unique code of the signal it is looking for, and by applying this code as a kind of filter to the stream of radio energy in the channel, it can recover the desired signal from the background of all the other signals.

6.4.5.1 Code Division Metaphors

CDMA can be a difficult concept to grasp; it may be helpful to resort to a couple of metaphors to convey some sense of how it works. For example, let us imagine a series of radio transmitters trying to communicate with a matched set of receivers. In a channelized systems (e.g., FDMA), we can conceive of each channel as a separate "pipe" and each message as a series of pellets (e.g., bits) passing through the pipe from transmitter to receiver. The pellets are all physically similar, but the use of separate pipes keeps each transmission from "interfering" with the others (see Figure 6.60). In a CDMA system, there is only one pipe, one channel, which is much wider than the narrow pipes used for FDMA. All of the transmitters are forced to use the same channel at the same time. Let us further imagine, however, that before each transmitter actually sends a message pellet, it "wraps" this message (or "bit") in a specially shaped package. One transmitter uses a spherical package. Another uses a cube-shaped package. A third uses a pyramid-shaped package, and so on. The shape of the package is unique for each transmitter. At the receiving end of the pipeline, each receiver uses a special "sieve" to filter only those messages with the properly

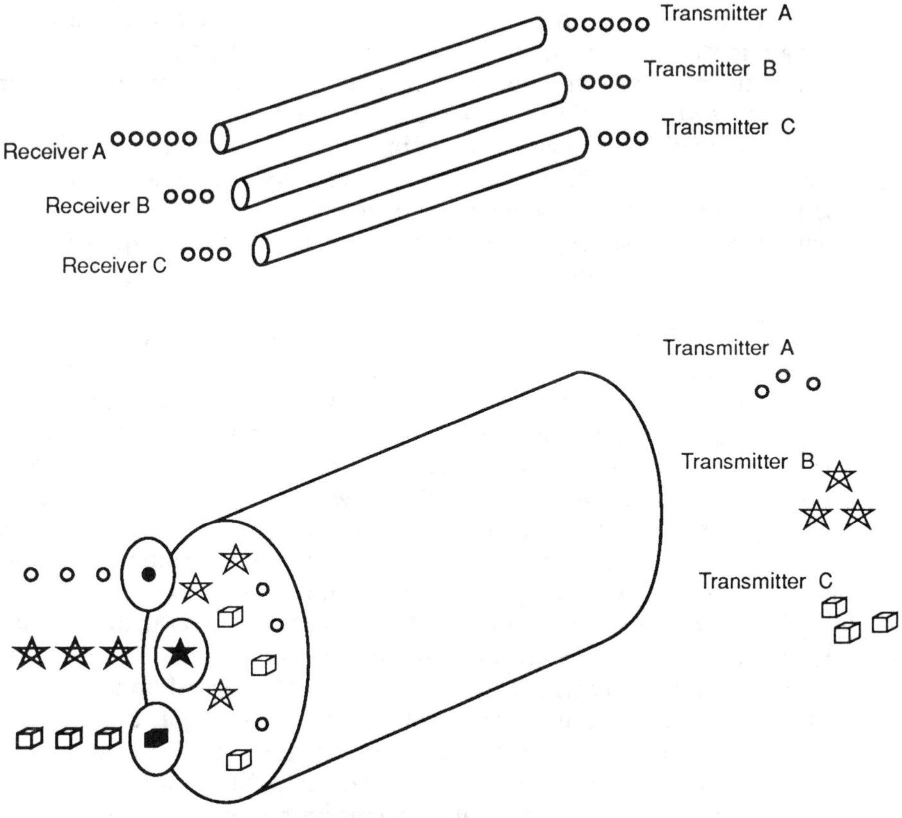

Figure 6.60 CDMA metaphor—"shapes" and "sizes."

shaped packages (see Figure 6.60). This is suggestive of how code division works. The shapes of the packages are the "codes"—a special extra layer of code applied to each message to enable it to be mixed with other messages in the same channel, and then recovered later by a receiver using the same code.

Our pellet example, however, does not convey the problem of CDMA interference very well. Let us now switch metaphors to extend the discussion and hopefully capture the concept of a self-interfering architecture, which is so central to CDMA. Imagine a medium-sized room in which there are a large number of TV sets arranged around the walls. Three of them are turned on. On one set there is a man speaking English. Another set has a man speaking in Japanese. The third TV has an announcer speaking in Arabic. Standing in the middle of the room is an individual (the receiver) who understands English but does not understand Japanese or Arabic. For him, the English-speaking TV is intelligible, and the others are simply "noise." In all likelihood, he will be able to understand and follow the English announcer fairly well. Meanwhile, we can assume that a second individual is standing next to him, who understands only Japanese. For him, the English and Arabic is simply "noise." Finally, an Arabic-speaker can "tune in" only the Arabic TV announcer and "tune out" the noise of the English and the Japanese. In this way, three individuals can all watch different television shows in the same room at the same time. This is analogous to the use of different codes ("English" versus "Arabic") to allow multiple users to share the same channel at the same time.

This second example suggests several additional issues that are quite important in actually making CDMA systems work.

1. *Channel capacity:* The reader can probably sense intuitively that the experiment of three TV sets speaking in three separate languages would probably "work"—that is, the listeners would actually be able to understand their corresponding announcers despite the background "noise"—but that it would work better under some arrangements than others. For example, how many more TV sets could we turn on in the same room? Could we operate 10 TV sets in the same room at the same time? Or 50 or 1000? Intuitively, we recognize that there is some upper limit, in the sense that the rising background of physical noise (sound energy in this illustration, radio energy in the actual case of CDMA) will ultimately become too large for the receiver to discriminate its desired signal from the noise background. We can also probably sense that this limit is not a hard limit, but a gradual degradation as additional TV sets are turned on. (It is often claimed that CDMA exhibits "graceful" degradation, but the example suggests that the lack of a hard limit does not imply that the gradual loss of quality is more pleasant or acceptable than simply being blocked. Would the user take a poor quality circuit immediately, or a higher quality circuit in 30 seconds? Two minutes? Ten minutes? These are questions that we do not have much experience with, because until now the world of telephony

has been a blocking world rather than a gracefully degrading world, at least for voice communications.)

2. *Orthogonal codes:* Imagine that there are three TV sets turned on, one speaking Spanish, one Italian, and one Portuguese—three very closely related and similar-sounding languages. In this case the challenge to the Spanish-speaking listener is probably a lot harder than in the case where the three languages are unrelated and quite different-sounding (such as English and Arabic). This suggests another design problem for CDMA systems: the proper selection of codes. To work best, a self-interfering CDMA system should use codes that are as different from one another as possible—specifically where the correlation between two codes is no higher than between two random numbers. Such codes are said to be *orthogonal* to one another. Designing appropriately orthogonal code systems is a subdiscipline of CDMA architectures.
3. *Power control:* It should also be intuitively obvious that even with two very different-sounding languages (i.e., orthogonal codes), the CDMA system works best if the TV sets are both equal in loudness, as perceived by the receivers standing in the center of the room. If the Arabic-speaking TV set is located right next to the English-speaking receiver while the English-speaking TV is located in the far corner, the noise level from the Arabic set will overwhelm the English-speaking receiver. Therefore, we must have some way of turning down the volume of the nearer TV and boosting the volume of the farther set, so that the loudness of both sets in the same to both receivers. Now imagine a "mobile application"—where the TV sets are on wheels, moving rapidly around the room. In this case, the power control must be highly dynamic, to ensure that the perceived volume level of all the sets is always approximately the same.

6.4.5.2 Code Division Techniques

Returning from the metaphorical plane, we find that the actual methods for implementing code division are complex and diverse. Indeed, CDMA is not a single technology at all, but an approach to building a multiple-user, multiple-access communications system that can be implemented in a number of quite different ways physically. The two most common forms of code division are:

1. *Frequency hopping spread spectrum (FH/SS):* In this approach, the total available spectrum is divided into a large number of frequency bands, rather like the frequency-denominated channels in an FDMA system. But instead of assigning each user to a single channel (for the duration of the call), his transmitter is instructed to "hop" from channel to channel in a very rapid sequence. Each user follows a different, unique sequence of hops. The sequence of hops is designed so that it minimizes the likelihood that any given hop will land on the same frequency at the same time as another user. Indeed, the reader may

have realized that the sequence of hops for any given user is in fact determined by an orthogonal code—the code division is implemented by varying the frequency-hopping sequence from user to user. Figure 6.61 shows how two orthogonal users can occupy the same wideband channel at the same time through FH/SS. Note that the code sequences do not guarantee that collisions will not occur (where both users may simultaneously occupy the same frequency and thus interfere with one another in a conventional sense); the code sequences merely limit the amount of mutual interference to a definable level.

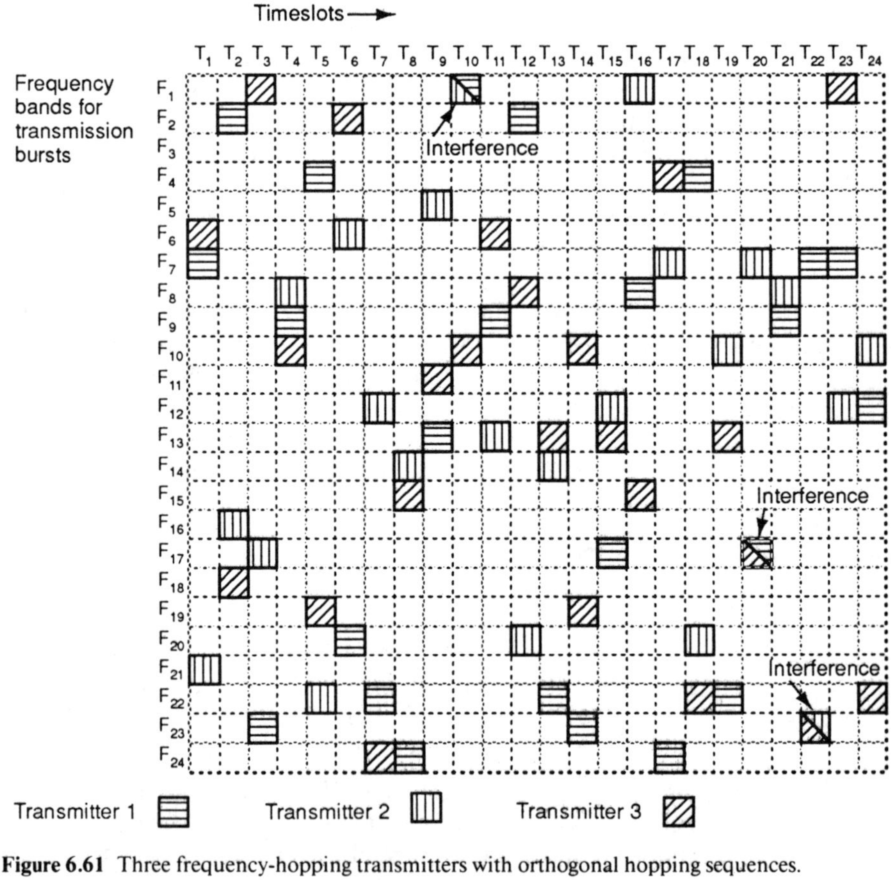

Figure 6.61 Three frequency-hopping transmitters with orthogonal hopping sequences.

2. *Direct sequence spread spectrum (DS/SS):* The other chief method of code division is even more difficult to explain in a short presentation. Essentially, it involves mixing the information-bearing signal with another digital signal which is generated at a much faster rate and over a wider bandwidth. This sec-

ond signal carries the unique orthogonal code. It appears, on its own, very similar to noise, and is often called a "pseudo-noise" sequence. When the signals are combined—the low-rate information signal with the very high-rate pseudo-noise signal—the result is a combined signal that also looks very similar to a noise signal, but contains the information signal embedded in its code (see Figure 6.62).

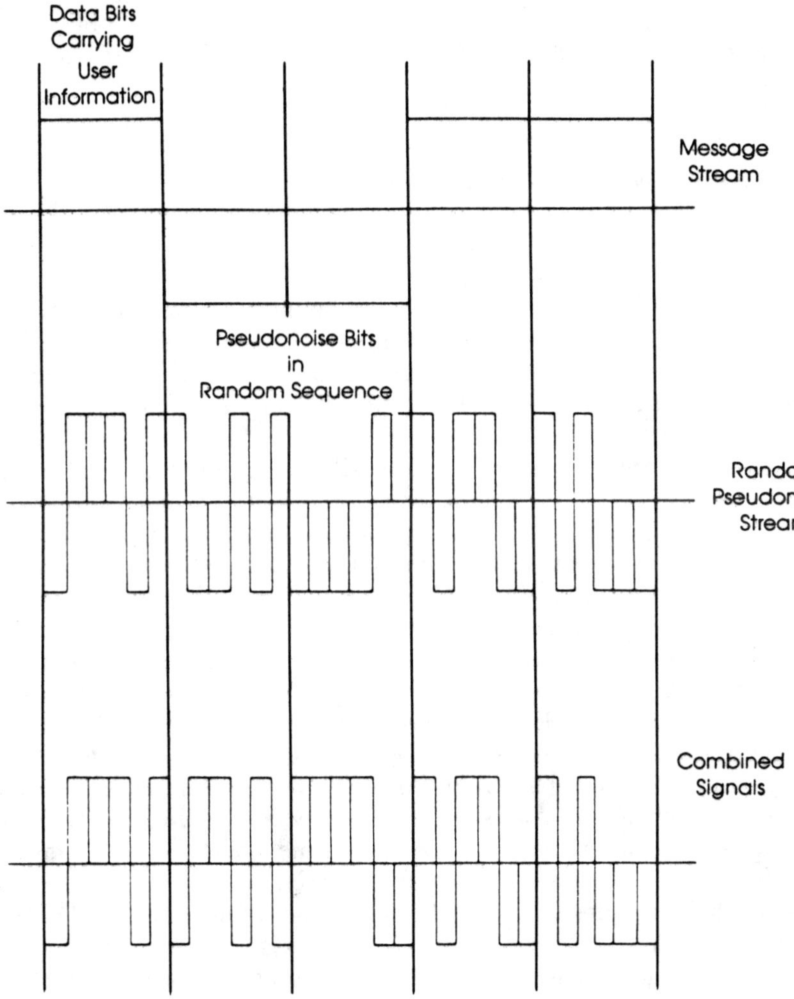

Figure 6.62 Direct sequence spread spectrum. (From G. Calhoun, *Digital Cellular Radio,* Artech House, Boston, p. 354.)

It is not possible here to describe CDMA techniques in detail. The reader is referred to a growing literature on the subject [105].

With respect to CDMA system characteristics, it is difficult to run through a quick list of salient features as we have done for FDMA and TDMA—for the simple reason that no one has ever built a CDMA system for a wireless access application. It is possible to envision many different kinds of systems, with a wide range of potential advantages. Yet it could be misleading to speculate on these issues until the outline of a workable CDMA telephony-oriented architecture becomes a little clearer, and I would not want to cast my own speculations in the guise of objective statements. In Section 6.4.8, I will offer some more personal views on the likely role for CDMA in the future.

6.4.6 Packet-like Techniques: Digital Speech Interpolation

6.4.6.1 The Concept of Packet Transmission

As noted in Chapter 5, wireless access generally involves the creation of *circuit-like connections* between users and network facilities. While there is no rigorous definition of a circuit, in general the term implies that the transmission path between an individual subscriber and the network is established at the beginning of the call or session, the path stays "up" for the length of the call, and it is dedicated to that subscriber for as long as the call lasts. It is also usually assumed that the circuit actually comprises a dedicated physical channel, such as a radio frequency or time slot. (This assumption may be qualified somewhat by the possibility of a "hand-off" in a cellular mobile system, however, where the physical channel may be changed in the course of the call. Nevertheless, the cellular circuit—viewed in logical or functional terms—is still dedicated to the caller for the length of his call.)

The assumption of a circuit-switched architecture reflects the voice traffic orientation of most wireless access applications. Among other things, voice traffic is very intolerant of delay; people do not like it. Also, holding times (the length of a voice session or telephone call) tend to be fairly short, averaging a few minutes or so (depending on the specific application). But the "capacity utilization" of the voice circuit is fairly high while the call is in process (as we shall see below, typically the human speaker is transmitting about 40% of the time). Under such circumstances, it makes sense to set up dedicated circuits for each session.

A very different philosophy underlies the design of many modern data communications networks. Early time-shared computer systems were made up of large numbers of "dumb" terminals requiring access to a set of central (shared) resources (e.g., a mainframe computer). The terminals tended to transmit data in short bursts. The sessions had longer holding times than voice phone calls, but actual channel utilization was much lower. There was a high percentage of idle time. In addition, the data traffic carried on these computer networks is tolerant of delay (within limits) but highly intolerant of transmission errors.

This type of traffic fostered a very different approach to network access. Instead of assigning dedicated circuits to each user for the length of the session, data messages were broken into small units, called *packets,* each with its own "address." Packet networks were set up consisting of a variety of different smart nodes. Each node is a device that may be capable of creating, sending, receiving, buffering or holding, transmitting, and/or reassembling these packets to recreate the original message (see Figure 6.63). The message is disassembled into many packets, which are routed by various paths through the network and reassembled at the receiving end.

Consider this analogy: Imagine mailing a 100-page book through the postal system, but instead of putting the whole book in one package, the sender tears each page out and mails it in a separate envelope. The 100 envelopes each carry the same address, and may in fact be routed by the Post Office over many different physical paths to the same ultimate destination. At the receiving end, the receiver can reassemble the book by using the page numbers, and if it finds that any pages have not arrived, it can dispatch its own letter to request that specific pages be resent. This is essentially what a packet network does by electronic means. In so doing, it is possible to construct a much more economical data network than would be the case for a circuit-switched architecture. Of course, the subject gets a good bit more complicated, and there are many different architectures for building a packet network [106]. Packet techniques are growing rapidly in importance as well, because they form the basis for local-area networks (LANs) and metropolitan-area networks (MANs), which operate at much higher speeds and vastly expanded functionality [107].

The ability to implement a packet architecture depends on the state of the technology and, fundamentally, on the nature of the traffic. Packet-switched networks are much more efficient than circuit-switched networks for the "bursty" transmissions characteristic of asynchronous data communications systems. Nevertheless, one characteristic of all packet networks today is still the introduction of delay—delay due to congestion and the statistical character of packet traffic, which is variable and generally does not allow for real-time applications such as voice or video.

6.4.6.2 Fast-Packet and Packet Voice

Traditionally, then, voice communications requires the use of a dedicated circuit, while bursty, delay-tolerant data communications can take advantage of the savings from packet architectures. On closer examination, however, it becomes apparent that voice traffic is also bursty. In fact, most voice traffic shows a pattern of alternating between short periods of speech and slightly longer periods of silence (the latter occurring in the form of pauses between words or syllables, as well as while listening).

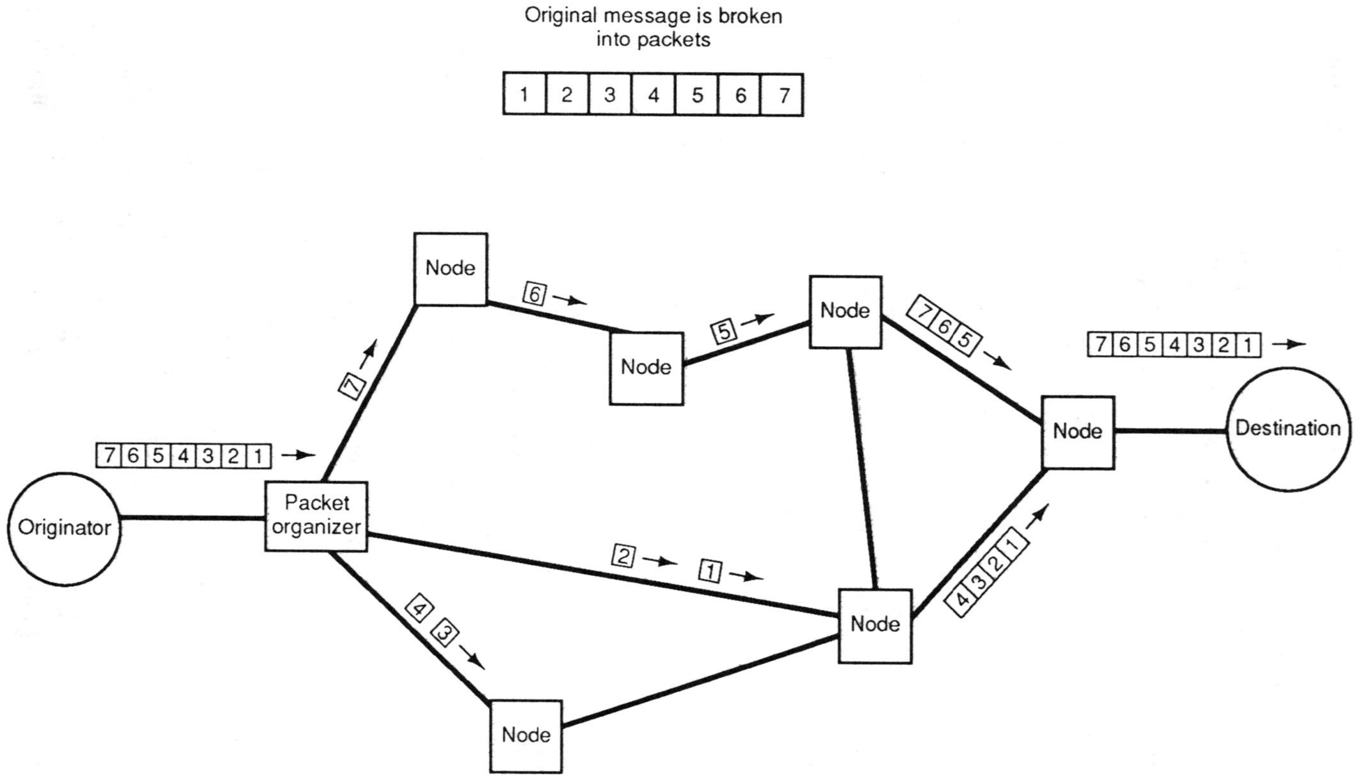

Figure 6.63 A packet network.

Studies show that even continuous speech comprises only about 65 to 75% actual speech production, after pauses for breathing, intersyllabic gaps, and normal hesitations associated with thought processes are taken into account [108]. And, of course, while one person is talking, the other is normally listening (i.e., not producing any speech at all). Certainly, this is true on average for any large population of callers. In fact, many studies have shown that the actual voice activity factor—the amount of active speech time per individual—is less than 50%. Voice activity factors of 35 to 40% have been reported [109].

This microscopically bursty nature of voice suggests that packet techniques might someday be applicable to voice communications, if delay can be controlled. In fact, the capabilities of modern technology are beginning to bring this prospect within view. As transmission systems improve and speed and bandwidth increase, interest has grown in the potential application of packet techniques to voice communications. To be suitable for voice transmission, the speed of the packet network would have to be increased to the point where delays are reduced to levels comparable to circuit-oriented architectures. Hence, the term *fast-packet,* which is sometimes used in this connection. A number of advanced packetized architectures are now emerging that anticipate handling voice traffic (as well as data) [110]. Indeed, some advocates of these approaches believe that early in the twenty-first century we may see the dominance of conventional circuit-switched networks challenged by so-called asynchronous transfer mode (ATM) switches based on such advanced packet concepts.

What would be the advantages of packet voice transmission? One of the most important potential advantages would be a savings in bandwidth, due to the fact that a packet voice system only transmits when the individual is actually speaking. When the individual is listening (i.e., not speaking), no packets are formed and no traffic is generated. *This means that a packetized transmission system could double (at least) the capacity of an appropriately sized multichannel transmission system.* The potential for such a gain is very attractive, obviously, for radio applications that must always seek to optimize the capacity of the available spectrum.

Another potential advantage is that packet techniques allow for the very efficient mixing of different types of traffic, from voice to computer data to video. This could allow efficient sharing of facilities for different types of services.

6.4.6.3 Digital Speech Interpolation

Commercial availability of true fast-packet or packetized-speech architectures lies well into the future. There is a related technique, however, that is readily applicable to wireless access applications in the near term. In fact, this technique was originally developed in the 1950s for voice circuits in which the transmission facility was very limited and expensive, which meant that channel capacity had to be highly optimized—the circuits of the undersea telephone cables crossing the Atlantic Ocean. In the analog domain, the approach became known as *time assignment speech interpolation* (TASI) [111]. In the 1960s and 1970s, the same principles were extended to digital communications. The digital version of this technique became known as *digital speech interpolation* (DSI), and was first implemented in satellite systems [112].

Simple speech interpolation is a rather straightforward enhancement, which has been implemented in a variety of applications and considered for others [113]. One author offers the following definition:

> Speech interpolation refers in general to the process of time sharing a multichannel transmission facility among a number of telephone conversations; individual segments of conversation, termed talk spurts, are given access to a transmission channel as soon as possible after their onset. In the process, N conversations are carried on M transmission channels where M is less than N and the ratio of N/M is referred to as the DSI gain. . . . [114]

As noted already, the normal DSI gain is about two, or a 50% reduction in the number of transmission channels required to carry a given (reasonably large) amount of traffic (see Figure 6.64).

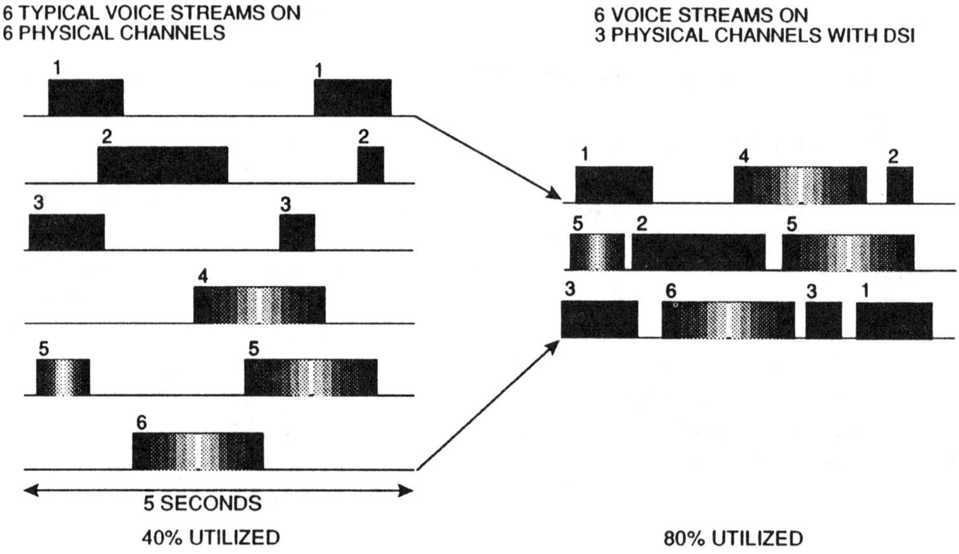

Figure 6.64 DSI compression. (From Hughes Network Systems.)

Let us examine the DSI process more closely. If we envision the actual speech activity on one-half of a radio-telephone circuit, we can see that typically there are short periods of speech, known as *talk spurts,* interspersed with periods of silence, known as *pauses* (see Figure 6.65). In fact, the average duration of a talk spurt is

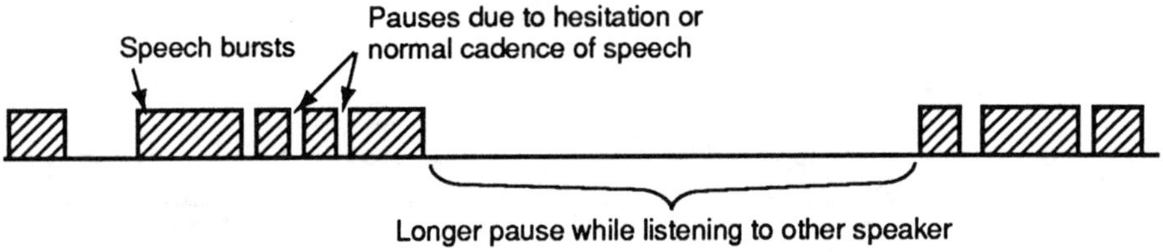

Figure 6.65 Talk spurts and pauses in a DSI channel. Typical pattern of speech and silence episodes for a single speaker in a two-way conversation. (From Hughes Network Systems.)

about 1.5 seconds, according to a number of studies, compared to about 2.25 seconds for the average pause (Figure 6.66). If we could find a second user whose talk spurts exactly coincided with the pauses of the first user, we would be able to interpolate the second user in the idle periods of the first user, and combine both users on a single channel with room to spare (see Figure 6.67). Unfortunately, no two users will show such a coordinated pattern, so DSI cannot be implemented quite this easily. The pattern of talk spurts and pauses generated by any individual speaker is too unpredictable. With just two users on one channel, we would often have cases where the second user would find the channel occupied and his talk spurt would be blocked or clipped (Figure 6.68).

But with a sufficiently large pool of available circuits—known as a DSI pool—it becomes possible to model the behavior of a large group of users statistically, such that we can assure with reasonable confidence that any individual user will be able to gain access to one of those circuits when necessary. Indeed, the problem of design-

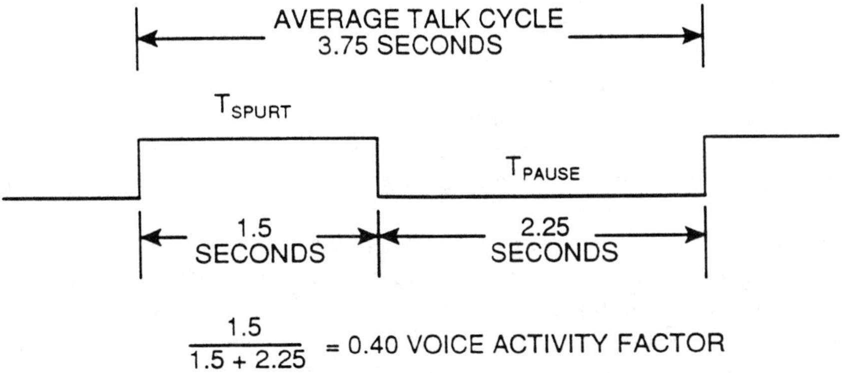

Figure 6.66 Average talk/pause cycle. (From Hughes Network Systems DSI Report.)

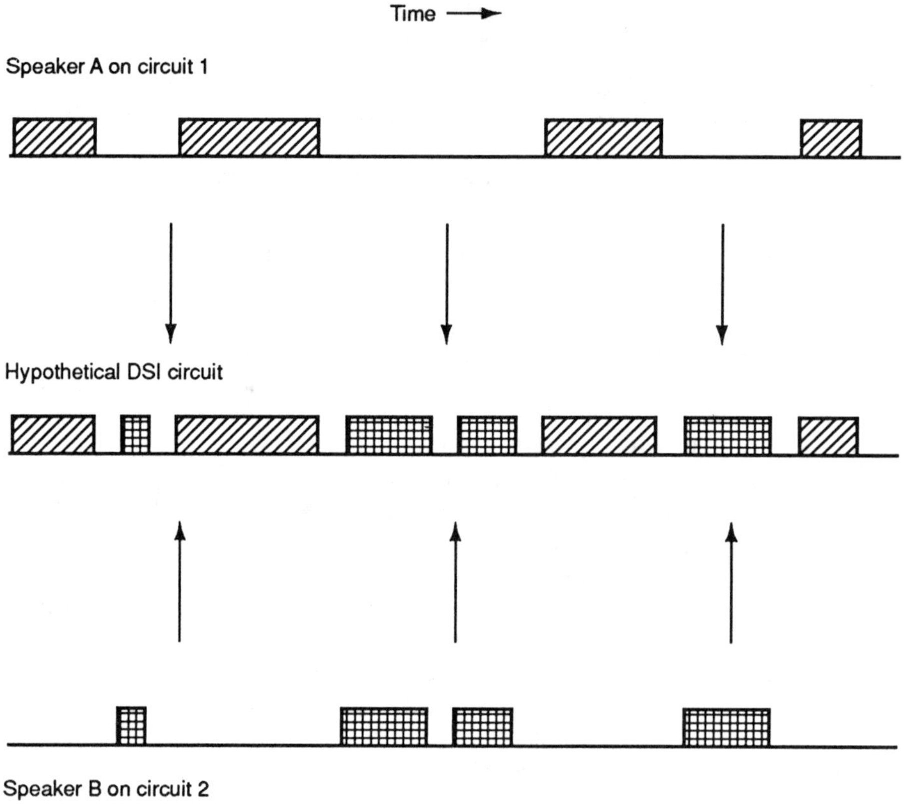

Figure 6.67 Idealized example of two perfectly matched users sharing a single channel.

ing the DSI pool is basically the same as that of designing any concentrated trunk group or statistical multiplexing process, except that it operates on a much shorter cycle than most trunking applications. A given number of circuits can support a larger number of users, each of whom generates traffic that is less than one full duplex channel's worth (see Figure 6.64).

In fact, one way to think of DSI is as a system where instead of setting up the circuit for an entire call, a circuit is set up for each talk spurt, which may be a syllable or a few words, and assigned to the user only for the duration of that talk spurt. The statistical principles that apply in calculating the gains from statistically multiplexing such discontinuous traffic sources onto a smaller number of channels are the same. Unlike capacity gains claimed for CDMA and other relatively unproven approaches, calculating the DSI gain is straightforward. DSI is really just another form of statistical multiplexing, a well-understood art. Or, as another author puts it, "this is a classical problem in the theory of queues" [115].

The basic technical elements for DSI are:

1. A *voice activity detection algorithm,* capable of distinguishing speech from other forms of background noise and capable of determining when talks spurts begin and end with a high degree of accuracy—this is probably the most challenging technical hurdle to overcome [116]
2. The *DSI pool,* with appropriate control slots and talk slots, properly sized to realize the maximum statistical gains
3. *Control procedures for rapidly assigning an available talk slot* to a user who is beginning a talk spurt—fast enough to avoid any loss of information (e.g., clipping) in the vast majority of cases

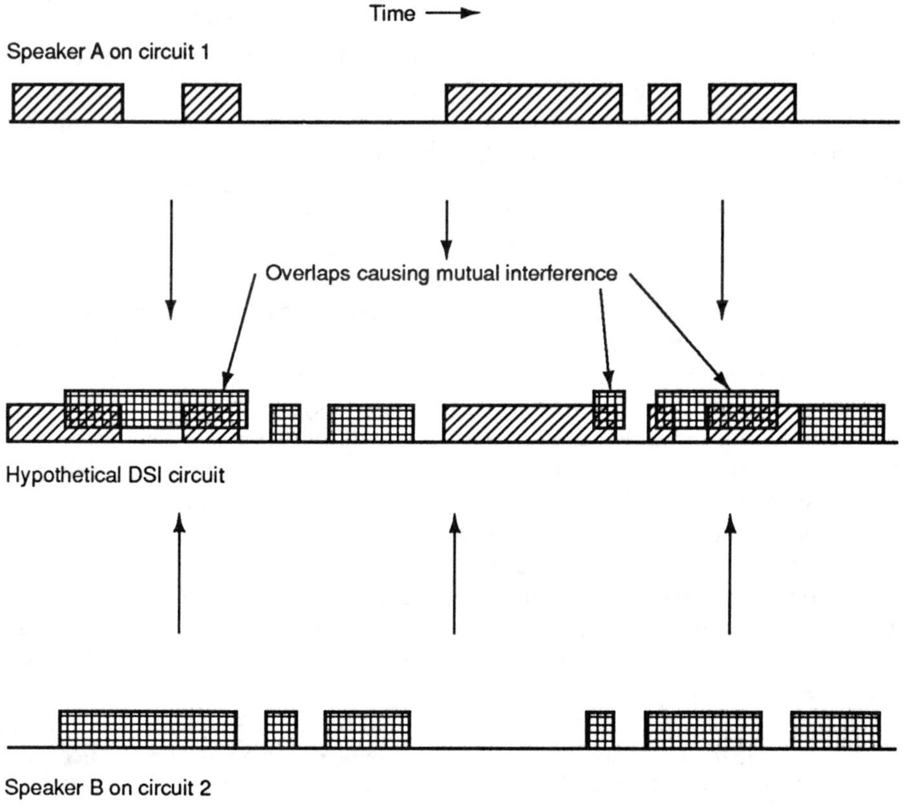

Figure 6.68 Clipping from actual speech overlaps.

4. A *demodulation algorithm capable of rapidly acquiring synchronization* on received talk spurts.

The application of DSI to wireless access was first proposed by a consortium of Hughes Network Systems (a subsidiary of General Motors), Alcatel, and IMM for an enhancement to the IS-54 TDMA standard for digital cellular radio, known as *E-TDMA* (for extended TDMA) [117]. In the system demonstrated in early 1991, each DSI pool comprises 12 radio-frequencies, and six time slots per frequency (as defined in the standard IS-54 framework), for a total of 72 circuits. Some number of these circuits are set aside for control—that is, for requesting talk slots and making slot assignments. The remaining slots are dynamically available to any user on demand at the beginning of a talk spurt (see Figure 6.69).

How does this system actually work? As we shall see, DSI is uniquely suited to block-encoded digital radio systems. First, let us consider the slot allocation process. It may sound like quite a challenge to process a slot request and make an assignment quickly and accurately enough that no speech data are lost. And yet the scheduling of talk spurts is paced primarily by the fact that such systems utilize block encoding of the voice data. Low-bit-rate voice coders take in a block of speech data, process it, hold the processed data in a buffer, and finally transmit the speech data in a burst when the appropriate time slot rolls around. This coding and buffering process typically involves delays of tens of milliseconds. Thus, there is a built-in delay before the first block of speech is actually transmitted. In the E-TDMA framework, this startup delay is about 40 milliseconds. It turns out that this is more than enough time for the subscriber terminal to send out a request to the base station for a talk slot and for the base station to process this request and transmit back the slot assignment. In fact, there is time for multiple transmissions of the slot request and slot assignment messages, to allow for the possibility that channel conditions may be poor—all while the voice coder is crunching away on its first block of speech data. The slot assignment request can be initiated as soon as the voice activity detector determines that a speech spurt has commenced, which occurs at the very beginning of the coding process (see Figure 6.70). Thus, by the time the subscriber terminal is actually ready to transmit the first block of voice data, the slot assignment has been made, usually with plenty of time to spare. As well, the system is designed so that the slot request and slot assignment messages are an order of magnitude more robust than the voice slot—so that in a mobile application the voice link will fail (from fading or other propagation problems) long before the control link fails. This prevents false assignments [118].

In the E-TDMA proposal, let us assume that there are 9 control slots, and thus 63 slots available for voice traffic. There may also be other types of nonvoice traffic, such as mobile assisted hand-off instructions (in the IS-54 cellular standard). These must be factored into the traffic calculations. The E-TDMA proposal can actually support about 120 users on those 63 slots, or nearly a doubling of capacity.

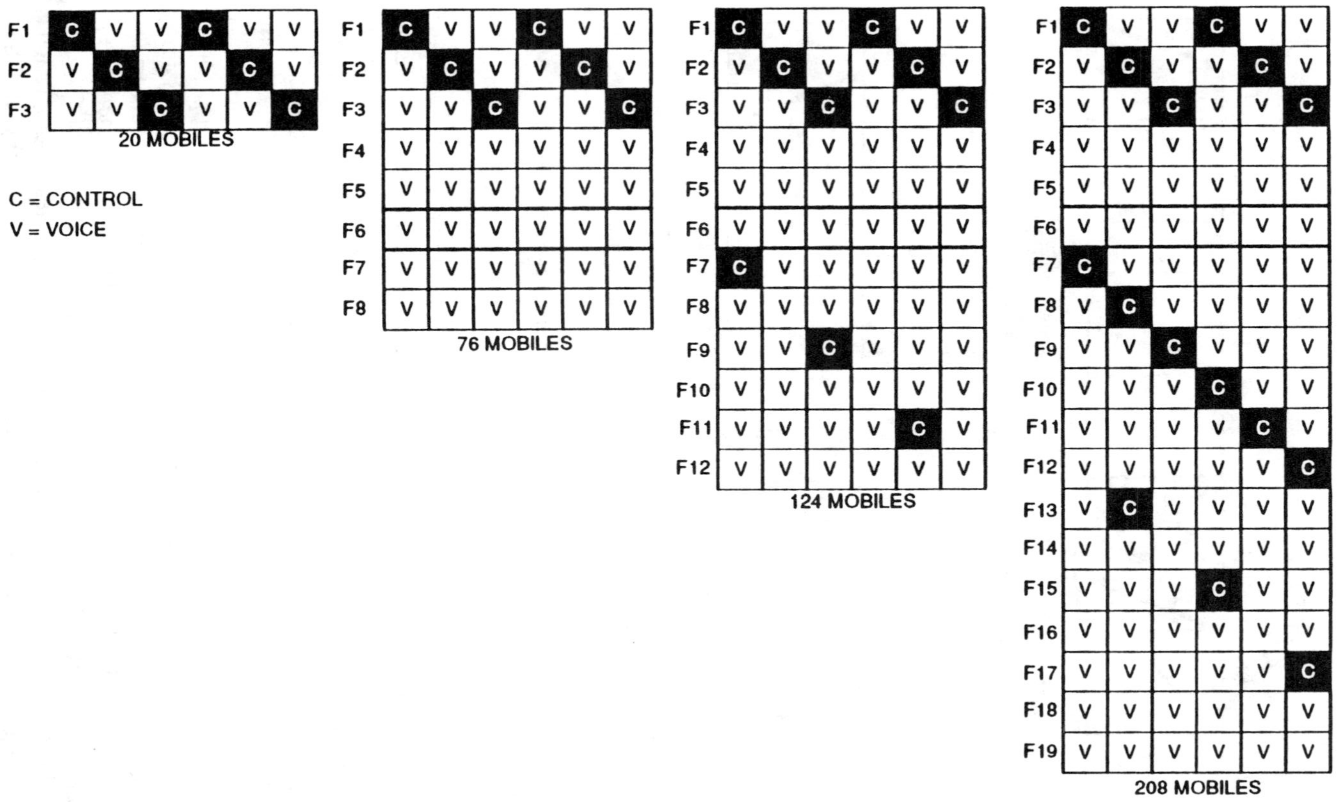

Figure 6.69 DSI pool. (From Hughes Network Systems.)

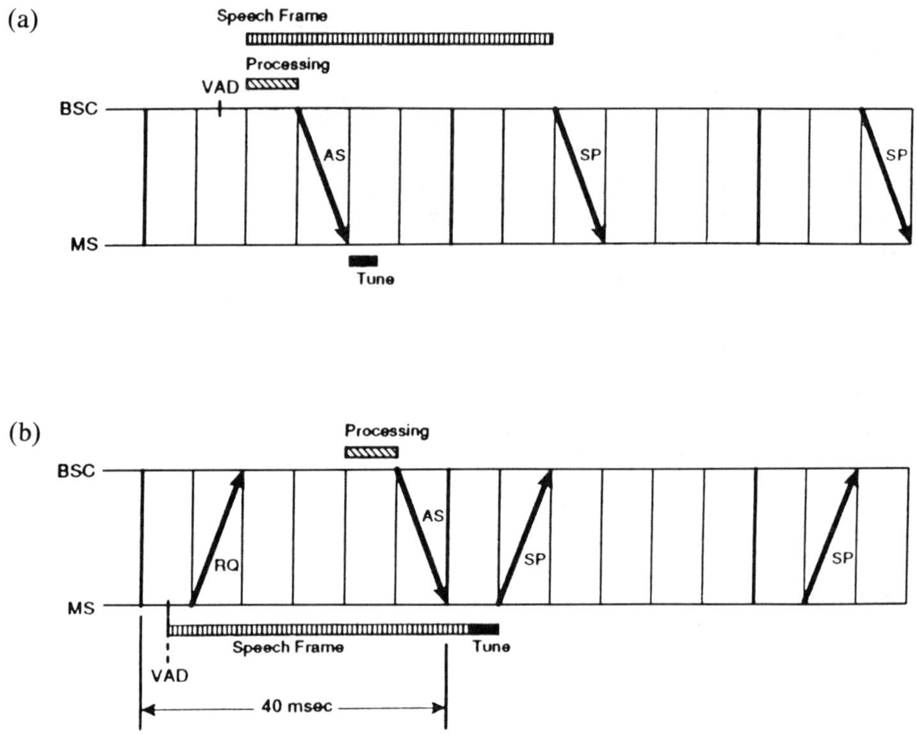

Figure 6.70 Talk-slot assignment request protocol; (a) forward channel assignment, (b) reverse channel assignment.

What happens if the 63 available slots are occupied and a 64th user begins a talk spurt and requires access to a talk slot? There are in general three types of outcomes in the event of congestion or the unavailability of a talk slot at the moment when a user requests one [119]:

1. *Freeze-out or clipping:* In this case, the first block of speech data may be thrown away, resulting in a front-end clipping of the speech; it is also possible to design the system to insert the clipping in the middle of the speech spurt, although this is generally less acceptable than front-end clipping [120].
2. *Variable delay:* It is possible to buffer the speech, without actually throwing away any of the blocks of data, by simply delaying the whole talk spurt by a fixed amount; the delay may vary from talk spurt to talk spurt when the system is highly congested [121].
3. *Variable speech coding rate:* Another conceptually interesting approach is the use of a variable coding rate, where as the system becomes congested the cod-

ing rate for all circuits (or possibly only certain circuits) is reduced in order to generate bandwidth for the overload; the possible disadvantage of this approach is its complexity [122].

A number of alternatives have been evaluated and, in general, it appears that in terms of economics, implementability, and performance, the best results are obtained from allowing a small amount of front-end clipping, possibly supplemented by a small amount of variable delay [123].

Let us examine this a little more closely in the context of the E-TDMA proposal. When a user tries to access a system that is temporarily fully loaded, a number of possible outcomes can occur. The first possibility is that while the voice coder is encoding the first block of User 64's talk spurt (which may take as much a 40 milliseconds or so, depending on the coder delay) one of the other users may stop talking and a talk slot may become available. Since the average talk spurt is about 1500 milliseconds long, and there are 63 users randomly starting and stopping their speech spurts, then we should expect that one speaker will stop transmitting—and a slot will open up—every 24 milliseconds or so when the system is operating under full loading. What if this does not happen? In that case, the first block of encoded speech from User 64 may be blocked and we wait another cycle (i.e., another 40 milliseconds while the second block is encoded) for a slot to become available. If so, User 64 starts transmitting, having "lost" the first block of data. In other words, the first 40 milliseconds of speech from User 64 were clipped. What does this do to speech quality? The answer appears to be that a single speech clipping event of less than about 200 milliseconds is generally not perceptible. Nevertheless, if a high percentage of talk spurts are clipped, there may be some perceived loss in quality [124]. In the E-TDMA system, the overall statistics are designed to ensure that no more than 2% of the speech spurts are clipped by 50 milliseconds, which is a recommended criterion by some previous studies [125]. The 2% level would only be reached at maximum system loading. Such a clipping level should produce no perceptible degradation in speech quality.

Therefore, one of the design parameters the operator can set in sizing the system is the number of talk spurts that are "allowed" to be clipped. It should be apparent that the amount of clipping is not fixed; it is a statistical parameter that can be controlled and modified by the operator. It is in this sense strictly analogous to blocking. But in perceptual terms, clipping and blocking are experienced differently, and the system design may exploit this to advantage if the clipping can be managed so that its effects are perceptually masked.

Advantages of DSI. DSI offers a number of advantages as an enhancement for conventional TDMA techniques:

1. *Capacity gains:* A properly implemented DSI should approximately double the baseline system capacity, with further gains realized from improved trunking efficiencies at the system level. In fact, the Hughes/IMM/Alcatel E-TDMA

proposal calls for about 2.5 times the baseline, or 15 times analog cellular capacity (see Figure 6.71).

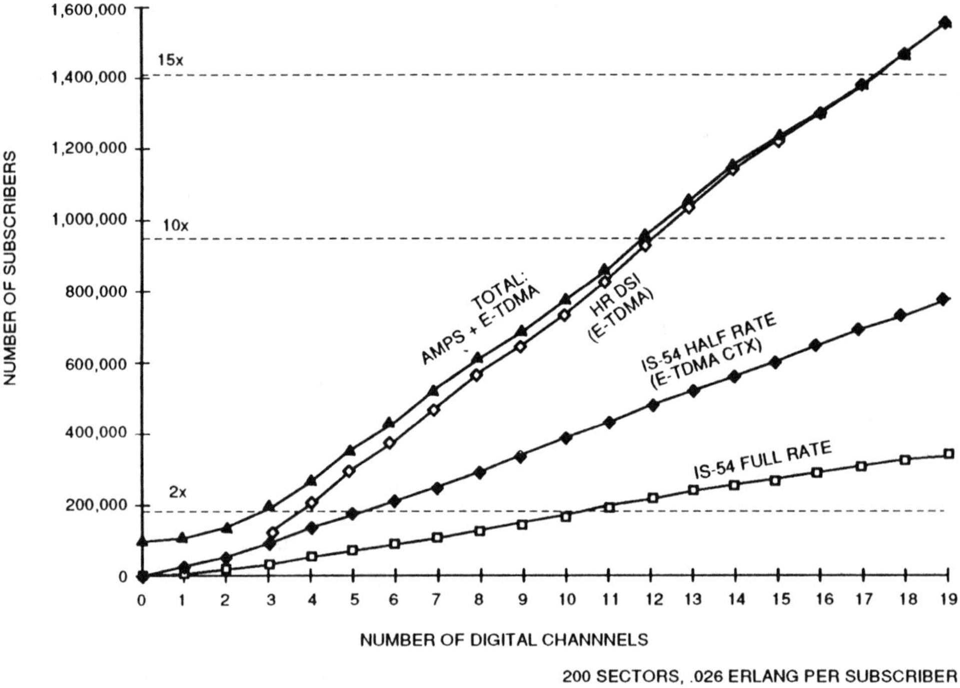

Figure 6.71 Capacity gains for DSI-enhanced TDMA over conventional analog (for AMPS cellular). (From Hughes Network Systems.)

2. *No change in RF emissions characteristics:* Unlike CDMA and channel-splitting FDMA, DSI does not alter the channel characteristics of the basic TDMA format. It does not affect frequency plans or cell-site placement.
3. *Implicit frequency hopping within the DSI pool provides some resistance to frequency-selective fading:* The DSI pools do not need to be made up of contiguous frequencies, and they can be designed to provide a modicum of frequency hopping that will enhance resistance to frequency-selective fades in much the same way as frequency-hopped spread spectrum systems. While DSI hopping *per se* is not as robust as a formal frequency-hopping scheme (as found in the GSM air interface, for example), it can help with the problem of a temporarily stationary antenna that is stuck in a deep fade (as when a car is standing at a stop light, for example).

4. *"Soft" failure and overload modes:* The system can be engineered to allow for a temporary overload condition without blocking callers. For example, if a traffic accident ties up a large number of people at a particular location and leads to a calling "hot spot" in a cellular system, it may be desirable to allow the system to "bend" in response to the radio-traffic bulge that occurs as people begin to call their office or homes to indicate they will be delayed. In a standard TDMA system or FDMA system this is not possible; excess traffic is simply blocked. In a CDMA system, extra traffic can, in principle, be accommodated, but the penalty is an increase in average interference levels in the system as a whole, which may lead to uncertain results for all users (as with many things about CDMA, the effects of overloading the system with voice traffic are difficult to predict and there is as yet no base of experience with CDMA voice systems to fall back upon). In a DSI/TDMA system, the result of overloading is an increase in the percentage of talk spurts that are clipped. This effect is thus felt at the individual circuit level, not at the system level as in CDMA. Once again, there is little experience yet with clipping involving coders of this type, but experience with PCM coder clipping suggests that there is considerable tolerance for clipping rates above 2% [126]. The same applies in cases where, for example, an equipment failure causes the loss of one of the channels in a DSI pool. The users in that pool would experience a slight increase in the clipping rate (if the pool were operating at maximum capacity when the failure occurred). Even calls in progress on the failed channel would not be dropped (although the speech spurt would be lost).
5. *Compatibility with TDMA standards:* DSI can be implemented without disrupting the standards that have been developed for TDMA architectures for many wireless access applications around the world. This allows for a lengthening of the lifetime of these standards in a dynamic technological environment.
6. *Straightforward implementability:* Unlike CDMA, DSI/TDMA does not require major technological breakthroughs. DSI principles are well understood from other applications stretching back to the 1950s. The voice activity detection algorithm is fairly straightforward [127]. The control structure of the DSI pool may seem complex, but with today's high-speed processors and frequency-agile radios, the implementation is not difficult.

It is worth noting that a somewhat broader conception of time-compressed speech techniques applicable to a variety of real-time and nonreal-time voice communications applications is emerging, of which DSI *per se* is a subset. One label suggested for this expanded concept is *dynamically managed voice* (DMV), and I believe the neologism is a valuable one.

> DMV is a generic term that encompasses systems utilizing speech activity detection . . . to statistically multiplex a number of speech sources onto a

> smaller number of transmission channels. Examples of DMV systems include: nonbuffered and buffered speech interpolators, store and forward and local area packet voice networks, integrated voice/data switches and networks, voice activated satellite systems, and voice messaging systems. [128]

This passage suggests how pervasive these techniques are likely to become, which bodes well for implementability.

All in all, DSI is a promising near-term enhancement for a number of wireless access architectures, including digital cellular radio.

6.4.6.4 Packet Reservation Multiple-Access Concept

A similar packet-oriented multiple-access concept has been advanced by David J. Goodman and his colleagues, which they designate *packet reservation multiple access* (PRMA) [129]. Essentially, PRMA is a "combination of time division multiple access and slotted ALOHA" [130]. In much the same way as the DSI system described above, PRMA begins with a time slot formatted channel. The slots in each frame are either available or reserved, on the basis of information broadcast from the base station on the previous frame. When a subscriber terminal begins its transmission (a voice detection algorithm is obviously required, in much the same manner as described in the previous section), it attempts to transmit during an available time slot. If it is successful, this time slot will be reserved for that user until the voice activity detector determines that the speech burst has finished. If, on the other hand, the initial attempt to gain access to the slot is blocked because of contention with other subscribers simultaneously attempting to do the same thing, then all contending users will back off and retransmit in a standard "ALOHA-type" manner (see Figure 6.72).

As I see it, the main difference between PRMA and the E-TDMA/DSI system described in the preceding section is the point of possible contention between users. In the E-TDMA system, there is first a circuit request (talk slot request) message, which is sent over a special control slot. It is possible that two or more users may block one another during the attempted transmission of this message. But once a talk slot has been assigned, there is no possibility of contention involving actual speech data from two users. The PRMA system uses a somewhat simpler talk slot assignment procedure: The subscriber terminal simply begins transmitting on any nonreserved time slot. If he is successful on the first try, the reservation procedure protects him from further contention. But there is a possibility that the first block of speech data may be lost due to contention. It is not clear to me whether this is a functionally important difference.

In most respects, these two systems should display similar strengths and weaknesses (or failure modes). I have not yet seen a formal comparison of capacity gains anticipated from these two techniques.

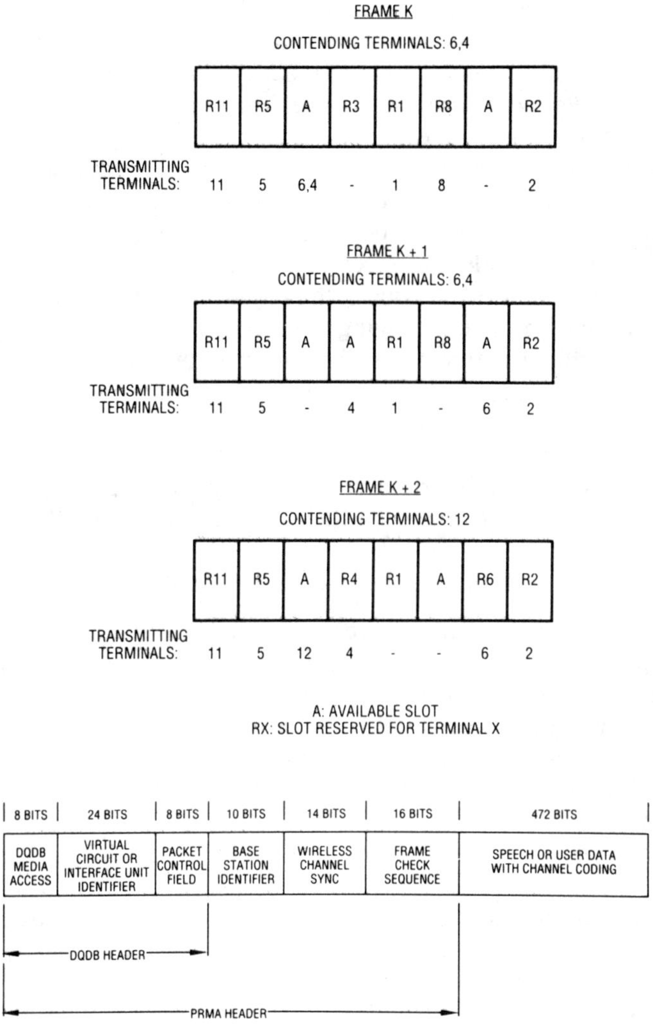

Figure 6.72 PRMA protocol operation. (From D.J. Goodman, "Cellular Packet Communications," *IEEE Transactions on Communications,* V. 38, No. 8, August 1990, p. 1277.)

6.4.7 Channel Access Techniques

The "other half" of a multiple-access radio system is the procedure for setting up a call, allocating a radio circuit (or a talk slot, in the DSI framework) to a user when he requests it. Normally, in the types of systems we are discussing here, the call setup is carried out over a special control circuit (which may be a separate time slot in a TDMA system). That is, the call setup interactions are kept separate from the actual voice transmission. In the context of wireline network terminology, we might refer to this as *common channel signaling,* although the term is not widely used in radio.

The problem the radio system must solve in providing channel access is that the user's calling behavior is not known ahead of time, and yet once a user initiates a call he demands relatively rapid access to a circuit—typically less than 500 milliseconds. Users are conditioned by the instant dial tone inherent in the dedicated loops of the wireline network. They expect the same from a multiple-access radio system, even though the use of nondedicated radio channels means that providing a rapid response is much more challenging.

The other problem is that the call setup channels are essentially broadcast-mode channels. Particularly in the reverse direction—that is, when subscriber units are attempting to initiate calls—there is a potential problem of contention: Two or more users may try to transmit a circuit request (or talk slot request in a DSI system) at the same time and hence interfere with one another. The access protocol must establish some way to control this problem if large user populations are to be supported.

To be effective, the channel access technique should be:

1. *Transparent:* It should not require the user to perform any unusual operation, beyond perhaps pressing a SEND button (and even that may be unacceptable in some applications, such as BETRS).
2. *Rapid:* As noted, access should be provided within 0.5 second, generally speaking, although once again there may be some leeway in some applications; where a SEND button is used, a longer period (up to several seconds) may be acceptable.
3. *Efficient:* The channel access protocols should be capable of handling a very large number of successful call requests without using up a lot of spectrum; it is generally felt, for example, that the protocols used in analog cellular radio are unnecessarily slow and wasteful of radio spectrum, and future access protocols will achieve much higher throughputs and support much larger numbers of users.

Although there are a large number of different techniques for accomplishing fast, efficient channel access, the subject generates far less controversy than the question of the multiple-access architecture. The field of access protocols is becoming very complex because it is the basis for today's LANs, MANs, and WANs which all use common channels for packet transmission. There is a large literature, which we will not review here [131].

Basically, the access protocols that have been used for radio fall into two categories: demand assignment (following the terminology of Sachs' taxonomy [132]) and random access models (sometimes called contention systems). Demand assignment systems include *polling* techniques, where the base station periodically interrogates each subscriber unit to see whether it wishes to transmit. Polling has been used in some of the early generations of microwave subscriber radio [133]. It is suitable only for very small subscriber bases, since otherwise the polling process would introduce too much delay in the access process.

Therefore, the vast majority of wireless access architectures today and in the foreseeable future use contention protocols for channel access, where any subscriber can initiate a call request on its own, without waiting for any signal from the base station. The simplest type of random access protocol is known as ALOHA, which has been widely described and analyzed [134]. ALOHA is, however, relatively inefficient in terms of throughput; therefore, a number of improvements have been developed, including *slotted ALOHA,* which is fairly widely used [135]. Another class of relevant protocols are known as *carrier sensed multiple-access* (CSMA), which is a fancy term meaning simply that unlike simple ALOHA systems, the subscriber unit first "listens" to the channel to see if it is idle before trying to transmit the call request. (The reader should not be misled by the CSMA terminology; this is not a method of organizing circuits in the radio spectrum like TDMA and FDMA, but simply an access protocol.) There are many variations on ALOHA and CSMA [136].

6.4.8 A Comparative Evaluation of Multiple-Access Architectures

Over the next 10 years, more ink will be spilled (and perhaps more tears shed) over the choice of access methods for wireless applications than over any other facet of the technology. In part, this reflects its relative importance in the hierarchy of design decisions. FDMA, TDMA, and CDMA represent broad frameworks within which specific circuit-level digital techniques (e.g., voice coders) can be modified "in software" to meet specific applications requirements or to stay current with the march of technology. It is also generally true that the different access techniques represent mutually exclusive alternatives. In other words, within any access framework, there is considerable flexibility; but differences between access methods tend to be embedded in a great deal of incompatible hardware. As well, changing access techniques usually involves changing the essential physical characteristics of the radio channel (such as its bandwidth), which alters RF performance characteristics such as C/I ratios, which in turn lead to new cellular geographies, new frequency plans, and so forth.

Access techniques will be debated as well because they represent the cutting edge of international competition among alternative standards and alternative applications. Many of the wireless access applications discussed in Chapter 5 are partially overlapping, and hence competitive. If a TDMA architecture and an FDMA architecture were to be in such competition with each other, we could expect to hear a great deal from each camp about the purported superiority of its approach. The North American cellular industry has already passed through one such phase, during the year or so prior to the January 1989 selection of TDMA as the access method for the digital cellular IS-54 standard [137]. Similar debates are raging as of this writing with respect to many private radio applications, as major manufacturers push for one access technique or another. Promoting a "preferred" access standard has become part of the competitive strategy for manufacturers, operators, and trading blocs.

Another reason that controversy over access methods is likely to maintain a high level of partisan interest is that there is still relatively little experience with digital radio in any of these applications. Large areas of technical uncertainty exist—and as nature abhors a vacuum, so too does the telecommunications industry rush to fill these areas of uncertainty with elaborate mathematical equations to provide us with an illusion of knowledge, in the form of models-which-are-true-if-the-assumptions-are-valid. This is particularly true for the more exotic CDMA architectures that are just beginning to be explored.

Under such circumstances, it is difficult to address the question of which access technique may be objectively superior, in a technical sense, for a given application, especially in a relatively brief and nonmathematical discussion. What follows, therefore, is a set of rather broad assessments that the reader may take (or leave) more or less at face value. I would defer to another author and to a very different kind of book any more detailed analysis.

1. *FDMA is a limited architecture which has been fairly fully exploited.*

   Almost all radio systems deployed up through 1990 (with the relatively minor exception of subscriber radio, BETRS-type systems, and some VSAT systems) have been based on FDMA, single-channel-per-carrier architectures. Essentially, the advantage of FDMA is that it allows trunking. But today we would assume trunking in virtually every application. It is part of the baseline, so to speak; indeed, as noted previously, most TDMA and even many proposed CDMA architectures actually combine frequency-division multiplexing as well. For example, the IS-54 standard for digital cellular radio is accurately designated an FD/TDMA system (or to be even more precise, an FDM/TDM, FDMA/TDMA system).

   Once multichannel trunking is part of the baseline, FDMA architectures have little to recommend them. They tend to be hardware-intensive, inflexible architectures that force major upheavals in order to take advantage of advances in circuit-level techniques. Essentially, the only way for FDMA to gain in spectral efficiency and capacity is by channel-splitting [138]. But channel splitting in a cellular system introduces many difficulties, including poorer cochannel interference performance with a high uncertainty as to just how much worse it may be [139]. Some have claimed that poorer RF performance (necessitating, for example, less efficient cellular reuse patterns) may in fact nullify the nominal gains from going to narrower channels in the first place [140].

Because FDMA hardware is not easily reconfigurable, it is also more difficult for FDMA systems to provide variable-bandwidth services to users who may want more (or less) than basic voice communications. If a lower bit-rate service (e.g., Group 3 fax or packet data) is desired, an FDMA system must still allocate an entire channel/circuit, which is wasteful. If digital services with bit rates higher than the standard voice rate are desired, the FDMA architecture is an awkward vehicle. As Raymond Steele has noted:

> A variable bit-rate service is difficult to accommodate when FDMA is used, although a variable number of FDMA channels could be assigned to satisfy the service requirements of a particular mobile user. However, it is very much easier to alter the number of TDMA time-slots (and thereby the data rates) associated with a particular RF carrier when the mobile user requests a particular service. [141]

FDMA systems also have the highest system infrastructure costs in most applications, mainly because of the requirement for duplication of all the RF hardware for every circuit. This is the root of the high system cost (and high airtime charges) of analog cellular [142].

Another argument that has been made on behalf of FDMA is that it is inherently a simpler architecture than, for example, TDMA, and that it might thus be capable of being brought to market more rapidly. This was probably true in the late 1970s and early 1980s; however, by the late 1980s the technical viability of TDMA systems had been demonstrated beyond doubt and this argument lost much of its force [143].

For these reasons, I believe that it is likely that most of the large-scale wireless access applications will probably not employ FDMA.

2. *Time-division duplex may well find an important niche in small-scale access systems, such as cordless telephony.*

   It may seem odd to refer to cordless telephones as "small-scale" systems, given that there are probably 5 or 10 times more cordless phones today than all other radio-telephony devices put together. Nevertheless, each cordless phone is a single-line device, operating on a single frequency, and the bulk of the market would seem likely to retain that structure. It would be very desirable to bring an improved, digital form of cordless telephony to the market to eliminate the many problems that plague today's analog sets. Moreover, from both a spectrum allocation perspective and for designing a cheap radio device, the ability to operate in a duplex mode on a single channel could be advantageous. TDD seems well suited to this specification.

Beyond residential cordless phones, some other small-scale applications may find TDD to be, in one form or another, a useful architecture. But for the big, mainstream wireless systems such as cellular radio, BETRS, and (probably) PCN, we will be playing some game other than ping-pong.

3. *TDMA is likely to prevail in the major applications for the 1990s.*

   Already, the standards for digital cellular radio systems in North America, Europe, and apparently Japan will all be based on TDMA. On the telephony side of the applications tree, BETRS and subscriber radio are TDMA-based. After evaluating other access methods, Bellcore advocated TDMA in its extensively researched PCN-type proposal [144]. Other systems such as the United Kingdom's version of PCN and the European cordless telephone standard known as DECT (which is really aimed at the wireless workplace application set) will also follow the TDMA trend.

   There are good reasons for this. First, TDMA is feasible with today's technology. Second, it offers a much more software-intensive framework that should allow for much more flexibility in integrating improved circuit-level techniques (e.g., voice coders). A good example of this is the anticipation of so-called "half-rate" voice coders within the framework of both the IS-54 and GSM digital cellular standards. The leap to the third generation of technology is already built in. There are, of course, limits to our ability to anticipate the future course of technology, but TDMA should be much more flexible than FDMA in a highly dynamic technology environment.

   TDMA also reduces system infrastructure costs dramatically. It does so in the same way that TDM works economic magic in the wireline world: by time-sharing the costly transmission equipment among many users.

4. *Packet-like techniques will gain in importance, with basic DSI leading the way.*

   While true, full packetized transmission systems for voice probably will not emerge until the twenty-first century, the exploitation of voice activity gains to double or more than double system capacity will undoubtedly play an early role. DSI will be a relatively straightforward enhancement of TDMA systems in many applications; indeed, as noted above, it has been announced by one manufacturing consortium for the first IS-54 applications [145].

What About CDMA?

There has been a flurry of interest in the applicability of CDMA techniques to digital cellular and PCN, which has captivated some of the media and enlivened industry discussions with a new and intriguing vocabulary. People who may have been uncertain about the meaning of "digital" and "analog" a few years back are now happily debating the merits of "pseudo-noise" and "orthogonal codes." The excitement has been stirred by dramatic claims for CDMA performance and, particularly, system capacity. The CDMA "capacity barometer" started initially at "forty times analog capacity" [146], and has ranged recently from "twenty times analog capacity" [147], dipped to "seven to ten times analog capacity" [148], and then rebounded to "fifty or more times analog" [149]. Such claims get attention, just as surely as would anyone who announced a car that average 300 miles per gallon of gas.

One of the hazards of attempting to deal in print with a highly topical subject is that today's interim conclusions may seem terribly naive or misguided in the not so distant future. This seems especially worrisome where CDMA is concerned, because it is such an unknown quantity. There is fundamental, sincere disagreement (as well as a lot of tendentious rhetoric) concerning almost every major claim made on behalf of CDMA today, and unfortunately many of the claims are such that we may only really be able to validate them by building and operating some pretty big CDMA systems for a fairly long period of time. It is quite possible, for example, that even a trial on the scale of the original Chicago AMPS trial, which encompassed a couple thousand customers and a dozen or so cells, and ran for several years back in the late 1970s, might well be insufficient to prove or disprove some of the claims now being made. To take one example, it has been suggested that a CDMA system could operate with a reuse factor of 1—that is, all the available frequencies could be used in all cells, and even in all the sectors within the cells, at the same type. If this were true, it would certainly be a boon to cellular system engineers. It would eliminate the tedious job of frequency planning. Yet if we think for a moment what it would really take to test this proposition, the scope of the experiment becomes quite daunting. After all, we are dealing with statistical factors, and statistics must be collected over a sufficient period of time to allow calculations of the right level of accuracy.

Thus, with some sense of trepidation (since no one likes to look ridiculous in hindsight), I offer the following observations, all of which seem fairly plausible as of this writing.

1. *There are no new laws of physics here.*

   As a number of authors (including some of the strongest CDMA advocates) have pointed out, the "theoretical" capacities for FDMA, TDMA, and CDMA are all about the same [150]. The relative advantage of one access method over another lies in (1) the economics of the implementation, and (2) the state of practical knowledge about the application. In other words, we should be cautious and critical regarding, for example, capacity claims for CDMA that are dramatically superior to any other technology.

   In the first place, many of the claims now being made for CDMA are simply specious [151]. And aside from partisan excesses, there are real uncertainties about whether given capacity claims represent worst case, average case, or best case numbers. As anyone who has ever built a product knows, there is a gap between the theoretical performance of a system on the blackboard, and the product specification that is handed to the factory. The latter incorporates countless large and small compromises with reality, including those based on the following:

1. Conservatism for any new design—we do not try to push a new technology to its outer limits in the initial applications; it is simply too risky for the manufacturing process and investment to follow this path; instead, a hefty margin for error is built in to be on the safe side.
2. If it is a standardized product, it must be manufacturable by a broad base of companies—the technology must be transferable and translatable to ordinary manufacturing environments, using ordinary procedures and ordinary people.
3. The technology and its implementation must be forgiving of slight deviations from specifications that are bound to creep in as manufacturers vie for cost advantages—if the entire system crashes when it hits a pothole, it is not really marketable.
4. It must be economical—high performance at any cost is acceptable in a military environment, perhaps, but not in the commercial market.
5. It must be based on a solid understanding of the technology and its applications within the industry that designs and builds it—the lack of applications experience in particular will hold back commercialization of "breakthrough products" unless there is an absolutely overwhelming and urgent market imperative (a rare situation—witness how long we have managed to live without HDTV).

Much of the industry recognizes that the claims have been overblown. Indeed, most conversations on the subject today begin with the question as to how much these claims are to be "derated" (in the words of one CDMA advocate). Some are more skeptical.

> The highest cell-site capacity estimates (factors of 20) are based on optimistic theoretical estimates of factors for highly complex spread-spectrum systems that incorporate multiple levels of power control, precise cell-site synchronization, and "soft hand-off." When it is noted that variations in power control with standard deviations of as little as 1 or 2 dB could result in capacity decreases by as much as 30% or more and that soft hand-off imperfections could result in similar decreases, the likelihood of realizing such high theoretical estimates is small. [152]

2. *No one knows yet what CDMA is capable of in these applications.*

In my earlier book, I called CDMA the "Great Unknown"—and I stand by this description. At least up through 1990, CDMA had never been used in a commercial telephony application. There are so many unknowns about its performance, some of which are so fundamental, that it is fair to say that no one really knows how far this architecture can be pushed within the economic and practical applications constraints of today's wireless applications. For all the wishful thinking and the optimistic views, there are some real concerns about whether CDMA will work at all (in the commercial sense) for many of these applications. As one authority has noted:

> The large-scale complex multiple-feedback dynamic system [which CDMA cellular and PCN proposals represent] also raises concern for overall system reliability and stability [!]. In order to achieve significant cell-site capacity, the system must be delicately balanced with tight feedback power control, and multiple cell-site and vehicle set interaction (soft handoff). Precise synchronization of radio ports (cell sites) to better than 1 microsecond, as required for soft handoff in these spread spectrum systems, is a particularly difficult problem for the many thousands of radio ports connected to many widely separated interconnected central office switching machines that would be needed to serve millions of users of low-power access systems. The margins and techniques needed to guarantee dynamic stability may further reduce the theoretically expected capacity. . . . [153]

It may be fairly observed that both time-division and code-division techniques were pioneered at the same time (during World War II) for the same purpose (secure communications) [154]. In a sense, they came out of the starting gate on more or less equal terms. From that point forward, however, what we have is a tale of two very different technologies.

On the one hand, time-division techniques went on to become the basis for the entire digital revolution in telephony, beginning with the introduction of T-carrier transmission systems in the 1960s [155] and encompassing the advent of digital switching, digital microwave, digital satellites, digital fiber optics—all based on the same time-division principles. Today in the United States there are hundreds of millions of TDM circuits, and the entire wireline core network will have converted to TDM by the end of the century. The telephone industry today can be called, without exaggeration, a TDM industry. The extension of TDM techniques to the sphere of wireless access is an evolutionary extension of this TDM empire.

By contrast, code division has never achieved commercial success in the world of telephony. In fact, there is to my knowledge not a single operational CDM telephony circuit anywhere in the world (unless it is classified). It has been used in certain military applications, where cost, capacity, and so forth are no object—and even in the military environment, CDM is a very small niche application. There may be a lesson in the failure of CDM techniques to penetrate the commercial market for more than 50 years—or it may simply be a matter of timing and availability of the enabling technologies (such as advanced DSP chips). That is another one of the unknowns.

Thus, the comparison of TDMA and CDMA is often a mixture of apples and pink elephants. There is simply too little industry experience with CDMA to deploy it as rapidly and as easily as many of its advocates believe. Even if all the technical unknowns could be resolved with the wave of a hand or the flutter of a viewgraph, the vast basis of applications experience that is required to sustain a new technology in a large-scale telecommunications application is simply not there (at least as of this writing in mid-1990). Keep in mind the lengthy gestation periods required for other new telecommunications technologies, such as fiber optics and digital switching, both of which took approximately two decades to move from concept and feasibility demonstrations to wide deployment supported by multiple vendors, backed by detailed standards, and so forth. For better or worse, I believe it is unlikely that CDMA radio systems will dramatically shorten this cycle.

3. *Finding out what CDMA is capable of will be expensive.*

TDMA and FDMA are circuit-modular systems. In other words, if a single RF channel of TDMA technology (e.g., IS-54) can be demonstrated to work, it is possible to extrapolate in a fairly straightforward manner to predict the performance of 10 channels, 100 channels, 1000 channels. In a channelized system (such as TDMA), the whole purpose of the multiple-access architecture is to keep individual channels separate from one another, in time or frequency or both. If the ability to separate the channels can be demonstrated, the question of how many channels are built is virtually trivial (from a technical feasibility standpoint).

CDMA is not circuit-modular. To demonstrate that a single CDMA mobile unit can communicate with a CDMA base station is of some value, but it says little about whether 10 units can share the same channel, or 1000. The capacity claims of CDMA advocates rest precisely on this unanswered question: How many mobiles can share the same channel at the same time, given that they are all mutually interfering with one another and the assumption of separateness cannot be used to extrapolate in a straightforward manner.

Indeed, almost every significant aspect of system performance in a CDMA architecture is subject to the same caveat. For example, it has been claimed that a reuse factor of 1 is possible for CDMA cellular and PCN systems [156]. If it proves to be true, it would be a tremendous boon for cellular operators. But I invite the reader to imagine what sort of test would be required to fully test this proposition under all the relevant traffic conditions.

Much the same sort of doubt hangs over the matter of power control, soft hand-off, "graceful degradation," and many other questions. These are aspects of performance that will reflect total system loading, total system traffic patterns, and total interactions among thousands or tens of thousands of mutually interfering mobile subscribers. Simulation will not be sufficient—because no one can agree on how many of these parameters should be simulated.

Obviously, to build and operate such a large system under these conditions of uncertainty represents an enormous financial risk for the operator that would undertake it. It may be that some operators would be willing to run these risks. But I believe that this is a large part of the reason why the application of CDMA to wireless access will be incremental. Early, less risky applications (e.g., in wireless LANs now reaching the market) will buy invaluable experience for reducing the uncertainties associated with subsequent steps.

4. *The CDMA mystique could be dangerous.*

   The idea of code division is so unusual, so counterintuitive, so clever, so revolutionary, at least to those educated to conventional channel-oriented radio system designs, that it possesses a certain intellectual allure—in the words of one author, a "natural beauty" [157]—which may captivate the unwary. Another commentator suggests that:

   > The mystique of spread-spectrum communications is such that commercial enterprises, as well as academia, are often attracted by the novelty and cleverness of the technique. [158]

   CDMA techniques may have an important role to play in at least some of the applications we have been discussing, at some point in time. But we should be careful not to fall victim to an intriguing idea simply because it is an intriguing idea. CDMA proposals must be subjected to the same test and scrutiny as any other architecture.

5. *The claim that CDMA can eliminate the need for spectrum regulation is too good to be true.*

   The regulators at the FCC and elsewhere are battered and bewildered by the competing claims on the finite radio spectrum. The idea that some new technology could emerge that might allow different applications and different users to coexist in the same spectrum at the same time, without the need for intensive regulatory oversight, presents an intense temptation toward wishful thinking. Just imagine that we could do away with licensing, with gerrymandered allocations, with tedious arguments over the relative merits of different services—and let the magic of code division and spread spectrum sort it all out. One can almost here the strains of the Beach Boys: *"Wouldn't it be nice if. . . ."*

   Unfortunately, the prospect of early retirement for our spectrum regulators is illusory. I believe that the claims of spread spectrum systems with respect to interference resistance have been misunderstood. It is not true that CDMA is more robust to interference than other systems; in fact, it may be more truly said that it is less robust to uncontrolled interference. CDMA is a *self*-interfering architecture that is able to engineer a carefully specified amount of interference in a tightly controlled system. When the interference environment is not well behaved, a CDMA system may have more trouble than a channelized system. I believe that the forthcoming experiments in spectrum sharing among unregulated CDMA users will establish the limits on this notion of interference resistance.

On the Other Hand. . . .

CDMA techniques do hold promise for some future wireless applications. It is likely that during the 1990s we will see the first commercial CDMA applications on a significant scale, probably not for mammoth cellular systems, but more likely (in my view) in more closed, tightly administered systems such as wireless PBXs and wireless LANs, and possibly some forms of PCN (although the processing complexity for CDMA militates against the early availability of portable terminals, in the opinion of some [159]).

Recall that CDMA emerged in the military environment, where the systems involved were generally small-scale (in terms of numbers of users) and where access to the system was strictly controlled. In prospective commercial applications, CDMA "works" by tightly controlling the interference environment that it creates. I believe this argues against an early success for CDMA in applications involving open access and large numbers of people. Cellular radio is such a system, where essentially any user can begin transmitting at any time, where new users are constantly entering the system, where the user terminal equipment is manufactured by dozens of different companies all around the world, following varying standards of manufacturing quality, with components of varying tolerances, etc.

On the other hand, in a wireless PBX or wireless LAN application, there will typically be a finite (small) number of users (dozens or hundreds—not hundreds of thousands), with tight access control (people will not be able to walk in off the street with their own radios and start transmitting and creating interference), with a central system administrator that can act as a "dictator" when the need arises and manage the system much more closely than the public network can normally be managed.

This is the kind of system that would seem to be more congenial to the needs of a CDMA architecture. In addition, since such systems operate over short distances in fixed environments (normally indoors), the RF propagation issues are mitigated. The power control problem that plagues a CDMA architecture in a big cellular system should become much less problematic for an in-building wireless PBX. The portables in such an in-building system should prove to be more congenial to CDMA than the 60 mph vehicles zipping through the Rayleigh environment at hundreds of fades per second.

In Summary

FDMA and TDD will be used in some small-scale applications, such as cordless telephony, very small wireless PBXs, etc.

TDMA systems, enhanced in some cases by DSI, will emerge as the preferred architecture for mobile applications and wide-area applications, especially those to be deployed in the 1990s.

CDMA will begin to show promise during the 1990s for applications that are:

1. In-building, or at most campus-wide
2. Relatively short-range, and low-power
3. Emphasizing portable terminals, rather than mobiles (portables should be easier from a system performance standpoint, although the complexity of the CDMA technique cuts the other way for a low-power, low-cost portable design)
4. Involving relatively small numbers of users per cell.

But TDMA will still probably compete (and perhaps dominate) in many of these applications as well.

The experience gained from these initial commercial applications of CDMA will illuminate possible future applications in more challenging environments such as third- or fourth-generation cellular.

NOTES

1. G. Calhoun, *Digital Cellular Radio,* Boston: Artech House, 1988, pp. 27 ff.
2. Daniel Noble, "The History of Land-Mobile Radio Communications," *Proceedings of the IRE, Vehicular Communications,* May 1962, p. 1413.
3. L. Lessing, *Man of High Fidelity: Edwin Howard Armstrong,* Philadelphia: J. B. Lippincott Company, 1956.
4. G. Calhoun, *Digital Cellular Radio,* Boston: Artech House, 1988, Chap. 4.
5. Federal Communications Commission, *Notice of Proposed Rule Making,* "In the Matter of Amendment of Parts 2 and 22 of the Commission's Rules to Permit Liberalization of Technology and Auxiliary Service Offerings in the Domestic Public Cellular Radio Telecommunications Service," Docket 87-390, released October 15, 1987; also see the resulting *Report and Order,* adopted October 13, 1988, and released December 12, 1988; see also various drafts of the digital air interface standard, known as IS-54, published by TIA.
6. See William C. Jakes, *Microwave Mobile Communications,* New York: Wiley, 1974, as well as any of several recent books by Dr. Bill Lee of Pactel Cellular, all quite good.
7. One of the best single-volume references in J. C. Bellamy, *Digital Telephony,* 1st ed., New York: John Wiley & Sons, 1982, which has also recently be reissued in a second edition.
8. This is actually a vast field. The reader with a specialized interest is referred to the many recent articles in the *IEEE Communications Magazine* (for example, the January 1991 special issue on switching systems and ISDN).
9. See N.S. Jayant and Peter Noll, *Digital Coding of Waveforms: Principles and Applications to Speech and Video,* Englewood Cliffs, New Jersey: Prentice-Hall, 1984, especially Chap. 1. For

example, "We shall also note . . . the inadequacy of SNR as a perceptually meaningful measure of digitized waveform quality. Very briefly, the inadequacy of the SNR has to do with the fact that the reconstruction error sequences in general do not have the character of signal-independent additive noise; and that therefore the seriousness of the impairment cannot be measured by a simple power measurement." (p. 6)

10. Michael J. Miller and Syed Ahamed, *Digital Transmission Systems and Networks,* Rockville, Maryland: Computer Science Press, Vol. II, 1988, pp. 41 ff.
11. William C. Jakes, *Microwave Mobile Communications,* New York: Wiley, 1974, Chap. 2.
12. *Ibid.,* p. 107.
13. *Ibid.,* p. 110.
14. Personal communications from Thomas Klos.
15. Donald C. Cox, Hamilton W. Arnold, and Philip T. Porter, "Universal Digital Portable Communications: A Systems Perspective," *IEEE Journal on Selected Areas in Communications,* Vol. SAC-5, No. 5, June 1987, pp. 764–773.
16. For a more detailed conceptual discussion of multipath problems, see George Calhoun, *Digital Cellular Radio,* Boston: Artech House, 1988, pp. 216–219; for a much more technical treatment, see William C. Jakes, *Microwave Mobile Communications,* New York: Wiley, 1974.
17. Jakes, *ibid.,* pp. 215–217.
18. SR Telecom company personnel report that in more than 10 years of applications worldwide now encompassing tens of thousands of subscriber lines, multipath effects have constituted a serious problem in only one case. IMM's experience with BETRS in the United States has been similar; to my knowledge, we have seen serious multipath effects at only one location involving several subscribers on a system in Silver Lake, Oregon.
19. See G. Calhoun, *Digital Cellular Radio,* Boston: Artech House, 1988, pp. 146–148, and associated references.
20. E.F. O'Neill, ed., *A History of Engineering and Science in the Bell System: Transmission Technology (1925–1975),* AT&T Bell Laboratories, 1985.
21. As good general texts on digital communications, I can recommend the following: John C. Bellamy, *Digital Telephony,* New York: Wiley, 1982; David R. Smith, *Digital Transmission Systems,* New York: Van Nostrand Reinhold, 1985; John C. McDonald, ed., *Fundamentals of Digital Switching,* New York: Plenum Press, 1983; Michael J. Miller and Syed V. Ahamed, *Digital Transmission Systems and Networks* (2 vols.), Rockville, Maryland: Computer Science Press, 1987.
22. G. Calhoun, *Digital Cellular Radio,* Boston: Artech House, 1988, Chap. 6.
23. E.F. O'Neill, ed., *A History of Engineering and Science in the Bell System: Communications Sciences (1925–1980),* AT&T Bell Laboratories, 1985, p. 530.
24. N.S. Jayant and Peter Noll, *Digital Coding of Waveforms: Principles and Applications to Speech and Video,* Englewood Cliffs, New Jersey: Prentice-Hall, 1984, pp. 9–12, also Appendix F.
25. See Bellcore, *Generic Framework Criteria for Universal Digital Personal Communications Systems (PCS),* FA-NWT-001013, Issue 2, December 1990, Sec. 3.5 and associated references on the definition of QDU; for a summary comparison of both measures, see N. Kitawaki and Hiromi Nagabuchi, "Quality Assessment of Speech Coding and Speech Synthesis Systems," *IEEE Communications Magazine,* October 1988, pp. 36–44.
26. N.S. Jayant and Peter Noll, *Digital Coding of Waveforms: Principles and Applications to Speech and Video,* Englewood Cliffs, New Jersey: Prentice-Hall, 1984, p. 660.
27. See Bellcore, *Technical Analysis Report: Ultraphone 100 Codec Subjective Voice Quality Analysis,* AR-NPL-001041, Issue 1, January 3, 1989. Note: The distribution of this document may be limited to Bellcore and authorized clients.
28. I am aware of one BETRS product (IMM Corporation) and one VSAT product (Hughes Network Systems) that have been sold commercially with low-bit-rate coders for a radio circuit. In both cases, a RELP-based coder was used, operating at bit rates between 14 and 16 kb/s.

29. G. Calhoun, *Digital Cellular Radio,* Boston: Artech House, 1988, Chap. 4.
30. Motorola's presentations for several years have called for such massive reallocations of the spectrum.
31. This has been the result of various studies of the future spectrum needs sponsored by or for the staff of the FCC. See, for example, the report entitled *Future Private Land Mobile Telecommunications Requirements,* prepared by the Private Radio Bureau in August 1983. The Sheriff's Department of Los Angeles has been particularly active in this crusade.
32. See G. Calhoun, *Digital Cellular Radio,* Boston: Artech House, 1988, Chap. 2, and K. Flamm, "Technological Advance and Costs: Computers vs. Communications," in R.W. Crandall and K. Flamm, ed., *Changing the Rules: Technological Change, International Competition, and Regulation in Communications,* Washington, D.C.: The Brookings Institution, 1989, pp. 36 ff.
33. John C. Bellamy, *Digital Telephony,* New York: Wiley, 1982, p. 273.
34. See Richard G. Saunders, "Ultraphone—Wireless Digital Loop Carrier System," *Proceedings of the National Communications Forum,* Vol. 42, No. 2, 1988, pp. 1860–1866; also see product specifications on the Ultraphone 100 system, available from IMM Corporation.
35. Donald C. Cox, "Personal Communications—A Viewpoint," *IEEE Communications Magazine,* November 1990, p. 12.
36. John C. Bellamy, *Digital Telephony,* New York: Wiley, 1982, Chap. 4, pp. 171 ff.
37. M.J. Miller and S.V. Ahamed, *Digital Transmission Systems and Networks, Volume 1—Principles,* Rockville, Maryland: Computer Science Press, 1987, p. 185 ff.
38. See summary of references in J.L. Flanagan, *Speech Analysis, Synthesis and Perception,* 2nd ed., New York: Springer Verlag 1972, pp. 4–8.
39. *Ibid.,* p. 6.
40. N.S. Jayant and Peter Noll, *Digital Coding of Waveforms: Principles and Applications to Speech and Video,* Englewood Cliffs, New Jersey: Prentice-Hall, p. 16.
41. See note 28.
42. See *Technical Analysis Report—IMM Ultraphone 100 Wireless Digital Loop Carrier System,* AR-TSY-001273, Issue 1, November 2, 1989. Note: The distribution of this document may be limited to Bellcore and authorized clients.
43. J.C. Bellamy, *Digital Telephony,* 1st, ed., New York: John Wiley & Sons, 1982, p. 113.
44. S. Millman, ed., *A History of Engineering and Science in the Bell System: Communications Sciences (1925–1980),* AT&T Bell Laboratories, 1984, Chap. 2.
45. Robert V. Bruce, *Bell: Alexander Graham Bell and the Conquest of Solitude,* Boston: Little, Brown, 1973, p. 36.
46. *Ibid.,* p. 5; also see J.L. Flanagan, *Speech Analysis, Synthesis and Perception,* 2nd ed., New York: Springer-Verlag, 1972, pp. 204 ff.
47. *Ibid.,* pp. 210 ff.; see also S. Millman, ed., *A History of Engineering and Science in the Bell System: Communications Sciences (1925–1980),* AT&T Bell Laboratories, 1984, pp. 99 ff.
48. The "dictionary" of symbols and sequences is as follows:
the * ing & and % of $
ed # it @ in ! be π
as ¢ sou ∞ st ¡ ic ≈
en ± tr - spe β od Ω
er £ at § on ¶ or ·
re Σ qu φ ch ç al å
ct √ el ¥ an ¿
49. N.S. Jayant and Peter Noll, *Digital Coding of Waveforms: Principles and Applications to Speech and Video,* Englewood Cliffs, New Jersey: Prentice-Hall, 1984, pp. 410 ff.
50. Robert D. Hoyle and David D. Falconer, "A Comparison of Digital Speech Coding Methods for Mobile Radio Systems," *IEEE Journal on Selected Areas in Communications,* Vol. SAC-5, June 1987, pp. 915–920.

51. N.S. Jayant and Peter Noll, *Digital Coding of Waveforms: Principles and Applications to Speech and Video,* Englewood Cliffs, New Jersey: Prentice-Hall, 1984, Chap. 11.
52. J.L. Flanagan, *Speech Analysis, Synthesis and Perception,* 2nd ed., New York: Springer-Verlag, 1972, especially Chaps. 6 and 8.
53. J.C. Bellamy, *Digital Telephony,* 1st ed., New York: John Wiley & Sons, 1982, p. 147.
54. Robert D. Hoyle and David D. Falconer, "A Comparison of Digital Speech Coding Methods for Mobile Radio Systems," *IEEE Journal on Selected Areas in Communications,* Vol. SAC-5, June 1987, pp. 915–920.
55. See J.-P. Adoul, "Speech-Coding Algorithms and Vector Quantization," in K. Feher, ed., *Advanced Digital Communications: Systems and Signal Processing Techniques,* Englewood Cliffs, New Jersey: Prentice-Hall, 1987, pp. 133–181; see also Allen Gersho and Vladimir Cuperman, "Vector Quantization: A Pattern-Matching Technique for Speech Coding," *IEEE Communications Magazine,* December 1983, pp. 15–21.
56. The two premiere texts on voice coding, which are nicely complementary to one another, are: J. L. Flanagan, *Speech Analysis, Synthesis and Perception,* 2nd ed., New York: Springer-Verlag, 1972; and N.S. Jayant and Peter Noll, *Digital Coding of Waveforms: Principles and Applications to Speech and Video,* Prentice-Hall: Englewood Cliffs, New Jersey, 1984. The Flanagan book is a classic, a work of enduring scholarship which exhibits a depth of historical understanding rare for a technical work. It is also a delight to read and browse through. As the title indicates, Flanagan deals primarily with the analysis-synthesis or parametric modeling approaches to speech coding. Jayant and Noll, on the other hand, deal almost exclusively with waveform coding, for which they provide a definitive treatment. These are the two pillars of the literature. Other, shorter discussions of the subject can be found in most of the general texts on digital communications, including John C. Bellamy, *Digital Telephony,* Wiley: New York, 1982, Chap. 3; Michael J. Miller and Syed Ahamed, *Digital Transmission Systems and Networks,* Vol. I, Rockville, Maryland: Computer Science Press, 1987, Chap. 6; David R. Smith, *Digital Transmission Systems,* New York: Van Nostrand Reinhold, 1985, Chap. 3. For short, readable, and fairly recent updates, see the following: R. E. Crochiere and J.L. Flanagan, "Current Perspectives in Digital Speech," *IEEE Communications Magazine,* January 1983, pp. 32–40; N.S. Jayant, "Coding Speech at Low Bit Rates," *IEEE Spectrum,* August 1986, pp. 58–63; Manfred R. Schroeder, "Linear Predictive Coding of Speech: Review and Current Directions," *IEEE Communications Magazine,* August 1985, pp. 54–61; Robert D. Hoyle and David D. Falconer, "A Comparison of Digital Speech Coding Methods for Mobile Radio Systems," *IEEE Journal on Selected Areas in Communications,* Vol. SAC-5, June 1987, pp. 915–920; Nobuhiko Kitawaki, Masaaki Honda and Kenzo Itoh, "Speech-Quality Assessment Methods for Speech-Coding Systems," *IEEE Communications Magazine,* October 1984, pp. 26–33.
57. Thomas L. Jones and William A. Kissick, "ACSB: What is Adequate Performance?," *Proceedings of the 37th IEEE Vehicular Technology Conference,* Tampa, June 1–3, 1987, pp. 492–497; see also *IEEE Communications Magazine,* "Single-Sideband Mobile Radio: A Review & Update," March 1979, pp. 25–28.
58. S.C. Gupta, R. Viswanathan and R. Muammar, Land Mobile Radio Systems—A Tutorial Exposition," *IEEE Communications Magazine,* June 1985, pp. 34–45.
59. *Ibid.*
60. F. de Jager and C.B. Dekker, "Tamed Frequency Modulation, a Novel Method to Achieve Spectrum Economy in Digital Transmission," *IEEE Transactions on Communications,* Vol. COM-26, May 1978, pp. 534–542; see also Flavio Muratore and Valerio Palestini, "Features and Performance of 12PM3 Modulation Methods for Digital Land Mobile Radio," *IEEE Journal on Selected Areas in Communications,* Vol. SAC-5, June 1987, pp. 906–914.
61. K.S. Chung and L.W. Zegers, "Generalized Tamed Frequency Modulation," *Philips Journal of Research,* Vol. 37, 1982, pp. 165–177; see also Muratore and Palestini, *op. cit.*

62. Sirikiat Ariyavisitakul and Tsutomo Takenchi, "A Novel Anti-Multipath Modulation Technique—DSK," *IEEE Transactions on Communications,* December 1987, pp. 1252–1258.
63. Subbarayan Pasupathy, "Minimum Shift Keying: Spectrally Efficient Modulation," *IEEE Communications Magazine,* July 1979, pp. 14–22; see also Muratore and Palestini, *op. cit.*
64. W.J. Weber, "A Bandwidth Compressive Modulation System Using Multiamplitude Minimum Shift Keying (MAMSK)," *IEEE Transactions on Communications,* Vol. COM-26, May 1978, pp. 543–551.
65. K. Murota and K. Hirade, "GMSK Modulation for Digital Mobile Radio Telephony," *IEEE Transactions on Communications,* Vol. COM-29, July 1981, pp. 1044–1050.
66. C.-E. Sundberg, "Continuous Phase Modulation in Cellular Digital Mobile Radio Systems," *Bell System Technical Journal,* September 1983, pp. 2067–2086.
67. Hiroshi Suzuki and Kenkichi Kirade, "System Considerations of M-ary PSK Land Mobile Radio for Efficient Spectrum Utilization," *The Transactions of the IECE of Japan,* Vol. E 65, No. 3, March 1982, pp. 159–165; Hiroshi Suzuki, "Canonic Receiver Analysis for M-ary Angle Modulations in Rayleigh Fading Environment," *IEEE Transactions on Vehicular Technology,* Vol. VT-31, February 1982, pp. 7–14; Hiroshi Suzuki and Kenkichi Hirade, "Spectrum Efficiency of M-ary PSK Land Mobile Radio," *IEEE Transactions on Communications,* Vol. COM-30, July 1982, pp. 1803–1805.
68. Yoshihiko Akaiwa and Yoshinori Nagata, "Highly Efficient Digital Mobile Communications with a Linear Modulation Method," *IEEE Journal on Selected Areas in Communications,* Vol. SAC-5, June 1987, pp. 890–895; S.A. Gronemeyer and A.L. McBride, "MSK and Offset QPSK Modulation," *IEEE Transactions on Communications,* Vol. COM-24, August 1976, pp. 809–820.
69. Yoshihiko Akaiwa and Yoshinori Nagata, "Highly Efficient Digital Mobile Communications with a Linear Modulation Method," *IEEE Journal on Selected Areas in Communications,* Vol. SAC-5, June 1987, pp. 890–895.
70. Sundberg, *op. cit.;* see also Justin C.I. Chuang, "The Effects of Time Delay Spread on Portable Radio Communications Channels with Digital Modulation," *IEEE Journal on Selected Areas in Communications,* Vol. SAC-5, June 1987, pp. 879–889.
71. S. Asakawa and F. Sugiyama, "A Compact Spectrum Constant Envelope Digital Phase Modulation," *IEEE Transactions on Vehicular Technology,* Vol. VT-30, August 1981, pp. 102–111.
72. John C. Bellamy, *Digital Telephony, op. cit.,* pp. 294 ff; for a proposed mobile/PCN application, see Raymond Steele, "The Cellular Environment of Lightweight Handheld Portables," *IEEE Communications Magazine,* July 1989, pp. 20–29.
73. Justin C.I. Chuang, "The Effects of Time-Delay Spread on QAM with Non-linearly Switched Filters in a Portable Radio Communications Channel," *Proceedings of the International Conference on Digital Land Mobile Radio Communications,* Venice, June 30–July 3, 1987, pp. 104–113.
74. J.C. Bellamy, *Digital Telephony,* 1st ed., New York: John Wiley & Sons, 1982, p. 272.
75. *Ibid.,* pp. 181–182.
76. Michael J. Miller and Syed V. Ahamed, *Digital Transmission Systems and Networks,* Vol. I, Rockville, Maryland: Computer Science Press, 1988, pp. 50–51.
77. *Ibid.,* p. 273.
78. *Ibid.,* p. 273.
79. M.H. Meyers and V.K. Prabhu, "Future Trends in Microwave Digital Radio: A View from North America," *IEEE Communications Magazine,* February 1987, pp. 46–49; see also T. Noguchi, Y. Daido, and J.A. Nossek, "Modulation Techniques for Microwave Digital Radio," *IEEE Communications Magazine,* October 1986, pp. 21–30.
80. Michael J. Miller and Syed Ahamed, *Digital Transmission Systems and Networks,* Vol. II, Rockville, Maryland: Computer Science Press, 1988, p. 37.
81. *Ibid.,* pp. 20 ff.

82. C. Sundberg, "Continuous Phase Modulation," *IEEE Communications Magazine,* Vol. 24, No. 4, April 1986, pp. 25–38; see also Miller and Ahamed, *op. cit.,* Vol. II, pp. 28 ff.
83. John C. Bellamy, *Digital Telephony,* 1st ed., New York: John Wiley & Sons, 1982, pp. 279 ff.
84. *Ibid.,* p. 297.
85. *Ibid.,* p. 292.
86. *Ibid.,* pp. 161 ff.; also M.J. Miller and S.V. Ahamed, *Digital Transmission Systems and Networks, Volume 1—Principles,* Rockville, Maryland: Computer Science Press, 1987, Chap. 3.
87. S. Millman, ed., *A History of Engineering and Science in the Bell System: Communications Sciences (1925–1980),* AT&T Bell Laboratories, 1984, pp. 422–424; for more recent developments see Shahid Qureshi, "Adaptive Equalization," *IEEE Communications Magazine,* March 1982, pp. 9–16; Thomas J. Aprille, "Filtering and Equalization for Digital Transmission," *IEEE Communications Magazine,* March 1983, pp. 17–24; T.A.C.M. Claasen and W.F.G. Mecklenbrauker, "Adaptive Techniques for Signal Processing in Communications," *IEEE Communications Magazine,* November 1985, pp. 8–19; and M. J. Miller and S. V. Ahamed, *Digital Transmission Systems and Networks, Volume 1—Principles,* Rockville, Maryland: Computer Science Press, 1987, Chap. 4 and references.
88. Shahid Qureshi, "Adaptive Equalization," *IEEE Communications Magazine,* March 1982, p. 13.
89. T.A.C.M. Claasen and W.F.G. Mecklenbrauker, "Adaptive Techniques for Signal Processing in Communications," *IEEE Communications Magazine,* November 1985, pp. 8–19.
90. S. Millman, ed., *A History of Engineering and Science in the Bell System: Communications Sciences (1925–1980),* AT&T Bell Laboratories, 1984, p. 424.
91. M.J. Miller and S.V. Ahamed, *Digital Transmission Systems and Networks, Volume 1—Principles,* Rockville, Maryland: Computer Science Press, 1987, p. 41; M. Miyake, T. Fujino, Y. Umeda and E. Yamazaki, "A Study of a Coded 8PSK Modem with a Viterbi Decoder Using Simplified Metric Calculation," *Proceedings of the Pacific Telecommunications Conference,* Honolulu, January 1985, pp. 279–286.
92. M.J. Miller and S.V. Ahamed, *Digital Transmission Systems and Networks, Volume 1—Principles,* Rockville, Maryland: Computer Science Press, 1987, p. 41.
93. For a short survey, see Vijay K. Bhargava, "Forward Error Correction Schemes for Digital Communications," *IEEE Communications Magazine,* January 1983, pp. 11–19; also, S. Lin and D.J. Costello, *Error Control Coding: Fundamentals and Applications,* Englewood Cliffs, New Jersey: Prentice-Hall, 1983.
94. Both Motorola and AT&T initially favored FDMA.
95. William C.Y. Lee, Remarks at the FCC Seminar on "The Future of Cellular Radio," Washington, D.C., September 2, 1987.
96. See Paul Clements, "BETR: Digital Radio in the Loop," *Telephone Engineer & Management,* August 15, 1988; also Richard G. Saunders, "Ultraphone—Wireless Digital Loop Carrier System," *Proceedings of the National Communications Forum,* Vol. 42, No. 2, pp. 1860–1866, 1988; also James F. Mullen, "Digital Radio Systems for the Local Loop Offer Rural Areas a BETRS Way," *Outside Plant,* July 1989.
97. Motorola, *Preliminary Narrow AMPS Specification,* November 1990.
98. See extensive references on SR Telecom and related systems in note 42 of Chap. 5 of this book.
99. See IS-54 digital cellular specifications, cited in note 4 of Chap. 5 of this book.
100. Bellcore, FA-TSY-001013, *Generic Framework Criteria for Universal Digital Portable Communications Systems,* Issue 1, March 1990. See also Donald Cox's many articles, including: "Portable Digital Radio Communications—An Approach to Tetherless Access," *IEEE Communications Magazine,* July 1989, pp. 30–40; "Universal Digital Portable Radio Communications," *Proceedings of the IEEE,* April 1987, pp. 436–477; and, with H.W. Arnold and P.T. Porter, "Universal Digital Portable Communications—A System Perspective," *IEEE Journal on Selected Areas in Communications,* June 1987, pp. 764–773.

101. The IMM Ultraphone system provides for dynamic variable slot assignments, with users operating with either 4-level or 16-level DPSK modulation, and 32 or 16 kilosymbols per second, respectively.
102. By IMM Corporation, along with M/A-COM (Linkabit), later Hughes Network Systems.
103. David G. Messerschmidt, "Digital Termination and Digital Signal Processing," in John C. McDonald, ed., *Fundamentals of Digital Switching,* New York: Plenum, 1983, p. 287.
104. Ray Simpson, presentation at WinLabs Conference, Rutgers University, Oct 19, 1990.
105. The field of CDMA is indeed shrouded in much mystery. I have tried to provide a summary of spread spectrum and CDMA concepts in my previous book, which is one of the least mathematical treatments available to the general reader. See George Calhoun, *Digital Cellular Radio,* Boston: Artech House, 1988, pp. 339–357. There are other references that may serve as tutorials on various aspects of CDMA, although most require a fair amount of prior technical knowledge. Probably the definitive work on spread spectrum at this time is the three-volume treatment by Marvin K. Simon, Jim K. Omura, Robert A. Scholtz, and Barry K. Levitt, *Spread Spectrum Communications,* Rockville, Maryland: Computer Science Press, 1985. The extensive historical chapter in Volume I is a real treat. Shorter references that are reasonably accessible, with an emphasis on commercial wireless access type applications, include George R. Cooper, Ray W. Nettleton, and David P. Grybos, "Cellular Land-Mobile Radio: Why Spread Spectrum?," *IEEE Communications Magazine,* March 1979, pp. 17–23; William Utlaut, "Spread Spectrum: Principles and Possible Application to Spectrum Utilization and Allocation," *IEEE Communications Magazine,* September 1978, pp. 21–30; Andrew J. Viterbi, "Spread Spectrum Communications: Myths and Realities," *IEEE Communications Magazine,* May 1979, pp. 11–18; Robert Eckert and Peter M. Kelly, "Implementing Spread Spectrum Technology in the Land Mobile Radio Services," *IEEE Transactions on Communications,* Vol. COM-25, August 1977, pp. 867–869; Charles E. Cook and Howard S. Marsh, "An Introduction to Spread Spectrum," *IEEE Communications Magazine,* March 1983, pp. 8–16; Robert A. Scholtz, "The Spread Spectrum Concept," *IEEE Transactions on Communications,* Vol. COM-25, August 1977, pp. 756–763; Raymond L. Pickholtz, Donald L. Schilling, and Laurence B. Milstein, "Theory of Spread Spectrum Communications: A Tutorial," *IEEE Transactions on Communications,* Vol. COM-30, May 1982, pp. 855 ff.
106. See, for example, M. Schwartz, *Telecommunications Networks: Protocols, Modeling and Analysis,* Reading, Massachusetts: Addison-Wesley, 1987.
107. *Ibid.;* also G. E. Keiser, *Local Area Networks,* New York: McGraw-Hill, 1989.
108. Bellcore, *Telecommunications Transmission Engineering, Volume 1—Principles,* 3rd ed., Bellcore Technical Publications, 1990, p. 398.
109. Paul T. Brady, "A Technique for Investigating On-Off Patterns of Speech," *Bell System Technical Journal,* January 1965, pp. 1–22; Paul T. Brady, "A Statistical Analysis of On-Off Patterns in 16 Conversations," *Bell System Technical Journal,* January 1968, pp. 73–91; P.T. Brady, "A Model for Generating On-Off Speech Patterns in Two-way Conversations," *Bell System Technical Journal,* September 1969.
110. See, for example, H. Ueda, I. Tokizawa and T. Aoyama, "Evaluation of an Experimental Packetized Speech and Data Transmission System," *IEEE Journal on Selected Areas in Communications,* Vol. SAC-1, December 1983; also C.J. Weinstein and J.W. Forgie, "Experience with Speech Communication in Packet Networks," *IEEE Journal on Selected Areas in Communications,* Vol. SAC-1, December 1983.
111. For the earliest discussions of speech interpolation techniques, see E.F. O'Neill, "TASI - Time Assignment Speech Interpolation," *Bell Laboratories Record,* Vol. 37, 1959, pp. 83–87; K. Bullington and J.M. Fraser, "Engineering Aspects of TASI," *Bell System Technical Journal,* March 1959, pp. 353–364; also see Bellcore, *Telecommunications Transmission Engineering, Volume 1—Principles,* 3rd ed., Bellcore Technical Publications, 1990, p. 56.

112. For early discussions of digital speech interpolation techniques, see K. Amano and C. Ohta, "Digital TASI System in PCM Transmission," *Proceedings of the International Conference on Communications,* June 1969, pp. 34.23–34.28; I. Poretti, G. Monti and A Bagnoli, "Speech Interpolation and their Application in TDM/TDMA Systems," *Proceedings of the Second International Conference on Digital Satellite Communications,* Paris, November 1972, pp. 348–357; J.A. Sciulli and S.J. Campanella, "A Speech Predictive Encoding Communication for Multichannel Telephony," *IEEE Transactions on Communications,* Vol. COM–21, July 1973, pp. 827–835; E. Lyghounis, I. Poretti, and G. Monti, "Speech Interpolation in *Digital Transmission Systems," IEEE Transactions on Communications,* Vol. COM-22, September 1974, pp. 1179–1189.
113. For more recent surveys, first see S. J. Campanella, "Digital Speech Interpolation Systems," in K. Feher, ed., *Advanced Digital Communications: Systems and Signal Processing Techniques,* Englewood Cliffs, New Jersey: Prentice-Hall, 1987, pp. 237–281. Other references include S. J. Campanella, "Digital Speech Interpolation," *COMSAT Technical Review,* Vol. 6, No. 1, Spring 1976, pp. 127–158; S. J. Campanella, "Digital Speech Interpolation Techniques," *Proceedings of the National Telecommunications Conference,* December 1978, pp. 14.1.1–14.1.5; Joseph Sciulli, "System Engineering Considerations in DSI Applications," *Proceedings of the National Telecommunications Conference,* December 1978, pp. 14.2.1–14.2.5; Yohtaro Yatsuzuka, "High-Gain Digital Speech Interpolation with Adaptive Differential PCM Encoding," *IEEE Transactions on Communications,* Vol. COM-30, April 1982, pp. 750–761; Michael S. Nakhla and David H. A. Black, "Analysis of a TASI System Employing Speech Storage," *IEEE Transactions on Communications,* Vol. COM-30, April 1982, pp. 780–785; Troy Malpass and Ray Lowe, "Adapting the CCITT 32 kbps ADPCM Algorithm for Operation at Variable Rates," paper available from Hughes Network Systems.
114. S. J. Campanella, "Digital Speech Interpolation Techniques," *Proceedings of the National Telecommunications Conference,* December 1978, p. 14.1.1.
115. *Ibid.,* p. 14.1.2.
116. J. A. Jankowski, "A New Digital Voice-Activated Switch," *COMSAT Technical Review,* Vol. 6, No. 1, Spring 1976, pp. 159–178; Yohtaro Yatsuzuka, "Highly Sensitive Speech Detector and High-speed Voiceband Data Discriminator in DSI-ADPCM Systems," *IEEE Transactions on Communications,* Vol. COM-30, April 1982, pp. 739–750.
117. This consortium was announced publicly in December 1990, with first equipment demonstrations (including DSI) in January 1991.
118. See Michael S. Nakhla and David H. A. Black, "Analysis of a TASI System Employing Speech Storage," *IEEE Transactions on Communications,* Vol. COM-30, April 1982, p. 780.
119. This analysis is drawn from Troy Malpass and Ray Lowe, "Adapting the CCITT 32 kbps Algorithm for Operation at Variable Rates," *op. cit.*
120. John G. Gruber and Leo Strawczynski, "Subjective Effects of Variable Delay and Speech Clipping in Dynamically Managed Voice Systems," *IEEE Transactions on Communications,* Vol. COM-33, August 1985, p. 806. Their comment: "The results of formal subjective testing . . . indicate that in terms of speech quality, FEC [i.e., front-end clipping] is preferred to MSC [i.e., Mid-speech Burst Clipping]. These results seem intuitively reasonable, since there is only one discontinuous amplitude transition due to FEC and two transitions with MSC. Also, unlike MSC, FEC is preceded by silence, so the listener is not anticipating the speech to follow."
121. See Michael S. Nakhla and David H. A. Black, "Analysis of a TASI System Employing Speech Storage," *IEEE Transactions on Communications,* Vol. COM-30, April 1982, pp. 780–785, who invent the interesting term "gap modulation" to describe the effects of variable delay. They describe the process as follows: "When a speech burst arrives, an active interrupt is sent to the [slot] assignment controller and simultaneously the speech burst is written into a buffer memory. . . . Subsequently, the speech burst is read out on the transmission facility without a loss. It has, however, experienced a delay, which could range from as little as 40 milliseconds to hundreds of millisec-

onds. Subjective tests have confirmed the tolerance of the ear to the variable delays introduced by the system, provided they remain within a limited range. For [a particular piece of DSI equipment they are describing], the requirement has been set that the variation in delay, or "gap modulation," shall have a probability of exceeding 300 ms for less than 10 percent of the time" (p. 780]) See S. J. Campanella, "Digital Speech Interpolation Techniques," *Proceedings of the National Telecommunications Conference,* December 1978, p. 14.1.3. Campanella comments on two different approaches to variable bit rate DSI. The first is a PCM-based bit-robbing type of voice system, where the number of bits in the PCM coder is reduced for all circuits as congestion increases. The second approach is based on delta modulation.

122. See S.J. Campanella, "Digital Speech Interpolation Techniques," *Proceedings of the National Telecommunications Conference,* December 1978, p. 14.1.3. Campanella comments on two different approaches to variable bit-rate DSI. The first is a PCM-based bit-robbing type of voice system, where the number of bits in the PCM coder is reduced for all circuits as congestion increases. The second approach is based on delta modulation.
123. See John G. Gruber and Leo Strawczynski, "Subjective Effects of Variable Delay and Speech Clipping in Dynamically Managed Voice Systems," *IEEE Transactions on Communications,* Vol. COM-33, August 1985, pp. 801–808. It is also interesting that CCIR has developed, quite some time ago, certain "standards" for clipping in DSI applications; see CCIR Doc. 4/29-E, "Proposed Amendment to Annex to Report 211-3: Speech Interpolation Technique in TDMA Systems," Japan, January 19, 1976.
124. See John G. Gruber and Leo Strawczynski, "Subjective Effects of Variable Delay and Speech Clipping in Dynamically Managed Voice Systems," *IEEE Transactions on Communications,* Vol. COM-33, August 1985, pp. 801–808; see also John G. Gruber, "A Comparison of Measured and Calculated Speech Temporal Parameters Relevant to Speech Activity Detection," *IEEE Transactions on Communications,* Vol. COM-30, April 1982, pp. 728–738.
125. For example, J. Campanella, "Digital Speech Interpolation," *COMSAT Technical Review,* Vol. 6, No. 1, Spring 1976, p. 129.
126. See references cited in notes 124 and 125.
127. K.S. Gilhousen, I.M. Jacobs, R. Padovani, A.J. Viterbi, L.A. Weaver and C.E. Wheatley, "On the Capacity of a Cellular CDMA System," to appear in a forthcoming issue of *IEEE Transactions on Vehicular Technology,* 1991. The draft of this paper states that "voice activity can be monitored, a function which virtually already exists in most digital vocoders, and transmission can be suppressed . . . when no voice is present." While this may be something of an overstatement, I agree that the implementation of a suitable voice activity factor should be fairly simple.
128. John G. Gruber and Leo Strawczynski, "Subjective Effects of Variable Delay and Speech Clipping in Dynamically Managed Voice Systems," *IEEE Transactions on Communications,* Vol. COM-33, August 1985, p. 801.
129. D.J. Goodman, R.A. Valenzuela, K.T. Gayliard and B. Ramamurthi, "Packet Reservation Multiple Access for Local Wireless Communications," *IEEE Transactions on Communications,* August 1989, pp. 885–890; D.J. Goodman and S.X. Wei, "Factors Affecting the Bandwidth Efficiency of Packet Reservation Multiple Access," *Proceedings of the 39th IEEE Vehicular Technology Conference,* San Francisco, May 1989, pp. 292–299; D.J. Goodman, "Cellular Packet Communications," *IEEE Transactions on Communications,* Vol. 38, No. 8, August 1990, pp. 1272–1280.
130. *Ibid.,* p. 1276.
131. Sonia R. Sachs, "Alternative Local Area Network Access Protocols," *IEEE Communications Magazine,* March 1988, pp. 25–45. This article makes an effort to provide a taxonomy for the bewildering variety of access protocols, and although it is oriented towards wireline LANs, much of the conceptual structure can be applied to radio systems. The article also contains a 154-item bibliography.

132. *Ibid.*
133. For example, the early generation of SR Telecom's equipment used polling for call setup. See Chap. 5 of this book.
134. G.E. Keiser, *Local Area Networks,* New York: McGraw-Hill, 1989, pp. 207–209.
135. For example, BETRS uses a slotted ALOHA for its call setup circuit.
136. Sonia R. Sachs, "Alternative Local Area Network Access Protocols," *IEEE Communications Magazine,* March 1988, pp. 25–45.
137. See George Calhoun, "Digital Cellular—A View from America," *Communications Systems Worldwide,* October 1988, pp. 20-25, as well as G. Calhoun, *Digital Cellular Radio,* Boston: Artech House, 1988, Chap. 11.
138. Motorola, *Preliminary Narrow AMPS Specification,* November 1990.
139. On December 6, 1990, in an industry conference in Arlington Heights, Illinois, Motorola introduced their concept for a narrowband analog "next generation" of cellular technology, based on channel-splitting. For one observer, the presentation served chiefly to highlight some rather important unknowns about system-level performance of the proposed narrowband channels in a multiple-cell system. For example, the degradation in C/I performance was variously stated as being between 0.8 dB (Motorola's claimed value, which would represent a relatively minor degradation) and as much as 6 dB, the textbook value for the static case, which would represent a truly massive loss of coverage and result in substantially inferior performance.
140. William C.Y. Lee, Sept 2, 1987, presentation at the FCC seminar on "The Future of Cellular Radio."
141. Raymond Steele, "The Cellular Environment of Lightweight Handheld Portables," *IEEE Communications Magazine,* July 1989, p. 26.
142. G. Calhoun, *Digital Cellular Radio,* Boston: Artech House, 1988, Chap. 4.
143. The Motorola N-AMPS proposal, however, still raises the issue of the time delays in reaching a working TDMA architecture as one important reason for deploying an interim round of FDMA narrow-channel systems.
144. Bellcore, *Framework Technical Advisory,* FA-TSY-001013, "Generic Framework Criteria for Universal Digital Portable Communications Systems," Issue 1, March 1990. The Bellcore report notes that "other schemes such as Frequency Division Multiple Access (FDMA) and Code Division Multiple Access (CDMA) have been studied but do not appear attractive for this application" (Section 3.4 of the cited Framework Advisory).
145. The extended TDMA (E-TDMA) proposal by Hughes, Alcatel and IMM, first announced on December 11, 1990.
146. This was the initial capacity claim in the first presentation of CDMA as a cellular alternative, by Irwin Jacobs of Qualcomm Inc. to the CTIA in 1989.
147. The standard claim included in presentations from Qualcomm Inc. during 1990 and 1991.
148. Personal communications with PacTel personnel; this was later adjusted in public to "6–8 times capacity increase" in comments by Craig Ferrill in January 1991, based on actual field experience.
149. Ali Salmasi (spokesman for Qualcomm, Inc.), in comments at the WinLabs Conference, Rutgers, October 19, 1990.
150. Notwithstanding his strong advocacy for CDMA, Bill Lee has been very conscientious about this in all his recent papers; e.g., William C.Y. Lee, "Overview of Cellular CDMA," presented at the Globecom '90 Conference, San Diego, December 2, 1990, p. 1: "It does not matter whether the spectrum is divided into frequencies, time slots, or codes, the capacity provided from these three multiple access schemes is the same."
151. For example, in one prominent article on CDMA for cellular radio, published in a respectable refereed journal (which shall here remain nameless), the capacity of a cellular CDMA system has been compared favorably against the capacity of the TDMA IS-54 standard—without clearly acknowledging that the CDMA numbers assume a voice-coding rate 50% lower than the calculations for

the TDMA standard. As well, CDMA numbers often assume that gains from exploiting the voice activity factor can only be realized through a CDMA architecture, which is not true (see Section 6.4.6 on DSI). Indeed, during 1989 and 1990, the North American cellular industry has been victimized by this sort of "numbers game"—where unstated assumptions, hidden in dense mathematics, have served to confuse and mislead many people.

152. Donald C. Cox, "Personal Communications—A Viewpoint," *IEEE Communications Magazine,* November 1990, p. 18; for more on the power control problem, see J.M. Holtzman, "CDMA Power Control for Wireless Networks," *Proceedings of the 1990 WINLAB Conference,* Rutgers University, October 19, 1990, pp. 264–273.
153. *Ibid.,* p. 18
154. For TDM background, see E. F. O'Neill, ed., *A History of Engineering and Science in the Bell System: Transmission Technology (1925–1975),* AT&T Bell Laboratories, 1985, Chap. 18; for CDM history, see Marvin K. Simon, Jim K. Omura, Robert A. Scholtz, and Barry K. Levitt, *Spread Spectrum Communications: Volume I,* Rockville, Maryland: Computer Science Press, 1985, Chap. 2.
155. E. F. O'Neill, ed., *A History of Engineering and Science in the Bell System: Transmission Technology (1925–1975),* AT&T Bell Laboratories, 1985, Chap. 18.
156. Irwin Jacobs, "CDMA for Digital Cellular," presentation at Supercomm, Atlanta, April 18, 1990.
157. William C.Y. Lee, "Overview of Cellular CDMA," presented at the Globecom '90 Conference, San Diego, December 2, 1990, p. 1.
158. A.J. Viterbi, "When Not to Spread Spectrum—A Sequel," *IEEE Communications Magazine,* April 1985, pp. 12–17. The irony is not lost. It may be that Dr. Viterbi has decided that long exposure to the CDMA virus has conveyed on him a sort of immunity, and thus he may rush in now where others fear to tread. Damn the torpedoes!
159. Donald C. Cox, "Personal Communications—A Viewpoint," *op. cit.,* p. 18.

Part IV

WIRELESS ACCESS: ECONOMICS AND ENGINEERING

Building a telecommunications network involves analyzing and solving two parallel problems: physical and economic. Both are often very complex. Viewed as a *physical* entity, a switched telephone circuit is a concatenation of many different types of transmission and switching facilities operating in diverse environments, which must carry a fragile signal instantaneously, and with fidelity, often across great distances. Viewed as an *economic* entity, a telephone circuit is even more abstract: a complex cost structure extended in time and space, comprising many elements, some shared, some dedicated, some traffic-sensitive, some part of the overhead, some new expenditures, many allocated from past investments. From end to end, each link must connect with the others in a predictable way, with tight performance requirements and strict accounting of costs—despite the fact that various elements of the circuit have been installed at different times, by different people, working for different companies, most of whom did not know one another or have the opportunity to interact directly. Viewed on the scale of the network itself, it is indeed daunting. Planning, deploying, and maintaining tens of millions of telephone circuits, involving hundreds of millions of circuit-segments (of which the access link is only a subset), poses an enormous logistical challenge. And keeping the books straight, understanding the cost structure so as to make intelligent investment decisions is, if anything, even more difficult.

To cope with this complexity, the telephone industry adheres to strict methods and procedures, honed and handed down over generations of experience. The planning of such systems is fundamentally a question of economic analysis of alternatives, or, as it is often called, *engineering economy.* The implementation of these plans focuses on developing specific methods and procedures for dealing with specific problems—site preparation, installation, testing, maintenance—often collectively referred to as *site engineering.*

The introduction of wireless access forces a reevaluation and redefinition of many of these procedures. Fortunately, in many respects, radio greatly simplifies the job. For example, radio economics are less variable and less contingent than wireline costs tend to be—which should simplify economic modeling. In implementation, radio systems can often be engineered in a much more streamlined manner as well. In Chapter 7, we deal with the basics of economic analysis for wireless access. In Chapter 8, we outline some of the engineering requirements for building and operating such systems.

Chapter 7

THE ECONOMICS OF WIRELESS ACCESS

7.0 "NO MORE COPPER!"

In 1980, near the beginning of my career in telecommunications, I had occasion to visit a group of network planners at one of the large "independent" telephone companies in the United States. They had just completed a major study of their outside plant, which accounted for about 42% of their total invested capital (and most of their day-to-day headaches). They had evaluated their deployment of assets and their cost structure. They had examined service trends and calculated growth opportunities. With the help of numerous "experts," they had made a sincere, diligent attempt to peer out into the uncertain future. Having done all this, they had reached a startling conclusion, which they cast in the form of a slogan: *No More Copper!*

In other words, they had come to believe that their business could not move forward on the basis of the existing access technology. For them, given their roles (as planners) and their training and experience, it was a straightforward matter of economics. Copper wire cost too much to deploy, maintain, and service. It was too inflexible, which meant that it tied up too much capital. Their state regulators allowed them depreciation lifetimes for copper access facilities that were so drawn out, and indeed so lacking in realism, that they found themselves paying for thousands of miles of copper plant that had long since ceased to function and had been taken out of service, yet was still being charged as depreciation against current earnings. They also realized that if, God forbid, someone were ever to try to do to them what MCI was beginning to do to AT&T—that is, to compete—the deadweight of those real and imaginary copper "assets" would be about as useful to them as a life preserver made out of granite. For all of these reasons, they had determined that they could not economically justify plowing another mile of copper cable in their system. *No More Copper!*

What made such a conclusion remarkable in 1980 was that at that time *there was no clear alternative to copper.* These intuitional planners were like those nineteenth-century astronomers who inferred the existence of the planet Pluto from perturbations in the orbit of Neptune—decades before Pluto was actually discovered.

These telephone company planners had posited the requirement for a noncopper-based access technology, without any real idea of what it might be. The reader should recall that in 1980, in the mainstream of telephony, radio systems of the sort we are discussing in this book were almost completely unknown. There had as yet been relatively little consideration given to fiber optics as a loop technology. "Divestiture" was not in the vocabulary. Even "digital" was a term of novelty for many veteran telephoners. If pressed to say how *"No More Copper!"* might have been implemented at that time, the planners would probably have pointed to the new digital loop carrier systems (SLC-96) just beginning to come on line. Loop electronics is indeed a way of avoiding, or at least postponing, the deployment of new copper facilities for at least a portion of the loop plant. But those planners also recognized that such systems had no effect on the bulk of their copper plant, which lay not in the feeder network, but in the distribution cable and the drop-line facilities, which were still dedicated copper wire-pairs. (Their company's exchanges served primarily nonurban areas.) Moreover, many of their feeder cables were not in good enough shape to support electronic pair-gain. And even where loop carrier systems could be applied, they were at best a band-aid. Feeder-multiplexer systems still depend on the existing copper facilities as the medium of transmission, and when those facilities eventually deteriorate, the telephone companies are confronted by the enormous costs of replacing the physical plant.

Thus, when the idea of wireless access was presented to these planners, albeit in a hazy and uncertain way, they seized on it. Their own calculations had created a receptiveness to the idea of providing access without using copper wire. Indeed, their response turns out not to have been unusual. Time and again over the past 10 years, I have seen engineers and planners respond to our presentations with the same alacrity. In that first instant of grasping the idea of wireless access, before they have begun to think about its problems and complexities, it has great persuasive power.

I have also seen, however, the pattern of second and third thoughts that begin to cloud over that first glow of enthusiasm. A now-predictable pattern of doubt begins to arise. Some questions relate to the performance of radio in an access role. Some have to do with how to install and service the new equipment, which may not be obvious when all the current procedures are based on the heretofore immutable tenet of a metallic loop. But the most difficult questions have to do with the costs of radio-based access. The longer we think about them, the more the differences between the cost structures for wireless access and wireline access become apparent. The more differences, the more work for the analysts and the more uncertainty about the results.

Soon it becomes clear that what is needed is a new analytical framework. Radio is so unlike wireline plant that the rules for costing out the alternatives are no longer valid. In fact, even after initial radio trials have demonstrated the technical and operational viability of radio-based access, the telephone company planners and engineers are often hesitant to use radio because they lack the tools and concepts to *prove* what they know intuitively to be true—that radio is much less expensive than wire for many access applications.

That last statement bears repeating: *Radio is much less expensive than wire for a very large number of access applications.* This conclusion is based on my experience in looking at wireline and wireless access costs in nearly 100 specified applications (and several hundred more that were analyzed in more abstract terms). I believe it is true despite the fact that many conventional analyses show different results. The reason for this discrepancy is that the conventional analyses are inherently biased toward wireline alternatives, and against radio. This is not because the people who constructed today's popular economic models were prejudiced against wireless solutions; it is simply that the industry still works within a framework of "economic engineering" that dates back decades and never anticipated the use of radio in the local access environment.

Consider one example. Wireline plant is often heavily weighted toward first-cost as opposed to life cycle cost. The reason is simple. If I want to install a feeder cable to serve a new housing development, I will plow in enough wire-pairs upfront to serve the entire future growth of the development (ignoring for the moment the option of using a digital loop carrier system). That is, if the ultimate size is 500 houses, I must install a feeder cable for 500 pairs, plus a reasonable margin for spares to extend the life of the cable facility for as long as possible. In short, I would plow in a 750-pair cable, even though there are only 100 houses actually occupied today and full occupancy may not be reached for 10 years (if ever). Despite the fact that most of the pairs are idle and do not generate any revenue during the early years of the project, I build the access network in this manner because the cost of the physical construction effort associated with laying the cable is so great that I only want to do it once. The feeder, which represents, say, 50% of the total investment, is paid for in year 1. Under such circumstances, if someone asks me how much the project costs, I would normally answer by citing the costs of year 1, the "first-cost." The telephone company budgeting cycle often tends to reinforce this emphasis on first-costs.

In contrast, the radio solution for the same service requirement would delay much of the investment until the later years of the project. The year 1 cost might be only 20% of the total, and the savings resulting from delaying the commitment of capital will be substantial (more quantitative examples are offered later in the chapter). Yet the standard telephone company analysis will tend to collapse the entire cost of the radio program into a single year and compare it strictly on a first-cost basis against copper. This is not a fanciful example; I have seen this done many times. The result is that a very significant element of the economic savings from wireless access—the ability to match investments much more closely in time with service requirements—is missed completely.

Even so, there are a number of indications of how cost-competitive wireless access will be. For example, recent Bellcore studies of copper loop costs have found that the average first-cost of a new loop in the United States in the late 1980s was about $1700 to 1800 [1]. I believe that by the mid-1990s, wireless access will be readily implementable at approximately the same first-cost per line. If $1800 is the average cost of a copper loop, then approximately half the loops cost more. Some significant percentage (5%?, 10%?) costs twice as much, which means that wireless access is immediately competitive on a first-cost basis. Recognizing that every 1% of the access base represents more than a million access lines in the United States alone, then even if we allow for the many advantages that copper enjoys today based on the ability to leverage the $100 billion or so of embedded investment in wireline facilities, it seems clear that wireless access will be the cheaper alternative in a very large number of loops in the United States.

Indeed, the very heart of the case for wireless access is economics. This assertion may seem to give short shrift to the mobile-oriented applications of wireless access, which offer capabilities that are simply not available from the wireline network. Nevertheless, with the partial exception of purely vehicular uses, most applications of wireless access in the PCN set, as well as all the applications in the telephony set, are explicitly or implicitly competing against wireline alternatives on a cost basis. If PCN turns out to be five times more costly than using a pay phone, the market for PCN will be substantially smaller than if the costs are about the same. There is admittedly some premium that some segments of the population will be willing to pay for portability, but the visions advanced by many PCN advocates of a seamless network based on portable wireless terminals replacing the wireline system will not come about until the baseline costs of the wireless access reach parity (more or less) with those of the copper-based network of today. And it is quite clear that BETRS, and other "wireless local-loop" applications, will be judged against prevailing copper (or fiber) alternatives first and foremost on the basis of the economic comparison (and perhaps also on the basis of bandwidth capabilities, an issue we shall examine in Chapter 9).

The purpose of this chapter is to outline a new framework for analyzing loop-plant alternatives, a framework that recognizes the dimensions of the significant difference between wireline and wireless access facilities. While the evolution of a complete, quantitative, and fully automated methodology for such an economic analysis will take many years and occupy many minds, it is possible to enumerate many of the basic applications principles at least in the broad conceptual terms appropriate to the level of discourse that I have tried to target throughout this book. For example, the concept of area-wide coverage, and the two-dimensional planning field that it opens up (as opposed to the essentially one-dimensional cable-route, which is the important unit of analysis for all wireline planning), can be conveyed and understood as a fundamental departure from traditional methods—despite the fact that the precise calibration and quantification of the trade-offs between cable costs and radio costs in an area-coverage model may take considerable time and field experience to develop. Some initial attempts have been made to assess and compare the costs of wireline and wireless access, but I regret to conclude that most of them are still conceptually inadequate in ways that this chapter shall hopefully clarify [2].

The focus of this chapter is admittedly oriented toward the telephony applications set, especially BETRS and wireless local loop. I believe this is justified for two reasons. First, it is in the realm of telephony applications that the comparative economics of wireless and wireline access have been most thoroughly studied. As noted in Chapter 5, the telephony service model of wireless access places little premium *per se* on radio's unique capabilities as a transmission medium, and reduces the comparison to its starkest (and clearest) economic terms. It thus avoids the more troublesome question of the "value-added" for the end-customer of radio's features such as portability, mobility, and the even more difficult (and usually ignored issues) of the "value-subtracted" for issues such as higher blocking, reduced privacy, increased risk of theft, poorer circuit quality, and other possible forms of degradation that are characteristic of at least some radio system implementations.

Second, as my comments hopefully have made clear already, the wireless loop or telephony application is the baseline for most of the wireless applications addressed in this book. Other than mobile radio (that is, phones in cars or other moving vehicles, for which there is no real wireline substitute), most wireless access applications involve some aspect of *substitution* of radio for wire, as well as (in some cases) the *enhancement* of the service by taking advantage of radio's unique capabilities. A wireless PBX competes against its alternative, a wired-up PBX. A residential cordless extension phone competes against its chief alternative, a wired extension phone (with perhaps a long extension cord). *This substitution is an economic substitution, first and foremost.* A cordless extension phone is nice to have, but millions of people continue to make do without one, in part because the value-added of limited portability does not offset (for them) the cost of purchasing the equipment. I believe that the prophets of "portability *über alles*" are prone to overstating their case. There are relatively few truly premium markets for portability, markets where consumers will pay very large sums for the portability feature. Of course it depends on which end of the telescope we are looking through; Motorola is selling what they consider to be a lot of cellular portables. But beyond Wall Street, and Rodeo Drive, and maybe the Strip in Las Vegas, I doubt there are vast numbers of consumers for a portable phone that costs more than a thousand dollars. Not when the corner pay phone still takes a quarter.

So the comparison of wireless to wireline access costs does constitute at least the first important baseline for the economic analysis. And it is a baseline that can be talked about in relatively objective terms. The more nebulous matter of the "economics" of additional features—such as portability, roaming, nationwide compatibility, call-in capabilities (an issue for CT-2 today), seamless hand-off—I bequeath to my successors.

The remaining sections of this chapter are organized as follows: Section 7.1 discusses some of the basic background assumptions that go into any economic analysis of access alternatives. Section 7.2 addresses the fundamental differences between wireless and wireline access configurations, with an emphasis on obtaining an accurate mapping of functionally equivalent elements between alternatives. Section 7.3

develops, piecemeal, some of the basic elements of a sound modeling approach to analyzing the costs of wireless access. Finally, Section 7.4 outlines a general program for conducting an economic assessment of the requirements for wireless access in a specific site or over a wide area encompassing several sites.

7.1 ECONOMIC MODELS FOR ANALYZING ACCESS PLANT DECISIONS

The telephone industry has developed a fairly standardized discipline for evaluating plant investment decisions, the aforementioned *economic engineering* [3]. In part this is done to justify major plant expansion or upgrade programs to regulators, and to prove that these decisions are reasonably "optimal" or at least "efficient" uses of capital in order to justify associated rate increases. Indeed, in the past, the financial accounting methods have often been specified in detail by the regulators, embodied in state-mandated depreciation schedules and other accounting methods as well as the FCC's Uniform Systems of Accounts [4]. As a result, the methods for economic analysis of new plant investment decisions in the telephone industry diverged considerably from the procedures followed in other industries.

These methods have often been criticized for producing artificial results and distorted decisions. For example, it is clear that the extremely long depreciation lifetimes imposed by regulators (in the interest of keeping rates down) have produced a huge surplus of "phantom" assets—plant that no longer exists, but is still carried on the balance sheet. These phantom assets correspond to a probable future writedown that overhangs the entire financial scenario, threatening the stability of the regulated system. It has also been argued, I believe convincingly, that traditional methods of analysis often do not favor the least-cost alternative at all, but instead encourage the selection of what might be called the "highest-cost-alternative-we-can-get-away-with." In other words, since revenues are essentially based on costs in the regulated segment of the industry, the goal of the rate-making "game" has been viewed by some as a question of how to increase the rate-base as much as possible. This phenomenon is known by many names—"goldplating the plant" is one of the more colorful I have heard—and it is undoubtedly a strategy that has been followed on occasion. I have spoken with more than one telephone manager who told me quite bluntly that even though radio might be cheaper than cable for a particular configuration, he would go ahead and plow in the cable because the state regulators accepted the use of cable and would not object to his padding the rate-base in this manner [5].

This is now changing, as telecommunications transitions from a regulated monopoly to a less regulated, more competitive industry structure. The underlying purposes of economic analyses are now changing. The goal is no longer solely to justify the reasonableness of a rate request to a state regulator. In the new era, alternative plant investments must be evaluated in comparison to counterstrategies that may be launched by a competitor. For example, a decision to plow "gold-plated" cable—which may have made sense in a monopoly framework—may be monumentally ill-advised if the competitive environment suddenly brings forth new competitors who choose less expensive plant alternatives (e.g., radio) to offer the same service, and hence are capable of underpricing the cable-bound operator.

It is likely, therefore, that future editions of the "Green Book" or the equivalent will reflect a somewhat different approach to modeling the economic analysis of plant investment decisions. However, it is beyond the scope of this survey to try to lay out a complete new model. The purpose of this section is to discuss the generic requirements of a successful model for economic analysis of access plant decisions, as a basis for looking more closely at the special requirements for successful modeling of wireless access alternatives.

7.1.1 Elements of an Economic Model

The building of models—simplified versions of some aspect of reality—could be called the core of the scientific process. There are many kinds of models, and even to categorize them would be a book-length undertaking on its own [6]. There are indeed many different ways even of talking about models and their components. To organize the presentation in this chapter, I have used certain terms here that are both arbitrary and of a fairly high level of generality. They have no religious significance. They only have value if they help to clarify the issues under discussion in this chapter.

For our purposes, we may view a model as a set of *structural relationships* (often expressed in the form of mathematical equations) between different sets of *variables.* It is customary to distinguish between an *independent variable,* the value of which is assigned by the model operator, and a *dependent variable,* the value of which is determined by the values of the independent variables and the structural relationships that comprise the model itself.

I would like to alter this terminology slightly. I believe there are three categories of variables that are of interest here: *assumptions, inputs, and outputs* (see Figure 7.1). We will consider each in turn.

Assumptions. Assumptions and inputs are both classes of independent variables, and it might be objected that there is no logical difference between them. Nevertheless, from a pragmatic standpoint, some independent variables are often *assumed* to have a fixed value for all cases, and which therefore may not even be recognized as variables capable of taking different values. I refer to these as assumptions. An example: Most telephony-oriented planners will tend to assume that access to a circuit must be granted immediately on demand (where "immediately" means, say, "in less than half a second"). If access is not granted within this short window, the call is blocked. Yet it is quite possible to design a system in which the delay associated with circuit

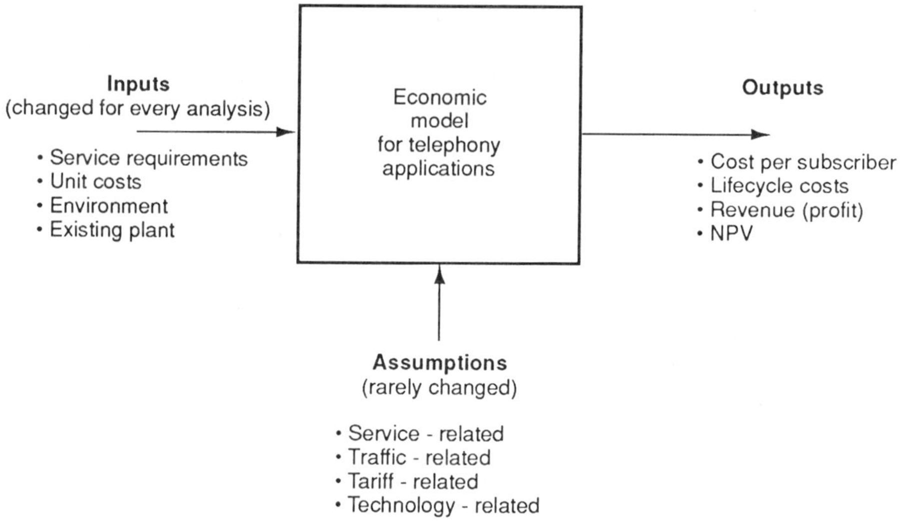

Figure 7.1 The basic economy model.

assignment is a variable; for example, some cellular systems use queuing, with a visible indicator at the subscriber terminal indicating how the queue is being counted-down—much like taking a number at a crowded fast-food restaurant. Another example: telephony engineers usually assume that once a caller has initiated a call, he will be free to continue to occupy that circuit for as long as he wants to. Once again, some cellular operators have instituted "time-outs" of, say, three minutes, after which a call is disconnected. This is done to cope with heavy loading during peak periods. It should be obvious that both of these "assumptions" are pregnant with economic consequences; indeed, the cellular innovations in both cases are strictly motivated by economic considerations (cost avoidance).

Another reason for assuming a fixed value for an independent variable is that it may be difficult to determine the actual value of that variable empirically. For example, we may not be set up to measure the actual amount and pattern of traffic generated by a given subscriber or group of subscribers. Of course, it is also possible to estimate (rather than simply fix) the values of such variables. In Section 7.1.3 we examine some of the common assumptions for modeling access economics.

Inputs. Some independent variables are widely recognized as being capable of taking on different values on a case-by-case basis. Examples of such inputs are the number of customers to be served, the price per unit for various types of equipment, and the hourly cost of labor for various categories of labor. In Section 7.1.4, we will look at the common categories of input variables for access economics.

Outputs. Outputs are the dependent variables that are considered to be the results of the model. "Cost per subscriber" is a typical output variable. The tricky part of specifying an output is to define it unambiguously. For example, "cost per subscriber" may mean "first-cost per subscriber" or it may mean "life cycle cost per subscriber" or it may mean "present value of life cycle costs per subscriber." In Section 7.1.2 we look at the outputs commonly sought by economic models of access decisions.

Structural relationships. These are the equations or formulae that relate assumptions and inputs to one another, along with certain truly fixed parameters, or constants, to derive the values of the outputs. In Section 7.1.5, we examine some of the common characteristics of the equations used in access economic models.

7.1.2 Outputs of Access Plant Studies

It is appropriate to begin by looking at the outputs of the economic analysis of access plant alternatives. These outputs define *what it is that we are interested in knowing,* and help us decide what other variables (i.e., assumptions and inputs) must be specified in order to determine the output values. Thus, although outputs are called dependent variables, they are only dependent in a logical sense. In a pragmatic sense, they are primary; the outputs (what it is we wish to know) determine the types of inputs (what we need to find out).

Outputs tend to be stated very simply, and often rather imprecisely. Some of the terms used include:

1. *First-cost:* This usually refers to the initial outlay of capital required for equipment, including installation and site-engineering costs, for a particular project [7].
2. *Lifecycle cost:* This refers to the total projected cost of the project, over the defined service life of the equipment or some other long-term basis. The term *life cycle* connotes that the equipment must be retired from service at the end of the period. This may not be a correct assumption for radio-based systems and wireless access applications, where subscriber radio terminals may be used for a considerable period of time, and then removed and replaced with wire, and subsequently reinstalled in another location, possibly after an idle period in inventory. These scenarios complicate the relatively straightforward interpretation of life cycle costs with which wireline plant planners are familiar.
3. *Cost per subscriber:* This is the most often cited output variable, and one of the least useful because it is underspecified. As noted above, the cost per subscriber specifies neither the numerator nor the denominator. Does "cost" encompass first-cost only or full life cycle cost? Does "subscriber" refer to only the initial subscriber population, or to some future expanded subscriber base that will be partially served by the equipment installed in a current project?

4. *Total project cost:* Again, the question as to whether project cost refers to first-cost only or life cycle cost must be addressed.
5. *"Capital" costs:* Some analyses focus on those funds that must be raised from external sources.
6. *Operating costs:* These are ongoing expenses that are paid from operating revenues, and may in some analyses be ignored or "swept under the rug."
7. *Cost per year:* The project cost broken down on a yearly basis, which may reflect both capital costs (new investments in plant) and ongoing maintenance costs.
8. *Return on investment (ROI):* Yes, Virginia, telephony is a profit-making business, and occasionally you see a return-on-investment calculation, in which future revenues are projected (as well as costs), and (usually) some appropriate discount rate is applied, to generate one of a various number of measures of ROI [8]. The use of such measures will obviously become more prevalent as we move toward value-added digital services and a competitive environment.

In summary, the emphasis in most analyses of access plant investments is still on costs, not revenues. As one standard economic engineering text concludes in its assessment of the alternative methods of analysis:

> In transmission engineering studies [i.e., loop plant studies, primarily] it is now found that the Present Worth of Expenditures [the well-known PWE] method is most easily applied and leads to the most useful results. . . . This method of analysis may be defined as the summation of cash flows for capital expenditures, net salvage, income taxes and operation costs (or savings) for a project after conversion to present worths at the appropriate rate. [9]

Note that this definition does not contain a reference to revenue. This reflects the still-pervasive monopoly environment for local exchange services [10].

Given this limitation, the best cost-type output variable for access plant studies, in my opinion, is life cycle cost per subscriber, where a salvage value can be used to reflect the "premature" retirement of wireless access equipment, which then becomes available for further use in other situations. In short-term deployments of wireless access (say, less than two years per site), it may be desirable to calculate a monthly or yearly charge for each subscriber unit and apply this in modeling these short-term applications.

It is very important to capture the time dimension of the cost picture; for this reason, first-cost studies are to be avoided. One of the chief difficulties in a life cycle or time-oriented analysis, however, is the question of how to handle staggered service requirements that share certain common equipment installed in year 1. If the future service requirements are uncertain (reflecting estimates of future growth, for example), the cost-per-subscriber measure may be cloudier. Nevertheless, "simple" first-cost studies are often "simply" wrong (as we shall examine below in the following sections).

Moving beyond cost-oriented models, the fuller picture provided by revenue-inclusive ROI-type models is certainly advantageous as the telephone industry becomes more competitive. ROI analysis must be done on an individual-circuit or individual-billed-customer basis, however; burying the analysis in averages and rate-bases is part and parcel of the wireline framework that a competitive industry must ultimately transcend. The industry may find it difficult, or even repugnant, to analyze the ROI for individual subscriber access circuits, because this will highlight negative ROIs associated with certain loops. If the truth were told, a lot of local access lines are money-losers. Focusing on this unpleasant fact may heighten the desires of some carriers to shed the "dogs" and concentrate on milking the "cows" (in the business school lingo of the 1980s). Nevertheless, this is the brave new world we are entering, where some loops will be much more profitable than others (the "cream," about the skimming of which we hear so much), and aggressive competitors will seek to exploit these opportunities.

7.1.3 Assumptions for Access Plant Studies

The assumptions are the relatively fixed features of our economic model. These tend to fall into three important categories:

1. Service-related assumptions, which describe the type and quality of service being offered
2. Traffic-related assumptions, which describe the individualized and aggregate behavior of the customers using the service
3. Regulatory assumptions, which specify certain parameters of the model based on fixed, external (if not "eternal") mandates handed down from state and federal regulators
4. Technology-related assumptions.

7.1.3.1 Service-Related Assumptions

Service-related assumptions define the nature of the telecommunications service being offered. Many of these are so basic that planners are hardly aware they constitute assumptions and not laws of nature.

For example, the assumption that *access opportunities are equal* among all customers (i.e., no one has priority to "bump" someone else off a circuit) is one of the service assumptions that is usually applied, but unstated, for telephony applications and probably for most others. (It might not be appropriate for some wireless PBX-type applications, however.)

Another service assumption, which is usually recognized as a variable at least in principle, is that the system should maintain uniform low blocking probabilities (say 1% or less). This assumption characterizes certain applications (BETRS) but not others (cellular radio). In fact, the cost of a particular application varies greatly depending on whether we assume a 0.5% blocking rate (typical of telephony applications) or a 5% blocking rate (probably typical of cellular systems today) (see Figure 7.2).

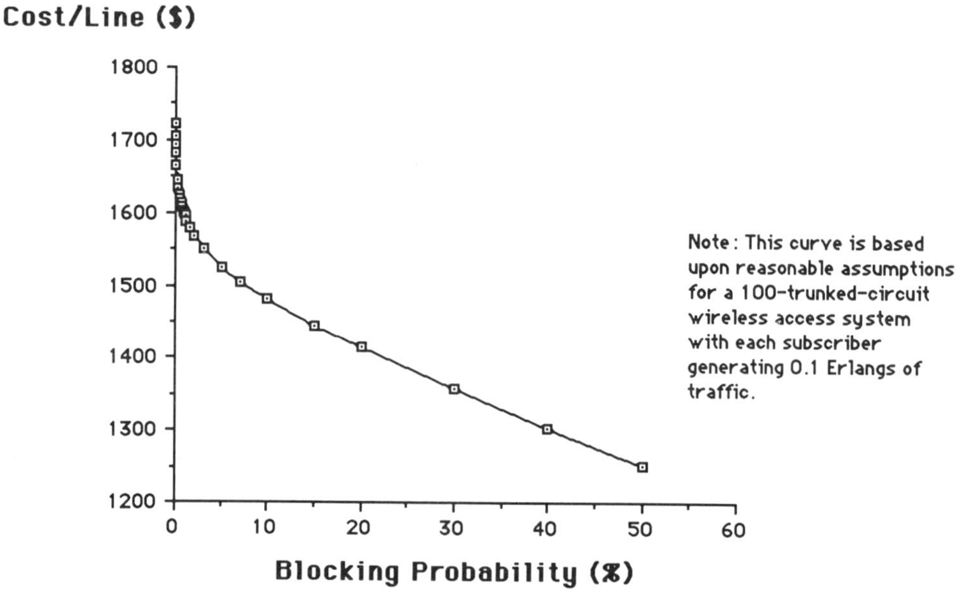

Figure 7.2 Abstract relationship between cost-per-subscriber and blocking probability.

Other service-related assumptions could include:

1. *Delay:* The acceptable time delay in setting up a call or allocating a circuit is often assumed to be fixed. Actually it is an engineerable parameter and can have a significant impact on system costs. Systems that are engineered to tolerate longer call setup delays during periods of high usage (including those that rely on queuing) will be less costly than those that require a very rapid access at all times.
2. *Error rates:* For digital services, the BER is often an important measure of service quality, and this also affects cost. For example, in a BETRS system it may be necessary to implement a more robust form of modulation to guarantee a BER standard for a marginal subscriber (whose voice quality may be quite acceptable otherwise), which in turn may reduce the capacity of the system and/or increase the cost to serve other customers. Other measures to decrease

BER for digital services may involve adding cost at the subscriber station, such as by using a higher gain antenna or a taller antenna mast.

3. *Nonvoice services:* As new services become more prevalent, this ought to appear in our economic models in a number of ways, such as in the potential difference in revenue generated per subscriber. Today's models may not be adequate. In cost-oriented models (such as PWE), the difference between one alternative that allows for enhanced digital services as well as voice (i.e., digital radio) and a second alternative that allows for voice only (pair-gain on existing copper) will be missed entirely because enhanced revenue opportunities are not considered.

7.1.3.2 Traffic-Related Assumptions

Other assumptions of the cost model concern the nature of the *offered traffic*—that is, the individual or collective behavior of subscribers using the service.

For example, how much traffic does each user generate? A typical telephony-oriented assumption is that each subscriber generates 0.1 Erlangs (i.e., 3.6 CCS) per busy-hour. Quite often, this assumption is not a real number at all (i.e., it does not reflect real traffic offered by a real population), but a textbook planning value that is assumed uniformly to save the trouble of analyzing actual traffic patterns on a case-by-case basis. Less frequently the traffic assumption may be based on real measured traffic. Once again, the cost of an application can vary depending on whether we assume 0.1 Erlangs (a normal textbook number) or 0.05 Erlangs (a measured value for rural telephone traffic in at least one environment [11]), or 0.026 Erlangs (the measured cellular average traffic as of 1990 in North America [12]) (see Figure 7.3).

Other traffic-related assumptions could include:

1. The pattern of talk spurts and pauses (important for DSI, PRMA and some CDMA system models, and costs)
2. Blocked-call retry behavior (the basis of arguments over the use of Poisson-B or Poisson-C, etc.)
3. Proportions of nonvoice traffic in the total offered traffic stream (nonvoice traffic often has different holding times and circuit usage characteristics)
4. Percentage of mobile (high-power) versus portable (low-power) subscribers in mixed systems, such as cellular radio
5. Percentage of vehicular (fast-moving) versus ambulatory (slow-moving) subscribers
6. The percentage of real-time traffic (e.g., ordinary voice telephony) versus non-real-time traffic (such as voice mail or alphanumeric messaging).

As the network becomes laden with more heterogeneous traffic streams (voice, data, video, real-time, delay-tolerant, etc.), the specification of different aspects of the offered traffic will become more important in economic analyses.

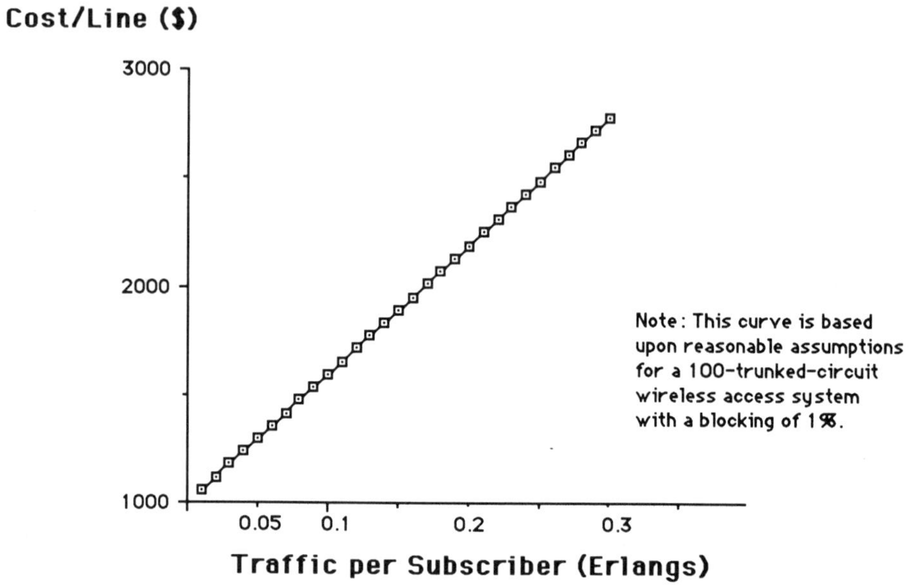

Figure 7.3 Abstract relationship between traffic and cost-per-subscriber.

7.1.3.3 Regulatory Assumptions

As a regulated business, telephony must in some cases adapt its economic models to the specifications of the regulators, reflected in certain assumptions or constraints [13]. Examples include the following.

Depreciation. State and federal regulators may mandate certain depreciation schedules. Typically these depreciation schedules assume artificial service lives for outside plant in order to keep annual depreciation charges down and temper rate increases [14]. Quite aside from the strategic and technological implications of long depreciation lifetimes (some of which we have touched on in Chapter 3), the effect is often to distort the economic model, sometimes severely. For example, radio systems—classified as electronics—are often depreciated on a much shorter lifetime than pole lines and buried cable. This is justified because the technological lifetime of any advanced technology will be shorter than the physical lifetime of a piece of copper buried snugly in the ground. But in a simple economic model, it means that higher annual depreciation charges will be levied against the radio alternative in the early years, and when present value is calculated, the result is to tip the balance against radio. This is a situation that I have actually encountered on a number of occasions.

Pricing. Actually, under the current regulated regime in the local exchange, the carriers lack much pricing freedom. The tariffs for many services are set by agreement with the state public utility commissions; they are negotiated with difficulty and changed infrequently. Moreover, the states enforce pricing based on average costs rather than marginal costs; in other words, the rates for any individual subscriber are not related to the costs of providing service to that subscriber, but to the total cost of providing service to all subscribers. The upshot is that most telephony economic analyses either ignore revenue and pricing as variables altogether (since they can be assumed to be equal for all options), or they assume only average pricing and not line-by-line pricing.

These practices are so ingrained that it may seem hard to imagine doing it any differently—until we recognize that in many nonbasic services, such as leased data service lines, pricing is currently based on the actual individual costs of each facility. Pricing for many enhanced services is not equal for all users, and the carriers apparently enjoy much greater latitude in setting and administering prices for these market segments.

7.1.3.4 Technology-Related Assumptions—The Copper Bias

More subtle are the assumptions about technology. The industry has been using metallic circuits for so long (since well before Alexander Graham Bell) that the copper bias is pervasive and pernicious. It compromises the accuracy of almost every cost analysis I have ever seen.

The following are a few of the more serious instances of copper bias that I have found in the economic models used to analyze access plant alternatives:

1. *The assumption that loop facilities are dedicated rather than shared.* Copper-biased models often assume that there is an identifiable local-loop facility for each individual customer that is 100% dedicated to that customer, and they may build this assumption into the structure of the model in a number of ways. For example, they may simply disallow "concentration" (i.e., trunking) in the loop. This is a bias inherited from the star network of the traditional local exchange (Figure 7.4). In fact, even future wireline networks (probably based on fiberoptics facilities) may use ring architectures, bus architectures, or other configurations in which the local access facilities are shared by many users. In a multiple-access radio system, the radio circuits are all trunked or shared resources. Of course some portion of the investment is still dedicated to the individual subscriber, even in a fiber or radio access network. The terminal costs and the drop-line to the home, as well as the in-building wiring and customer-premises equipment, are still dedicated (unshared) in many cases. In other configurations, radio subscriber terminals may support two or more subscribers. The model should be flexible enough to allow for the analysis of various levels of concentration, delay, and grade of service in the loop itself.
2. *The local access network is assumed to have a "utility-like" cost structure.* A utility-like cost structure is one in which the common costs are high and mar-

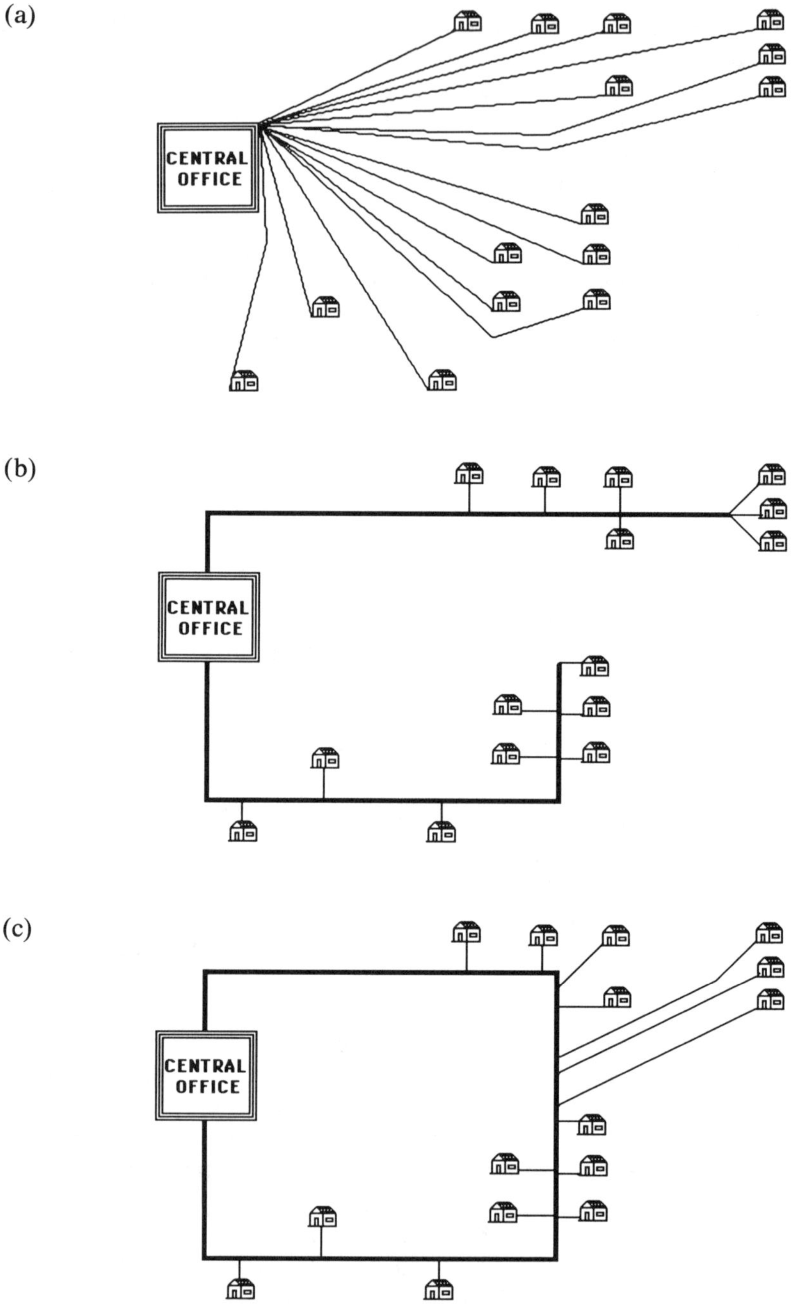

Figure 7.4 Alternative local wireline network typologies; (a) pure star, (b) bus, i.e., cable route, (c) ring.

ginal costs per subscriber are low. An example might be a subway line. The cost of building the subway is enormous. The line must be completed before a single passenger can be carried. Once it is operating, however, the cost of carrying the second passenger is very small. Such systems often are said to approximate a natural monopoly, because the initial costs of setting the system up are so great that the market can sustain only one operator. Therefore, the monopoly may be formalized and subjected to regulation as a public utility. Examples of operations with such a cost structure include water companies, rail transport, electric companies, gas companies, and (traditionally) telephone companies.

This may at first seem to be contradictory with the first assumption. Yet copper-based telephone systems do exhibit both a high percentage of cost allocated to dedicated (as opposed to shared) facilities, and yet the marginal cost of building these facilities per subscriber is usually low compared to the common or shared costs. Both can be true because of the way a telephone cable route is constructed. The cable consists of many individual wire-pairs, each of which is dedicated—hence the circuits are dedicated, not shared. But most of the cost of the cable is in the common costs associated with laying the cable: the labor, the right-of-way, the planning and engineering, and so forth.

3. *Loop costs are assumed to rise as distance increases from the central office.* This is one of the tenets of copper-based access networks (and indeed all star networks based on wireline facilities, whether copper or fiber). The longer the loop, the higher the cost. It should be obvious that with radio costs are essentially distance-insensitive. The model should be capable of varying the relationship between customer location and cost.
4. *Loop costs increase as population density decreases.* This is another very common manifestation of the copper bias in telephony economics, the assumption that rural costs are inevitably higher than urban costs. This is no longer (necessarily) true when we consider wireless access alternatives.
5. *Most large projects are "front-end loaded" in the time-distribution of costs.* In other words, copper-biased models often assume that first-cost so dominates the model that all other costs during the lifetime of the project can simply be treated as a factor applied to first-cost, such as the present value of maintenance (so much per year per loop, discounted). Such models thus boil down to comparisons of first-costs, since the factors applied are equal for all alternatives (another shortcut). Despite the widespread lip-service paid to life cycle modeling, I have found that very few models really handle life cycle costs in any other fashion than as a fixed factor applied to first cost.

Radio is very different. The staging of an investment in time to match service requirements much more accurately can lead to large differences in life cycle costs between wireline and wireless alternatives. However, these are not simply "present value" factors or the like—they are real, substantive phases in the service program.

6. *Zero or negative "salvage" value for loop facilities.* Wireline facilities are not easily redeployable, and many models assume that they have no salvage value, or even a negative value if the cost of removing the wireline facilities at the end of their service life exceeds the salvage value of the copper. In radio-based systems, on the other hand, "salvage" (with its connotations of commodity scrap) is not really even the appropriate concept. A system may be installed, used for a considerable period of time (months or years), and then redeployed elsewhere. The value of the equipment may be considerably greater than zero at the end of a particular project.
7. *Intensive planning overhead is required for any facilities expansion.* Planning for a new cable route is quite an undertaking. Service needs must be forecast for as much as 20 years into the future [15]. Rights-of-way must be negotiated. The logistics of construction dictate that construction workers and equipment must be marshalled in force, and because these resources are needed for many projects, each project requires careful scheduling and management to avoid affecting the overall program. Weather is important, which introduces a seasonal dimension to outside plant construction. Thus, there is a heavy overhead of planning, engineering, and management associated with new outside plant. This is often reflected in the economic model by a heavy overhead charge that is applied to direct costs for the project. It is not uncommon to see overhead charges as high as 50%.

   Radio is very different. It requires very little planning, right-of-way negotiation, or construction team scheduling. The costs associated with planning, engineering, and management are likely to be much lower (as we will discuss below), and the overhead charges should reflect this.

We could find other examples of the copper bias that infests today's economic models of telephone systems. The message here is simply that if you find any of these assumptions hard-wired into your economic model, they should be carefully examined. Loop access economics is about to undergo a revolution (not only because of wireless access, but also because of new fiber-based loop architectures), and it is time for the planners and economists to get ready for it.

7.1.4 Inputs for Access Plant Studies

While the variables we refer to as assumptions tend to be held constant across many different cases, the values of certain variables are likely to change with every analysis—in effect, they specify each case and differentiate from one project from another. Some of these inputs are obvious, such as the number of customers to be served or the prices for various types of equipment. Other inputs are somewhat less obvious, and may be overlooked by many simple models.

7.1.4.1 *Existing Plant*

The most important input variables—and those most often underspecified—are those that characterize the state of the pre-existing outside plant in a particular exchange. It cannot be stated too strongly: *The status of the existing plant is probably the single most important factor in determining whether wireless access is the preferred alternative from an economic standpoint.*

In the simplest case, there is no outside plant at all—a region completely without wireline telephone service. The cost comparison is straightforward, and is based mainly on the equipment and materials prices from vendors (for first-cost, at least). But when there is existing plant, the analysis becomes more complicated.

Consider a situation in which a certain amount of growth is anticipated within the boundaries of a rural exchange. On the one hand, an existing feeder cable may contain enough idle pairs to allow for new customers to be connected directly by means of short drop wires (see Figure 7.5). Or, even if there are not enough physical pairs, the existing feeder may support pair-gain equipment to derive the required circuits electronically in the feeder portion. In both these situations, a wireline solution may make the most economic sense, because the additional cost of adding new subscribers is small *given the state of the existing plant.* (In both cases, however, the economics are also affected significantly by the simplest of all configuration variables: the physical proximity of the feeder route to the new subscribers, which dictates the lengths of the drops.)

Now consider a different scenario. The existing feeder is in poor condition. It cannot support loop electronics for pair-gain. Moreover, it will have to be completely replaced in a few years, which will force a major new investment to maintain service for the existing customers on the route. Finally, we could assume that it is not well located to serve the new customers. In short, the existing plant here will weigh against a wireline solution.

In practice, I have found that—as one of our installers put it very succinctly—"when you have to dig, radio wins." The cost of plowing new cable is so onerous that wireless access becomes the economically superior solution. I would go so far as to say that in the majority of rural situations, were it not for the ability to piggyback small incremental investments on the existing plant, wireless access would always win the economic contest.

7.1.4.2 *The Physical Environment*

The nature of the terrain is important. For wireline alternatives, the distance-sensitive costs of laying cable can vary by a factor of up to 10 or more depending on the nature of the soil, rock, hills, and water obstacles. For radio alternatives, obviously

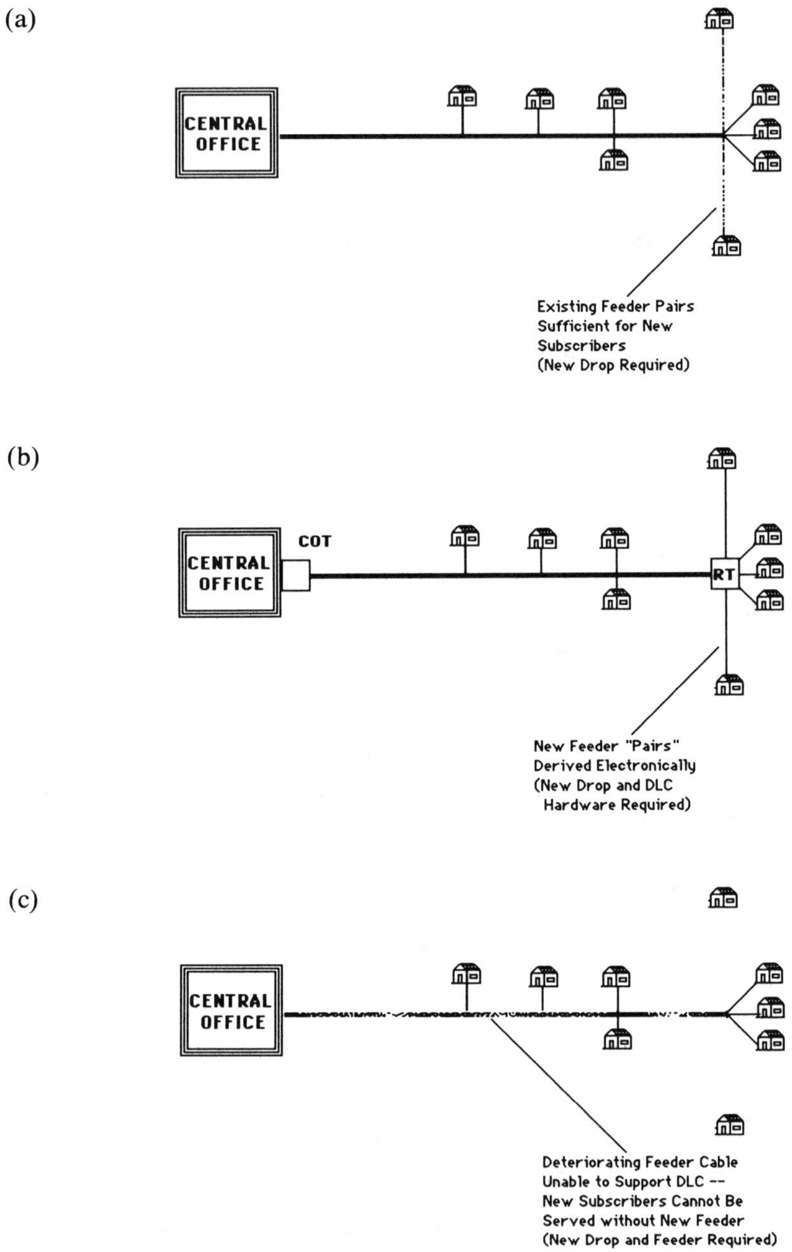

Figure 7.5 Cost of copper-based access as a function of the status of existing feeder plant; (a) the least expensive, (b) more expensive, (c) the most expensive.

any terrain features that affect propagation are important to take into account. A good engineering model should provide terrain-sensitive inputs to the economic analysis (see Chapter 8).

Climate is also a cost factor, particularly for maintenance. Climate may also (at some frequencies) affect radio propagation, limiting range and increasing costs for wireless alternatives.

7.1.4.3 Spatial and Temporal Characteristics of Service Requirements

While every project plan specifies the number of customers to be served, not all plans indicate the location and timing of each requirement. This is particularly true where "future growth" is estimated. Simple annual percentages may be used, with no real effort to examine expected loop lengths or other distance-related characteristics.

While future growth may not be known, it is possible to run sensitivity analyses on a project to evaluate the effects of different scenarios on costs of different alternatives. What happens if the projected growth occurs faster or slower than the baseline forecast? What happens if the average distance of new subscribers from the central office is greater than expected?

The costs of wireline alternatives tend to be insensitive to different scenarios for the spatial and temporal characteristics of service requirements—for the simple reason that (as already noted) wireline costs are heavily front-end loaded. Whether the projected growth occurs in year 3 or year 5 has little overall effect on the cost of a wireline project. On the other hand, because radio access investments can be matched much more closely with actual service needs, the costs of a wireless access alternative may change considerably across different scenarios. A radio alternative that costs the same as a wireline alternative where future growth is projected for year 3 of the project may be substantially cheaper if the growth does not occur until year 5.

7.1.5 Structural Relationships for Access Plant Studies

The structure of the model is embodied in the equations that relate the independent variables (inputs and assumptions) to one another and to certain fixed constants or coefficients so as to yield a value for the dependent or output variables. In some cases, the calculation is simple. Some variables are statistical in nature and may be represented by a probability distribution (e.g., the Poisson distributions for call attempts). There are no easy generalizations about the structure of economic models, except perhaps to say that they are usually oversimplified as a result of numerous shortcuts that have gained acceptance over the years as a practical way of making life easier for the planners who use them.

This is a sore point, perhaps, yet it must be confronted. The telephone industry prides itself on the sophistication of its analytical methods. This pride is justified in many areas, but not in the field of loop plant analysis. Indeed I may say, as a general observation, that *the modeling practices currently followed by the telephone industry for loop plant studies will have to be expanded considerably in scope, and refined in detail, to deal with the potential trade-offs raised by the emergence of wireless access alternatives and the need to assess these alternatives in a competitive environment.* In the design of most other complex systems, from mass transit systems to computers to airplanes, modern engineering practice makes liberal use of simulation and intensive analysis. By contrast, the plans for the expansion of local telephone networks are amazingly primitive and imprecise considering the scope of the investment. Many of these plans today are drawn up literally with pencil and paper, with but a single alternative costed out, on a first-cost basis only, and no calculations more demanding than multiplication or division. Quite often hundreds of thousands or millions of dollars are spent with no real analysis at all. "When the ground thaws, we just fire out and start plowing cable," as one telephone company planner put it. "It all goes into the rate-base anyway." The planning processes followed by cellular engineers are if anything even more empirical.

Ten years from now, the technical and economic modeling of access decisions will be much more of a science. Flexible, sophisticated computer models will replace the seat-of-the-pants practices that prevail today. The sorts of issues we are discussing at a conceptual level in this chapter will have been organized into a far more specific set of tools and techniques. For now, however, we can only touch on a few issues related to the structure of these models.

7.1.5.1 Fixed and Variable Costs

The concept of fixed and variable costs, or (in what is essentially the same distinction) common (or shared) costs and per-line or per-subscriber costs, is fairly straightforward. Many of the equations that we find in the economic model consist of equations of the form:

$$R = AX + BY + C$$

where, for example:

R is the total project cost,
X is the number of subscribers,

Y is the number of radio channels,
A is the equipment cost of the per-subscriber equipment (such as line cards or subscriber stations),
B is the equipment cost of per-RF-channel equipment (such as transmitters, receivers, etc.), and
C is the cost of the common equipment (such as the power supply for the base station).

The idea is simple: some costs are shared by all users, some are shared by groups of users, and some are borne by a single user. This is true of almost every sort of economic system, at some level. This is the sort of equation that I have used frequently, and successfully, to derive a rough estimate of equipment costs for wireless access projects. Yet as obvious as this sort of relationship is, many telephony models today do not address wireline costs in this manner. In fact, wireline costs tend (for reasons relating to the historical shortcuts mentioned previously) to be stated in more esoteric terms, such as cost per foot of cable, which may vary somewhat with the gauge of the cable. But the assignment of costs to specific users is not the common practice, and wireline models do not usually follow this type of structure.

7.1.5.2 Statistical Factors

Telephony gave birth to modern statistics [16]. Yet it is surprising how primitive the grasp of statistical factors often is in loop plant models. This is probably because in the environment where all loops are dedicated facilities, there was little need to apply statistical methods to the access decision. This will change, however, with wireless access where loop facilities are shared.

For example, in the previous example, where I want to estimate the equipment costs of a wireless access project with an equation such as $R = AX + BY + C$, there is actually an unspecified, but crucial relationship—the relationship between the number of subscribers (X) and the number of RF channels (Y). How many subscribers can share one channel? How many channels are necessary to provide service to a given number of subscribers? It is probably clear that to answer these questions, a number of other variables must be specified, including the grade of service or blocking factor (service assumptions) and the average amount of traffic offered by each customer (traffic assumptions). The relationships between these variables are complex and involve the use of complex statistical formulas. The tendency in today's models is to sidestep these complexities and make certain gross simplifications. For example, in the cellular industry it is customary to simply fix a "loading factor" of, say, 20 subscribers to each radio circuit. Similarly, in telephony models, the "concentration ratio" is often invoked. As traffic becomes more heterogeneous (mixing various types, including synchronous and asynchronous data, voice, and even video or imaging), the shortcuts will become less useful.

7.1.5.3 Iteration, Optimization, and Simulation

Most economic analyses of access plant in the telephone world today are *ex post facto* justifications of a single alternative that has already been selected. In a few cases, a few alternatives may receive cursory analysis. It is for this reason that today's primitive hand methods have survived in the field [17]. As we shall see below, the more complex multiple-facility access environment of the near future will require models that can analyze and compare dozens or hundreds of slightly different alternatives.

Consider, for example, an exchange serving 500 customers. What mix of cable and radio is appropriate? Which configuration is cheapest? Such network design problems are not often amenable to deterministic solutions for wireline networks. The famous "traveling salesman" problem is a good example, and a relevant one: How do we construct the least cost (shortest route) network to serve a given large number of customers? This is too complex to be solved directly. As well, the points in the traveling salesman's network are themselves undergoing dynamic changes in time. There are new services, new competitors to be considered. There are, in effect, several traveling salesmen, competing for the most lucrative customers and whoever gets there first captures that customer and he drops out of the other salesmen's networks (for the time being).

What we must begin to recognize is that the access network is no longer a static graph of points. It is a fluctuating grid of revenue opportunities and service requirements. What will really give one salesman (operator) an edge over his competitors is the ability to plan his networks through simulation, where models are used not to develop a single definitive solution, but to evaluate dynamically a large number of scenarios. We need an iterative model that can automatically evaluate a large number of variations and search for the optimal, least-cost solution.

This may all sound revolutionary, but it is not. These sorts of techniques have been in use in planning the core network for many years. It is now time to extend this sophistication to the loop or access environment.

7.2 WIRELESS AND WIRELINE ARCHITECTURES: THE BASICS

To perform a valid economic analysis, we must be able to accurately and realistically compare very different plant configurations. This involves a sort of mapping of one exchange model, and one type of local loop, onto a different exchange and a different loop configuration. By "mapping," I mean that we must establish the relationships of functional equivalence between different elements of the wireline and wireless access networks. Unless this is done correctly, it will be impossible to perform a valid cost comparison.

In this section we will look at what I shall call the "standard" models for wireline and wireless access, and discuss the mapping problems that confront the model builder. We shall also make a brief mention of some of the alternative configurations for both wireline and wireless systems.

7.2.1 A Standard Wireline Access Architecture

The standard configuration for a wireline access network is based on a star architecture (see Figure 7.4).

7.2.1.1 Cost Structure of the Star Network

There are two points of overriding importance concerning the cost structure of a star network: a high proportion of dedicated (as opposed to shared) costs and highly distance-sensitive costs.

High Dedicated Costs. In what we may call the *pure star network,* there is a dedicated loop facility for every subscriber. From the telephone instrument all the way to the line card at the switch, nothing is shared with any other subscriber. In the purest case, this could extend to setting an individual line of poles or buried wire to each subscriber. For economic reasons, the pure star is almost never used in permanent outside plant installations. There is always an effort to construct facilities in a manner that allows for sharing of as much of the cost as possible, or, said another way, to minimize the duplication of costs and facilities. This goal is realized in part through the application of new technologies, especially digital loop carrier in the feeder portion of the loop, and in part through the age-old practices of cable-route design (see next section).

Nevertheless, the economics of star networks in the access environment are inherently poor, as we have discussed in more general terms in Chapter 2. The high percentage of dedicated facilities, combined with the very low utilization of those facilities by any individual subscriber (usually less than 1% utilization on a time basis), makes for a very expensive access network.

Distance-Sensitive Costs. It should be intuitively obvious that the cost of any given loop in a star network is related in a fairly direct manner to the length of the loop, which is of course a function of the distance of the subscriber from the central office. Indeed, in the abstract, it is often said that the cost of a loop rises "more than linearly" as a function of distance (see Figure 7.6). The steps in the graph represent the cost of additional loading coils for long loops, as well as the cost of larger gauge cable. It is also empirically true that long loops tend more often to be found in regions of difficult terrain, which leads to increased construction costs.

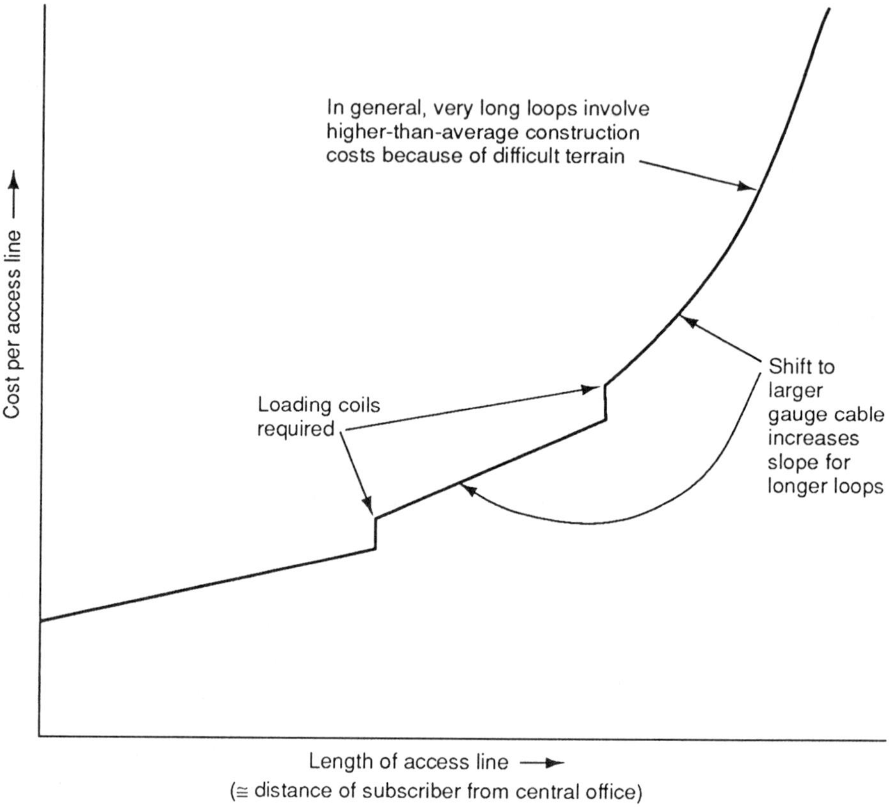

Figure 7.6 Length of access line *versus* cost-per-access line.

In practice, the situation is complex, for reasons that become clear in the next section. Air-mile distance of a subscriber from the central office is less important than the distance from the nearest cable-route or cross-connect point. Nevertheless, the minimization of total loop length in the system has been historically a very important factor in shaping the current network. The placement of switches and the outlines of today's exchange boundaries reflect the principle of loop-length minimization.

Trade-offs. In practice, we find that there is a trade-off to be made between these two factors. The percentage of shared costs can be increased by the creation of cable-routes (see next section) at the cost of increasing the total number of circuit-miles in the system (since the individual circuits no longer follow the most direct route to each customer). In general, the loop-length minimization criterion is applied to determine where exchanges should be located and where their boundaries should be defined. Within the exchange, the goal of reducing duplication of costs through cable-route planning tends to be dominant.

7.2.1.2 The Standard Exchange Configuration: Cable-Routes

As noted above, a true star configuration is too expensive for most actual access applications. Normally a *modified star network* is built, based on the aggregation of individual circuits into *cable-routes* (see Figure 7.7). Cable-routes make economic sense because the cost of today's wireline circuits is dominated not by materials but by labor and other costs associated with installation (including poles or conduits). It saves money to construct facilities in such a way that these costs can be shared by as many circuits as possible, over as much of their length as possible.

The aggregation of circuits into cable-routes has many implications for network architecture and economics. It defines the hierarchy of loop segments (discussed in the next section) (see Figure 7.8). For example, it becomes clear that maintenance, upgrade, and replacement of cable facilities will affect groups of subscribers located in geographical proximity, in what the telephone industry now refers to as *serving areas* [18] (see Figure 7.9). A logical extension of this is the concept of a *car-*

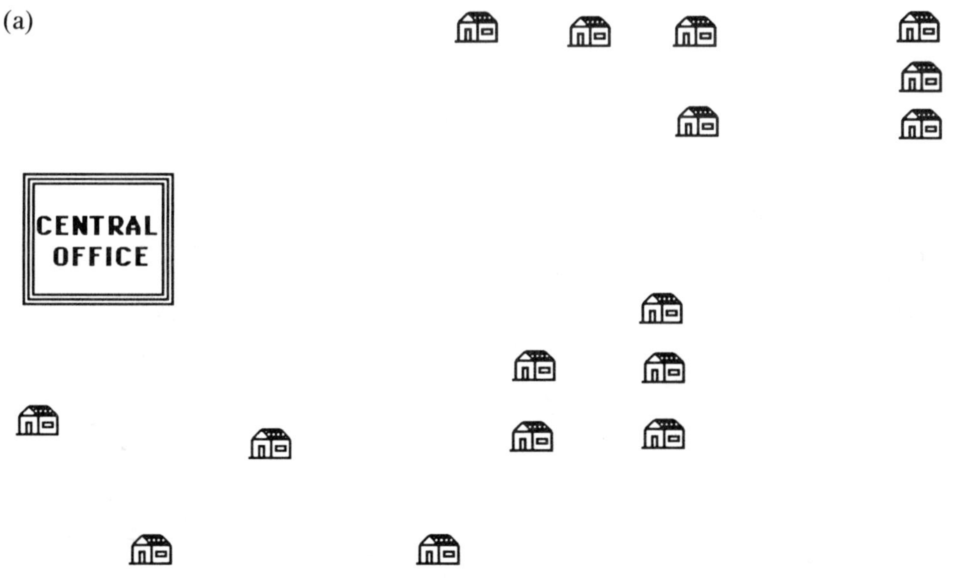

Figure 7.7 Cable-route local networks; (a) before cable, (b) after cable.

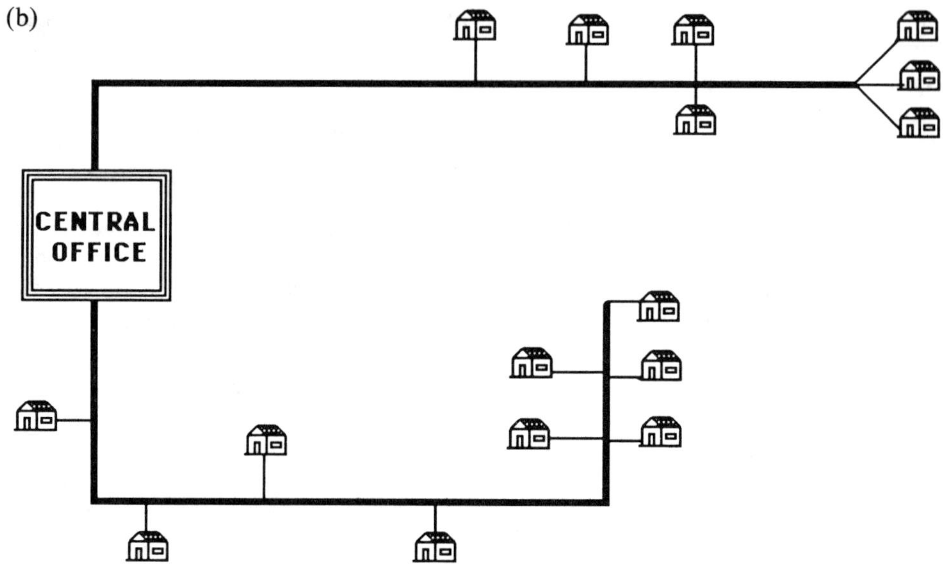

Figure 7.7 continued

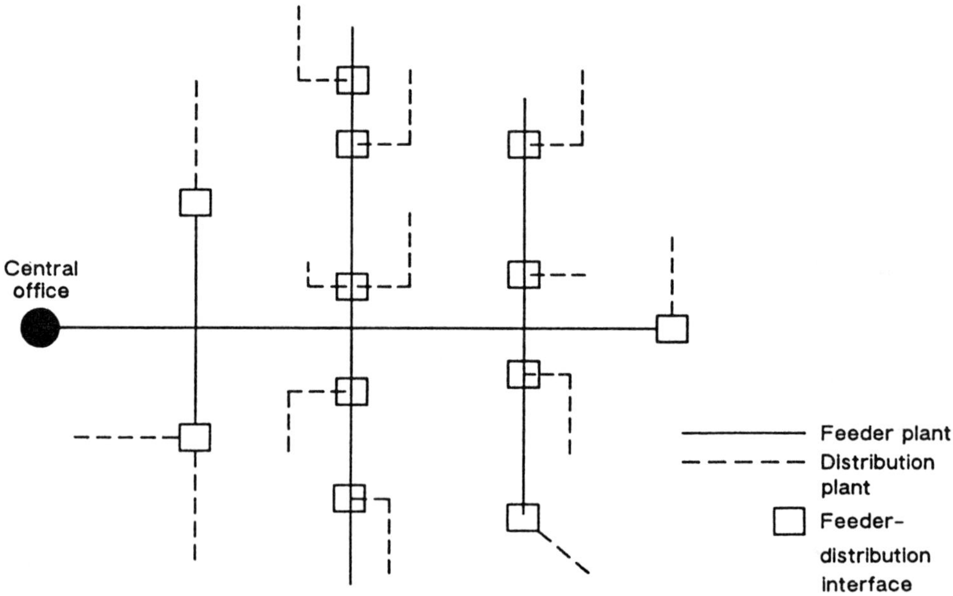

Figure 7.8 Cable-route servicing areas. (From Bellcore, *Telecommunications Transmission Engineering Volume 2: Facilities,* Third Edition, 1990, p. 90.)

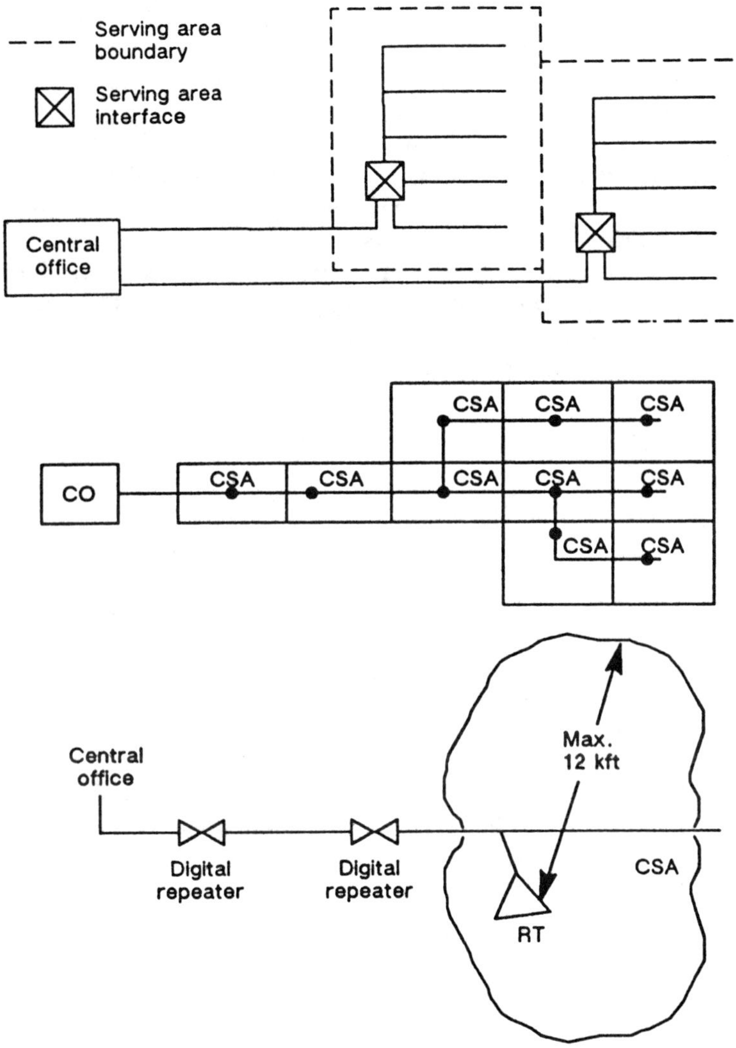

Figure 7.9 Schematic of servicing areas. (From Bellcore, *Telecommunications Transmission Engineering Volume 2: Facilities,* Third Edition, 1990, pp. 93 and 95.)

rier serving area, where digital loop carrier is applied to serving area aggregations [19], and even to the *ISDN serving area,* where digital subscriber services are offered to a particular zone of an exchange defined by the cable network [20].

The serving area and the cable-route become, in effect, the major unit of analysis for access economics, midway between the exchange and the individual circuit.

Unfortunately, the prominence of the serving area creates complications for the economic modeling of access plant decisions, because serving areas are site-specific, vary widely in size, age, and other relevant details, and cannot be handled abstractly very well. Unlike per-circuit cost elements, which are easily abstracted to "dollar-per-subscriber" (sometimes compounded with a distance factor stated as "dollar-per-meter"), and also unlike shared central office cost elements, which are abstracted as "dollar-per-exchange," the serving area or cable-route-related costs are stated as "dollar-per-*X*-subscribers" where *X* varies from case to case and over time. As well, cable-route costs are based on a more complex per-meter cost equation, which varies depending on the type of cable facilities employed. Many models tend to ignore the serving area and cable-route-related cost structures in the wireline exchange altogether. The results can be grossly inaccurate.

7.2.1.3 The Standard Loop

Let us look a little closer then at the structure of a typical loop in the standard wireline configuration, especially since this is often the unit of analysis for simplified economic models.

The standard wireline loop is shown schematically in Figure 7.10. There is a *feeder* segment (usually a large, multipair cable), which is connected to *distribution segments* (smaller gauge cables), which in turn are connected to individual *drop segments* (single or double twisted-pairs connecting to individual subscribers). The connection between the feeder segments and the drop segments is today normally accomplished by means of a *cross-connection* device, which defines an administrative boundary sometimes called the *serving area interface.* This allows for planning, administration, maintenance, and provisioning to be carried out separately for the feeders and the distribution segments, which reduces costs considerably over the older dedicated plant methods (in which a single physical facility was used for the entire length of the loop) [21]. Digital loop carrier technology is normally applied

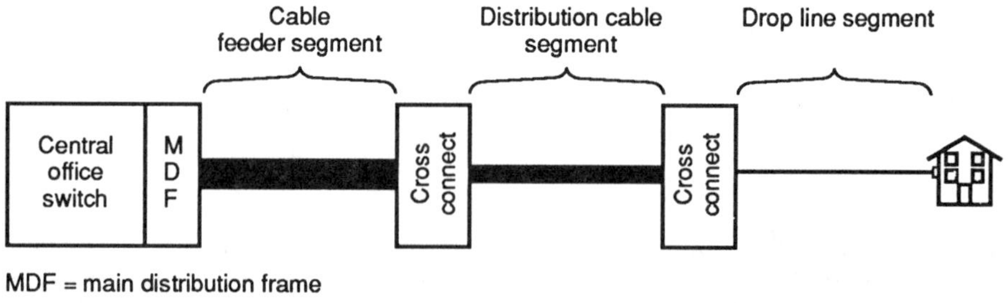

Figure 7.10 Components of the standard wireline access circuit.

today only to the feeder segment, although "fiber to the curb" and similar concepts may take this out into the distribution segment. (The "curb" is shorthand for the interface between the distribution segment and the drop segment.)

As far as economics is concerned, the costs per foot or per meter will be different for each segment, depending on the gauge of the cable and the construction procedures. For a given subscriber, 100% of the drop segment costs will be allocated to his circuit. However, the distribution segment costs will be divided over a number of active and future subscribers, and the feeder segment will generally be shared by a very large number of subscribers. Moreover, there is usually substantial unused capacity, the cost of which must be borne by today's subscribers. Finally, these costs—other than the drop costs—are quite "lumpy." Removing a single subscriber from a cable-route does not reduce the telephone companies' cost of the feeder and distribution segments. Cost as a function of the number of subscribers is a step function, with arbitrary and contingent steps, which means that modeling the cost of alternatives can be arduous.

7.2.1.4 Alternative Wireline Loop Architectures

The copper-based star network is ubiquitous in the circuit-switched, voice-oriented telephone networks of today. There are, however, other network configurations that are emerging from data-oriented applications, such as local-area packet networks, which could be quite different in terms of their cost structure.

For example, many LANs use ring or bus architectures, which come in many varieties [22] (see Figure 7.5). These networks reduce the length of physical wireline facilities needed to serve a given population. One could imagine a local exchange served by a few giga-band fiberoptic rings, with short individual drops (this is indeed one of the "fiber-to-the-curb" concepts), and much less physical plant than today's networks (see Figure 7.11). LAN-like switching might even be performed on the ring itself for local, intraring traffic, removing this traffic from the switched network altogether, which could allow further savings. Such advanced architectural concepts blur the distinctions between switching and transmission. Of course, there would be added costs associated with the interface multiplexers or packetizers that connect each subscriber loop to the shared ring. And the technology for transporting voice over such a packetized facility with adequate delay control is still in the future. Nevertheless, it is possible that with the realization of fast-packet or asynchronous transfer mode architectures, it may at some point become possible to implement these new concepts for conventional telephony applications, significantly reducing wireline costs. It is even possible that the changes in wireline economics resulting from such new architectures could alter some of the basic assumptions that I have made in this book regarding the relative economics of wireline and wireless access (i.e., those derived from the proposition that wireline access can only be provided by a star network). Yet it is also clear that it will take quite some time before these options become available.

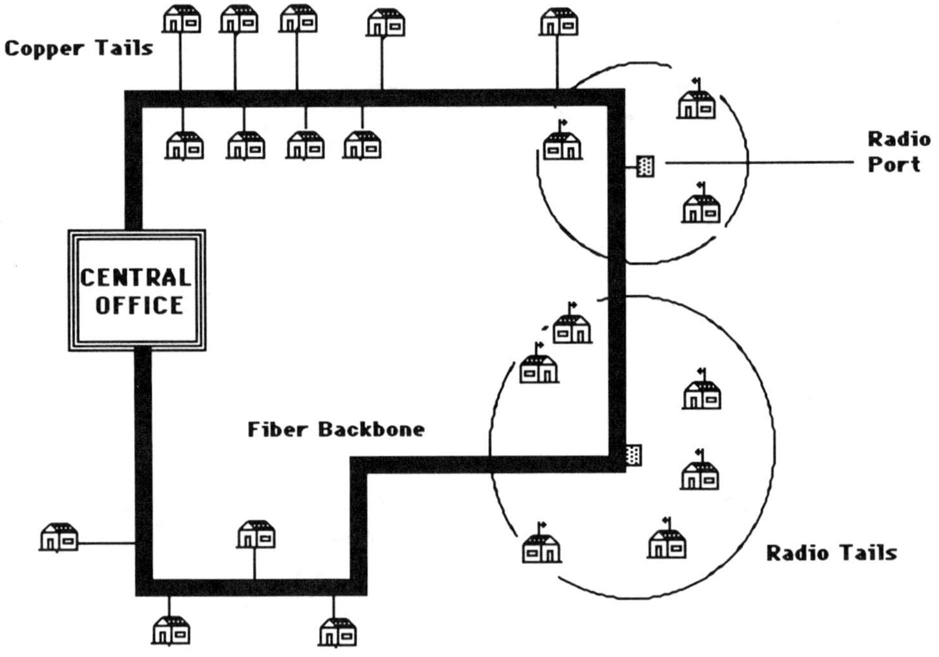

Figure 7.11 Alternative wireline loop architecture: fiber ring with copper drop lines and "radio tails."

7.2.2 A Standard Wireless Access Architecture

Wireless access configurations are much simpler than the wireline model, largely because of the several additional degrees of freedom that radio introduces as a transmission medium. We assume here a BETRS or cellular-type wireless architecture.

7.2.2.1 The Exchange Configuration: Area Coverage

The standard wireless exchange configuration consists of a radio coverage area, or cell, served by a central base station that is connected to the network (see Figure 7.12). The radius of this cell may range from 50 feet or so, to tens of miles (hundreds in some satellite proposals). However, it is important to realize that the coverage area is not defined by distance, but by signal levels and cochannel interference considerations. For terrestrial base stations, this means that the coverage reflects the terrain.

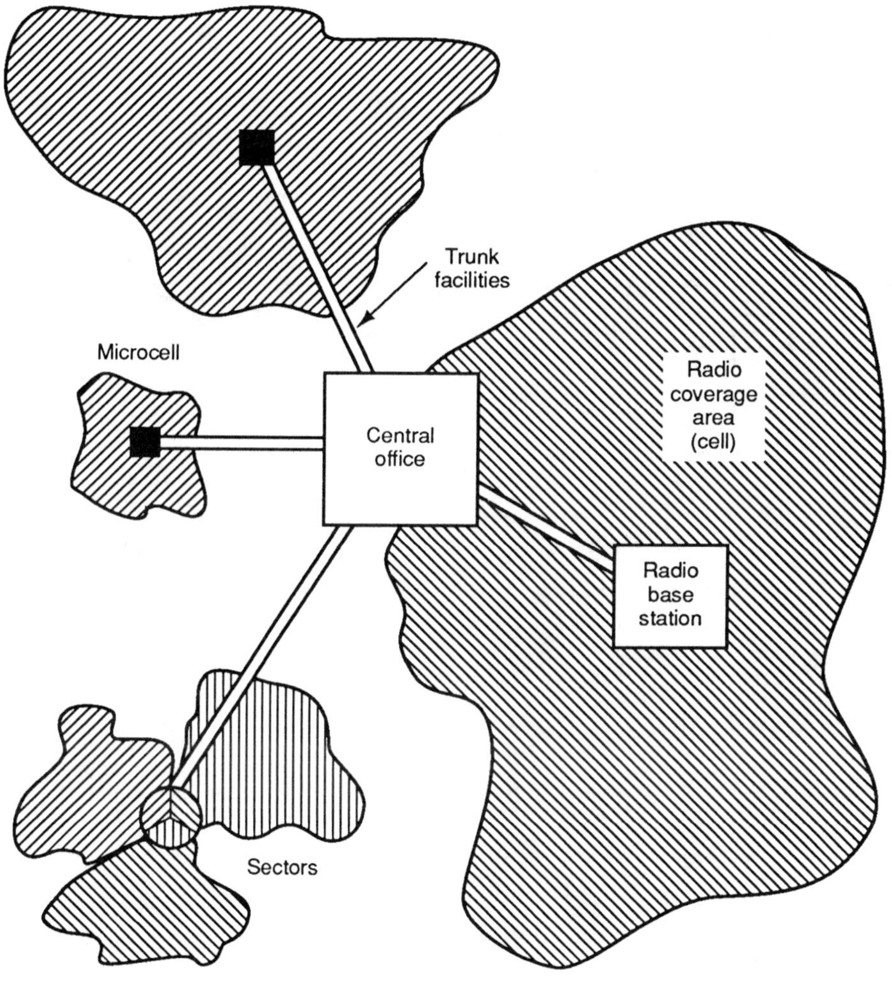

Figure 7.12 Radio coverage area.

In the simplest form of a wireless access cell, subscribers are handled independently of one another; we do not care (for economic modeling) how they are situated or clustered, as long as they are located within the cell.

A more complex model may be necessary for some applications. For example, if a group of subscribers is located close together, it may be more economical to serve them with a multichannel or "cluster" radio terminal with wire connections to the homes (see Figure 7.13). In SR Telecom's subscriber radio planning model, this is extended to the concept of *outstation serving areas* [23] (see Figure 7.14).

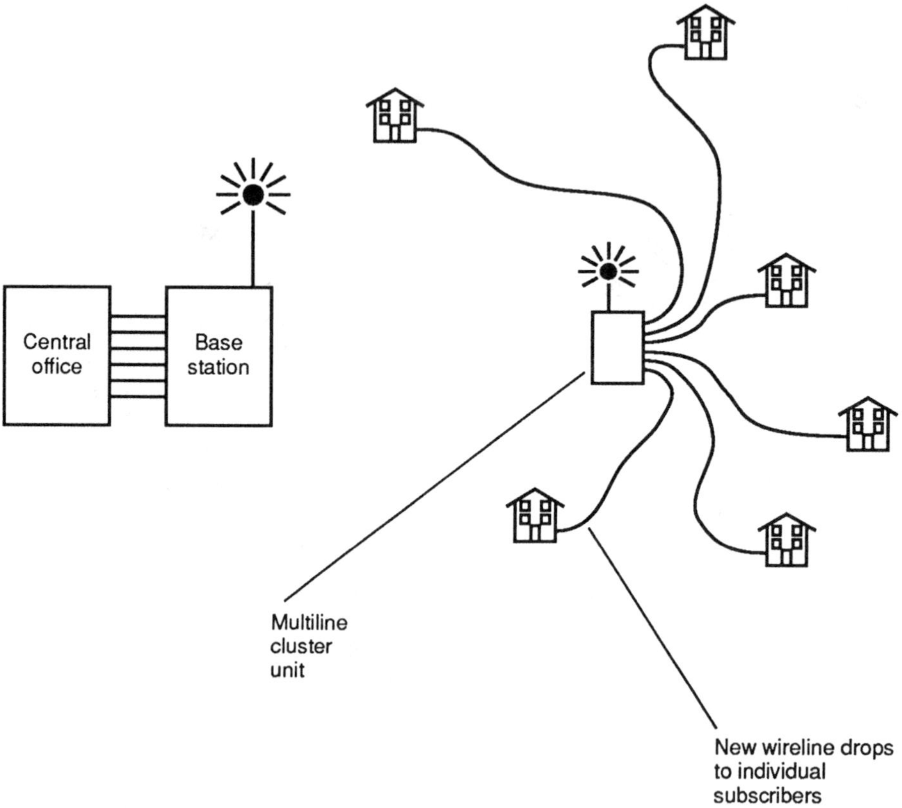

Figure 7.13 Cluster configuration.

Secondly, in some applications (e.g., BETRS), it is desirable to distinguish between a *core coverage zone* and a *fringe coverage zone.* These distinctions are drawn on the basis of actual signal strength rather than distance. The core zone is where signal levels are consistently high enough to allow, for example, a lower power subscriber unit (or perhaps a portable unit), while in the fringe areas additional processing and/or power may be necessary.

Nevertheless, the cost of wireless access circuits *within the cell* is in general both distance-insensitive and location-insensitive. That is, *it does not matter where the subscriber is with respect to the base station or with respect to other subscribers;* the cost of serving him will be about the same. (This statement may seem to understate the many difficulties associated with terrain-effects on radio propagation; note, however, that large-scale terrain features, which may create blockages, are taken into account in the definition of the coverage area, and smaller variations can be handled

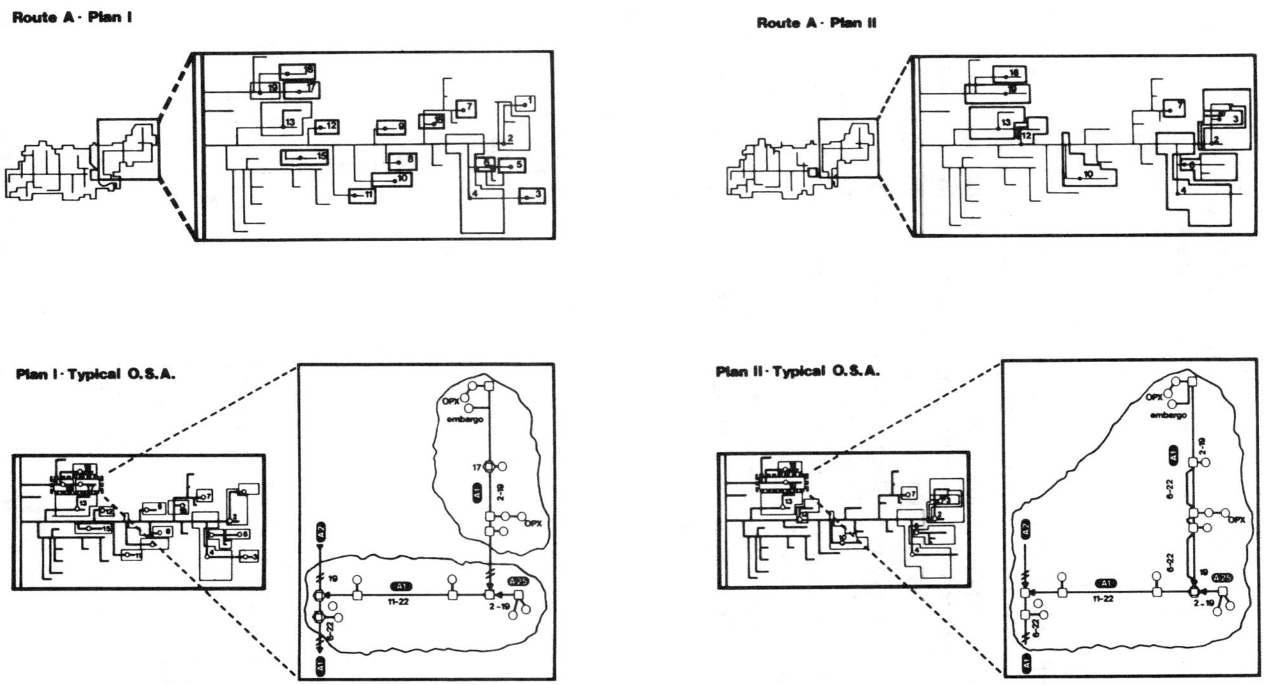

Figure 7.14 Outstation serving areas. (From SR Telecom.)

by the use of appropriate margins. This at least has been my experience. The point is that wireless access is in this respect very different from wireline access, where distance and clustering patterns are the defining parameters of the exchange network.)

7.2.2.2 The Wireless Loop

The longitudinal structure of a standard wireless loop (e.g., cellular or BETRS-type) is shown in Figure 7.15. Essentially, there are four segments. First, there is a concentrated digital facility to connect the radio base station to the switch; this facility may be wireline, fiber, or microwave radio, and may adhere to one of several network-standard digital formats (T1, E1, SONET). Second, there is the radio base station, which converts the signal from the incoming network-standard digital format to the selected digital air interface. Third, the subscriber transceiver receives the radio transmission and reconverts the signal from the air interface format to either analog or network-standard digital form. Finally, there is sometimes a length of wire to connect the terminal to the CPE device(s). As noted in Chapter 5, for many applications the CPE and the subscriber transceiver are integrated in a single unit.

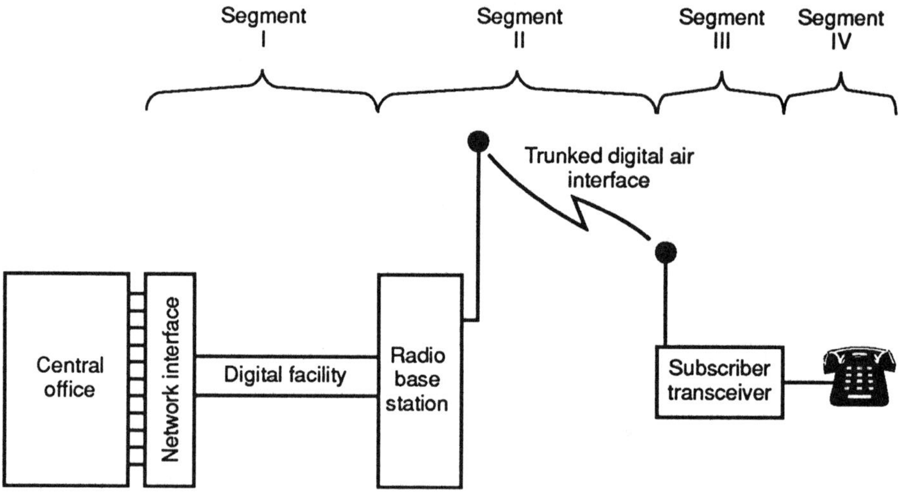

Figure 7.15 Components of the standard wireless access circuit.

Variations on the standard model (e.g., cluster-units or outstations, and repeaters) must also be considered.

There are as yet no standard names for these segments of the wireless access circuit. It is tempting to apply wireline terms such as "feeder segment" and "drop segment" to various parts of the wireless loop. This sort of naming, however, implies a certain mapping of functional equivalencies, which is a more complex subject than may first appear. We will return to this question in Section 7.2.3.

7.2.2.3 The Cost Structure of a Wireless Network

The cost structure of a wireless access system is much simpler than for wireline, which makes modeling it much easier. In fact, in most economic comparisons that I have performed, the wireless alternative could be analyzed very quickly—often in a matter of minutes, given a few simple inputs—while the wireline alternative required a great deal of time and many simplifying assumptions before a valid comparison could be rendered.

Distance Insensitivity. As we have already touched on, the most important economic characteristic of wireless access is that the costs to serve a given subscriber are essentially distance-insensitive. Distance from the base station or central office to the subscriber has no effect on the cost of serving that subscriber. (When one reaches the stage of picking at nits, it may be admitted that there are certain additional costs associated with very remote subscribers living in fringe coverage zones. It may be necessary to install a higher subscriber antenna mast, or a higher gain antenna, or to operate in 4-level modulation instead of 16-level, which reduces system capacity slightly and increases cost. In my experience, however, such "extras" rarely amount to more than 5 to 10% above the normal cost per subscriber, and do not therefore alter the basic observation that radio is essentially distance insensitive.)

Shared and Dedicated Costs. To build a valid and flexible cost model, we must be able to separate shared costs (which must be allocated over a number of subscribers) from per-subscriber or dedicated costs. Actually, in my experience, it has been useful to subdivide the costs into five categories:

1. *Common equipment costs:* This is equipment required for a wireless access system that does not vary as a function of the size of the system (e.g., power supplies, antenna, central processor for system control).
2. *Traffic sensitive equipment costs:* This covers equipment to cost of which varies as a function of the traffic, which is a function of the number of subscribers and the average traffic statistics; the most important category of traffic-sensitive costs are associated with the RF channel hardware.
3. *Per-subscriber equipment costs:* Equipment in this category must be purchased one unit per customer; examples include the subscriber terminal (for cellular and most BETRS applications) and the line card (if there is one) to interface to the central office.
4. *Ancillary materials and labor:* This refers to the power supply for the subscriber, the subscriber antenna, the pole, and so forth. It may be dedicated to a single customer or shared by several customers.

5. *Overhead charges:* Normally these charges are applied as a standard percentage or rate against the total value of a project.

The mapping of different types of equipment and other charges into these five categories may differ from application to application. Figure 7.16 shows a fairly standard mapping, based on a cellular or a BETRS model.

Clustered versus Distributed Systems. As we have also noted, in some configurations and some applications, the clustering of subscribers may be important. The two configurations shown in Figure 7.17 would lend themselves to different equipment specifications, with different costs. To compare clustered and distributed wireless alternatives properly can be a challenge to the model builder's persistence. Aside from differences in equipment cost, the use of a clustered alternative changes the mapping of some equipments items into the five cost categories identified in the previous paragraph. A clustered approach would restate Figure 7.16 as shown in Figure 7.18.

The effects of clustering, and the potential savings or cost penalty involved, are highly dependent on the specifics of a particular site and a particular application. As a general rule, clustering will not affect common or traffic-sensitive costs, and should affect per-subscriber terminal costs only slightly if at all. *Clustering may reduce, often dramatically, the ancillary costs* (by sharing a single power supply, a single pole, etc., across several customers). However, *clustering increases, often dramatically, the cost of the drop facilities* (see Figure 7.19).

7.2.2.4 Alternative Wireless Loop Architectures

The model we have developed here follows the BETRS or cellular pattern. It should be clear from the survey of wireless applications in Chapter 5 that there are many other configurations—for example, nontrunked radio links or dispatch systems—that would be modeled differently. Certainly a VSAT system is a very different animal.

One interesting alternative involves the marriage of radio and fiber optics, using a fiber-based ring or bus to support a series of radio base stations or ports as already shown in Figure 7.11. This hybrid holds much promise, I believe, and we shall return to it in Chapter 9. It must be admitted, however, that such designs are still rather speculative, and the actual implementation of such architectures probably lies well into the future.

7.2.3 Alternative Mappings between Wireless and Wireline Architectures

We are viewing radio as a substitute, or functional equivalent, for some portion of a wireline access circuit. Yet it is perhaps not entirely obvious how many different substitutions may be envisioned. That is part of the versatility of radio; it can replace

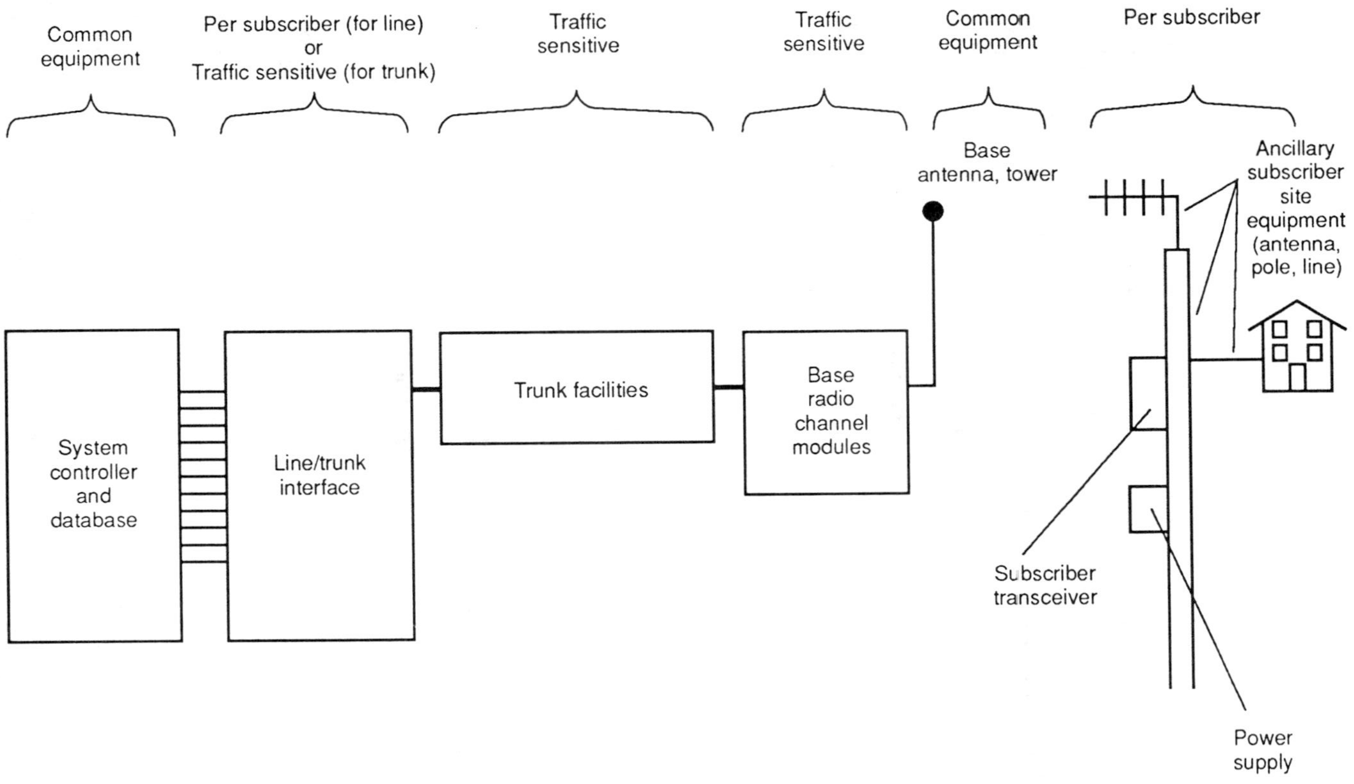

Figure 7.16 Cost categories for wireless loop components.

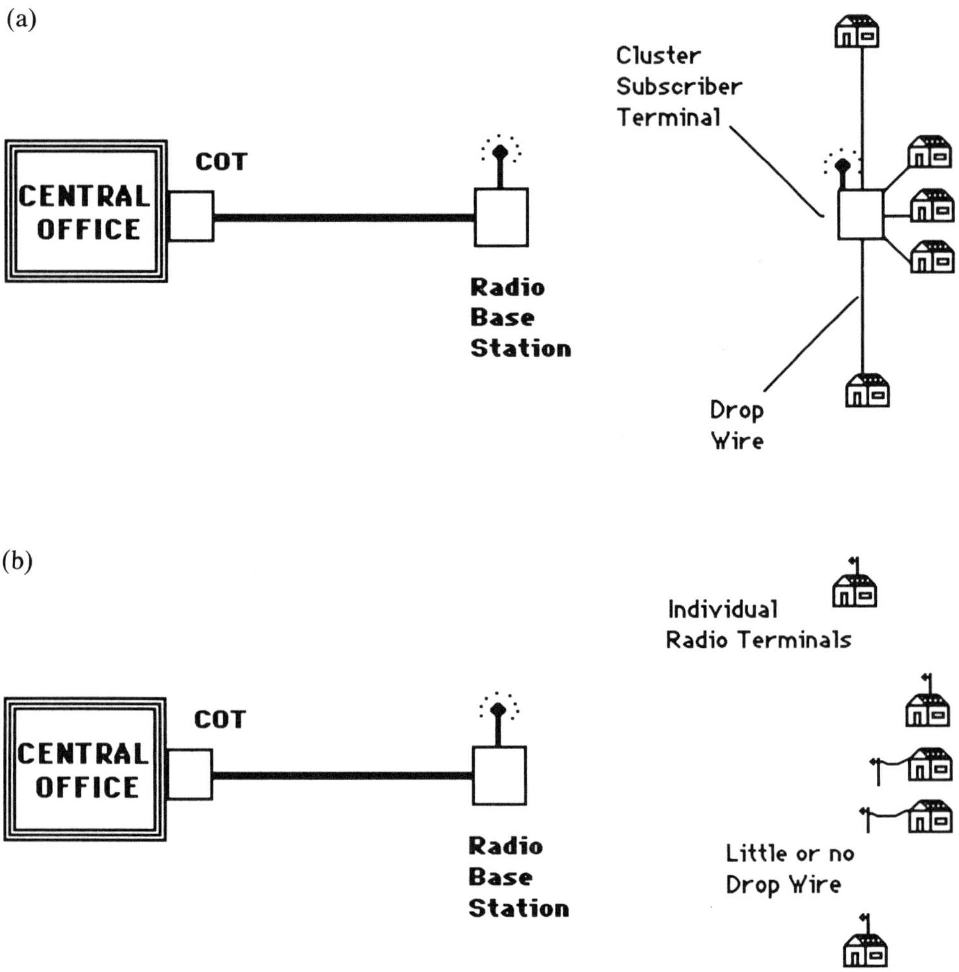

Figure 7.17 Clustered *versus* distributed service plans with wireless access; (a) clustered, (b) dispersed.

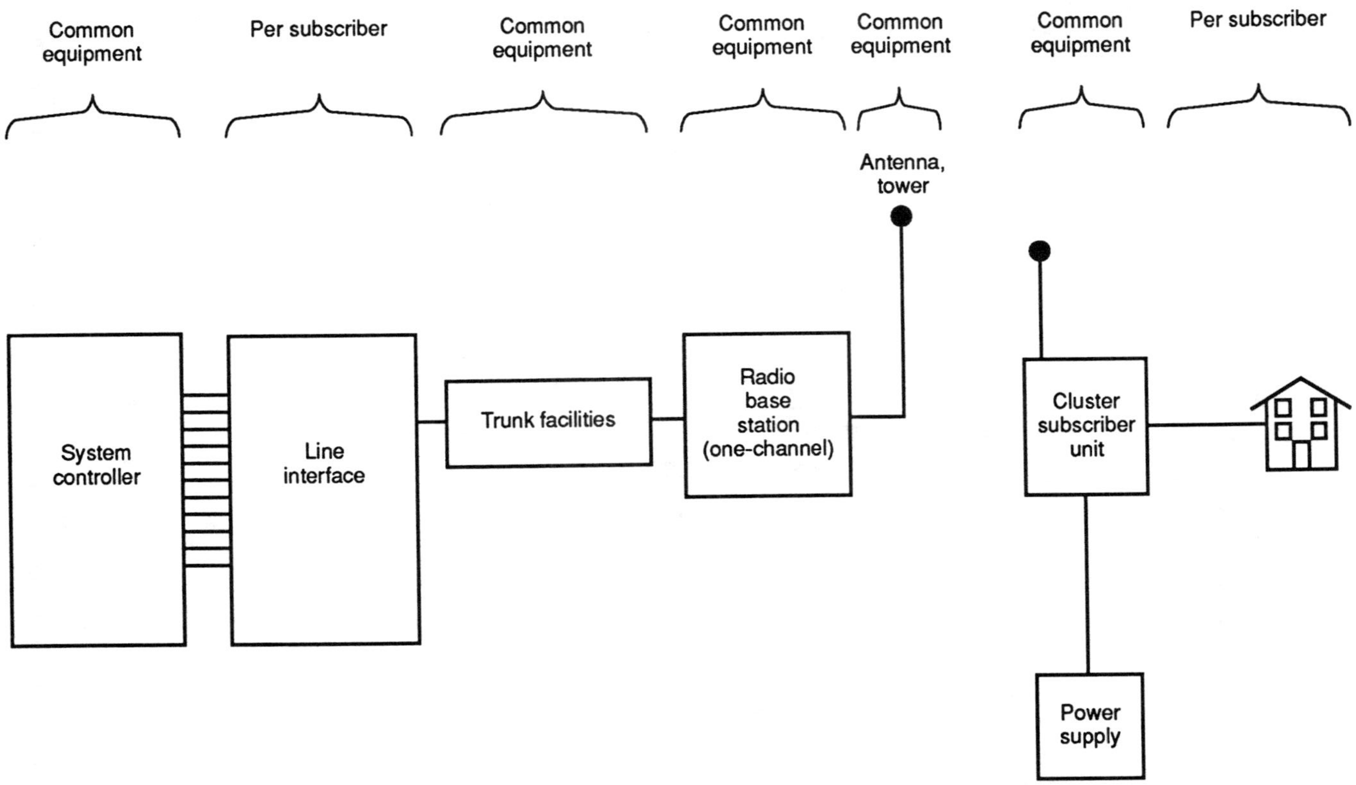

Figure 7.18 Cost categories for wireless loop components (clustered case).

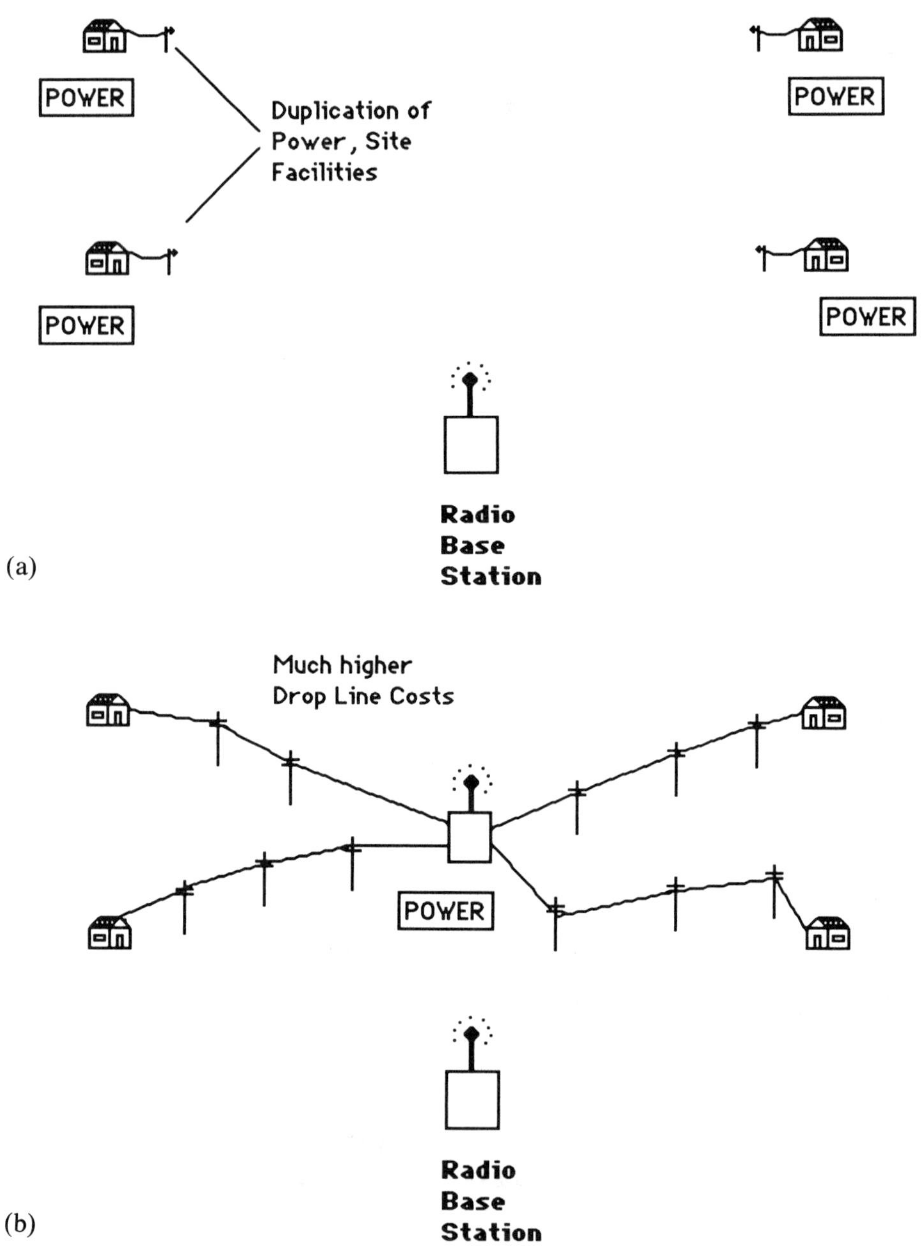

Figure 7.19 Clustering increases drop-wire costs; (a) dispersed service plan, (b) clustered service plan.

almost any segment, or all segments, of the circuit. But it is also a trap for the unwary model builder, who may not recognize the limitations he imposes, unconsciously, on the scope of the application.

7.2.3.1 Drop Replacement Only: "PCN"

The least ambitious mapping would substitute radio for the drop segment (and indeed, for only part of the drop) of the wireline circuit (see Figure 7.20). The radio base station, often called a *radio port* in such applications, is often envisioned as a small unit mounted "on a lamp post" or a street-side utility pole, radiating over a short distance. In modeling drop replacement, the cost of radio equipment is compared against the cost of the wire drop facilities only; the feeder and distribution facilities remain the same for both alternatives.

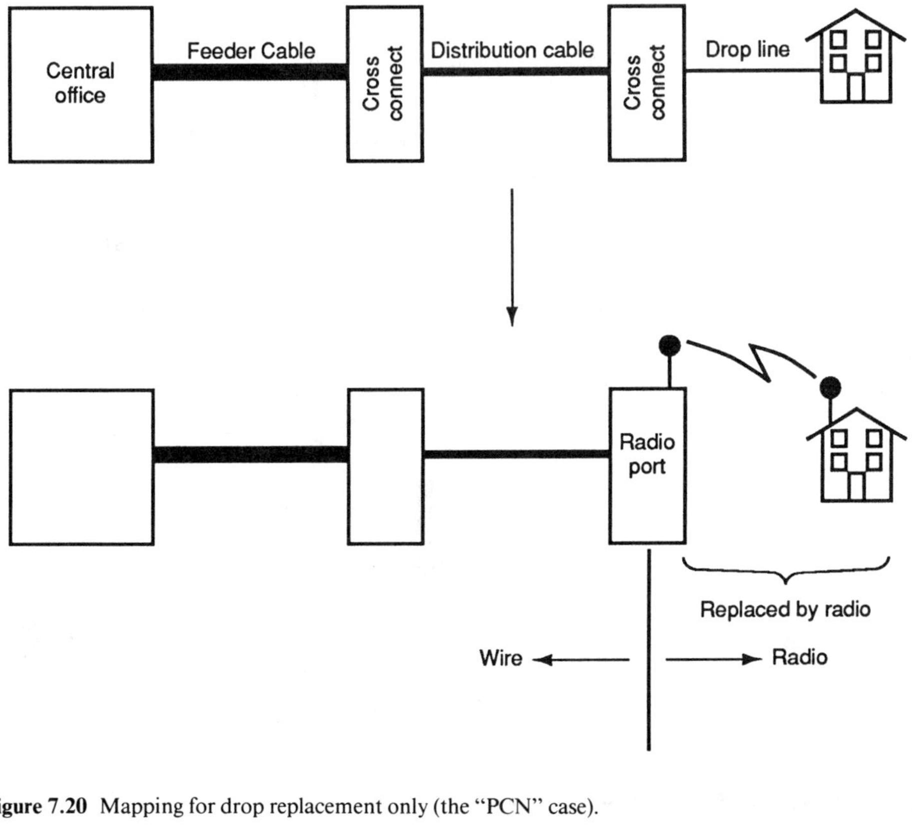

Figure 7.20 Mapping for drop replacement only (the "PCN" case).

As such, drop replacement is the most difficult case to prove-in for radio. As we discuss below, a round number for the cost of drop wire is $1.00 per foot installed [24]. In rural areas where drops of half a mile or more are not uncommon, radio may be justified for drop replacement alone. But short urban drops are typically only a few hundred dollars of first cost, which is difficult to beat with radio at the current state of technical development. However, when maintenance and life cycle costs are factored in, drop replacement by radio becomes more attractive.

It also appears that a number of major U.S. carriers are taking a hard look at drop replacement strategies for the long term. Drop replacement is the application envisioned by some PCN configurations, such as the Bellcore PCS proposal [25] (see Figure 7.21). It also underlies the recent NYNEX wireless loop proposals [26] (see Figure 7.22). Discussions of "fiber to the curb" and "radio tails" also blend in with drop replacement scenarios [27].

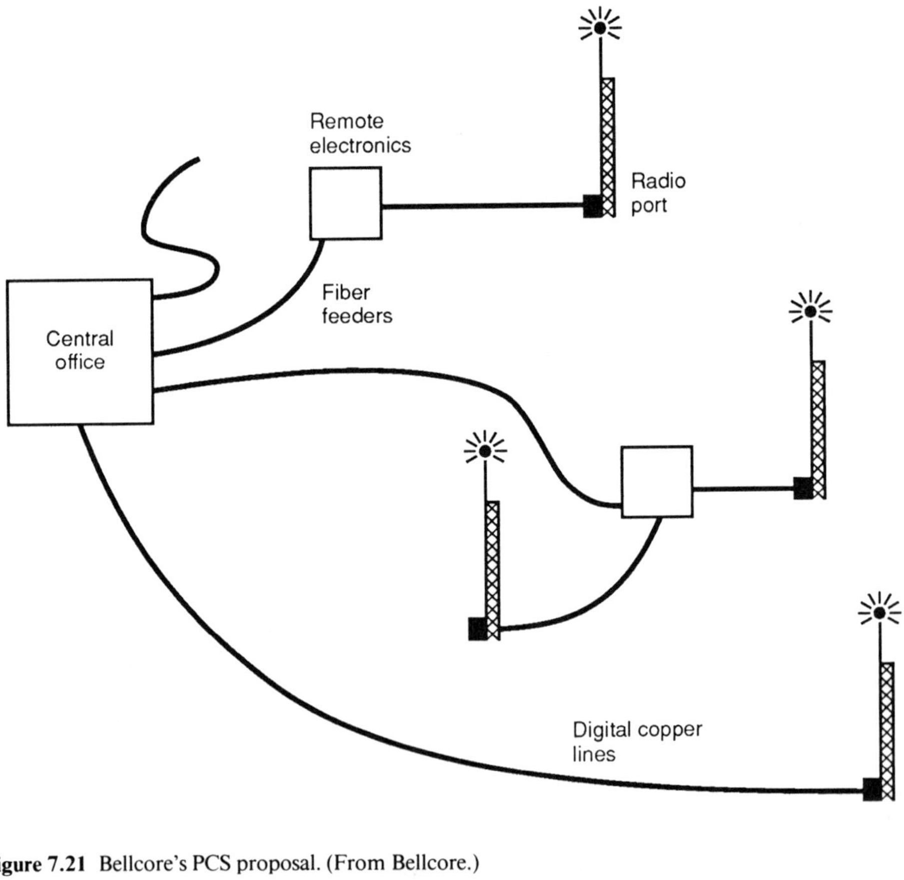

Figure 7.21 Bellcore's PCS proposal. (From Bellcore.)

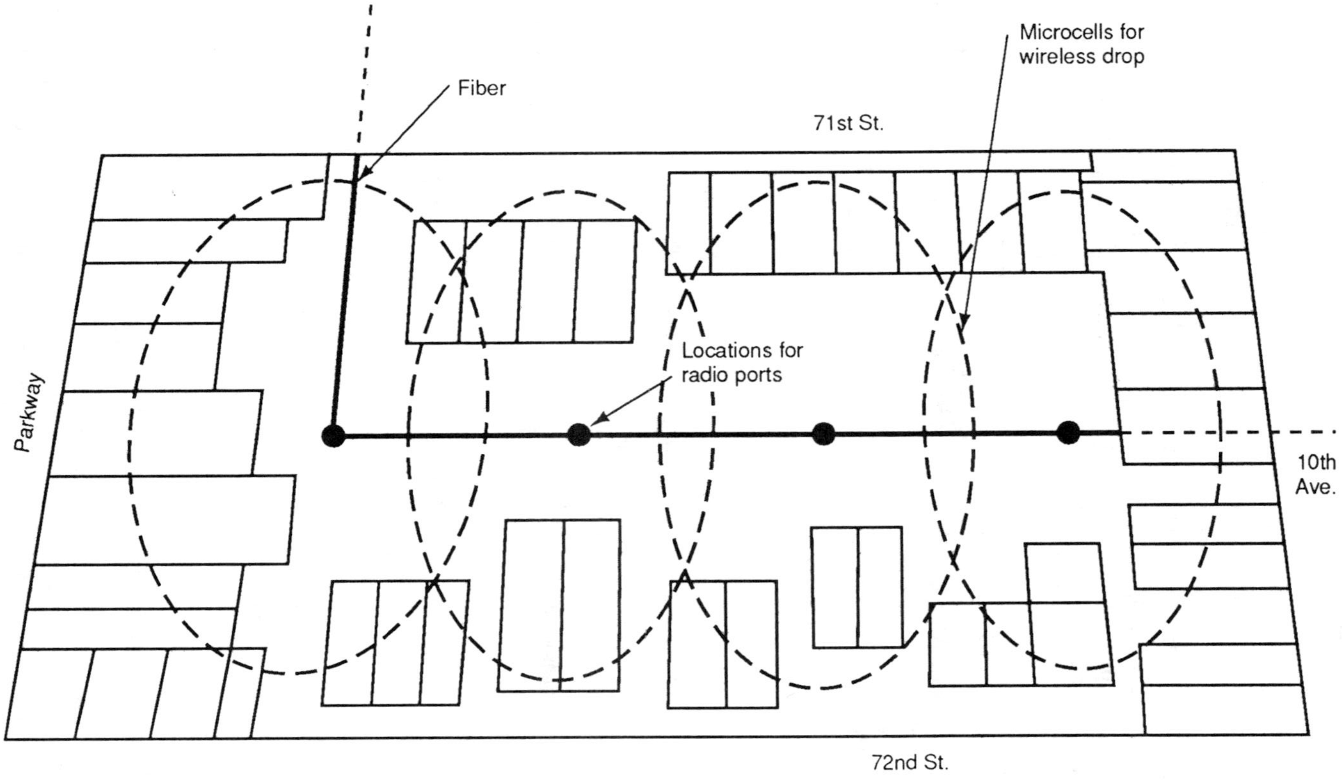

Figure 7.22 Depiction of an urban wireless local loop site. (From NYNEX RFI.)

7.2.3.2 *Distribution Replacement: Wireless DLC*

A more common view of the mapping of wireless access into today's exchange environment accepts radio as a replacement for both the drop and the distribution portion of the network (see Figure 7.23). The base station in this scenario is located at the interface between the feeder segment and the distribution segments, or, in other words, at the point where the remote terminal of a digital loop carrier system would normally be positioned. I have called this the *wireless digital loop carrier* scenario (see Section 5.3.4.2).

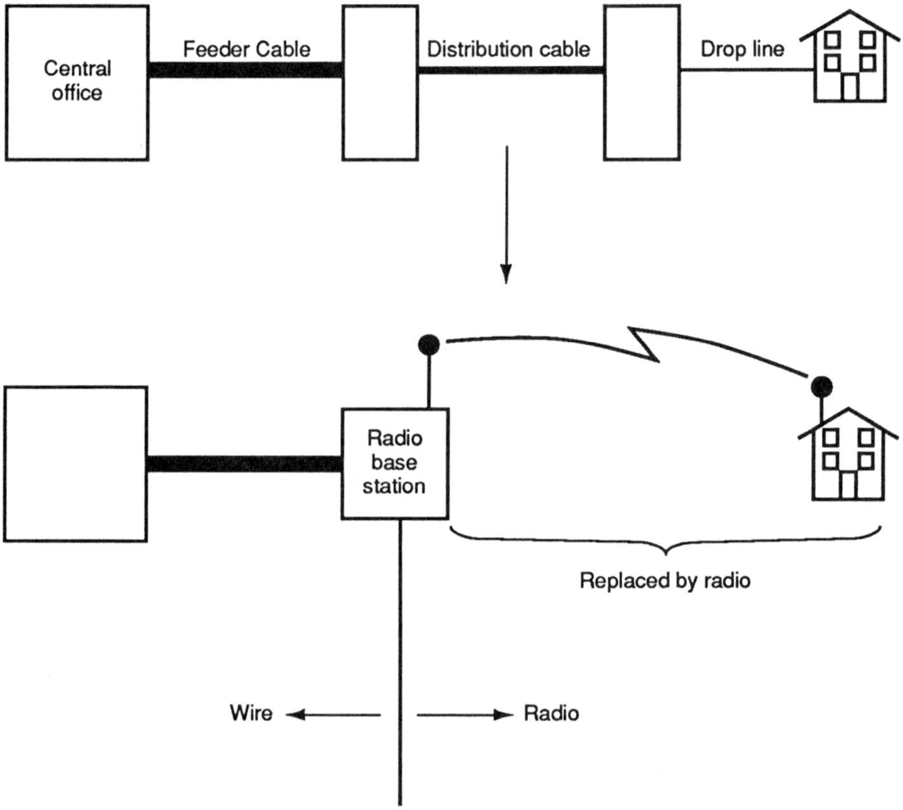

Figure 7.23 Mapping for drop and distribution replacement.

Wireless DLC is obviously easier to prove-in economically against wire—because more wire is being replaced. This is in fact the typical BETRS application. What may not be as obvious is that most cellular networks are organized in a similar manner. The radio base station is connected to the mobile switch by means of a trunk route (normally based on digital carrier), which serves the same function as the feeder in the wireline network.

7.2.3.3 Feeder Replacement: Radio Carrier Serving Area

The third scenario maps radio onto the feeder segment as well as the distribution and drop segments (see Figure 7.24.) With the acceptance of radio as a feeder replacement vehicle, the full area coverage concept can be realized, because a single radio system can substitute for multiple feeder/distribution systems in an exchange (Figure 7.25). I have labeled this a *radio carrier serving area* (see Section 5.3.4.3). The RCSA is very powerful in terms of the economic comparison against wire facilities. Indeed, it becomes possible to consider replacing the entire outside plant in an exchange.

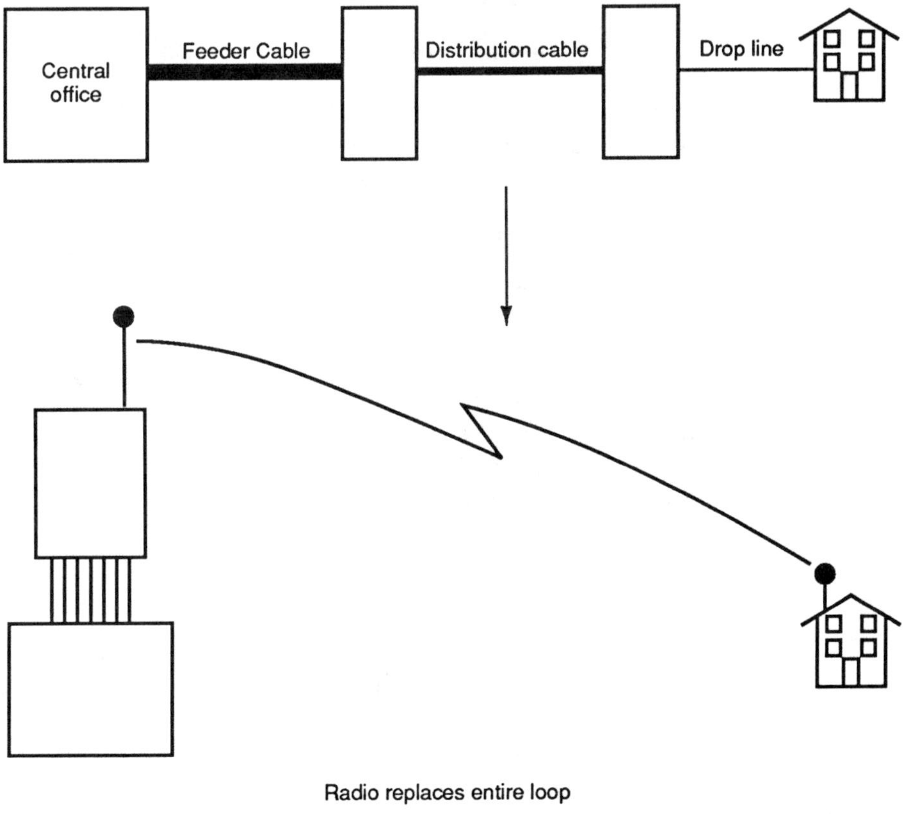

Figure 7.24 Mapping for replacement of feeder, distribution, and drop segments.

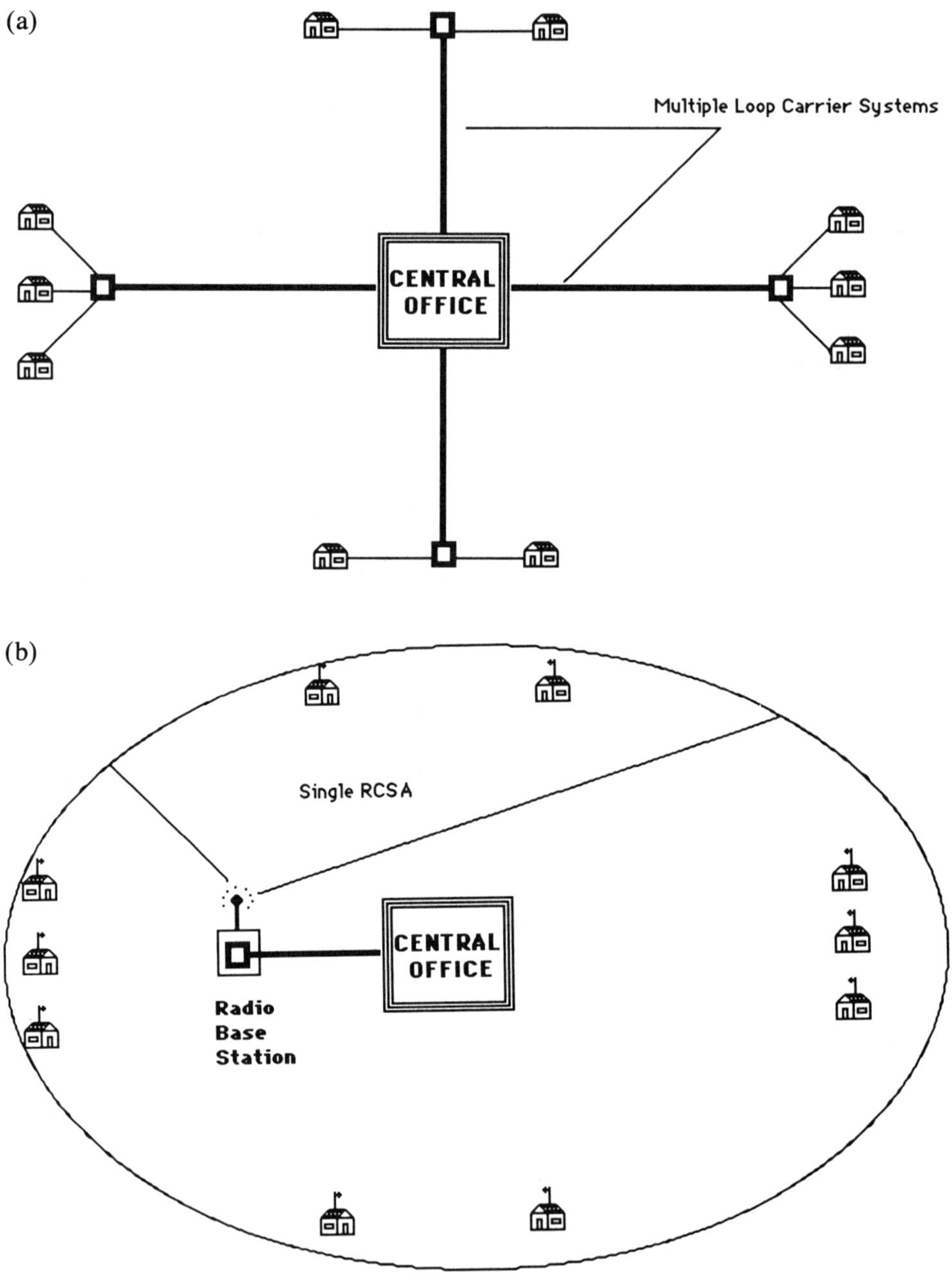

Figure 7.25 Radio carrier service area; (a) wireline system with loop carrier, (b) system with radio carrier.

It is also more apparent with the RCSA that the substitution for existing plant may take place in a gradual, even piecemeal, manner. The RCSA serves as a kind of safety net, or backstop, for a deteriorating outside plant system. We will examine the very positive economic implications of this below in Sections 7.3 and 7.4.

A brief editorial: I view the RCSA as the basic unit of analysis for wireless access. Drop replacement and wireless DLC are both valid concepts, but place unnecessary constraints on the planner. In effect, they both tie the radio system down to a particular fixed point in the access network. The RCSA, on the other hand, addresses the entire exchange or, more likely, several exchanges. I believe that in modeling the RCSA should be the base case, with drop replacement and wireless DLC viewed as special, limited applications.

7.2.3.4 Exchange Replacement/Consolidation: Wireless Overlay Network

The next step is fairly obvious: Once we realize that an RCSA typically covers multiple exchanges, there is no reason to limit the application to a single exchange. Indeed, a single BETRS-type RCSA covers several thousand square miles and up to a dozen or more wireline exchanges. It becomes possible to substitute radio for some of these exchanges (see Figure 7.26). Thus, the mapping of radio now covers even the exchange costs (or at least the fixed costs of that exchange). We refer to this as a *radio overlay network* (RON) [28].

(a)

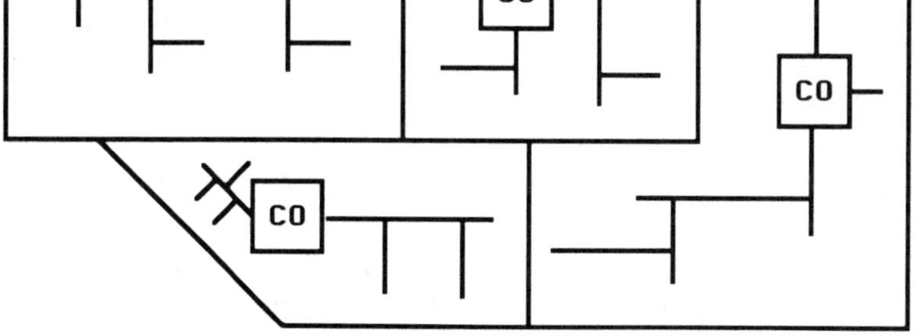

Figure 7.26 Exchange consolidation by means of wireless access; (a) today's wireline exchanges with feeder routes, (b) elimination of exchanges by radio carrier serving area.

(b)

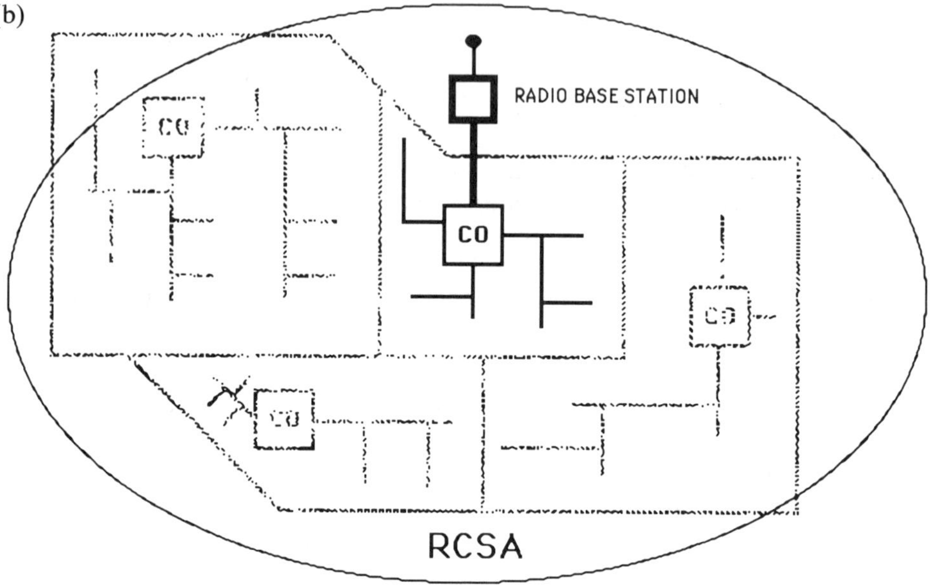

Figure 7.26 continued

The RON is another very powerful concept in my experience. It offers the prospect of reversing to a degree the long trend toward distributed switching. It also opens up interesting possibilities for digital services overlays (see Section 7.4).

There are other mappings. For example, radio may be used to replace only the feeder portion, with essentially a point-to-point configuration and extensive cable distribution from the remote outstations. This indeed is the standard model for subscriber radio (Figure 7.27). There is also the cordless extension phone, where radio replaces a portion of the "inside wiring" on the customer's premises.

What is important is that we understand and make explicit the particular mapping of functional equivalencies between wireless and wireline for any particular model of a wireless access application.

7.3 SIMPLIFIED MODELS OF WIRELESS ACCESS COSTS

In this section we develop a series of simplified analytical models, each illustrating a particular aspect of the cost comparison between wireline and wireless access alternatives. These models are not ready to use and need to be adapted to the circumstances and objectives of a particular operator's planning regime. Our purpose

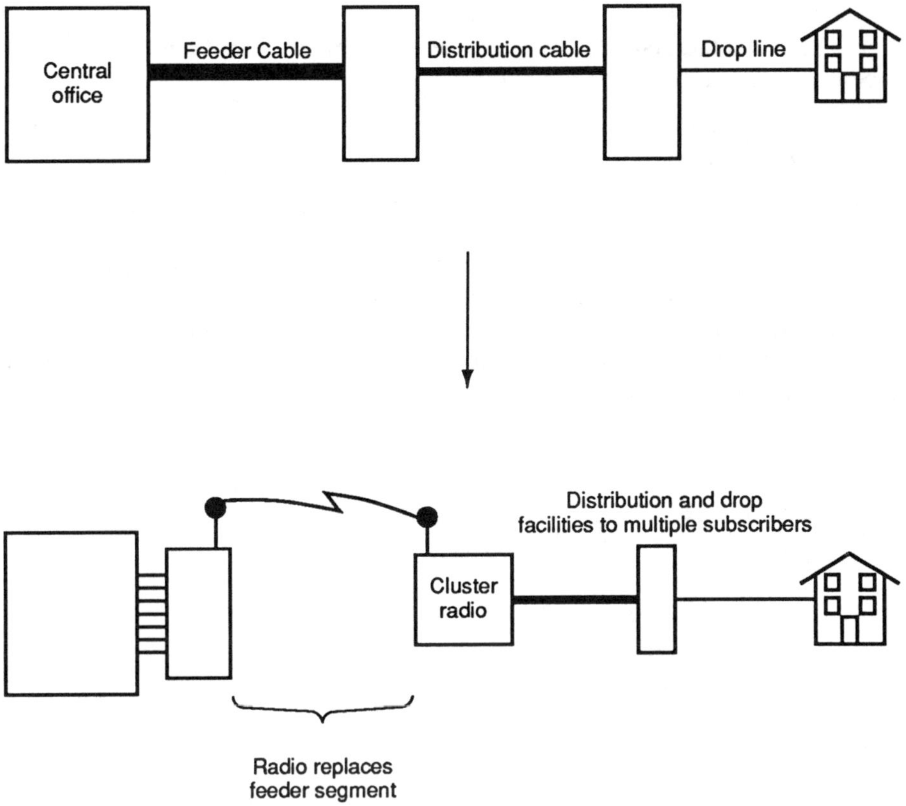

Figure 7.27 Mapping for feeder replacement only.

here is to elucidate some of the important principles that such models must incorporate.

7.3.1 Modeling the Spatial Dimension I: Cable *versus* Radio

As noted above, one of the most analytically important differences between radio and cable is that *cable access plant incorporates a per-unit-of-distance cost component, while radio does not.* Also, since radio provides area coverage, it adds a degree of freedom for the planner in the sense that we can be much less concerned with the spatial configuration of subscribers. Thus, a cable-route model incorporates (often implicitly) a population density function—often stated in terms of subscribers per route mile—while radio is much less sensitive to density and can often ignore it altogether. These two factors—distance and density—are the most obvious differentiators between wireline and wireless access.

7.3.1.1 Cost and Distance

Per-Unit-of-Distance Costs. When we overlay the distance function for wireline access costs (Figure 7.6) and for wireless access, we see that (at least in an abstract sense) there is a point at which the two curves cross, beyond which radio is cheaper than wire (see Figure 7.28). There is no standard terminology for this concept yet, but most analyses recognize the importance of this crossover point, often referred to as the *radio prove-in distance.* It is obvious that as the cost curves for either alternative shift, the crossover point will change.

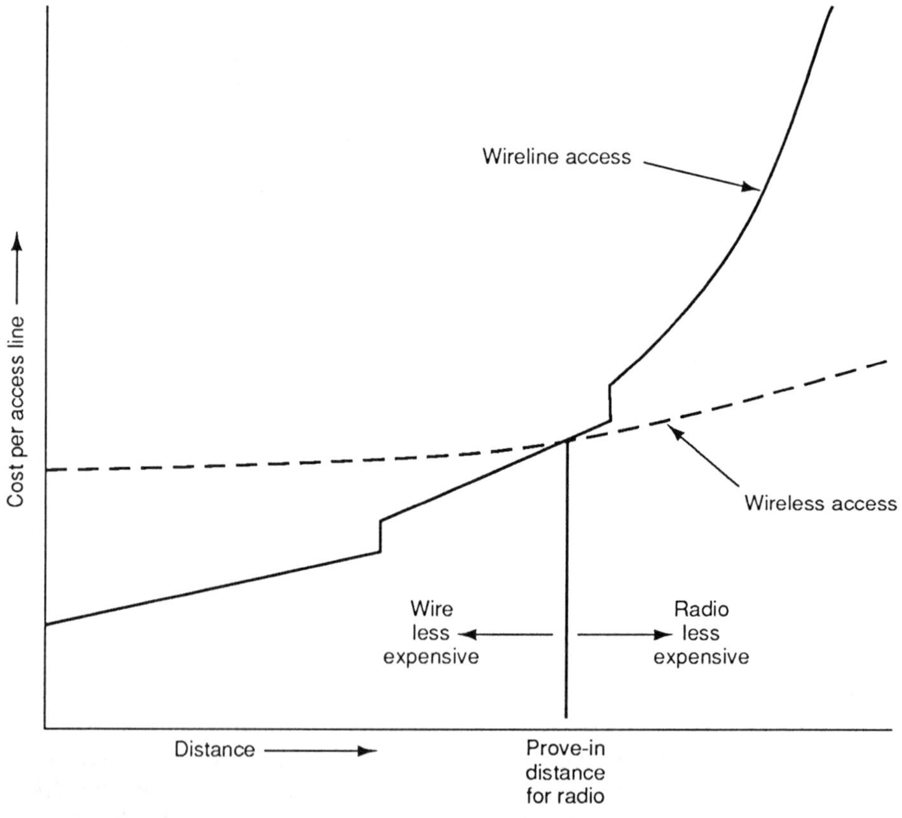

Figure 7.28 Wireline and wireless access economics.

The prove-in distance remains an abstraction, however, since it does not describe individual loops or cable-routes, but may be regarded as being true only as an average for a large number of loops. *There are too many uncertainties associated with the wireline cost/distance curve.* The specific wireline cost curve in a given exchange depends greatly on the specific configuration of the existing plant, especially the grouping of loops into feeder-routes, since the cost-per-unit-of-distance is higher for feeder segments than for drop segments, but the cost-per-line-per-unit-of-distance is the reverse—higher for drop segments than for feeder segments. It is also difficult to factor in the costs of loop electronics alternatives as a function of distance.

Prove-in distances have been calculated on several occasions. Sometimes, it has been used as an input, the value of which is assumed and varied for sensitivity analysis. Crossover points of 15 kilofeet and up have been proposed (see the studies cited in Sections 5.3.5.2 and 5.3.5.3). Despite the popularity of this concept, I am skeptical of most of these analyses. In fact, based on my experiences in the field, I would argue that while the prove-in distance concept may have some value as a general diagnostic indicator, it is very difficult to calculate for real exchanges and may not even be meaningful in any practical planning sense.

Frequency Distribution of Loop Lengths. The other important variable for the estimation of wireline costs—again dealing with large aggregates of loops—is the frequency distribution or probability distribution of loop-lengths. A number of studies [29] indicate that this distribution typically, and not surprisingly, follows a lognormal type curve, where most loops are short but there is a long tail to the right representing the small percentage of very long loops in the system (see Figure 7.29). The dynamics of this distribution are also interesting; there are indications that as loop lengths tend to increase (see Chapter 2), a disproportionate amount of the growth is taking place among the longer loops. One study disclosed a trend toward a bimodal distribution, with a second peak at around 18 kilofeet (the natural interpretation was that this represented the development of a suburban ring around the small towns that made up the exchanges in this population [30]).

The implication of such frequency distributions is that for a fixed or even increasing cost/distance function the total cost of the access plant in the exchange is disproportionately weighted down by the unusually high costs of a few long loops. Many in the telephone industry seem to be aware of this; as we noted in Chapter 2, in one instance a major carrier has produced an informal estimate that 3% of its loops create 30% of its costs (including maintenance as well as first costs).

Cost/Distance Indices. Thus, if we recognize the abstractions involved, we can still develop a potentially useful *index of wireline loop costs* for a given exchange by a formula similar to the following:

$$\text{Cost of n loops in exchange} = \sum_{J=1}^{n} L(J) * D$$

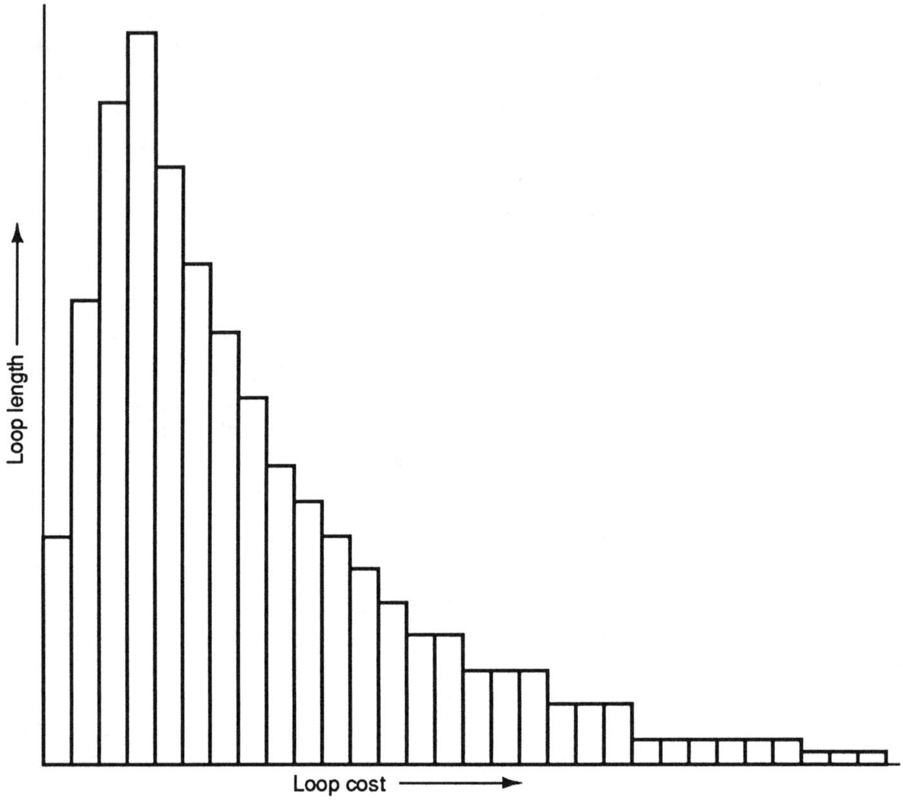

Figure 7.29 Lognormal loop length distribution.

where

L = Length of Loop J, and
D = the cost-per-unit-of-distance factor applied to wireline plant.

A more sophisticated analysis of the cost of a single wireline loop would reflect the different costs-per-unit-of-distance for different segments:

Cost of the Jth wireline loop =

$$\frac{\text{No. of Meters in Feeder segment} * \text{Feeder Unit-Cost per Meter}}{\text{No. of First Year Customers Sharing the Feeder}}$$

plus

$$\frac{\text{No. of Meters in Distribution segment} * \text{Distribution Unit-Cost per Meter}}{\text{No. of First Year Customers Sharing the Distribution Cable}}$$

plus

No. of Meters of Drop Wire * Drop Wire Unit-Cost per Meter

It is further assumed that these per-unit-of-distance costs are loaded with all appropriate overhead, labor, materials, and so forth. If it is necessary to break down these units into their component costs, the cost formula becomes even more formidable.

Note that these complexities attach primarily to the wireline half of the analysis. The radio analysis is much simpler, because it does not incorporate a distance factor:

Cost of a Wireless Loop =

Cost of the Subscriber Station

plus

$$\frac{\text{Cost of Shared Radio Base Station Equipment}}{\text{No. of Subscribers on Wireless Access System}}$$

The shared costs of the wireless access system are dependent on traffic, as we shall discuss in the next section. Note also that this simplest radio loop formula assumes a single-access-line subscriber station. If clustered subscriber stations (i.e., stations serving more than one customer, connected by separate drops) are assumed, the distance factor reenters the wireless access cost formula, but only for the drop portion:

Cost of a Clustered Wireless Loop =

$$\frac{\text{Cost of Subscriber Station}}{\text{No. of Subscribers in Cluster}}$$

plus

No. of Meters of Drop wire * Drop Wire Unit-Cost per Meter

plus

$$\frac{\text{Cost of Shared Radio Base Station Equipment}}{\text{No. of Subscribers on Wireless Access System}}$$

It is a common error in wireless access analyses to omit the distance factor for drop wire costs when calculating the cost of a clustered radio alternative compared to a single-line alternative [31]. Clustered systems, such as subscriber radio [32] and clustered BETRS systems [33], are often advantageous, but they do increase loop costs for the additional drop wire.

7.3.1.2 Cost and Population Density

The relationship between loop costs, loop architecture, and population density has not been thoroughly analyzed, to my knowledge. Intuitively, it seems obvious that wireline architectures show increasing costs per line as population density decreases. It also seems plausible that radio will prove-in more easily for lower density situations than higher density ones; however, demonstrating the precise analytical relationship is difficult, and once again most of the complexity adheres to the wireline case.

Cost/Density Functions for Wireline Systems. The chief problem in mapping the relationship between wireline cost and population density is the intervening variable describing the type of network architecture employed. We may illustrate the impact of this variable by considering the two extreme cases—a pure star network and a pure bus network. In the former, there is a completely dedicated facility to every subscriber, while in the bus architecture there is a shared transmission path connecting all subscribers with the minimum of cable.

Consider an area comprising a 8 square miles on a side, or 64 square miles in all, divided into an even grid of 16 square-mile sections with subscribers located at the corners of each section and the central office in the center of the area (Figure 7.30). (We may think of this as a farm region in rural Nebraska.) It is possible to provide wireline service to these nine subscribers (including one who lives at the central office itself) with a fairly efficient star network, using 38.6 miles of cable facilities. We can also serve the subscribers with a bus network (leaving aside the question of transmission technology for the moment) requiring only 32 miles of cable. In this relatively low-density situation, both alternatives are about equally efficient, with only about a 20% edge to the pure bus over the pure star.

Let us increase the density by assuming that the size of the sections is reduced by half (see Figure 7.31). Now the star network requires over 93 miles of cable, while the best bus network needs just 48 miles. In fact, there are many bus networks that come close to the minimum. We can see now that the bus networks approximate the layout of a feeder-route system.

If we increase the density again (Figure 7.32), we find that the star network needs 277 miles of cable while the bus network needs only 80.

The impact on cost is equally dramatic. If we assume a standard cost per mile of cable facilities (obviously an abstraction, but an illustrative one), then as the density increases from 0.14 subscribers per square mile to 1.27 subscribers per mile, the cost of the star network falls by about 20% per line—but the cost of the bus network falls by almost 75% per line (Figure 7.33). If we increase the density to 4.52 subscribers/mile, the cost per line in the star network declines by only an additional 5%, while the cost per line in the bus network drops by another 50%, to less than $1000 per line.

(a)

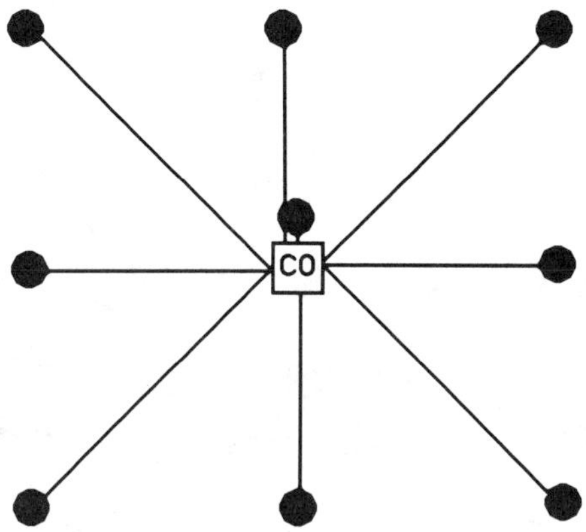

(b)

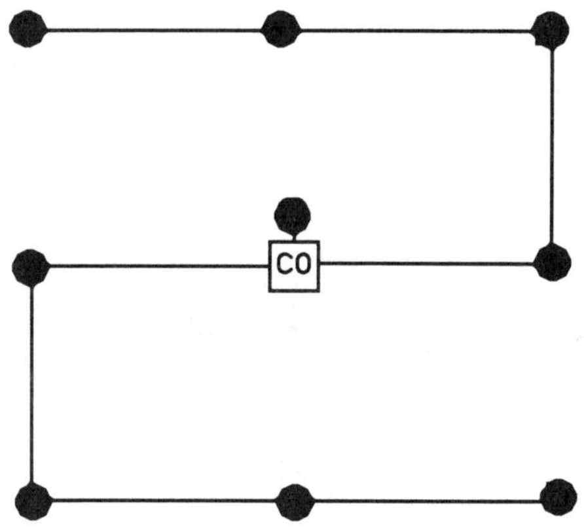

Figure 7.30 "Pure" alternative loop architectures—low density; (a) star configuration, (b) bus (cable route) configuration; "bus" = 83% of the plant mileage of "star."

What is the relevance of such abstract network alternatives? After all, no one ever builds a pure star network—it would be economically stupid—and no one can ever build the perfect bus network, because the real world intervenes.

The relevance is that the bus defines the best we can achieve with a well-designed cable-route system. The actual savings of real cable-routes is certainly not the full potential of the perfect bus architecture. For a variety of reasons, there is still a substantial portion of the total cost located in the drop segment—perhaps as much as 50% of the total cost in rural exchanges. And while the feeder segments are bus-like in cost structures, the drop segment remains star-like. Moreover, the cost per

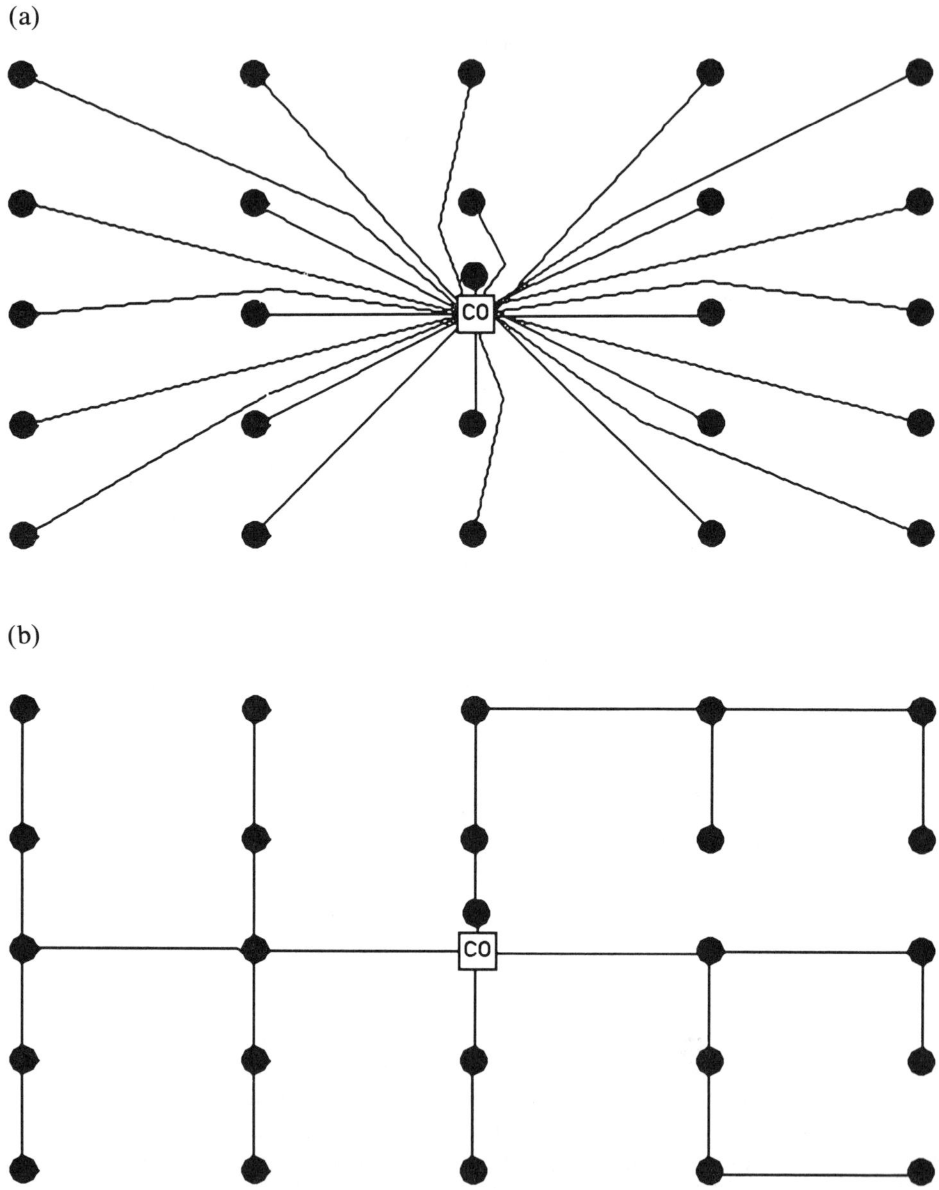

Figure 7.31 Alternative loop architectures—high density; (a) star configuration, (b) bus (cable route) configuration.

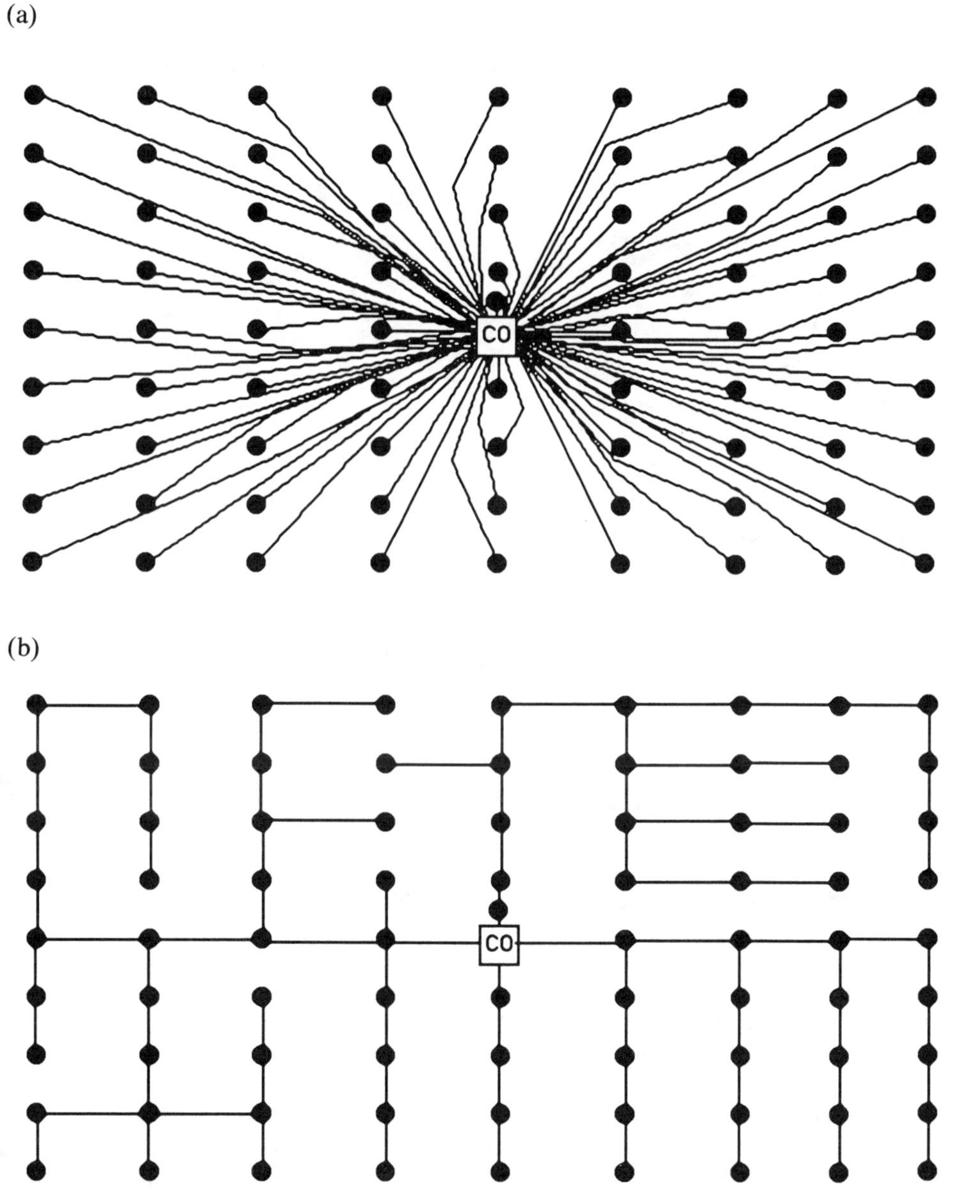

Figure 7.32 Alternative loop architectures—high density; (a) star configuration, (b) bus (cable route) configuration.

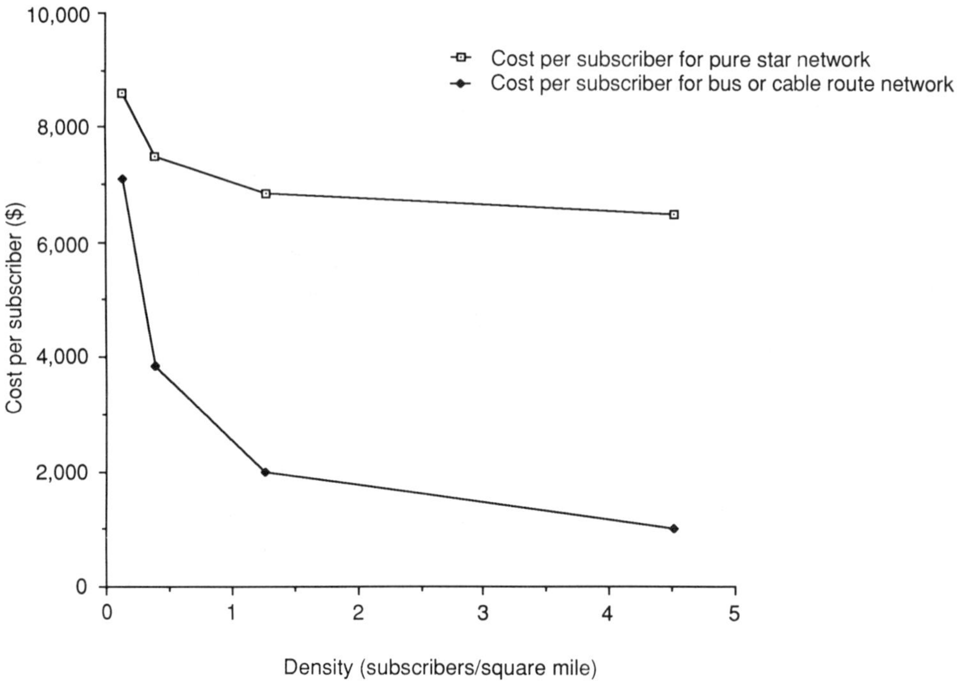

Figure 7.33 Cost-per-subscriber for wireless access as function of population density.

unit of distance is higher for the bus-like feeder portion. Finally, the availability of feeder-routes is always nonoptimum, subject to the constraints of terrain and the availability of rights-of-way. All in all, it doubtful that more than 50% of the potential savings from a bus-like architecture can actually be realized in today's wireline systems.

We can now understand the powerful drive toward feeder-route architectures, which attempt to render the access network as bus-like as possible. These simple illustrations show how much more can be gained from this approach in a high-density environment than in a low density environment.

Cost/Density Functions for Radio Access Systems. The cost/density relationship for wireless access systems is much simpler, especially if we assume single-line subscriber stations. We do not have to worry about configuration. Instead, the effect of increasing density is to spread the cost of the common radio equipment over a larger number of subscribers, driving the cost per line toward an asymptotic value that represents the cost of the subscriber station together with the traffic-sensitive portion of the base station (see Figure 7.34).

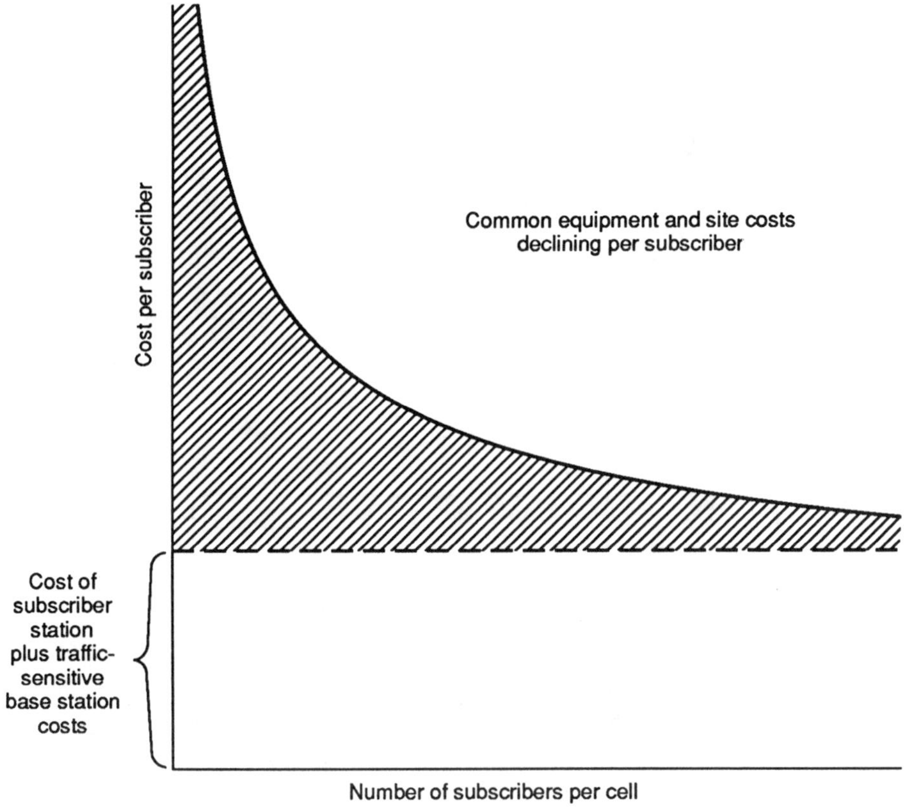

Figure 7.34 Cost-per-subscriber for wireless access as a function of the number of subscribers.

This cost/density curve dominates radio applications in different ways, depending on which variable is seen as independent from an engineering perspective. In one approach, perhaps characteristic of BETRS applications today, there is a fixed cell size; increasing subscriber density can help bring the cost per line down to where an application proves-in over wire. This may seems counterintuitive at first; we may suspect that the lower the density, the more favorable for radio. However, very small systems (i.e., serving a small number of subscribers over a large area) do not share the common system overhead costs across enough users to bring down the cost per line to be competitive with wire. Often, therefore, the challenge is to find additional subscribers within the radio carrier serving area, to drive the cost per line down to where the entire system proves-in (Figure 7.35).

In the second approach, density is held constant and the cell size is decreased (or increased) to achieve a desired cost per line. This may be the case in a fairly dense,

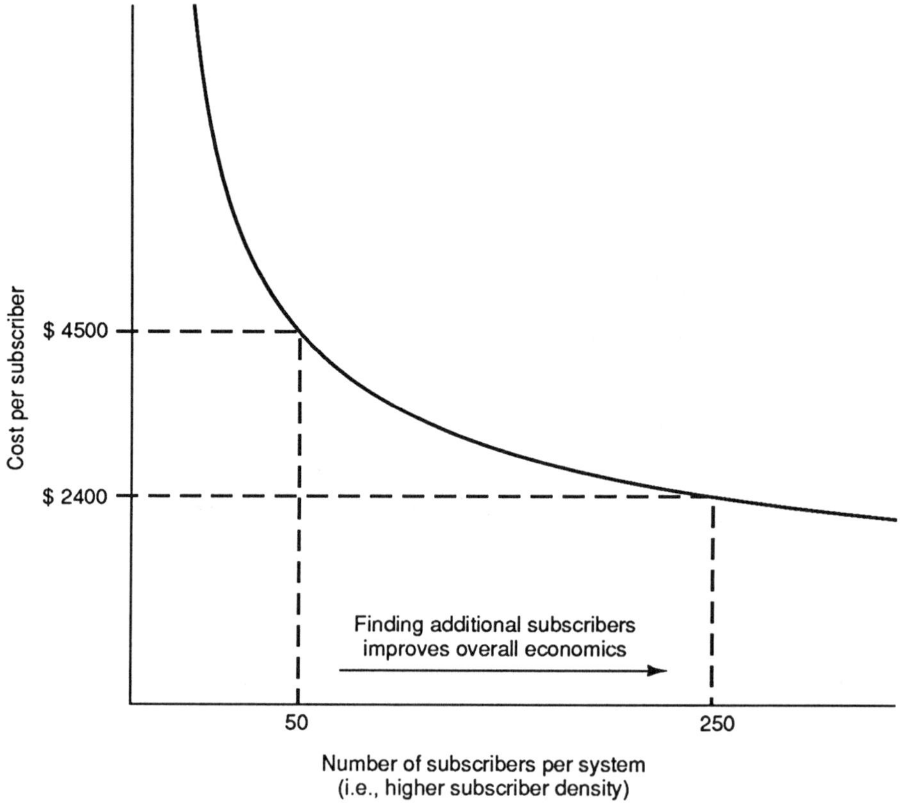

Figure 7.35 Cost/density relationship with fixed cell size.

urban setting, where the cost per line for radio with large cells is so favorable—i.e., nearly asymptotic and well below the wireline cost threshold region—that we may decrease the cell size in order to facilitate reuse and higher overall system-wide capacity. This will drive up the cost/density curve in the reverse direction, so to speak, but it may be possible and desirable if we can gain capacity without raising the cost per line significantly (see Figure 7.36).

Comparisons. The comparison of wireline and wireless alternatives as a function of density comprises about 99% perspiration (on the wireline case) and 1% respiration (on the wireless case). So far, there are no models that really help automate what can be an arduous procedure. I might suggest, as a general approach, the following:

1. The costs for the wireline case can be bounded by calculating two scenarios: a pure star network, which is fairly easy, and an efficient bus network (we need

not worry about finding the optimum since a slightly suboptimum network, relatively easily found, will serve just as well in this procedure).

2. Derate the bus approach by 50%—i.e., split the difference, roughly, between the star and the bus.
3. Calculate the radio cost per line based on the maximum feasible density (usually, 100%), and evaluate sensitivity and degree of asymptotism by backing off the density of subscribers to the point at which the costs are approximately equal.
4. This should provide a general idea of the comparative economics; it is probably best viewed as a diagnostic step, however, leading to a more specific modeling process, such as that discussed in Section 7.4.

Clustering and Cost/Density Functions. The clean, simple radio analysis unfortunately becomes rather messy when we take into consideration a clustering option, in

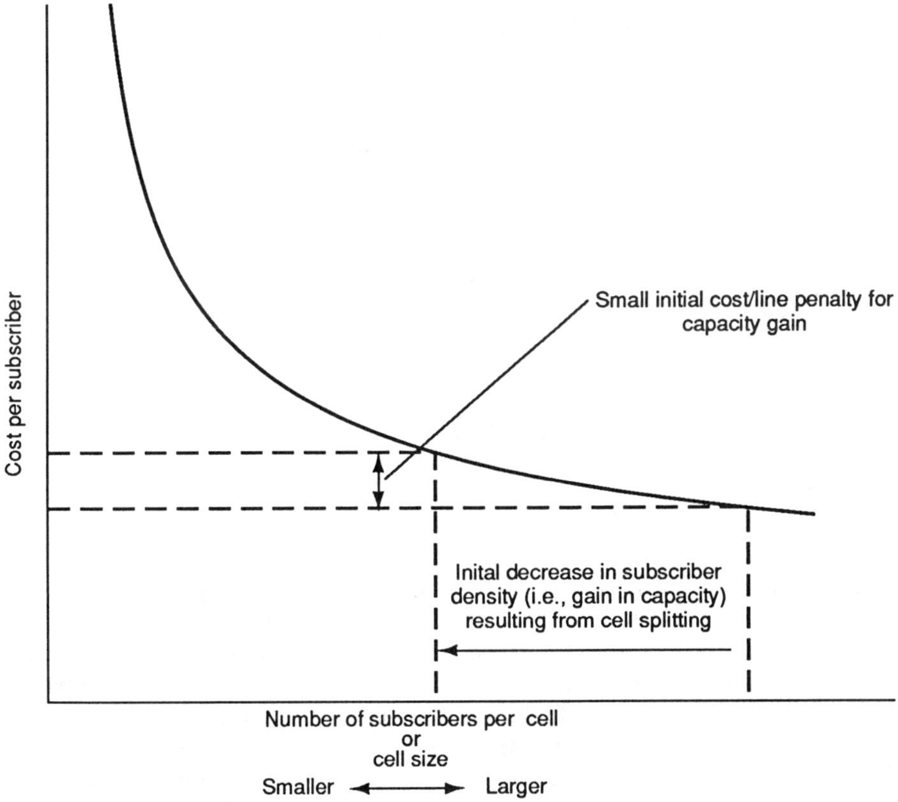

Figure 7.36 Cost/density relation with decreasing cell size.

which a group of subscribers may be served by a single subscriber station. In effect, clustering reintroduces the wireline configuration model—in miniature—at the remote end of the circuit. A cluster of, say, 24 subscribers must be connected to the cluster station by means of conventional drop wire facilities. This raises once again the question of the density and the costs of drop wire versus the cost of additional radio hardware.

Consider the scenarios shown in Figure 7.17, in which customers may be served by either individual subscriber stations, each with its own pole, power supply, etc., and a (negligibly) short drop (Configuration A), or by a cluster terminal, with a single pole, power supply, and so forth, but with the addition of significant lengths of drop wire to connect the individual residences to the cluster radio terminal (Configuration B). The costs are:

Configuration A (Single Line Units):
Cost of each subscriber access facility =

$$\text{Shared Base Station Costs (same for both cases)} + \text{Subscriber Terminal Cost} + \text{Ancillary Costs (Pole, Power, etc.)}$$

Configuration B (Clustered Units):
Cost of each subscriber access facility =

$$\text{Shared Base Station Costs} + \frac{\text{Cost of Cluster Terminal Equipment}}{\text{No. of Subscribers Served by Cluster}} + \frac{\text{Ancillary Costs (Pole, Power, Etc.)}}{\text{No. of Subscribers Served by Cluster}} + (\text{No. of Meters of Drop Wire} * \text{Drop Wire Unit-Cost per Meter})$$

The important observation is that drop line costs are increased by the use of clustered systems, and this may offset the savings from sharing subscriber radio equipment and other ancillary equipment. Indeed, since drop wire costs are commonly estimated at $1.00 per foot in the rural United States, it is clear that a cluster should probably be located within 1000 feet or so of all homes served by it in order to be cost-justified, since in our experience the savings from clustering the subscriber radio hardware, exclusive of the drop, rarely exceed $1000 per line (and usually less than half of that) over single-line applications. Of course, in rural areas outside the United States where the labor costs associated with drop wire are lower, this "cluster threshold" may be increased. (Although typically in such situations the use of very long drops is best avoided, to minimize the amount of plant exposed to scavenging.)

Note that the provision of two or more subscriber access lines to a single customer location is a special case of clustering, where the availability of subscriber transceiver equipment that can be shared is almost always advantageous (since there is no increase in drop wire facilities).

7.3.2 Modeling the Spatial Dimension II: The Problem of "Critical Mass"

Wireline plant tends to require a large "critical mass" of customers to justify the high fixed costs associated with feeder-cable facilities. For example, consider a housing development under construction about 5 miles from the nearest central office; it will ultimately contain 500 homes. Assume the cost of the feeder-cable is $250,000, fully loaded. Further, let us assume that the individual lots are being sold and built one by one, with an average drop cost from the feeder interface of $600 (see Figure 7.37). Yet at the end of year 1, only 20 tracts have been built and occupied. If the feeder is constructed now, the cost will be $12,500 per paying customer, plus another $600 for each drop, or over $13,000 per line. Since the telephone company's rates are predicated on its average cost per line of $1700 (an assumed number), the feeder is a money-losing proposition. Either the phone company builds it anyway, to fulfill its

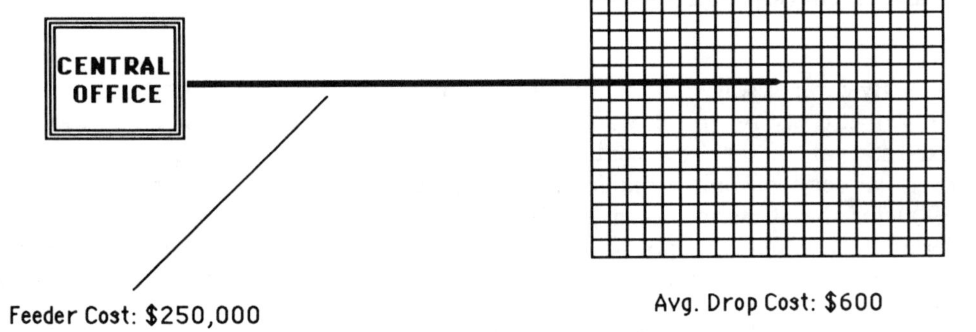

Figure 7.37 Serving of a new housing development.

commitment to universal service, and loses about $9 for every $1 of revenue—or it postpones building the feeder until it would be more fully loaded. When does it become a profitable project? A little arithmetic shows us that not until it is nearly half completed does the project go into the black. The project suffers from high fixed costs (see Figure 7.38).

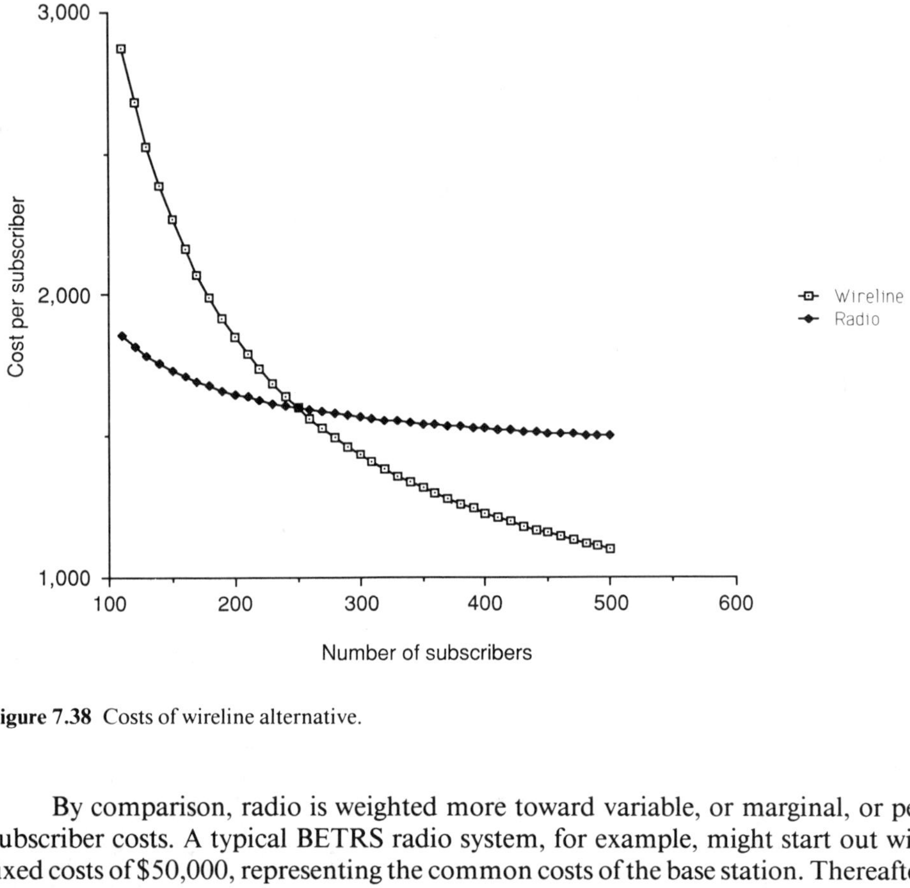

Figure 7.38 Costs of wireline alternative.

By comparison, radio is weighted more toward variable, or marginal, or per-subscriber costs. A typical BETRS radio system, for example, might start out with fixed costs of $50,000, representing the common costs of the base station. Thereafter, every subscriber would be added at a marginal cost of $1400, representing about $900 for the subscriber terminal, $100 for the ancillary equipment, and $400 for the traffic-sensitive portion of the base station. (All the numbers used here are hypothetical, but the approximate order of magnitude and cost structure is realistic.)

Under these assumptions, the radio cost curve is asymptotic to about $1500 per line, compared to an asymptote of about $1100 per line for the wireline alternative. However, until the housing development reaches its 250th subscriber, the cost per line for radio is less expensive than for the high fixed cost wireline solution. And the advantage of wireless access is especially pronounced for very low numbers of subscribers in the early stages of the project. For example, the cost of wireline for the first 10 subscribers is more than $25,000 per access line, compared to about $6400 for radio.

This brief example illustrates the importance of assessing accurately the fixed cost components of wireline and wireless alternatives. The actual recognition of these cost differentials will probably occur in the time dimension (see Section 7.3.5).

The critical mass problem of wireline plant crops up in many different guises. For example, whenever there is a question relating to the degree of "fill" or capacity utilization of a feeder-route (and housing developments are often problematic) this issue arises. It also arises when feeder upgrades are considered. In large measure, the rate base philosophy itself has been inspired by the critical mass issue, which is simply another way of saying that telephony has a utility-like cost structure.

7.3.3 Modeling the Spatial Dimension III: Consolidation of Feeders and Exchanges

7.3.3.1 Feeder Consolidation

The same example can also be used to illustrate the economic significance of the area coverage ability of radio-based access. Consider an exchange with four feeders identical to the one described in the previous section (see Figure 7.39). In the wireline model, each of the feeder-routes comprises its own economic universe, with its own critical mass problem. However, a radio solution that covers all four feeder-routes can collect its smaller critical mass four times faster. In our example, the wireline feeder solution crosses our hypothetical profitability threshold of $1700 at around 225 subscribers per route. The radio access system covering four cable-routes reaches the same cost per line at about 40 subscribers per route. Thus, we may find ourselves in a situation where radio cannot be proved against any single feeder line, but by using the power of radio to aggregate them into a serving area covering multiple feeder-routes, we can prove-in radio at lower densities. This is the economic basis of the RCSA described earlier in this chapter and also in Chapter 5.

The same principle can be extended even further. A single RCSA may cover multiple wireline exchanges, each with its own network of feeders (Figure 7.40). Now the critical mass for the radio system can be collected from an even wider area. Using the same figures as in the previous examples, we find that if we assume an RCSA covering four exchanges, each with four feeder-routes, *radio can reach the $1700-per-line threshold level with only 10 wireless access subscribers per feeder (i.e., per serving area).*

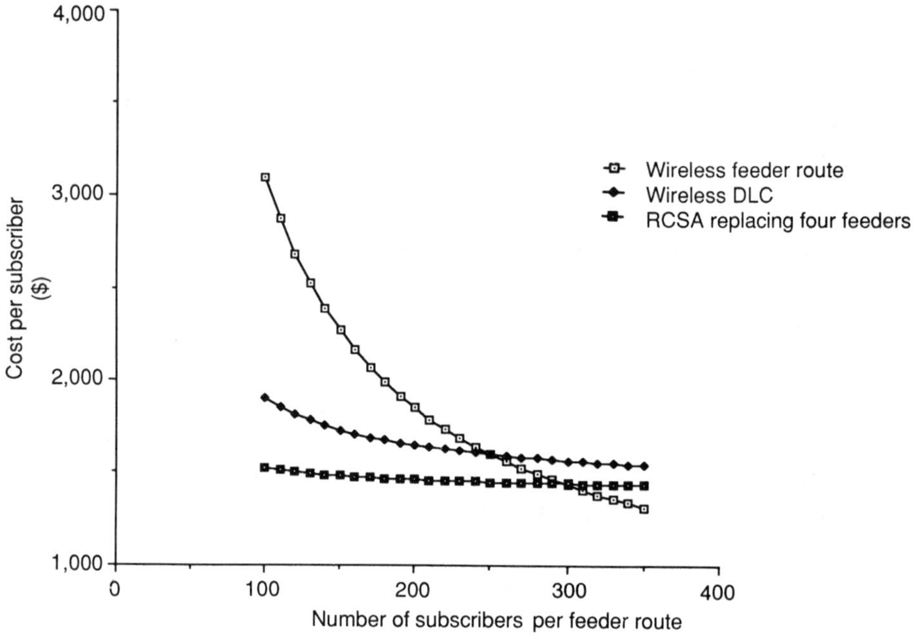

Figure 7.39 Multiple feeder routes *versus* radio carrier serving area.

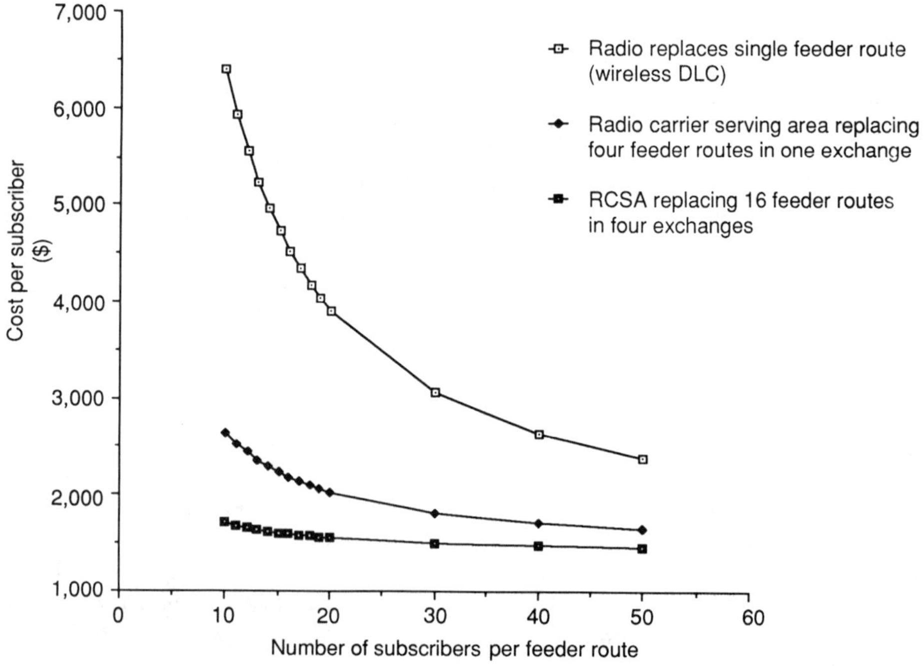

Figure 7.40 RCSA covering multiple exchanges.

7.3.3.2 *Radio "Safety Nets"*

Moreover, the telephone company planner has considerable latitude in selecting which subscribers to place on radio. This feature of wireless access may be particularly valuable when radio is being used as a "safety net" to capture customers who are "dropping off" a deteriorating cable network.

Consider, for example, a situation in which a deteriorating cable is losing pairs and has reached the point where there are just enough pairs to serve the remaining customers. If one more pair is lost, the entire cable will need to be replaced (or possibly DLC may be applicable; however, in such situations the cable is normally too deteriorated to support electronic pair-gain). With a radio safety net, individual subscribers can be selectively removed from the cable system and served by radio subscriber stations; meanwhile, the cable can stay in place to serve the remaining customers. The life of the cable plant is extended, which results in savings through deferrals of expenditures (see Section 7.3.5).

7.3.3.3 *Elimination of Exchanges*

Finally, the availability of wide-area wireless access may permit the elimination of some exchanges altogether, with the savings of all common equipment, overhead, and trunking costs for the switches taken out of service (Figure 7.26). There are a number of cost savings that can be realized from such consolidations. For example, maintenance is often reduced, especially if the exchanges that are eliminated are based on aging, electromechanical switches. A reduction in travel time to remote and normally unmanned central offices has also been cited as an important benefit [34]. It is often the case that these seemingly indirect savings are the decisive in the selection of wireless access.

7.3.4 Modeling the Spatial Dimension IV: Beyond Basic Services

The ability to use radio to "pick off" customers, singly or in whole cable-routes, over a wide area and essentially without regard to their spatial orientation to one another (as long as they lie within the RCSA), combined with radio's lower critical mass and its ability to collect that critical mass from a wide area, suggests that radio-based access can be used to introduce new advanced services in a much more flexible and economical way.

Consider a hypothetical plan to introduce basic rate ISDN services (144 kb/s) in a given area. Let us assume that the existing copper plant, including the feeder, is adequate for voice transmission but unable to handle the ISDN transmission. Therefore, a complete replacement of the feeder portion (and in many cases, the drop portion as well) will be required. Let us assume that a fiberoptic feeder system (e.g.,

SONET) is to be used (see Figure 7.41). The telephone company planner now confronts another critical mass type problem: He must find enough ISDN customers, conveniently located together in the same cable serving area, to support the cost of upgrading the feeder to ISDN capability. If, as is more likely, the potential ISDN customers are scattered around the exchange and their density on any given feeder-route is too low, the ISDN facilities will not be upgraded. An impasse has been reached. In fact, the only people likely to receive the new services any time soon are those living in the highest density areas such as major cities, in concentrated business/industrial zones, or major corporations who provide the critical mass in a single lump. Small businesses, most residential customers, and almost everybody living outside the largest metropolitan centers will be excluded from the information age party by the critical mass character of wireline economics.

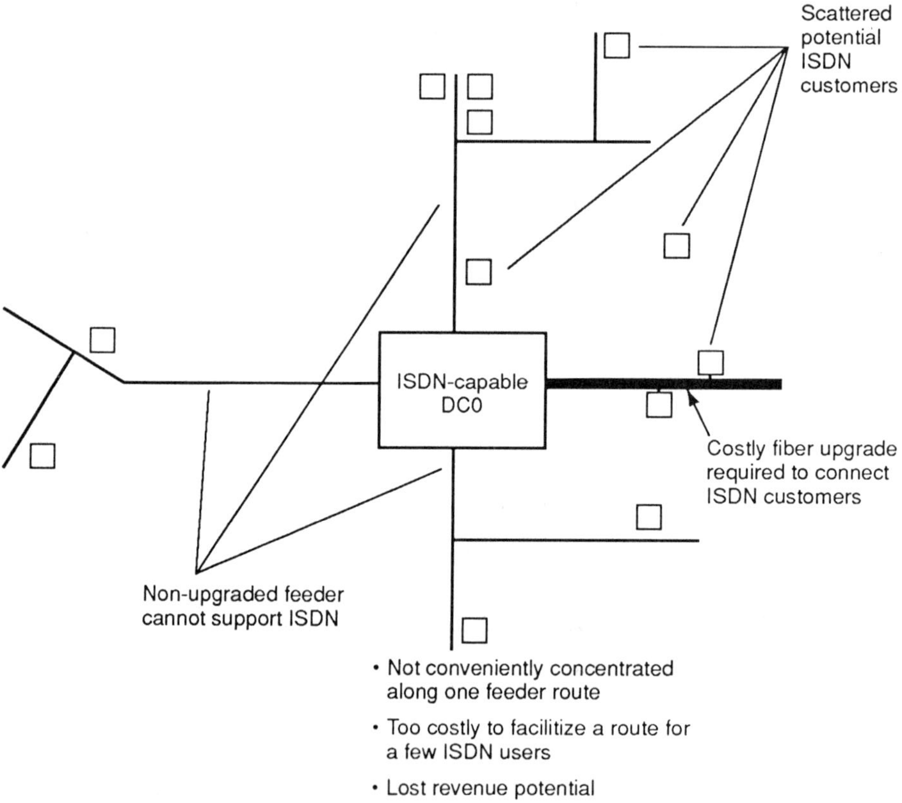

Figure 7.41 Fiber-based ISDN upgrade.

The reader should readily perceive how area coverage and RCSAs—using advanced digital radio—can transform this picture. Because radio has a smaller critical mass to start with, and because it can collect its critical mass from such a large area—encompassing potentially many exchanges—it can rapidly achieve a per-line cost that will allow the prove-in of radio-based ISDN to almost any area (see Figure 7.42).

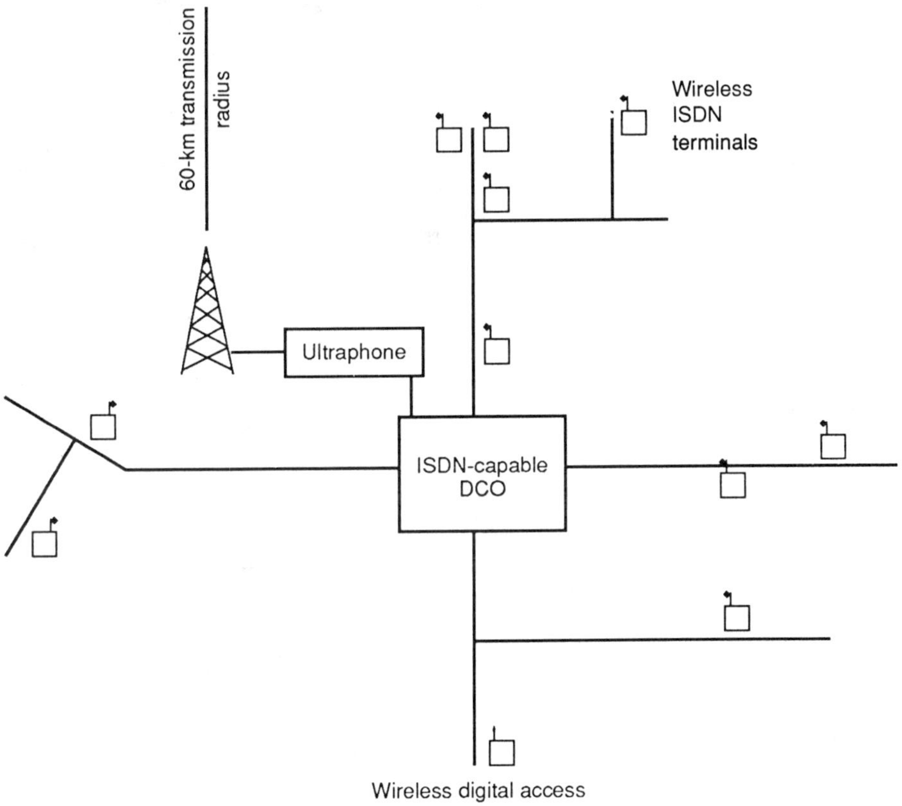

Figure 7.42 Wireless implementation of ISDN.

Moreover, the radio ISDN terminals (which might differ from the standard voice and low-data-rate terminals) represent a flexible, recoverable investment. If ISDN really catches on, and enough customers in a cable serving area do sign up, then the fiber facility can be emplaced and the radio ISDN terminals can be redeployed elsewhere. There are some limits to this scenario. For example, true broadband-ISDN and even more bandwidth-hungry services still on the drawing board for switched video, digital HDTV, and the like probably will not be easily handled on radio systems of the kind we are discussing here. However, data rates up to and beyond 144 kb/s and even up to T1 rates or slightly higher can probably be accommodated fairly readily.

The importance of this analysis for economic modeling is that a value should be attached to the ability of radio to handle new revenue-generating digital services. This is a plus that conventional cable solutions often cannot match.

7.3.5 Modeling the Time Dimension of Access Economics

There are a number of economic effects that can best be viewed as a function of the time dimension, and some of these have a very significant impact on cost comparisons between wireline and wireless access alternatives. We will examine four aspects of the time dimension of access costs:

1. Investment modularity and present value
2. Operating expenses (e.g., maintenance)
3. Inflation (and deflation)
4. Salvage value.

7.3.5.1 Modular Investments and Present Value

Radio-based access systems are inherently more modular than wireline systems, as a rule. That is, a greater proportion of the total cost is comprised of per-line cost elements (e.g., subscriber terminals and traffic-sensitive base station RF equipment) than shared or common cost elements (such as the cost of the 500-pair feeder-cable in our example). This can allow a closer matching of costs to service requirements and, in particular, a deferral of a larger portion of the total project cost until the later years of the project. If we combine this ability to defer costs with a standard *present value* concept, we can discount future investments relative to present investments. Thus, in principle, an alternative that requires me to invest \$1 million in year 1 is more expensive than a functionally equivalent alternative that lets me invest \$500,000 in year 1 and \$500,000 in year 2. The difference is roughly equal to the amount of money I can earn on that second \$500,000 by putting it in a bank for a year or, to be more precise, by investing it for a year in the most lucrative vehicle available that still provides me the liquidity when I need it. (Since the present value concept is so well known, we shall not review it here. The reader is referred to the standard Bell System references on engineering economy [35], or, for that matter, to almost any finance textbook.)

To examine the importance of this modularity, let us develop the example used in the previous sections. We again assume a 500-unit housing development. The wireline service alternative utilizes a 500-pair cable (we ignore the need for extra pairs as a margin or cushion, thus understating the wireline costs somewhat). The cost of the cable, which is incurred entirely in year 1, is $250,000. The cost of connecting each customer by drop wire to the serving area interface is an average of $600 and is incurred only in the year service is actually required. (Again, this distribution network is greatly simplified; for example, there are no distribution cables, only drop lines, which has the result of understating the wireline costs somewhat.)

The radio alternative requires an initial outlay of $50,000 for the common equipment at the base station (a year 1 expenditure). Subsequently, subscriber station costs and traffic-sensitive base station equipment costs (e.g., the RF channel elements)—which we have assumed to total $1400 per subscriber—are incurred in the year service is initiated for a given subscriber.

Let us look at three scenarios for the growth of the subscriber base.

Scenario 1: A front-loaded scenario in which half of the circuits are installed in the first two years, more than 80% by year 10 (the project midpoint), and the rest by year 20; this may be viewed as a fairly realistic scenario.

Scenario 2: 50 subscribers are installed every year for 10 years.

Scenario 3: 25 subscribers are installed every year for 20 years.

The capital required for the wireline alternative in all three scenarios is $550,000, compared with $750,000 for the radio alternative. The timing effects show up, however, when we apply an 8% discount rate to the three different investment streams. Figure 7.43 shows the effects of an 8% discount factor. The undiscounted premium for radio in this example shrinks from 36% to about 20% for scenarios 1 and 2, and about 3% for scenario 3 (the most attenuated deployment) (see Figure 7.44). Of course, the results in practice will vary with actual equipment costs and project characteristics. This simple example illustrates, however, that the greater the percentage of the project occurring in the later years of the project, the more advantageous the wireless access alternative appears to be.

The choice of the discount rate itself can also change the preferred alternative. Figure 7.45 shows the discounted cost of both the wireline and radio-based alternatives over a range of discount rates. Note that with a 10% discount rate (often recommended by telephone textbooks [36]) the radio alternative under scenario 3 of our example becomes less expensive than the wireline alternative.

7.3.5.2 Operating Expenses: Maintenance Costs

Another consideration that is important for the time dimension of the economic analysis is the possibility that different technological alternatives may give rise to dif-

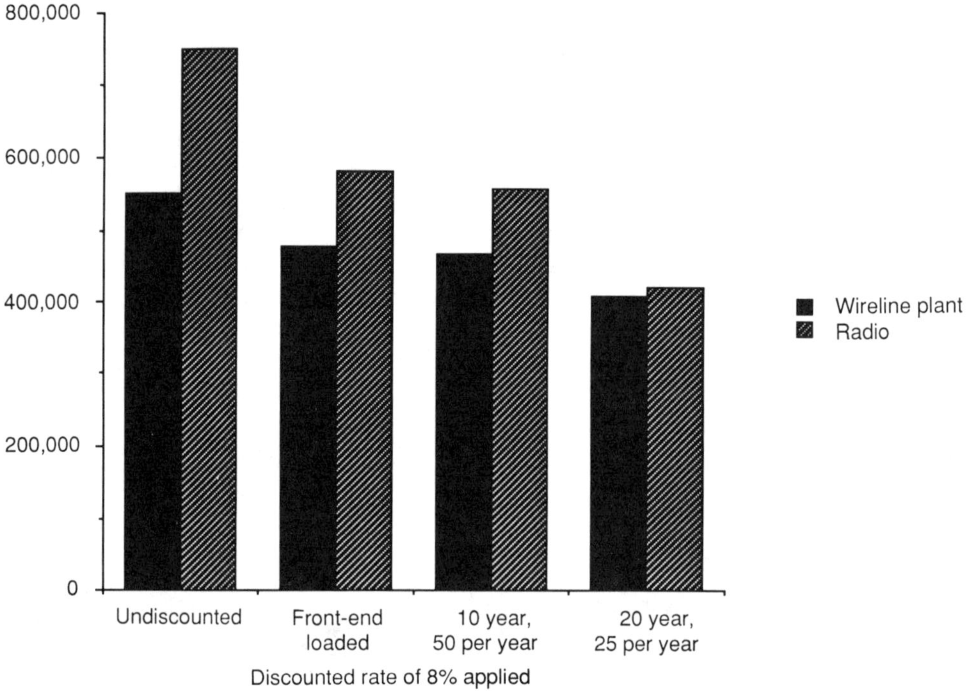

Figure 7.43 Wireline and radio access costs at various discount rates.

ferent levels of operating expenses. These are ongoing costs that are associated with operating and maintaining the access facilities.

One of the more important ongoing operating costs is *outside plant maintenance.* Despite the recognition of the importance of maintenance as a cost factor, and an extensive conceptual literature on the subject [37], it is difficult to obtain hard cost figures for maintenance associated with loop or access facilities. There is general agreement, however, that the loop plant generates a disproportionate amount of maintenance expenses [38]. In the world of rural wireline loops, I have seen a figure of 5% of first-cost quoted as an estimate of annual maintenance expense per access line. In other words, if the cost of the loop is $1700, the projected annual maintenance (at least in rural areas) is about $85. This certainly seems reasonable [39].

There is an intuitive feeling on the part of many of those who have worked with wireless access (especially BETRS) that maintenance for radio systems should be considerably lower [40]. Certainly, anecdotal information from quite a number of BETRS sites indicates a dramatic reduction in the number of complaints [41]. In one case, a company involved in a BETRS program estimated maintenance cost for radio loops of about 2.4% of first-cost per access line [42]. This was based on their experience with other types of radio systems, and was about half of the rate they normally used for wireline loops in rural areas. Experience with both BETRS [43] and microwave subscriber radio [44] shows that very high mean times between failures (MTBF) can be achieved, despite the tendency for such systems to be implemented in extremely hostile environments. Moreover, our experience with BETRS subscriber station equipment is that the mean time to repair (MTTR) is about 20 minutes in the worst case (i.e., complete replacement of the subscriber unit, exclusive of "windshield" time).

As experience is gained and quantified further with wireless access maintenance costs, this differential needs to be factored into the economic analysis. To get an idea of the importance of this factor, the example we have been using was adjusted to reflect a maintenance charge of 5% per year of embedded first cost for wireline alternatives, and 2.5% per year for radio. The result, not surprisingly, is to bring the cost of the two alternatives much closer; indeed, in scenario 3, radio becomes signif-

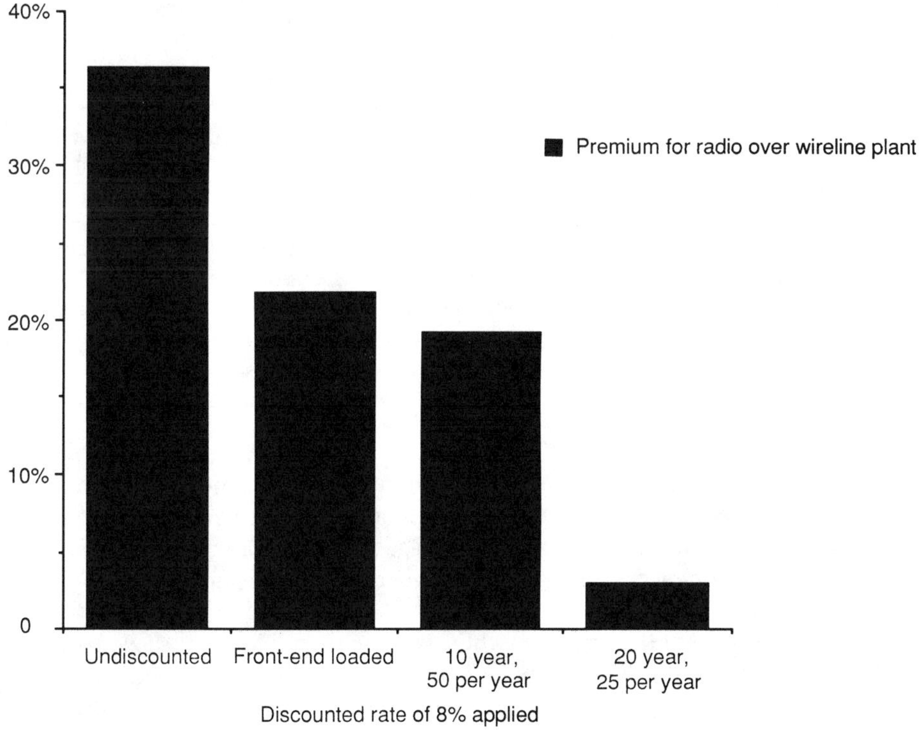

Figure 7.44 Premium for radio under various scenarios.

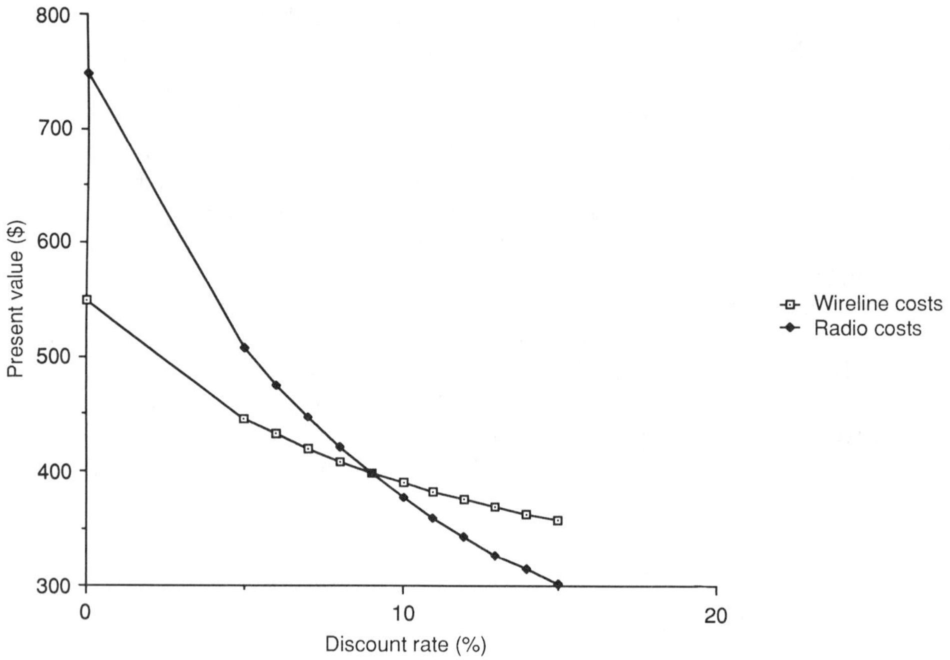

Figure 7.45 The effect of discount rate upon wireline and radio access costs.

icantly less expensive than wireline (see Figure 7.46; note that an 8% present value discount is applied to all scenarios). Moreover, *the fact that maintenance can amount to nearly 50% of the total life cycle costs* (for wireline) suggests how important it is for the analyst to get a handle on this cost component.

7.3.5.3 Inflation and Deflation Effects

Another time-dimensional effect is the projected rise in the unit rate of certain cost components. For example, the hourly rate for labor can be projected to increase over time. So can the cost of most bulk materials (e.g., the world price of copper has more than doubled in recent years). Thus, for a 20-year project, it is important to factor in inflation for certain cost elements.

It is also important to recognize, where we can, that different cost elements may exhibit different rates of inflation. Indeed, in some cases, we can confidently project a rate of *deflation.* For example, the cost of line cards in digital switches has fallen by more than 50% in the past decade. The cost of cellular radio terminals has dropped by at least two-thirds since 1984. Optical fiber transmission systems are projected to decrease in unit costs by 14% per year between 1985 and 1995 [45]. In general, we have come to expect that high technology equipment will fall in price, often very rapidly. I have seen many forecasts of equipment sales for the cellular industry, and there is not a single one that does not anticipate continued reductions in hardware prices (except for discontinuities in the trend where a new technology base, such as digital cellular, is first introduced).

This is a very important difference between copper-based wireline access, which is based on low-technology, commodity-type materials that may be expected to show rising prices, and wireless access, which is based on digital electronics that will undoubtedly show falling prices for a considerable period of time. In fact, the difference is likely to be so great that we should be careful not to understate it. In the past 18 years, for example, the cost of digital loop carrier hardware has fallen by a factor of about 15, while the cost of ordinary copper loops has risen by a factor of 3

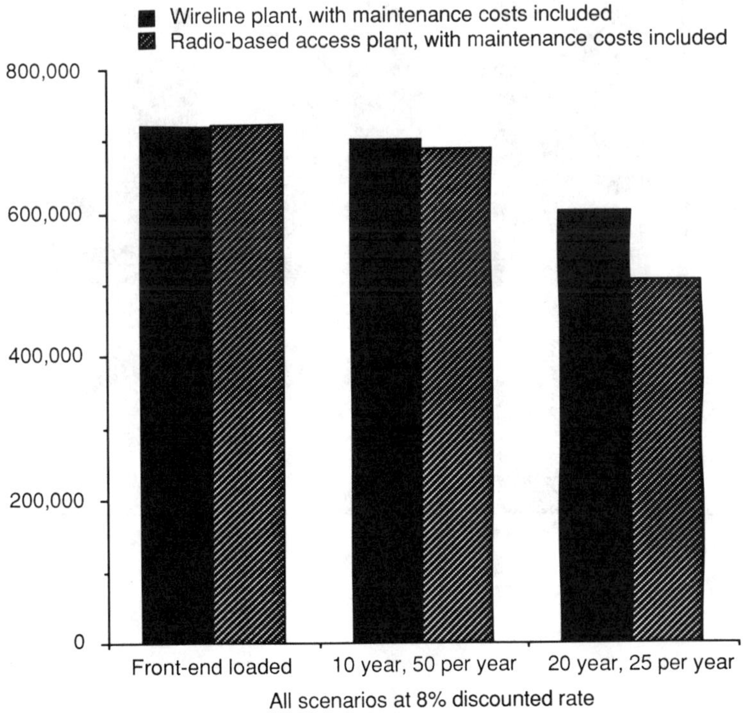

Figure 7.46 The effect of maintenance costs.

[46]. If we could peer into the crystal ball for the next 18 years, it would not be surprising to find a similar reduction in the cost of radio-based access hardware combined with a similar increase in copper costs.

Let us compare the three scenarios we have been analyzing when we now add the assumption that wireline per-line access costs (essentially, the drop wire materials and labor) will show a rate of inflation of 3% per year, and that radio-based access costs will show a rate of deflation of 3% per year. The results are shown in Figure 7.47 (which also combines the maintenance effects and an 8% present value discount). Note that in all cases examined radio is now the cheapest alternative.

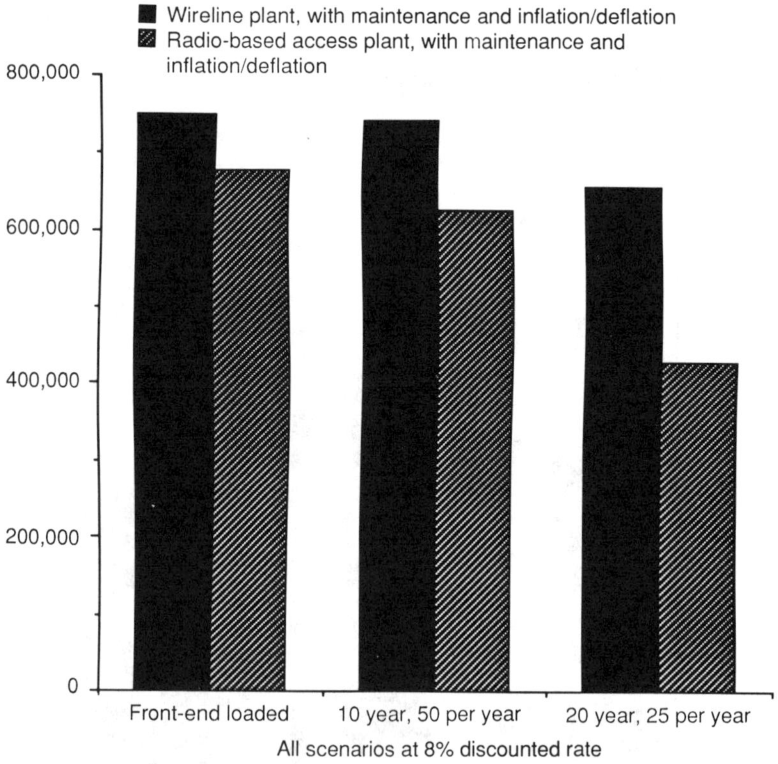

Figure 7.47 The effects of inflation/deflation.

The effects of timing variables on the economic analysis are important in absolute and relative terms (see Figure 7.48). In absolute terms, the effects of adding maintenance and inflation alone more than double the discounted cost of the wireline alternatives we have examined here. In relative terms, the time variables cause a transposition in our preferences: Radio becomes (for these cases) considerably more attractive for wire—for the entire project, it may be emphasized, not just for the initial phases as we examined in Section 7.3.2.

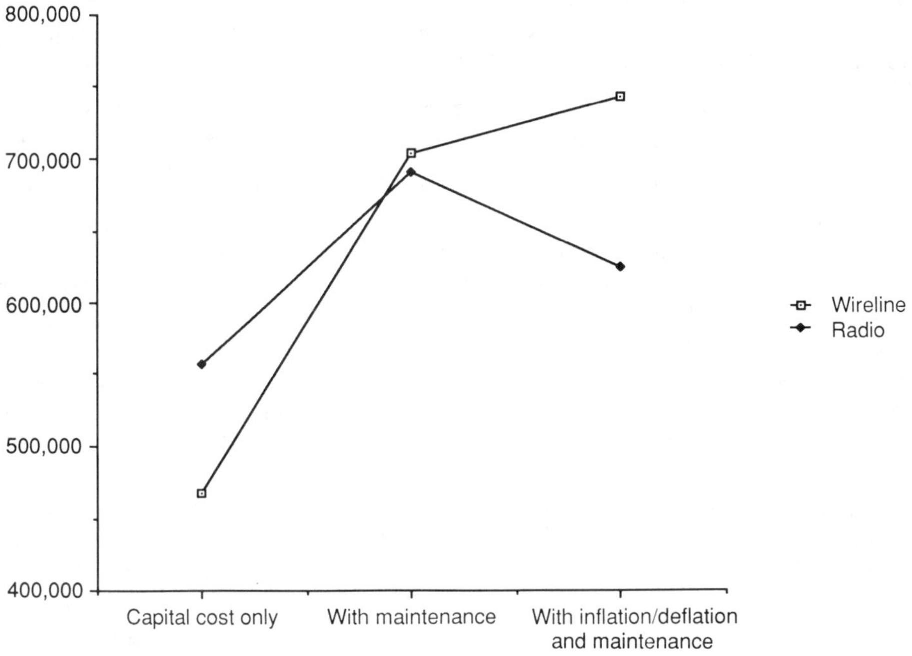

Figure 7.48 Summary of wireline *versus* radio access costs under various scenarios.

7.3.5.4 Recovery Value and Stranded Investment

Radio terminals, unlike wireline plant investments, usually have a substantial recovery value at any time prior to the actual end of their service life. With wire, the net value of abandoned plant is usually considered to be zero; that is, the salvage value of the copper plant is about equal to the cost of removing it [47]. With a radio system, if a subscriber is taken out of service for some reason, the subscriber radio terminal can usually be removed and redeployed [48].

In practice, some of the BETRS applications I have been involved with are motivated chiefly by this consideration. For example, in one of the fringe suburban areas of Los Angeles, a major independent telephone company has been using radio since 1987 to provide temporary service (up to two years) to homes being built beyond the reach of the cable infrastructure. As the "frontier" of construction moves out, the radio systems can be redeployed. Removal times are down to less than 20 minutes and reinstallation times are less than one hour [49].

The effect of adding back the recovered value of radio subscriber terminals taken out before the end of their service life and used elsewhere is to accentuate further the relative advantage of radio over wire. It is, in effect, an insurance policy against the possibility of a *stranded investment.* As such, the value of radio is augmented in situations where there is either:

1. a high degree of *uncertainty about growth* forecasts or
2. a likelihood that the number of access lines in an area may actually *decline.*

To look at the quantitative significance of radio as an antidote to stranded investment, let us consider two modifications of scenario 1 (the realistic scenario). In scenario 1B, we assume that the housing development grows to nearly 400 homes, and then a decline sets in and the population of lines in the feeder-route is affected by attrition and becomes only 250 lines at the end of the period. In this rather grim scenario, which might correspond to an "oil bust" in Wyoming, wireline investment is assumed to be stranded (i.e., zero value recovered), but radio terminals are removed and 50% of their deflated value is recovered. (Both modified scenarios here assume full effects of maintenance and inflation/deflation as described in the previous section.)

In scenario 1C, there is simply no growth after the first two years. Out of a potential 500 access lines, only 250 are placed in service and no more are added or subtracted after that. This scenario is probably more common, and represents either a gross planning error on the part of the telephone company planners who forecast the requirements for a 500 access line serving area, or perhaps a failed housing development.

The results are shown in Figure 7.49. Both scenarios exacerbate the disadvantage of wireline plant compared to radio. The percentage differences are shown in Figure 7.50. In the standard scenario 1 (with maintenance and inflation/deflation), the surplus cost for wireline is about 10% higher than radio-based access. In the growth-and-decline scenario, the difference rises to 17%, and in the no-growth scenario, wireline facilities cost almost 25% more than radio-based access facilities. (All figures assume the a present value discount factor of 8%; higher discounts would increase the percentage difference between wireline and radio.)

A recent article surveying the historical efforts of the telephone industry to plan the loop facilities notes that "one of the greatest difficulties in planning the installation of loop cables over the years has been the uncertainty of the timing and location of future demand for services" [50]. I believe that in the appropriate analytical framework, in which different contingency scenarios are assessed to show the sensitivity of the results of potential uncertainties in the forecast, radio will emerge as a strong alternative in a much wider range of applications.

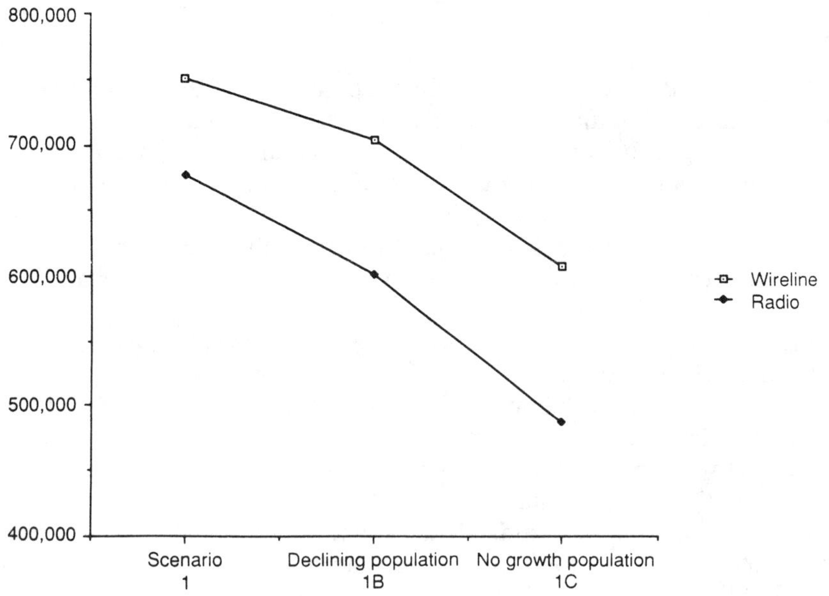

Figure 7.49 The effect of salvage value.

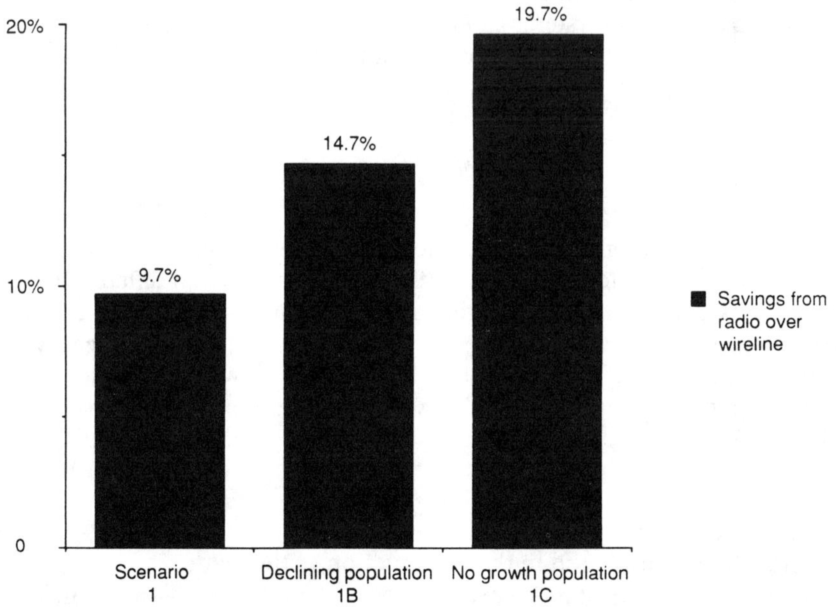

Figure 7.50 Percentage difference in wireline *versus* radio-based access.

7.3.6 "Intangibles:" Flexibility and Response Time

I recall having a discussion with one of the top planners at the headquarters of one of the regional Bell holding companies in which he advised me that when selling wireless access equipment to the operating divisions at the state and district level, a wireless system supplier should not allow itself to be "bullied" by the local planners on the issue of first cost, or even equipment cost *per se*. "The greatest savings that your equipment will bring to them is that it will allow them to avoid the horrendous overtime charges they have to pay today for a lot of the wireline installations they do."

It is true that running wire, setting poles, and burying cable involves a great deal of planning and logistical overhead, especially in light of the public's expectations for virtually instantaneous initiation or restoration of service on request [51]. To meet response time goals, telephone companies are often forced to spend well above their normal rates to pay for overtime and other charges.

I classify these as an "intangibles" not because they could not be quantified, but because normally it is so difficult to obtain the data that telephone company planners simply do not factor it into the specific costs for a given project. Instead it is buried in the surprisingly high overhead rates assessed against direct (measurable) costs for materials and labor.

In time, the proper analysis of wireless access economics will force telephone planners to deal more accurately with these intangibles—in large part because, once again, radio-based access systems should result in dramatic reductions in these overhead expenses. We will touch on a couple of these issues very briefly, although the absence of data makes it difficult to suggest modeling strategies.

7.3.6.1 Overhead

The overhead charge for a project is often applied as a fixed percentage on top of direct costs (like equipment and labor). It is handled somewhat differently by different companies. Under the FCC's Uniform System of Accounts [52], a number of "clearing accounts" are set up to handle *operating costs* that cannot easily be allocated to a single project. These include vehicle and other work equipment expenses, engineering, and plant supervision [53]. Also, when defining the first-cost of a project (which is normally capitalized), the AT&T handbook provides the following definition:

> The more important first-cost items [include]:
> Material: All material used in the construction of the plant . . .
> Installation: All direct labor costs, as well as motor vehicle and incidental expense.

Miscellaneous Loading: Supervision, tool expense, general expenses, social security taxes, and relief and pensions.

Engineering: All engineering time and associated costs. [54]

Other than materials (which is usually straightforward), all of these categories can present problems for planner seeking a fair comparison between wire and radio in loop applications. For example, in one BETRS project I was involved with, the telephone company planners insisted on assigning the *10 installation man-hours per subscriber,* regardless of whether it was a wire case or a radio case. Ten hours was felt to be accurate on the basis of historical experience for wire loops, but it was clearly an overestimate for radio loops (where many companies have found that installation takes about 1 hour per subscriber, unless the situation is very unusual). Since each of those man-hours, loaded, costs about $50 in the planner's model, the effect was to eliminate several hundred dollars of potential advantage for radio over wire in that application.

Planning and engineering overhead is also quite often overstated for radio as a result of reliance on wireline rules-of-thumb. It can be argued, in fact, that once a wireless access system is installed and operating, the marginal planning overhead for new subscribers is virtually zero. Assuming that the telephone company maintains a small inventory of subscriber terminals, a request for service by a customer can be met very rapidly and without any particular planning effort. (Consider for example how little planning is involved for a cellular carrier in adding a single new customer; quite often the carrier has to do nothing except enter the subscriber's authorization in the database.) Of course, there is considerable planning associated with the initial construction of a wireless access system, as well as with any significant expansion in coverage (although expansion of capacity within a cell by adding RF channels is quite straightforward). But the actual costs of such planning should be used, and new rules-of-thumb developed from them, rather than simply slathering the whole project with layers of wireline-type overhead charges (which can sometimes reach 50% of the total cost of the project).

7.3.6.2 Overtime

The issue of overtime raised by my Bell System friend is another valid point of difference between wire and radio in the loop. The telephone industry is used to having to fix things in the field, under all sorts of difficult conditions. A repairman may have to worry not only about the problem in the equipment, but about darkness, rain, lightning, hornets, ice, and so forth. And he has to perform his repairs (or sometimes new installations) under conditions of urgency, to meet the service outage objectives set by the industry (and often enforced in a global sense by the state regulators).

In contrast, the standard procedure for restoration of service in a BETRS-type system, for example, is to swap the troubled subscriber terminal for a good one taken from a service inventory. The bad unit is returned to a depot or engineering center (or to the manufacturer) for diagnosis and repair. Radio thus virtually eliminates all but the most basic maintenance or testing in the field. As mentioned above, the MTTR has been found to be about 20 minutes (not including windshield time).

The ability to restore (or initiate) service so rapidly, while minimizing environmental hazards, allows the telephone company to plan for a far larger percentage of repair within normal business hours. The advent of self-diagnosing radio terminals often means that the telephone company personnel are aware of a problem well before the customer and allows them to plan accordingly.

Automated loop testing, of course, is making inroads into the wireline environment as well. But the nature of intelligent radio terminals, combined with the ease and speed of restoring service, should allow for a significant reduction in the cost per trouble call (in addition to reducing the number of trouble calls).

7.3.7 Summary

Radio is often the least expensive access facility. But proving this is not always easy. It is fairly simple to analyze the economics in "virgin" installations (i.e., large areas where there is no existing plant of any kind); generally, first-cost considerations alone will suffice to justify radio-in-the-loop. But realizing the full economic benefit of the radio alternative in situations where there is already an extensive wireline access network in place requires a more supple and detailed analytical framework. In particular, we must draw out some of the hidden or implicit assumptions of the wireline plant model—assumptions as trivial as the idea that loop plant has no salvage value, or as profound as the realization that radio can eliminate the exchange boundaries altogether (see Section 7.5).

In this chapter we have touched on some of the subtleties of wireless access economics and in some cases suggested ways of thinking about them analytically. Building these suggestions into hard modeling techniques will occupy many researchers in the coming years; it is a task that is unavoidable. That today's modeling techniques are so rudimentary is perhaps best seen as one of the lesser symptoms of a regulated monopoly, and especially one that has been locked into a static access technology for 100 years. Why sweat the details of costing specific access configurations when it is all going to be plowed into the rate base anyway? Why be concerned with the arduous tabulation of specific costs for a specific subscriber or a specific service when everything will be averaged over millions of lines?

Competition and the emergence of technological and service alternatives will change all this. The methodology of access economics is bound to undergo its own revolution, echoing the technology revolution represented by both digital radio and fiber optics. Modeling economic alternatives will become a more precise and a more painstaking endeavour—and this will call forth more sophisticated tools for automating large segments of the process.

Let us summarize some of the more important concepts we have touched on in this chapter. The major differences in cost structure that wireless access introduces include:

1. Improving the inherent economic efficiency of the loop plant
2. Increasing the economic (as opposed to spatial or temporal) flexibility of the loop plant
3. Increased freedom from spatial constraints in meeting service requirements
4. Increased freedom from timing constraints in meeting service requirements
5. Ease of actual deployment
6. Options for redundant and overlay networks.

7.3.7.1 Greater Economic Efficiency

As we have stressed in Chapter 2 and elsewhere, the fundamental diseconomy of the access plant is the extremely poor efficiency of loop circuits (measured variously, e.g., traffic per dollar of investment) compared to the circuits in the core network. This is true even for access networks that have been fully optimized for the spatial and temporal distribution of subscribers. This inefficiency is rooted in the dedicated copper loop and the star network configuration. It is exacerbated by conventional wireline practices that almost always overbuild in the loop plant, resulting in a significant idle (i.e., completely unused) investment.

Radio changes the balance of shared and dedicated facilities. Compared to standard wire loops (100% dedicated facilities), radio loops may have anywhere from 30 to 90% of the facilities costs shared by many customers. The capacity utilization rises from 1% or so for a typical wireline loop to 30 to 70% for the shared portions of wireless access circuits (especially the RF channel equipment) [55].

Radio also virtually eliminates the need for overbuilding in the access network. The low marginal cost of adding new subscribers and the rapidity of service provisioning means that the standard practice of burying a large number of spare pairs in a cable facility or doubling up on drop lines to anticipate possible future two-line-per-household requirements can be avoided.

The trend toward sharing a larger portion of the access network is making itself felt in the planning for fiber-based access systems, as well. In some fiber-to-the-curb projects, the fiber ring is shared all the way to the drop-line interface. Certainly, the use of pair-gain systems (both electrical and optical) in the feeder portion of the network also takes advantage of the opportunity to improve efficiency through sharing of facilities. In this respect, *the use of radio is simply part of the trend toward radically more efficient loop networks.*

One important implication, which I would like to reemphasize, is that since trunking will now occur in the loop itself, telephone planners must be prepared to conduct traffic analysis and size the access system based on the statistical methods already employed in most plant decisions in the core network. Unfortunately, I am aware of a number of telephone companies that still adhere to the doctrine of "no concentration" in the loop. Indeed, "concentration" has little meaning for star networks based on copper facilities. There is no opportunity for concentration in such networks, at least until loop electronics come in, and even with pair-gain, the cost of overbuilding the electronics to engineer for zero concentration is a small increment to the initial investment [56].

7.3.7.2 Greater Economic Flexibility

Economic flexibility may be defined (very generally) as the ease with which the system operator can match his investment program and his cost structure to the service requirements of his subscriber population. Since subscriber requirements tend (for purposes of this discussion) to change in small increments, economic flexibility is related to the "lumpiness" of the capital requirements imposed by the particular technology in question. For example, an electric utility that relies entirely on nuclear generators for power production has less economic flexibility (because of the enormous cost of building each new generating plant) than another electric company that relies on a larger number of smaller and individually less costly coal- or oil-fired generators. A third utility that utilizes cogeneration, conservation incentives, and small hydroelectric plants to supplement its main generating stations may have even more economic flexibility.

Wireless access should offer greater economic flexibility in many applications. In most cases, where two alternatives—wireline and wireless—have the same total cost, a higher proportion of the costs of the wireline alternative will be upfront costs. Laying a cable or wiring a building is a difficult and expensive step that must be taken before a single customer can be connected. By comparison, a radio-based access system usually has a smaller cost for the initial common equipment associated with the base station, and a lower cost for the infrastructure associated with expansions in system capacity.

It is also difficult in most cases to take out wireline plant (at least to do so and realize any savings) if subscriber requirements decline or change for whatever reason. In contrast, radio can be taken out of service in one area and redeployed to another area much more easily. The ability to match system capacity (roughly corresponding to dollars invested) more closely with changes in demand is another indicator of the greater economic flexibility of radio (see Figure 7.51).

It is interesting that the initial deployment of DLC technology in the 1970s was premised in part on the increased flexibility it accorded the telephone company planner. As one recent account has noted:

> A common practice during this period was to use DLC systems as a temporary relief measure on long-loop cable routes that had relatively slow growth rates.

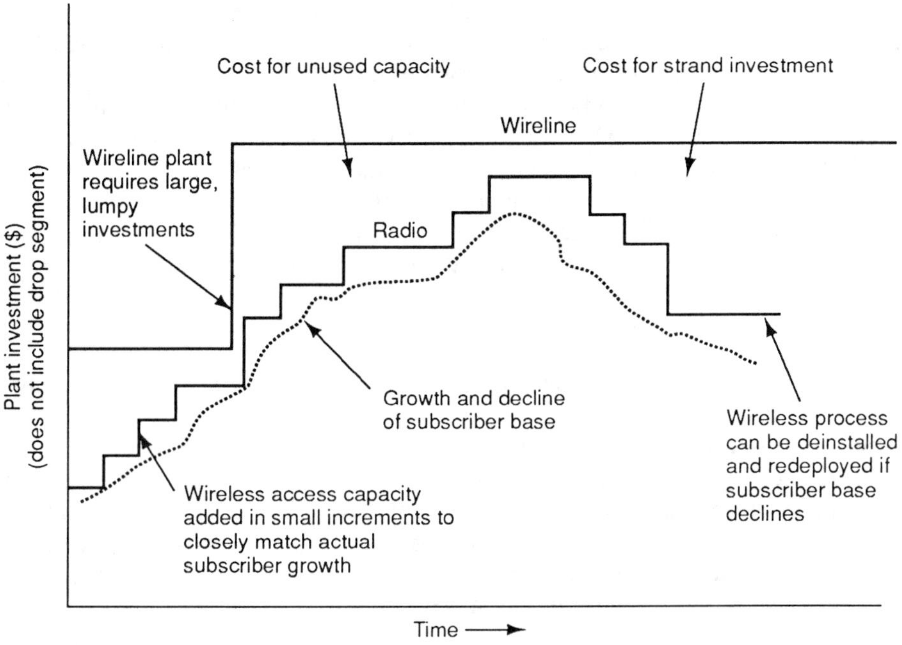

Figure 7.51 Relative "lumpiness" of wireline and wireless access investments.

> This was made possible by placing a 40-line system in a nearly full cable to gain the equivalent of about 36 additional pairs of wires. . . . If the growth along the route due to new demand and requirements for upgrading party lines was less than six lines per year, it would be possible to defer construction of a new cable for more than six years, at which time the DLC systems could be removed and used elsewhere. If the cable route was long enough, the savings achieved through such a deferral more than compensated for the cost of installing, maintaining, and later removing the DLC system when growth in demand justified the addition of new cable. The approach also proved to be a hedge against an uncertain forecast of demand. . . . [57]

Thus, we may expect that the flexibility of radio will generate significant savings for today's operators.

7.3.7.3 Freedom in Space: Area Coverage versus Cable Route

We have stressed the many ways in which radio opens up the planning "game" (in the game theory sense) from a one-dimensional, traveling-salesman type problem to an area-coverage problem in two dimensions. In a practical sense, I believe this is the most striking difference about radio-based access for planners accustomed to the many spatial constraints of planning for wireline plant. It is true that a radio access plan must develop a good radio propagation model to determine the area coverage for the system (see Chapter 8). Once this coverage area—the RCSA—is known, however, the planner need not be concerned about where the subscriber is located within that area, the distance of the subscriber from the base station, nor the relative position of one subscriber with respect to other subscribers [58]. Indeed, for many applications, it does not matter (for economic planning purposes) whether a particular subscriber is stationary or mobile.

7.3.7.4 Freedom in the Time Dimension: Life Cycle Economics

We have also stressed the importance (in our examples in Section 7.3.5) of the fact that radio accords us freedom in the time dimension. Investments can be matched in time much more closely to service requirements. Life cycle costs (including field service, maintenance, and repair) should be lower. The planning overhead associated with the need for accurate forecasts of demand many years into the future is greatly diminished by the nature of radio-based access plant.

7.3.7.5 Flexible Deployment of Wireless Access

Radio is more easily deployed than wire. In principle, a wireless loop (whether long-range BETRS-type or short-range PCN-type) should be as easy to install as a cellular phone. This allows the operator to plan to meet emergency, temporary, seasonal, or backup requirements without the cost and overhead of wireline plant.

7.3.7.6 Redundant Networks

Radio is well suited to providing overlays for new services that are complementary to existing wireline plant. For example, if enhanced digital services are to be offered in an area currently served by older electromechanical switches, these services can be made available initially through a radio overlay network from a single regional digital switch, which allows for such services to be offered to interested customers much sooner than if all the switches in the area must be replaced (see Figure 7.52).

This example suggests the even more general concept of multiple-access networks that can, in effect, serve as redundant backups to one another. Today, the telecommunications user enjoys redundancy in almost every segment of the network except the loop. There are multiple links and heterogeneous facilities in the inter-exchange network that mean that vital service can be restored (often almost instan-

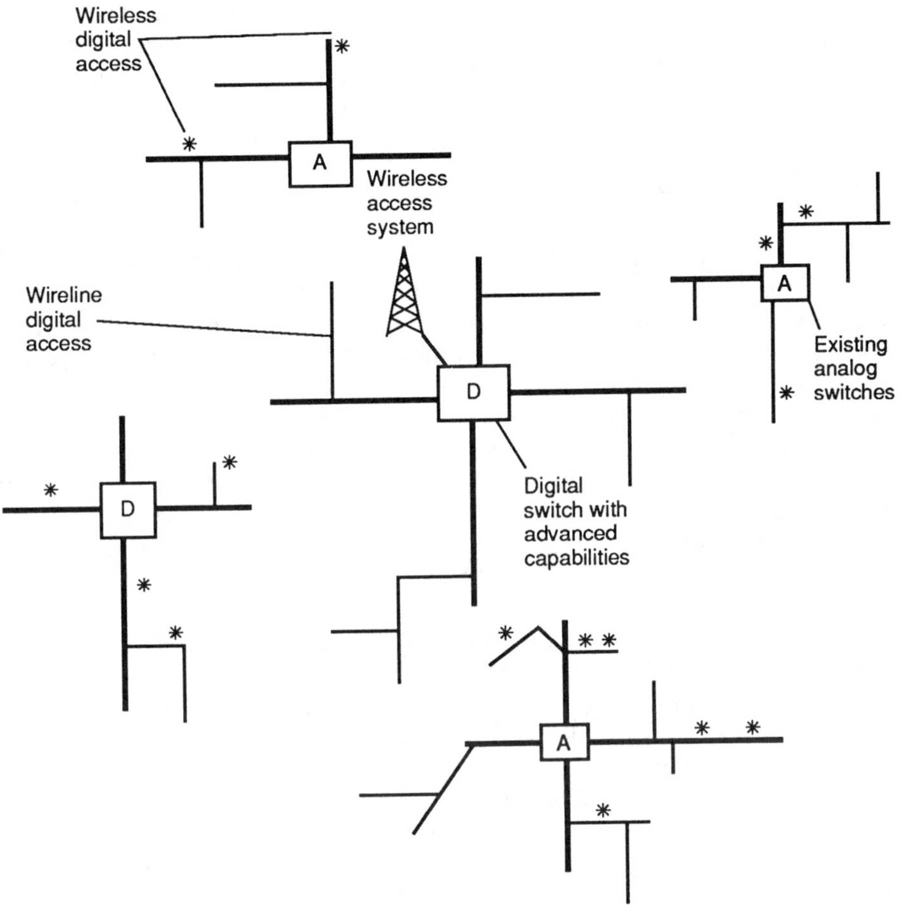

Figure 7.52 Radio overlay for delivery of advanced digital services.

taneously) in the event of an outage. But the last mile of the access circuit is still the nonredundant weak link for most networks. The use of radio as a safety net has already been described, mainly with reference to voice services, in Section 7.3.3.2, where we were concerned with the ability of such safety nets to allow operators to avoid, postpone, or optimize the scheduling of painful plant overhauls necessitated by the slow or catastrophic deterioration of wireline plant. Such safety nets may be of even greater interest to customers who need to ensure high circuit reliability end-to-end and would use radio as a backup capability for vital links.

These examples also indicate, however, just how difficult it may be to quantify fully (and thus model) all of radio's economic advantages in the loop environment. Undoubtedly, our models will improve by increments, and this brief survey would seem to indicate that the process will take some time.

7.4 PLANNING A WIRELESS ACCESS PROGRAM

If we accept the likely premise that radio is not the access technology of choice for every single subscriber—in fact, that many different access technologies will coexist and complement one another in the public network for a considerable period of time—then we face a very complex planning process. Indeed, up until this point we have discussed the "wireline alternative" and the "radio alternative" as though they are always distinct and exhaustive. In fact, in almost every setting I have analyzed, whether it was a virgin application or one involving considerable interaction with existing plant, there was actually a range of alternatives, which might best be viewed as different mixes of radio and cable. More recently, even fiber is beginning to enter the equation as a loop technology. So the real question faced by planners is not "Which is cheaper, radio or wire?" but "What is the most cost-effective mix of radio, cable, and fiber in the access network for a given situation?"

The complexity of this question can overwhelm even a fully computerized modeling process. In principle, there are dozens, or hundreds, of solutions (in the mathematical sense of the word) for any given exchange. And when we realize that radio easily transcends the traditional exchange concept (see Section 7.5), the solution set multiplies further. If we look at what may seem like a reasonable midlevel planning question (i.e., one that is between the tactical and the grand strategic), such as "How should wireless access be applied in the state of Pennsylvania?", the set of considerations and possible configurations is too large to handle directly. (At least that has been my experience working with telephone company planners at this still-early stage of wireless access deployment.)

In this section we examine briefly some of the steps that can be taken to help cut the planning problem down to size and get the process started. We shall look at three basic stages in the planning process:

1. *Screening* of individual applications (using the wireline exchange as the current unit of analysis)
2. Analyzing the *radio/cable mix* fr a given exchange
3. Looking outward from the selected exchange at *overlay* possibilities.

7.4.1 Site Screening: Indices of Wireless Access Requirements

We shall begin by taking (essentially at face value) the wireline exchange as the unit of analysis for planning purposes. The exchange represents the ubiquitous administrative unit in today's telephone network, and it is based on an *area* concept that is at least roughly compatible with wireless access applications principles discussed in this chapter and in Chapter 5.

The initial problem facing the telephone planner is thus to sort individual exchanges into two categories: those for which wireless access is likely to be a significant, or even dominant, element in the access network, and those for which wireless access is likely to play a more minor role. The boundary between these two categories is, of course, quite fluid as technology improves (mainly on the radio side), as equipment costs change, and as the existing wireline plant ages. The initial sort is not intended to be a hard categorization, nor a permanent one. It is simply a "first cut" procedure to reduce the overall analytical program to a manageable size. The specific target of the first sort may be to select only the 10% of exchanges where wireless access appears to offer the greatest benefits (or 5%, or 25%). These best case applications can then be addressed in the first stage of the program, while later applications (where the prove-in may be more difficult, or may require more detailed analysis) can be addressed later.

This screening process should be designed so that it can be carried out rapidly, without a great deal of expensive data collection, by planners who do not necessarily have (and as a practical matter generally do not have) specific knowledge of the individual exchanges. Since the major operators in the United States today often encompass hundreds or thousands of exchanges, it is imperative that the first step be capable of being carried out without detailed site-specific information.

The approach that I have found most useful in this regard is the development of certain wireless access *indices.* These are simple criteria that can be applied singly or weighted as elements of a combined general index of the likely requirement for wireless access in a given exchange. The indices are predictive; that is, they predict that a given exchange will probably be a good candidate (or not) for wireless access applications. These predictions must be verified, however, by site-specific analysis (see next section) before it can be said with certainty that a particular exchange is a good wireless access target.

7.4.1.1 Individual Indices for Wireless Access

Individual measures may be used as indices for the likelihood of finding a good wireless access application. The specific quantitative thresholds can be set based on company norms (e.g., the exchanges in the system in the highest quartile for average loop length), or on external or industry standards (e.g., exchanges rank-ordered in terms of the percentage of loops over 18 kilofeet in length). Considerable work still needs to be done to fine-tune most of these measures.

Among the indices that I have found to be useful are the following:

1. *Loop length:* This is one of the clearest indicators, if the data are available (which is not always the case). Exchanges with more than the average number

of long loops will tend to be found in rural areas, and will tend to have high loop costs (especially for replacement).

2. *Population density:* If loop length is not available, population density (or better yet, subscriber density) per square mile is a very good surrogate that is readily available. In one study it was concluded that wherever subscriber density was less than 25 per square mile, there was a high likelihood of finding a good wireless access application [59]. In my experience, this is probably a good number for screening purposes, but it should be recognized that there are many viable wireless access applications in exchanges with higher densities, where other wireless access indicators may be strong.
3. *High growth rate:* Areas of high growth in the base of access lines will often be good candidates, either for temporary wireless access (as in the Victorville, California, example described previously) or for permanent wireless access. High growth rates increase the likelihood of finding large forecasting uncertainties, which can lead to overbuilding and stranded wireline plant.
4. *Declining population:* It may seem paradoxical, but areas of population decline may sometimes be good wireless access targets as well. The problem of stranded investment is likely to be more acute in such areas, and this may sensitize local managers to the need to use plant that is less vulnerable to this result for those new applications that do occur. For example, the replacement of a feeder-cable by a radio system may make more sense in an area where further population declines are expected (which would produce more stranded investment in the future) than in an exchange where the population is expected to remain stable.
5. *High "churn" in the base:* "Churn" is often defined as any sort of turnover in the base, including new requests for service and terminations of existing service. In principle, an exchange may have a fairly stable average population and yet experience a high rate of churn as many people move in and out of the area.
6. *Large seasonal population shifts:* An access line in a summer resort area may earn revenue for only a few months of the year. A wireline loop in such a situation may be viewed as virtually a semi-stranded investment, since its cost is the same as a full-service line but much of the time it lies idle and unproductive. A radio loop can in principle be redeployed from a summer resort region to a winter resort region (although admittedly not all operators may experience optimally complementary applications). Nevertheless, the existence of a high percentage of seasonal lines is a fairly strong indicator.
7. *High percentage of party lines:* Of the several million party lines still existing in North America (as of 1991), most are found in rural areas, which are conducive to BETRS-type wireless access applications. Moreover, the remaining multiparty lines are the residue of decades of single-party conversion programs, and we can generally assume that they represent the highest cost,

most intractable segment of that original population. Thus, an exchange with a lot of party lines is probably a good candidate for wireless access applications.

8. *High maintenance costs per loop:* The advantages of radio in reducing maintenance and operating expenses indicates that exchanges with unusually high per-line maintenance costs are often good candidates for radio. These data may not be easy to obtain, however.
9. *Trouble reports:* A related measure is the number or rate of trouble reports, or service outages (variously defined), for a given exchange. Since exchanges with deteriorating cable plant are often ripe for wireless access applications, a higher-than-average rate of customer problems can be a good indicator.
10. *Age of plant:* If the data can be developed at this level, plant age may be another good indicator, for the same reasons.
11. *Held orders:* A high number of held orders is probably indicative of some fundamental difficulties in serving the area with conventional plant.

7.4.1.2 Composite or Weighted Indices

Any one index used to screen the total population of exchanges with a bias toward a particular type of application. *Loop length* and *subscriber density,* for example, will tend to select rural BETRS-type applications. The age of a plant or maintenance costs will generate many urban candidates as well (such as the "brownstone" example cited in Chapter 5). Seasonal demand will select for a different set of exchanges located primarily in resort areas. Since there are so many different applications, the best approach may be to use a composite measure in which the individual dimensions are scored and weighted in some appropriate manner.

7.4.2 Single-Site Analysis: Evaluating the Radio/Cable Mix

Once a particular exchange has been identified as a likely candidate for a wireless access application, it becomes necessary to examine the specifics of that site in some detail to determine whether the prediction of the index is correct, and whether there is some peculiar factor that may limit the ability to use wireless systems in that area. This more detailed evaluation involves (1) analyzing any unusual constraints that may exist in that area and (2) assessing the status of the existing plant in greater detail. Assuming that neither of these factors eliminates the exchange from further consideration, the next step—and probably the most important single step in the whole process—is to evaluate the range of radio/cable mixes to as fine a level of detail as possible.

7.4.2.1 External Constraints on the Use of Radio

There are two main external constraints that may limit or even prohibit the use of radio-based access in a particular exchange, even when the economics are favorable:

1. *Frequency availability:* There are always limits on the availability of the radio spectrum. For example, the BETRS allocation in the United States today encompasses a total of 94 radio channels, with most unavailable in urban areas. Thus, a wireless application in a dense urban setting is likely to be constrained by spectrum availability today (1991). Future rule-makings will undoubtedly improve this situation, probably in a dramatic fashion. Nevertheless, frequency limitations are always an important consideration.
2. *Terrain difficulties:* As we shall discuss in Chapter 8, terrain and the limits it imposes on radio propagation and hence on radio coverage are the primary engineering considerations in determining engineering (as opposed to economic) feasibility. In most cases, at this early stage in the analysis, it is not necessary to examine a detailed propagation model (although as these models come on line more and more they will be brought into the process at an earlier stage). However, if the site is located in a particularly mountainous region, it may be very desirable to perform such a detailed propagation analysis and compare the expected coverage area with the location of known service requirements. This may result in the elimination or downgrading of the radio option for all or part of the subscriber population in the target exchange.

7.4.2.2 Assessing the Existing Plant

Based on our comments in previous sections of this chapter, it should be clear that another important potential constraining factor is the status of the existing wireline plant. One of the most important characteristics is the age of the plant. If the outside plant is brand new and undepreciated, the prospect of a large writedown may offset the savings to be realized from the use of radio. If the feeder-routes have all recently been upgraded to fiber optics, it may be difficult to justify removal of these facilities.

The specific location of feeder facilities is also important, as well as the existence of unused capacity already in place. For example, if the exchange is experiencing rapid growth in the base of access lines, it makes a considerable difference whether this growth is taking place within existing (wireline) carrier serving areas or outside of them. In other words, if growth can be accommodated by deriving new pairs electronically from feeder facilities in good condition, supplemented by relatively short drop lines from the serving area interfaces, then it may be hard to prove-in a radio solution. On the other hand, a feeder in bad condition, without excess capacity, coupled with a need for longer drop lines or new distribution cable to reach outside the carrier serving area—any of these may augur well for a radio-based access alternative.

7.4.2.3 Optimizing the Radio/Cable Mix

After a given exchange has been qualified as a candidate for wireless access applications, the next step is to examine the range of possible cable/radio configurations. The important point here is that in a situation where there is existing plant in place that is not totally deteriorated, *the best option is usually to use radio for some customers and to continue to use cable for others.*

This raises the question: Which customers should be placed on radio? The logical answer is that we should find those whose conversion to wireless access will yield the greatest savings. We might look first at the highest cost customers as the candidates for conversion and generally this would lead us in the right direction. However, the cost for one customer may be affected by the alternative chosen for another. For example—and this is the most common case—by placing one customer on radio, thereby removing him from an ailing feeder, we may be able to free pairs to serve other customers without a complete feeder replacement. The more general question is: What configuration results in the lowest overall access costs?

Let us consider a hypothetical rural wire center with several deteriorating feeder-routes and several hundred current customers, plus a variety of projected requirements (party-line upgrades, possible growth, held orders, second-line requirements, etc.). Basically, for existing customers there are four options:

1. Serve them with radio—in which case they are of course removed from the wireline network.
2. Regrade or replace the feeder to provide/continue service over new dedicated pairs.
3. Serve them with dedicated pairs remaining on the feeder after other customers have been put on radio-based access.
4. Serve them with electronically derived pairs.

Of course, for new customers or party-line upgrades we face the same choices, but with the added requirement for constructing new drop-line facilities. If we add in the various critical mass type constraints for each system (i.e., pair-gain, feeder, or radio), it is clear that solving this problem for even a small wire center is probably computationally very difficult.

We must fall back on heuristic procedures and approximations. One approach I find attractive was invented by Paul Clements [60]. Essentially, it assumes that the best candidates for conversion to radio will be those customers with the longest loops, at the far end of the feeder-route, and it develops a simple, iterative computer model for finding an approximately optimum mix of radio and cable. The basic steps are as follows.

First, the costs are calculated for the two polar options: the all-cable solution (usually with pair-gain assumed to be exploited) and the all-radio solution. These become the two vertical axes of a graph (Figure 7.53). The next step normally begins with the all-cable option; a small number of the most distant customers (usually 10 or so) are removed from the cable system and assumed to be placed on a radio access network. The total cost (i.e., for the cable and radio systems together) is calculated for this option and plotted on the chart. Then another small group of customers are removed from the cable, placed on radio, and the cost is calculated, and so forth. Normally, a U-shaped curve develops as shown in Figure 7.53. The lowest cost option is usually neither the all-cable option nor the all-radio option, but a mixed scenario.

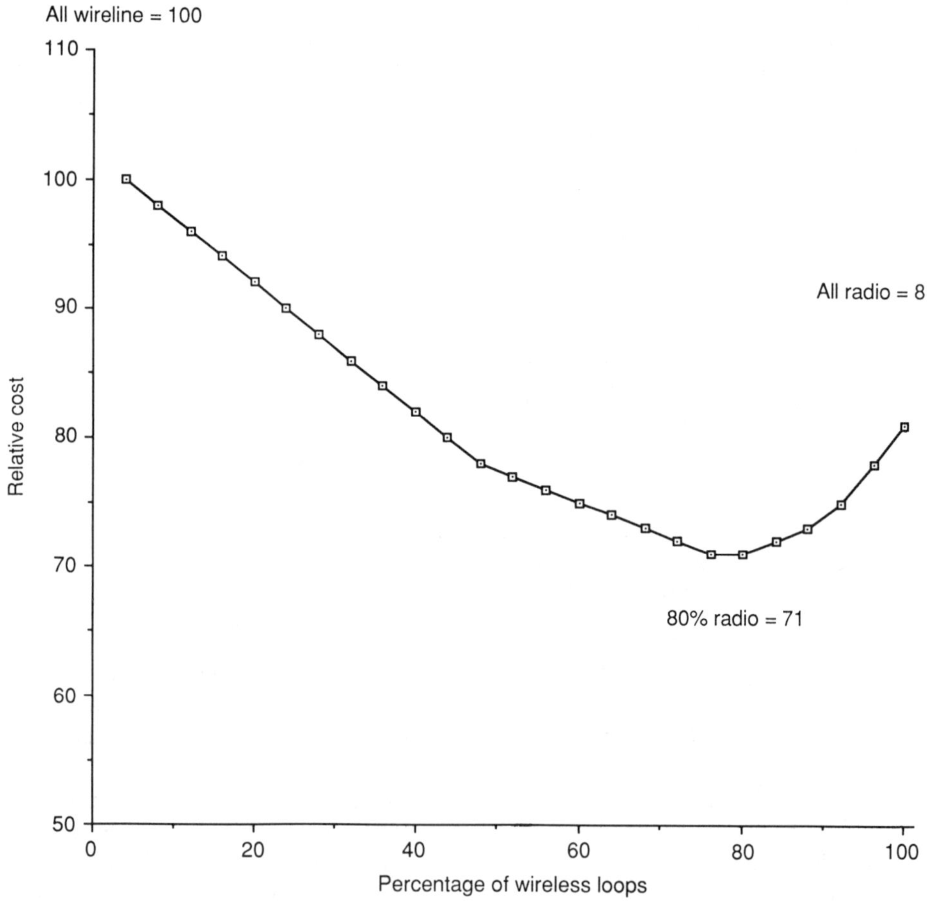

Figure 7.53 Radio/cable alternatives.

There are different ways to manipulate this kind of model. For example, instead of picking the most distant customers first for radio conversion, we can select those belonging to some class of applications, such as those requiring new drop wire (e.g., party-line upgrade, held orders). Several different curves may be plotted (see Figure 7.54). These analyses do not require excessive computation, and can easily be handled on a microcomputer. It is not too difficult to develop an interactive capability along these lines.

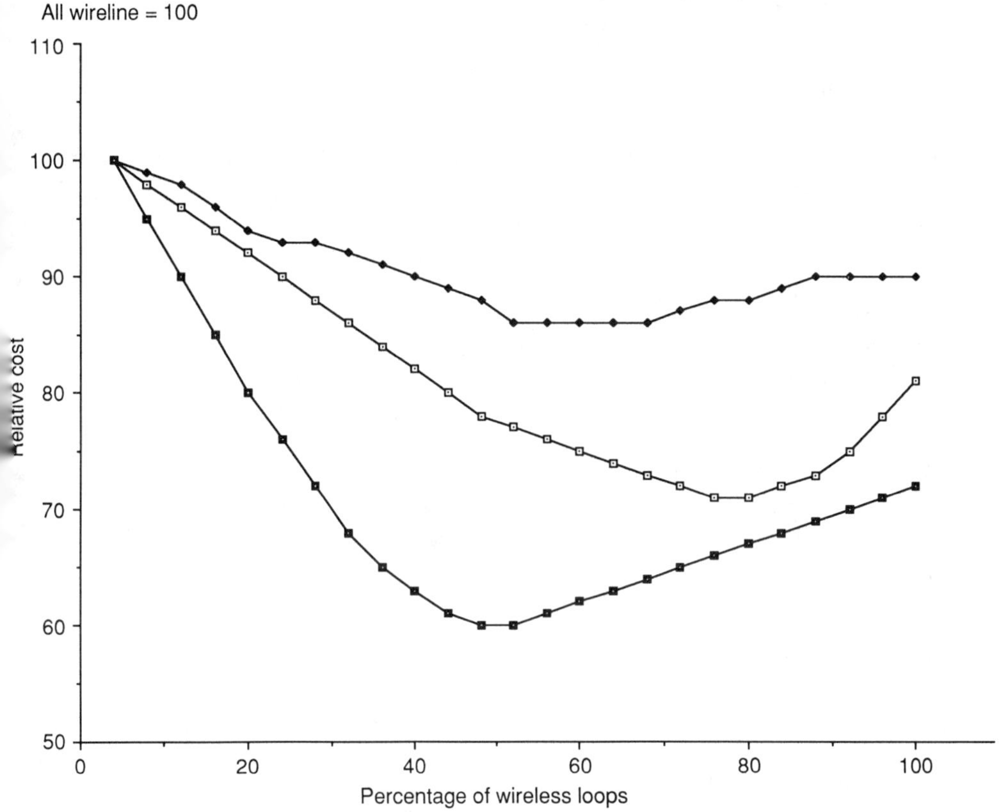

Figure 7.54 Radio/cable curves for various replacement strategies.

This analysis carries the economic study about as far as it is feasible to take it, short of beginning actual site engineering. In the next chapter, we shall further considerations that emerge when we actually develop a site plan.

7.4.3 Area Analysis: Radio Overlay Networks

The detailed analysis of radio coverage involves us in the transition from economics to engineering (Chapter 8). There are, however, a number of specific cost considerations associated with the definition of radio coverage areas and overlay networks, which may be incorporated into the economic analysis. These include:

1. Siting of radio base stations relative to switching facilities
2. Administration of multiple exchanges from a single wireless access system
3. Definition of network interface points and facilities
4. Definition of interconnection facilities.

7.4.3.1 Siting Radio Base Stations

Wireless access systems are set up to allow the radio base station to be separated from the switch/network interface point. The two are normally connected by some standard transmission facility such as a T-1 link. This provides engineering flexibility (see Figure 7.55). The location of the radio base station transmitter is constrained primarily by site engineering considerations, such as the effects of terrain on radio propagation, and the availability of suitable power and shelter facilities (see Chapter 8). Nevertheless, there are economic implications and in some cases the economics of a particular configuration may be the determining factor.

The low-cost option will usually call for the radio transmitter to be colocated with the switch, which provides the interface to the external network (see Figure 7.55). This normally avoids the requirement for another shelter and minimizes or eliminates the transmission facilities in the system. However, the switch may not always be well situated for radio propagation, and in multicellular systems most of the base stations will still require interconnecting transmission links.

7.4.3.2 Administration of Overlay Networks

The administration of multiple exchanges from a single RCSA may require, or permit, the redefinition of numbering plans and calling areas. New software may be needed for different call accounting procedures. In general, this should add a relatively small amount of cost per line (less than 1%) to reflect the cost of software enhancements.

7.4.3.3 Network Interface or Interconnection

I use the term *network interface* to denote the connection between the wireless access system and the external network (see Section 5.1). We are especially interested in the interface to the public switched telephone network (PSTN). In the cellular world (and elsewhere), this function is often called *interconnection,* which I find an ambiguous (but perhaps inescapable) term.

(a)

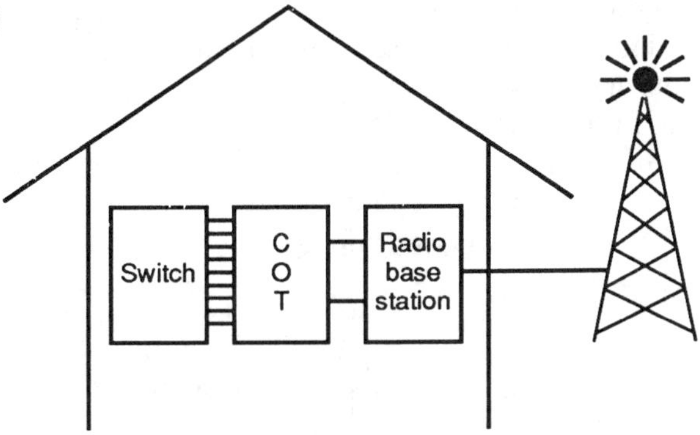

(b)

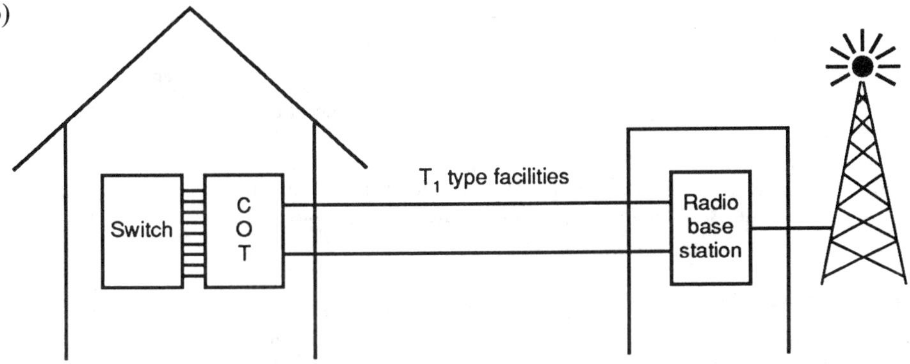

Figure 7.55 Colocated and remote options for radio carrier station placement; (a) colocated base station, (b) remote base station.

There are four aspects of the network interface that have potential economic importance:

1. *Interface type* (i.e., the point of interconnection within the switching hierarchy)
2. *Interface level* (i.e., the nature of the circuits that cross the interface)
3. *Interface location* (i.e., the physical location of the interface)
4. *Interface facilities* (i.e., the hardware).

Interface Type. The interface between the wireless and the wireline networks can occur at different points relative to the standard PSTN switching hierarchy. For example, the wireless access network controller (switch or switch-adjunct processor) may attach to or be part of a PBX, which in turn attaches to a Class 5 central office (what the cellular world calls a *Type 1 interconnection*). Alternatively, the wireless access network controller may attach directly to a Class 5 switch, or to a Class 4 or tandem switch. Said another way, the network controller itself may appear to the network like a DLC, a PBX, a full central office, or some hybrid of these (see Figure 7.56). The economics of these alternatives will vary (especially operating costs). The reader is referred to the literature for further details [61].

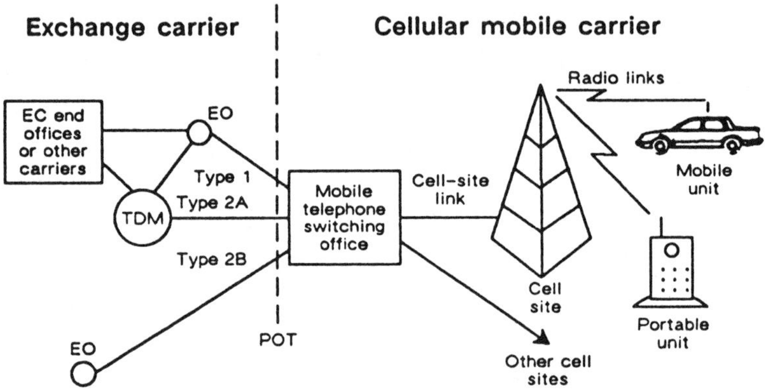

Figure 7.56 Alternative interconnection points to the PSTN. (From Bellcore, *Telecommunications Transmission Engineering Volume 3: Networks and Services,* Third Edition, 1990, p. 470.)

Interface Level. The circuits that cross the interface point may vary from analog 2-wire voice frequency line appearances (common with BETRS hardware today) to digital DS0 trunks as specified in the so-called "TR-8" interface for DLCs [62], which are also common in modern cellular switches, all the way to higher level digital transmission formats (such as a SONET format). The type of interface circuit will of course have an impact on equipment cost, although the cost per customer is usually small relative to the total cost of the wireless access circuit [63].

Interface Location. Occasionally, the physical location of the interface points may have economic implications. In small networks, or those serving fixed subscribers, the interface points are normally defined by the existing wireline network infrastructure. In larger multicellular networks, the choice of one or a few interface points versus many interface points can have a potentially significant economic impact on the cost of the intercellular trunking facilities. A distributed switching network will reduce the backhaul costs associated with getting the traffic from the cell-sites to the switching and network interface points. Some cellular manufacturers have developed switching architectures to take advantage of this [64].

Another optimization along these lines is the option for intracellular traffic to be handled without the necessity of routing the traffic out of the cell to a central controller, which can sometimes be accomplished with an *intercom-type function* (in such cases, however, there is usually no call record so that the charge is on a flat-rate basis rather than a measured service basis).

A third approach uses local switching over wire for calls between subscribers connected to the same cluster-radio terminal in either a BETRS unit or a subscriber radio system (i.e., calls between subscribers within the same outstation serving area). SR Telecom has implemented this feature, which effectively removes a good deal of traffic from the radio channel altogether [65].

Interface Facilities. The actual equipment used to accomplish the interface may also vary, from a simple channel bank (the cheapest alternative), to a DLC central office terminal with time slot interchange capabilities [66], to a full-fledged digital switch. The allocation of costs may be somewhat tricky in the face of these various alternatives, and oversimplifying the analysis should be avoided.

7.4.3.4 Intercell Trunking Facilities

The cost of transmission facilities linking the cells in a multicellular system is a traffic-sensitive cost. In today's star-configured networks, the only real variable (once the distribution of switching functionality has been taken into account) is the bit rate of the voice traffic on these intercellular links. Since the digital generation of cellular radio utilizes low-bit-rate coders (8 and 16 kb/s with overhead), some savings can be realized by transmitting over the intercellular links at the lower bit rate. In North America, none of these interfaces has been standardized, and as of this writing, the plans of the various manufacturers are still fluid. There are certainly problems with converting the wireline network to handle low-bit-rate voice traffic; indeed, there may be cost penalties that will more than offset the prospective savings. In an era of "unlimited bandwidth" over fiber, the potential savings from low-bit-rate backhaul may be smaller than anticipated.

The architecture of the multicellular network is also a potential factor in reducing the cost of backhaul facilities. A ring-type network to link the radio ports in an urban PCN-type application could result in significant savings over a star configuration. Such architectures are still in the future.

7.5 "NO MORE EXCHANGE BOUNDARIES"—THE IMPACT OF COMPETITION

The wireline landscape today is divided into thousands of individual exchange areas. In the distant past, the telephone exchange boundary represented the natural equilibrium between the tendency to seek minimum loop lengths, on the one hand, and the optimal sizing of switches on the other hand. In that sense, the exchange reflects the primordial switch/transmission dichotomy we have touched on in previous chapters. Beyond that, however, the exchange boundary has acquired over the years a degree of importance that goes beyond any mere technical equilibration. It is now the molecular unit of the telephone business. The exchange area is the basis for the service plan; its boundaries harden into political and economic barriers reinforced by tariffs and prohibitions. All programs, all technologies, must be mapped into the exchange jigsaw.

The advent of wireless access will spell the end of exchange boundaries as we have known them. Radio coverage areas for BETRS typically overlay several exchanges, and it makes no sense to artificially restrict the availability of wireless access to customers who happen to be located in the exchange area where the base station is located. In fact, it may be desirable to locate the radio base station outside of the target area. This situation came up several years ago in the state of Mississippi. An area coverage analysis (see Chapter 8) determined that the best way to serve the customers in one particularly problematic exchange was to locate the base station over the border of an adjacent exchange. There was a tower already available at this location, with power, road, and good propagation. It was superior to any site actually in the target exchange itself (primarily because a costly tower would have to be erected). The project was not authorized, however, because the carrier in question could not rationalize the idea of transmitting photons across an exchange boundary. What a pity.

Another common conundrum involves the wireless service plan for several exchanges located next to one another. It is almost always the case that a single RCSA can cover several contiguous exchanges, and (if need be) the customers can still be segregated for billing or other purposes in the system database. Nevertheless, most carriers are still reluctant to implement such multiexchange RCSAs, and would prefer instead to install one per exchange. This adds to the expense, of course, especially when multiple towers and power plants must be constructed.

The coming of competition will put an end to such fastidiousness. Radio systems that can reach from one company's territory into neighboring exchanges owned by other carriers can be a potent competitive weapon. Wireless access will be the perfect "foreign exchange" system for many situations [67]. It will allow carriers to offer wide-area overlay networks for specialized services. As we discuss in Chapter 9, I believe that this overlay capability will become one of the distinctive features of the network in the twenty-first century, nd radio will be its primary vehicle.

The economics of wireless access are just beginning to be understood. Most of the work so far has focused on comparisons within the wireline framework, and as such we are still constrained in our thinking to wireline terms of analysis, such as "feeders" and "exchanges." We are still being forced to prove that the airplane can be used as a ground transport vehicle by taxiing around the runways. The liberating character of wireless access will really be felt only when we have left these concepts behind and soared into the perpendicular dimension that radio opens up for us.

NOTES

1. See the Bellcore references cited in Chapter 9 of this book, pertaining to loop costs.
2. For example, there is a Bellcore study—Bellcore Special Report SR-NPL-000676, *Comparison of the Economics of Radio and Conventional Distribution for Rural Areas,* Issue 1, April 1987—which is best characterized as well-intentioned but preliminary. The title itself suggests the nature of at least one important limiting assumption, namely that the use of radio is appropriate only for remote rural areas. There is, of course, nothing rural *per se* about wireless access, as more recent proposals for the use of radio in urban settings, and indeed the deployment of CT-2 systems and proposed PCN franchises in the United Kingdom, clearly indicate. The best effort so far at reducing the economic analysis of wireless access to a simple comparative model is the work done by Paul Clements at IMM Canada in 1987 and 1988. This work is unfortunately unpublished, but much of this chapter draws on it.
3. The now-classic, albeit obsolescent, reference work in this area is the Bell system manual, *Engineering Economy: A Manager's Guide to Economic Decision Making,* 3rd ed., prepared by the AT&T Construction Plans Department, and published by McGraw-Hill (New York, 1977). The first edition of the "Green Book" was issued in 1952 and the second edition in 1963. Other synchronous references include E. L. Grant and W. G. Ireson, *Principles of Engineering Economy,* 5th ed., New York: The Ronald Press Company, 1970; and American Telephone & Telegraph Company, *Telecommunications Transmission Engineering, Volume 1: Principles,* 2nd ed., published by AT&T, 1977, especially Chapter 11. This last has been recently updated as Bellcore, *Telecommunications Transmission Engineering, Volume 1: Principles,* 3rd ed., Bellcore Technical Publications, 1990. The multi-edition, collective authorship character of these references, which summarize dozens of more specific Bell System studies and guides, suggests the pervasive nature of this discipline under the monopoly framework. This is now changing.
4. Federal Communications Commission, "Uniform System of Accounts for Class A and Class B Telephone Companies," *Rules and Regulations, Title 47, Code of Federal Regulations,* Part 31, Washington, D.C.: U.S. Government Printing Office, 1987; also see Bellcore, *Telecommunications Transmission Engineering, Volume 1: Principles,* 3rd ed., Bellcore Technical Publications, 1990, p. 367; and AT&T, *Engineering Economy: A Manager's Guide to Economic Decision-Making,* 3rd ed., New York: McGraw-Hill, 1977, Chaps. 3 and 7.
5. Academicians who have studied regulated utilities are often startled to make this discovery. For example, see J.C. Bonbright, A.L. Danielsen, and D.R. Kamerschen, *Principles of Public Utility Rates,* 2nd ed., Arlington, Virginia: Public Utilities Reports, Inc., 1988, p. 356: "In a seminal paper Harvey A. Averch and Leland L. Johnson gained imperishable fame [!!!] by developing the provocative thesis, commonly known as the Averch-Johnson effect, that traditional rate-base and rate-of-return regulation biases a regulated firm, as compared to an unregulated one, towards more capital-intensive modes of production. The reason is that purchased capital becomes part of the utility's rate-base upon which an allowed or approved rate of return may be earned."
6. Russell L. Ackoff, *Scientific Method: Optimizing Applied Research Decisions,* New York: John Wiley & Sons, 1962.
7. See Bellcore, *Telecommunications Transmission Engineering, Volume 1: Principles,* 3rd ed., Bellcore Technical Publications, 1990, p. 368; also AT&T, *Engineering Economy: A Manager's Guide to Economic Decision-Making,* 3rd ed., New York: McGraw-Hill, 1977, pp. 136 ff.
8. American Telephone & Telegraph Company, *Telecommunications Transmission Engineering, Volume 1: Principles,* 2nd ed., published by AT&T, 1977, p. 292.
9. *Ibid.,* pp. 292, 294; see also AT&T, *Engineering Economy: A Manager's Guide to Economic Decision-Making,* 3rd ed., New York: McGraw-Hill, 1977, pp. 291 ff; also R.L. Freeman, *Reference Manual for Telecommunications Engineering,* New York: John Wiley & Sons, 1985, pp. 1373 ff.

10. It is interesting to note, however, that in the just-released update to the Red Book, the definition of Present Worth of Expenditures (PWE) has been modified to include a revenue factor; see Bellcore, *Telecommunications Transmission Engineering, Volume 1: Principles,* 3rd ed., Bellcore Technical Publications, 1990, p. 378.
11. This figure was provided by the staff of Alberta Government Telephone, in connecting with a BETRS trial in 1987–88. Another study of rural loops, by United Telephone, yielded a measured figure of 2.3 CCS/main station, or about 0.64 Erlangs; see United Telephone, "Technological Substitution Study for Rural Copper Loops," 1985.
12. This has been cited in internal cellular operators' reports as a "typical" metro-urban value. Since all such data are shrouded in secrecy (for competitive purposes), a dose of sodium chloride is indicated.
13. Bellcore, *Telecommunications Transmission Engineering, Volume 1: Principles,* 3rd ed., Bellcore Technical Publications, 1990, p. 379, which notes that even the output variables can be "distorted": "Net Present Worth of Expenditures (NPWE) represents the present worth of revenue requirements. Regulators often prefer to express project value in terms of revenue requirements. In fact, NPWE is 'unique' to a regulated business. It was developed to give project information from the regulators' (and the customers') point of view.".
14. See Bellcore, *Telecommunications Transmission Engineering, Volume 1: Principles,* 3rd ed., Bellcore Technical Publications, 1990, pp. 368–9: "If the problem conditions do not give an indication of the life of the plant, life must be estimated based on past experience and engineering judgment. Even here, *study life will rarely coincide with accounting life."* [Emphasis added]
15. See the discussion in G.T. Hawley, "Historical Perspectives on the U.S. Telephone Loop," *IEEE Communications Magazine,* March 1991, pp. 24–28.
16. M.D. Fagen, ed., *A History of Engineering and Science in the Bell System: The Early Years (1875–1925),* AT&T Bell Laboratories, 1975, pp. 541 ff., and pp. 923 ff.
17. Undoubtedly there will be some who may feel my characterization of the telephone industry's analytical methods is harsh or unfair. Certainly, there are a great many sophisticated tools in the hands of planners at Bellcore and in some of the headquarters staff groups at the major operating companies. But at the level where plant decisions are often made, namely the exchange or district level, the methods are very much paper-and-pencil, seat-of-the-pants, and rule-of-thumb oriented (to mix several metaphors).
18. Bellcore, *Telecommunications Transmission Engineering, Volume 2: Facilities,* 3rd ed., Bellcore Technical Publications, 1990, pp. 93 ff; see also G. T. Hawley, "Historical Perspectives on the U.S. Telephone Loop," *IEEE Communications Magazine,* March 1991, pp. 24–28, for a discussion of the historical origins of the serving area concept.
19. *Ibid.*
20. *Ibid.,* pp. 110 ff.
21. *Ibid.,* pp. 90 ff.
22. Bellcore, *A Compilation of Results from Recent Bellcore Studies on Fiber to the Home (FTTH),* SR-TSY-001468, Issue 1, December 1989, Bellcore, Chap. 5.
23. For SR Telecom references see note 42 in Chap. 5.
24. The aforementioned United Telephone study cites a figure of $1.29/foot as the installed cost of distribution facilities from the far end of the feeder to the customer; see United Telephone, "Technological Substitution Study for Rural Copper Loops," 1985, pp. 2–3.
25. Bellcore, *Generic Framework Criteria for Universal Digital Personal Communications Systems (PCS),* FA-NWT-001013, Issue 2, December 1990.
26. NYNEX Enterprises, "Request for Information on Radio Loop Access," RFI No. 546-NSC-10B, October 12, 1990.
27. Peter Purton, "Radio over Fiber?" *Telephony,* November 1989. See also Chapter 9 of this book and the other references cited there.

28. Paul E. Clements, "Upgrading Rural Access Services through the Application of a Digital Radio Overlay," internal publication, IMM Corporation, 1987.
29. See for example N.C. Hightower, "Economic FO System for New Residential Services," *Telephony,* March 17, 1986, pp. 44–56; see also the many REA studies of loop lengths; one of the best of these types of studies I have seen was performed by United Telephone in the mid-1980s, although it was never published, to my knowledge.
30. United Telephone study, see previous note.
31. In 1988, Alberta Government Telephone evaluated a number of radio applications involving both clustered and single-line alternatives, and it was in the course of these studies that the importance of increased drop wire costs for clustered systems became clear.
32. See Section 5.3.2.
33. IMM Corporation and Rockwell both offered clustered BETRS-type systems in the early 1990s.
34. Scott Chesbro, "A New Era for Navajo Phone Service," *Telephony,* February 6, 1989.
35. See, for example, Bellcore, *Telecommunications Transmission Engineering, Volume 1: Principles,* 3rd ed., Bellcore Technical Publications, 1990, Chap. 11; also AT&T, *Engineering Economy: A Manager's Guide to Economic Decision- Making,* 3rd ed., New York: McGraw-Hill, 1977, especially Appendix 5A.
36. Bellcore, *Telecommunications Transmission Engineering, Volume 1: Principles,* 3rd ed., Bellcore Technical Publications, 1990, pp. 366 and 384, cites a 10% rate as the "actual composite cost of capital."
37. *Ibid.,* Chap. 22; also Bellcore, *Telecommunications Transmission Engineering, Volume 2: Facilities,* 3rd ed., Bellcore Technical Publications, 1990, Chap. 23.
38. See Bellcore, *Telecommunications Transmission Engineering, Volume 2: Facilities,* 3rd ed., Bellcore Technical Publications, 1990, p. 635: "The high rate of station movement and the complexities of feeder and distribution cable layouts make loop maintenance difficult. . . ."
39. In IMM's "wireless city" application with GTE, the wireline maintenance costs were stated to be $110 per loop per year.
40. See K. Flamm, "Technological Advance and Costs: Computers vs. Communications," in R.W. Crandall and K. Flamm, eds., *Changing the Rules: Technological Change, International Competition, and Regulation in Communications,* Washington, D.C.: The Brookings Institution, 1989, p. 50: "Precise calculations [of operating costs] will vary with the exact technology examined, the length of haul, and the cross section, but significant reductions are definitely associated with use of microwave radio. Microwave radio has much lower operating costs than carrier on coaxial. . . ."
41. For example, in videotaped interviews with outside plant engineers involved in the installation of a number of early BETRS systems, much lower maintenance requirements were often cited as a key benefit of wireless access. The companies involved included GTE, Pacific Telecom, and other independents. In my experience, the "maintenance differential" is quite pronounced, although I have never seen a nonproprietary quantitative analysis.
42. The company in this case was Alberta Government Telephones, for a 1987 BETRS project at Etzikom, Alberta.
43. IMM Corporation calculations show an MTBF in excess of 15 years for BETRS subscriber terminals.
44. SR Telecom reports an MTBF for their entire system in excess of 14 years.
45. Bellcore, *Telecommunications Transmission Engineering, Volume 1: Principles,* 3rd ed., Bellcore Technical Publications, 1990, p. 743.
46. G.T. Hawley, "Historical Perspectives on the U.S. Telephone Loop," *IEEE Communications Magazine,* March 1991, p. 25.
47. Bellcore, *Telecommunications Transmission Engineering, Volume 1: Principles,* 3rd ed., Bellcore Technical Publications, 1990, p. 369.
48. I am indebted to Kris Shelton for providing this insight.

49. The site in question is near Victorville, California. For further information, contact IMM Corporation.
50. G.T. Hawley, "Historical Perspectives on the U.S. Telephone Loop," *IEEE Communications Magazine,* March 1991, p. 26.
51. *Ibid.,* pp. 24–28.
52. Federal Communications Commission, "Uniform System of Accounts for Class A and Class B Telephone Companies," *Rules and Regulations, Title 47, Code of Federal Regulations,* Part 31, Washington, D.C.: U.S. Government Printing Office, 1987.
53. See AT&T, *Engineering Economy: A Manager's Guide to Economic Decision-Making,* 3rd ed., New York: McGraw-Hill, 1977, p. 45.
54. *Ibid.,* p. 138.
55. The range of capacity-utilization figures depends on many things, and can be calculated by standard traffic statistical methods. True loop applications will always have significant unused capacity if for no other reason than that most telephones are not used very much at night. The advent of new services that take advantage of idle periods at night to transmit other forms of data may push these figures higher.
56. I am aware of a number of telephone companies, for example, that have explicit policies against any "concentration" in the loop at all, even where DLC systems make this a perfectly reasonable option.
57. G.T. Hawley, "Historical Perspectives on the U.S. Telephone Loop," *IEEE Communications Magazine,* March 1991, p. 25.
58. As we have noted, these statements are not quite categorical. We are concerned about subscribers operating at the fringe of the coverage area who may need special attention. We are also concerned for some applications about making sure that two subscribers are separated by some minimum distance, which is usually very small (in BETRS applications, for example, it is typically well under 1/100th of 1% of the system transmission radius). In my experience, however, these are truly *de minimis* considerations.
59. The study was conducted in 1984 by Booz Allen Hamilton, a consulting firm, for United Technologies.
60. Paul Clements developed this procedure while working for IMM in 1987-88, in connection with the Alberta Government Telephone program for "digital loop replacement radio," Alberta Government Telephones, *Technical Specification for AGT Rural Individual Line Service (ILS) Digital Loop Replacement Radio (DLRR) System,* June 13, 1986. See Section 5.3.3.4.
61. Bellcore, *Telecommunications Transmission Engineering, Volume 3: Networks and Services,* 3rd ed., Bellcore Technical Publications, 1990, pp. 466 ff.
62. Bellcore, TR-TSY-000008, *Digital Interface Between the SLC-96 Digital Loop Carrier System and a Local Digital Switch,* Issue 2, August 1987, and supplements. The "next generation" of trunk interface designs will be based on a fully integrated DLC, as specified in Bellcore, TR-TSY-000303, "Integrated Digital Loop Carrier System Generic Requirements, Objectives and Interface," Issue 1, September 1986, plus numerous revisions and supplements.
63. For example, in BETRS-type systems today (1991), the 2-wire VF line interface contributes less than 5% of the total "cost per line" of the system.
64. The most advanced distributed architecture is the CD-900 system manufactured by SEL Alcatel, which is based on the System 12 switch. I am indebted to Ken Giesken, now of Hughes Network Systems, for background material on the System 12, which he helped to develop.
65. See references on SR Telecom in note 42 of Chap. 5.
66. For example, in BETRS or subscriber radio.
67. Indeed, cellular radio operators are already exploring these service options.

Chapter 8
ENGINEERING FOR WIRELESS ACCESS

8.0 "MAKING IT WORK!"

In this era of dazzling technology, there is often a peculiar juxtaposition of the marvelous and the mundane. We may find a digital radio-telephone—which is actually a specialized and very powerful computer—lashed together with a nineteenth-century vintage power supply and bolted to a wooden pole. CMOS circuitry competes against fungus, ice, and grime. Sophisticated electronic systems are encumbered with lightning arrestors, grounding rods, heat shields, and insect-proofing. Indeed, there are two camps within the technical community. There is the laboratory—the design environment—and the field—the user environment. For every technological balloonist ready to soar through the air there is a muddy-shoed pin-sticker of an outside plant engineer ready to bring it all back down to earth. And we find that every problem has two faces. For every elegant "solution in software," there is an equal and opposite solution involving a chain saw, or a backhoe, or a cherry-picker, or some other instrument of internal combustion.

In this chapter we examine some of the challenges of taking wireless access technology into the field and making it work. We are *not* talking about field testing, which is properly a phase of design and development. Field testing has to do with evaluating technical performance under real-world physical conditions. In contrast, "making it work" is primarily a challenge of designing for the user (operator or end-user, it makes little difference). The human being who has to install, operate, and maintain any telecommunications system becomes the ultimate filter, the common denominator that the technology must subserve. The system designer must anticipate that the person to whom the responsibility falls for engineering a particular installation may not have a very advanced knowledge of the underlying technology. (This is perhaps especially true for radio systems, which have not been widely used in telephony heretofore.)

As a result, the telephone industry often forces the adoption of a *black box* approach to packaging the technology. This is a product design philosophy in which the technological genie is confined within its opaque cage, out of sight and mind, with an interface to the real world that is as absolutely simple as it is possible to make it. Typically, for a wireless access subscriber terminal, there may be a dc power input, an RJ-11 telephone jack, an electrical ground, and a standard coaxial cable connector to the antenna. A couple of LEDs for status or power indicators may be our only window to the inner workings of what is an extremely complex digital system (far more sophisticated by some measures than the typical microcomputer) [1]. Imagine building a computer, or an automated machine tool, with such a simplistic user interface! Yet that is the standard practice in telephony.

The black box thus defines a known set of performance parameters, which allow for a fairly standard approach to engineering and installation, diagnostics, and maintenance. We *assume* the technical parameters. "Making it work" thus comes down to a matter of designing *methods and procedures* that streamline the deployment of these black boxes.

In my experience, the challenges of defining these methods and procedures for wireless access fall into three major areas:

1. Definition of service requirements and objectives
2. Prediction of radio propagation, area coverage, and link characteristics
3. Subscriber site-engineering and installation procedures.

Each of these tasks involves a degree of novelty that poses difficulties for the traditional industry. These are the most important areas in which it will be necessary to develop extensive new methods. There are, of course, many other aspects of engineering a particular system, but most of the others can be adapted relatively easily from existing procedures developed for other types of equipment. (For example, the growth of DLC technology has forced the industry to adapt to the idea of deploying electronic systems in the loop environment; fortunately, wireless access does not have to fight this battle again.)

I draw this distinction because it will help us focus in this brief chapter on the most challenging topics. There are many facets of deploying a wireless access system that, while they involve considerable effort and require considerable care, do not present a real creative challenge to the industry. For example, installing the base station and network interface equipment for a BETRS-type system is not very different from installing the electronics for a digital loop carrier or microwave radio system. On the other hand, installing a subscriber terminal (e.g., a BETRS-type terminal) is a very different operation from installing a wireline loop.

In the following sections of this chapter, we will examine each of these areas.

The scope of this chapter, once again, must fall short of providing a definitive handbook. Ultimately, of course, the procedures discussed here will have to be reduced to a rigid and detailed format. For example, the subscriber installation (for BETRS terminals) will be cast in the so-called *TOPS* format to suit many of the Bell operating companies [2]. This sort of material is becoming available, at least in preliminary form, from some manufacturers [3], from the Rural Electrification Administration (REA) [4], as well as from certain Bellcore references [5]. The goal here is more modest: to offer a kind of conceptual checklist of some of the more important issues facing the wireless access engineer in these three important areas (i.e., service definitions, RF engineering, and subscriber installation).

I must also defend, again, my reliance on the BETRS-type application as the model for much of this chapter. There are several reasons for this. First, there is very limited industry experience with wireless access deployment other than BETRS and related wireless loop type applications, for which at least the groundwork has begun to be laid. Some might argue that the cellular industry has undergone an extensive systems deployment process during which a great many systems have been, after a fashion, "engineered." Nevertheless, the cellular operators have yet to codify most of the deployment procedures to anywhere near the degree required. With respect to service issues (see Section 8.1), I have found that the cellular industry is still remarkably loose. (I recall one major operator who advised its customers that service to a particular area would improve in the autumn when the trees shed their foliage.) As far as RF engineering is concerned, the operators still rely in many cases on outside engineering consulting firms to develop the actual implementation plans. The models and procedures commonly used by the cellular industry to predict coverage and RF link quality are, unfortunately, of rather coarse resolution [6]. And, of course, in the cellular business the operator does not perform (in most cases) the subscriber installation. This is left to the local garage mechanic or car stereo outlet, with predictably unsatisfactory results that cellular operators often complain about.

Second, BETRS applications involve much higher service standards (as we shall see in the next section), which means that all methods and procedures must be much more exacting. I am not sure whether it is 1000 times harder to guarantee 99.99% availability (for BETRS) than it is to promise 90% availability (for cellular), but it often seems like it.

Finally, the BETRS application defines, I believe, the superset of wireless access engineering issues. PCN and other portable-oriented (or CPE-oriented) applications involve eliminating or radically simplifying many of the engineering issues. For example, as noted already, the installation issue (at least for the subscriber terminal) vanishes in the PCN application. Moreover, the assumption that a subscriber is mobile (i.e., capable of moving at all) means that we can be much less concerned about the degree of resolution in predicting radio coverage, compared to fixed applications (see Section 8.2).

8.1 SERVICE REQUIREMENTS FOR WIRELESS ACCESS

To engineer a real-world telecommunications system, it is necessary to specify the *type, quality,* and *availability* of the service being offered to the subscriber. These service objectives define the technical performance requirements that the system must be engineered to meet.

8.1.1 Types of Service

There are many different categories of telephone service within today's wireline network, and of course wireless networks will extend the range of services to include mobile and portable services not currently available. The reader is referred to one of the standard Bell System references for a complete catalog of service options [7]. In this section, we will touch on some of the most basic service categories that apply to most of the important wireless access applications.

8.1.1.1 Voice Services

The specification of telephony voice services constitutes an elaborate literature [8]. Actually, any voice service comprises two subservices: a *voice circuit* (also called a *voiceband channel* [9]), for the transmission of actual speech, and a *signaling set,* for transmitting control information of various kinds [10]. For purposes of defining wireless access requirements, it is important to be specific about the nature of both the voice circuit and the signaling the system is expected to support. There are several broad alternatives.

Analog Voice Circuits. An analog voice circuit is usually defined by a 3-kHz frequency bandwidth, which is normally specified as 200 to 3200 Hz [11]. The standard analog circuit is also expected to support DTMF signaling and dc signaling (hookflash). It also provides access to the switch (dial tone) immediately on going off-hook. The analog circuit is still almost universal for loop applications.

Although we have defined wireless access as a digital transmission facility, it is perhaps surprising that many of the service requirements for certain wireless access applications are derived from the analog voice circuit specification. This is especially true of basic exchange radio (e.g., BETRS), where the goal is to mimic an analog wireline loop circuit.

DS0 Digital Voice Circuits. The other standard definition for a voice circuit is the 64 kb/s PCM-based digital circuit, known as a *DS0* in the North American hierarchy. (The European version of this standard is slightly different.) It is widely found in trunking applications in the interexchange and long-distance environment, as well as (less frequently) in the feeder portion of the loop where digital loop carrier has been used. It is, however, still virtually unknown in end-to-end loop applications. In general, the 64 kb/s digital circuit has been designed to mimic exactly the characteristics of an analog voice circuit.

In wireless access applications, the standard DS0 circuit will not be used very often. It is highly inefficient to use 64 kb/s voice coding for a radio system; it requires too much bandwidth, much more than a comparable analog FM radio-telephone circuit. The only wireless application where the DS0 circuit has been used is in so-called thin-route microwave subscriber radio (see Chapter 5).

Note, however, that a 64 kb/s bearer channel (not the same as a DS0 voice circuit) will be important in the ISDN service (see Section 8.1.1.2).

Low-Bit-Rate Voice Circuits. The spread of wireless access will introduce low-bit-rate voice (LBRV) coding into the network on a broad scale for the first time [12]. A LBRV circuit may be defined as a digitized circuit with a net bit rate of 32 kb/s or less. Likely rates for standardization purposes are 32, 16, and 8 kb/s. Under the IS-54 digital cellular standard, a half-rate coder with a gross rate of 6.5 kb/s is currently being specified (1991). Although such coders are often designed to provide service that is nominally equivalent to wireline (analog) quality, their use defines a telephone circuit with very different characteristics from those found in standard analog and DS0 voice circuits. This will pose numerous challenges to the telephone industry.

For one thing, the analog characteristics of the output signal of an LBRV circuit may no longer match the characteristics of an all-analog voice circuit, with respect to parameters such as frequency response. The new coders may respond in unusual ways to standard test procedures [13].

Second, signaling may be disrupted. Some LBRV coders have difficulty passing DTMF tones and other signaling required for ancillary services such as a coin pay phone.

Third, LBRV will usually result in additional transmission delay. The total delay is a composite of several factors, including framing (for TDMA systems), but the LBRV voice coders now being developed for wireless access can be expected to introduce processing delays of tens of milliseconds [14]. In the (false) light of analog experience, this may be seen as a potential problem (i.e., echo); however, the use of software-based echo cancellation is already a standard for BETRS wireless access systems [15], which should allow telephone engineers to tolerate moderate processing delays (as long as the total delay remains below the threshold of perception, which is usually felt to be several hundred milliseconds).

There is no question that LBRV results in a different type of telephone circuit than conventional 3-kHz analog transmission or DS0 64 kb/s digital transmission. Nevertheless, the economics of wireless access are highly dependent on the use of extremely spectrum-efficient voice circuits, which require LBRV. The evolving standard for a wireless access voice circuit will have to adapt to the requirements of these coding systems.

Ancillary Voice-Oriented Services. There are quite a few enhanced or ancillary voice-oriented services, which differ from basic voice services chiefly in the enhanced forms of signaling employed. A good example is coin telephone service. Coin phones come in a bewildering variety and have proliferated over the years as different manufacturers and operators have invented different signaling schemes, all of which are workable within the framework of the robust and forgiving analog channel. Unfortunately, accommodating these many schemes on a much more tightly designed LBRV channel can pose severe difficulties for designers of wireless access equipment.

Our experience has been that coin operation is one of the most troublesome features to engineer into an LBRV channel.

There are other ancillary services, including PBX and CENTREX trunks, 4-wire lines, and so forth [16]. Some of these may present difficulties in an LBRV wireless access framework. Nevertheless, by the application of appropriate software modules (as described in the following section) and occasionally specialized terminal interfaces, it is possible to produce a satisfactory wireless solution for these requirements.

8.1.1.2 Data Services

Over the past 30 years, a cornucopia of data traffic has begun to pour onto a network originally designed only for voice service. Much of this traffic is "opportunistic"—in the sense that it involves only a certain amount of specialized CPE (e.g., a computer modem) and makes no demand *per se* on the telephone network itself. This is referred to as *voiceband data,* which simply takes the voice channel (i.e., the 3-kHz analog voice channel and its digital DS0 counterpart) as it finds it. More recently, specialized networks (which may be physical or virtual networks) have been developed specifically to handle *direct digital transmission,* using packet transmission or other arrangements that optimize the transmission for the characteristics of data traffic (as opposed to voice). Finally, we are beginning to see the emergence of integrated voice and data networks, of which the most important is the ISDN concept.

The advent of wireless access networks will have an impact on all three types of data services. Engineering a wireless access system will mean thinking out the adaptations of data service concepts to the restrictions of the radio channel.

Voiceband Data. Voiceband data transmission assumes the 3-kHz bandwidth of the standard analog (or DS0) voice channel, and adapts the data transmission protocols to the constraints and opportunities presented by this channel. Typically, this means using a modem to convert data signals generated by a computer, for example, into an acoustic analog of some sort, which can then be transmitted over the 3-kHz circuit. The simplest modems use short tones at different frequencies to convey ones and zeros in a digital bit stream. (Indeed, DTMF is itself a form of tone-based signaling that can be used for data transmission.) More sophisticated modems take advantage of both frequency and phase information, and may use multilevel coding to increase the information density of the signal. The same approach has been developed for transmitting images (facsimile, or fax) using similar acoustically oriented signals. By following this general strategy, fax and modem manufacturers have been able to achieve bit rates of 2400, 4800, and even 9600 b/s on a fairly routine basis over ordinary voice circuits. With special lines, even higher rates are possible. Indeed, fax and data transmission, at least up to 2400 b/s, are now usually accepted as a standard POTS requirement [17].

The advent of wireless access complicates this situation, primarily because of the use of LBRV coding and narrower bandwidth channels. Sophisticated voice coders are capable of achieving telephone quality voice transmission in an 8 or 16 kb/s channel *precisely by optimizing for the characteristics of the human voice* (see Section 6.3.1). To the extent that the channel is optimized for a voice-like signal, it tends to become less friendly for modem tones and fax transmissions. Indeed, some low-rate voice coders will have difficulty even passing standard DTMF tones for overdialing (to access voice mail systems, credit cards, etc.).

The problem is not, therefore, that a reduced bandwidth *per se* inhibits data transmission at these rates, but that the channel is no longer a general-purpose acoustic pipe with a flat 3 kHz of frequency bandwidth. The solution is to design in the flexibility in the software to detect and process different types of signals in different ways. If a fax tone is detected, a fax algorithm is switched in; if a modem tone is detected, a voiceband data algorithm is used. By reconfiguring the radio channel by means of software, it should be possible in most cases to support today's voiceband services even over the narrower bandwidth channels of a wireless access system.

As such, this is a matter of basic technology, a concern for the system developer. It may affect the engineer's job, however, by forcing him to make choices about which types of voiceband equipment shall be supported. There is a wide variation in the performance of nominally equivalent fax machines from different manufacturers, for example. There is a potential cost penalty in requiring wireless access systems to adapt to the many acoustically oriented voiceband protocols that have developed over the years. It may be that in some applications, such as PCN, a narrower specification may be acceptable. In other words, there may be a narrower specification for a PCN-compatible fax unit, and the operator may hold in abeyance the more general requirement that a PCN device handle any fax machine that currently works on the wireline network. It is true that this may cause inconvenience and even impose a cost on the end-user; the alternative, however, may be that all wireless access users are penalized by the cost of adapting to a large number of minor variations in the performance of fax equipment.

Direct Digital Transmission. Direct digital transmission occurs when the data are transmitted in an electrical rather than acoustical form. A DS0 64 kb/s channel is actually a form of direct digital transmission, with electrical pulses used to encode the digital information of the signal being transmitted (see Section 6.1.3). Of course it is possible to transmit computer data (or imaging data) in the same manner. Different types of network arrangements have been developed for handling direct digital transmission of nonvoice traffic, either on a synchronous basis (e.g., 56 kb/s DDS lines) or an asynchronous basis (packet data systems).

Wireless access systems, which are based on digital transmission, generally find direct digital transmission a far more congenial form in which to handle nonvoice traffic. In principle, translating a 64 kb/s wireline data circuit to a 64 kb/s digital radio circuit is straightforward; the voice codec is simply bypassed and the RF carrier is modulated directly with the electrical digital carrier input. Packetized transmission is also easily accommodated, especially by TDMA and CDMA architectures.

The chief complication (in my experience) has been that since LBRV coders are designed to withstand high BERs, a wireless access system that is implemented for voice service as its primary function may be engineered for a BER level that is quite reasonable for voice (say, 10^{-4}) but too high for some data applications. It may be necessary, therefore, to add extra fade margin or to engineer the subscriber station to a higher standard if data traffic is expected to play a significant role in a given subscriber installation. (For example, in the BETRS architecture, a voice circuit may be perfectly adequate at 16-level modulation, whereas a data circuit may require 4-level modulation to assure the required RSL and BER criteria.)

Integrated Voice and Data Networks. Instead of either force-fitting data traffic to the Procrustean bed of the voice channel, or overbuilding physically separate data networks that are generally unable to carry voice traffic at all, future network planners are moving toward the Integrated Services Digital Network (ISDN). ISDN will be capable of handling voice and data within a flexible framework that in principle avoids doing violence to the needs of either type of traffic.

Briefly, ISDN would offer two generic "channels" —each of which would support a variety of specific services. Voice and synchronous data would be provided by a 64 kb/s *bearer channels* to the end user. The nominal standard configurations would offer either *2 B channels*—the *basic rate interface* (BRI), designed for a residential or small business user—or *23 B channels*—the so-called *primary rate interface* (PRI), intended for larger organizations. Asynchronous data would be handled on the so-called *D channel*—a single 16 kb/s packet channel for the BRI or a 64 kb/s packet channel for the PRI.

The extension of ISDN to radio-based access systems raises a number of interesting questions. Let us consider here the BRI only, since it is likely that this will be the more general service offering in the loop access environment. The "strict constructionist" view of the ISDN BRI is that it calls for a full 144 kb/s (2B+D) to be delivered to the end-user. In the wireline environment, it is anticipated that ISDN BRI will be provided over essentially the same access facilities as today's analog voice circuits—copper pairs. Indeed, the ISDN BRI is premised on taking advantage of the fact that a copper wire-pair has a bandwidth much wider than the standard 3-kHz voice circuit.

In a radio system, however, transmitting 144 kb/s to a single user will result in a relatively inefficient use of the radio spectrum, especially where it is only being used for voice transmission. If we accept a LBRV standard of 16 kb/s for a telephone-grade voice circuit, then a wireless ISDN channel would occupy nine times the spectrum of a single voice circuit. There are significant economic implications in following a strict constructionist approach.

An alternative approach to ISDN over radio, which we might call "loose constructionist," would be to redefine the BRI functionally around an LBRV voice channel, as some combination of the following:

1. Two (or more) telephone-grade LBRV voice channels
2. The capability for a 64 kb/s bearer channel for synchronous data
3. A 16 kb/s D channel.

A different "quasi-BRI" circuit could also be imagined: For example, a wireless access BRI could comprise an 80 kb/s channel, capable of offering a 64 kb/s bearer channel and a 16 kb/s D channel, plus the ability to offer multiple 16 kb/s voice channels with either the B or the D channels (but not all three together). This would be a workable radio adaptation that would maintain data-rate continuity with the wireline implementations of ISDN.

An even more radical approach is one that has been proposed by a number of authors, in which B and D channels would be themselves redefined, and even renamed (in some proposals these become C and I channels, operating at much lower bit rates) [18].

Other wireless access concepts, notably that being proposed by Bellcore, essentially sidestep the ISDN issue for now. This may indeed by the most appropriate stance to take at this point, since ISDN is a heavily wireline-oriented service concept and many of its economic and service assumptions need to be carefully examined in light of radio's unique constraints and opportunities. Moreover, since wireless access is by definition a digital service, the fundamental need for the integration of voice and data facilities and services has already been addressed, in principle. What needs to be worked out now is a radio-friendly approach to providing integrated access that accomplishes the following:

1. Interfaces with wireline integrated networks in some reasonable manner (but does not assume the necessity of cloning wireline protocols).
2. Provides a platform for voice, synchronous data, and asynchronous data in a convenient package to the end-user.
3. Offers some degree of "bandwidth on demand" capability.

This task is still very much in its early stages. I will confine my discussion here to the observation that ISDN service is likely to involve a specialized terminal (also true for the wireline version, of course), such that there may well be one sort of air interface specification for voice and basic wireless access services and another air interface for wireless ISDN. As with the wireline ISDN, the responsibility for integrating multiple levels of service can be laid on the network—in this case, the radio interface itself can be designed (if we decide to do so) with sufficient flexibility to enable both levels, that is, both subscriber populations, to coexist in the same spectrum.

8.1.1.3 Special Services

There are many types of telephone service that go beyond the basic voice channel. The category of "special services" encompasses a potpourri of different offerings, with more being developed all the time. As the standard Bellcore reference notes:

> Special services are *special* in that they generally require engineering and operations treatment beyond that applied to ordinary services in terms of transmission, signalling, maintenance, and customer use. They are installed and removed in response to a specific customer order, as opposed to message trunks that are provided for general public use. . . . [19]

The key to this definition is that special services require special engineering: "Special services are laid out on a per-circuit basis; individual-circuit engineering is one of the major features that distinguish special services" [20]. Special service circuits also generally require more stringent transmission standards [21].

For purposes of this discussion of wireless access, it is useful to divide special services into three groupings:

1. Services based on different forms of signaling to support different types of CPE
2. Enhanced services that provide access to various intelligent network features
3. Enhanced services that are based on radio's unique capabilities as an access medium.

Alternative Signaling Formats. The wireline network today supports a number of variations on the basic loop voice circuit. For example, a coin telephone offers a somewhat enhanced set of options that are implemented by means of one of several different signaling formats. (Note that, although Bellcore considers coin service to be a basic service [22], from the standpoint of a wireless access system it is perhaps best regarded as a special in the sense of the definition provided above, for reasons we shall touch on shortly.) Trunk circuits (such as PBX or E&M trunks) may appear on the loop side of the access point in some cases and may require different signaling. Over the past century, a variety of access alternatives has grown up, spawning a diverse set of signaling systems [23]. Many of these systems are clearly obsolescent—but there are still large numbers of lines and facilities in place in the network that rely on them. Even dc signaling (for pulse dialing and hook-flash) is still in wide use, several decades after it was technically superseded.

The wireless access engineer faces a choice as to whether to require 100% backward compatibility between a new radio-based access system and existing plant and services or to accept some compromises for the sake of overall economy. It is difficult to accommodate all of these older forms of signaling in a wireless access circuit. In many cases, it may be best to take a "clean slate" approach; for example, cellular systems do not plan to interface to pulse dial telephones. On the other hand, BETRS-type wireless access systems—which are often installed in rural exchanges—are usually required to support dc signaling. Yet if a $1000 radio terminal is going to be installed at a subscriber location, why not spend another $50 for a DTMF telephone? Or simply require the customer to upgrade his CPE? Another example is the coin telephone problem to which we have already alluded. Rather than try to adapt a sophisticated digital radio technology to a 40-year-old coin signaling format, it may be cheaper all around to simply replace older coin phones with modern "smart" coin phones that can interface electronically to a wireless access terminal.

Intelligent Network Capabilities. The last 10 years have seen the beginnings of a new class of enhanced services, based on modern digital switching systems, which bring so-called "intelligent network capabilities" to the ordinary end-user. Current examples of such services include call forwarding, three-way calling, call tracing, calling number identification, and even 800 and 900 services. There are many other services in development or in the early stages of deployment [24].

Wireless access systems should strive to be compatible with these future-oriented services [25]. The difficulty here arises from the fact that the specifications for many of these new services are still being developed—usually without any explicit reference to wireless access. By the same token, complete specifications may not be available to wireless access system designers or engineers. This complicates the mutual adaptation that must take place. (One example is automatic number identification, which requires on-hook transmission and may be difficult to implement within the control structures for many radio access architectures today.)

Unique Radio-Based Services. The emphasis in this section so far has been on maintaining transparency or equivalence with the wireline network. There are also totally new services based on the unique capabilities of a radio-based access system. We are still in the very early stages of this process. A number of very creative proposals have been hatched, but few have yet reached the stage of formal specification. It is instructive, however, to glance at a few of the ideas offered in Bellcore's Framework Advisory on PCS:

1. *Telepoint-Like Service* will allow origination of voice calls from within the serving range of a public radio port. . . . In addition to providing low-price one-way calling, the service can be used as an "invisible" public pay phone, especially in crime- and vandalism-prone areas.
2. *Neighborhood Alert Service* is a public safety service. . . . A portable unit provided to the subscribers enables them to alert their neighbors, or the local authorities, in an emergency.
3. *Medical Alert Service* is essentially a "reverse paging" service. The subscriber carries a pager-like device that enables the subscriber to signal someone in the event of a medical crisis. The location of the port selected by the pager-like device is identified by the network intelligence. . . .
4. *Child Locator Service* combines several aspects of traditional paging services and the unique aspects of PCS. When interrogated, a Child Locator Service

> unit identifies its wearer's location in terms of the proximity to the accessed radio port. . . . [26]

As some of these examples suggest, the ability of radio-based access systems to provide information about the location of the customer will undoubtedly suggest many interesting new services. From an engineering standpoint, it is probably too early to anticipate these types of services in the shaping of field-oriented procedures for wireless access, but a degree of flexibility should be maintained in the face of continuing technological and product evolution.

8.1.1.4 Summary: Mutual Adaptation

The challenge in engineering a wireless access system is one of mutual adaptation. One the one hand, the radio system should be designed to support as many of today's services as possible. On the other hand, this may increase the cost of the system significantly. Some of the services available today represent fossilized vestiges from the distant past, solutions that have for the most part been superseded. The cost of forcing a fundamentally modern technology such as spectrum-efficient digital radio to maintain full backward compatibility with such antiquated signaling protocols should be carefully weighed. And many of the newer, futuristic service concepts emerging from the intelligent network programs for the next century do not yet recognize the unusual requirements and technical constraints of wireless access.

The broad goal should be for wireless access to provide transparency to all important network services, and where possible, to ensure full, literal equivalence of those services with their wireline counterparts. There are many individual cases, however, in which the principle of equivalent service must be examined intelligently in light of spectrum limitations, the vagaries of the mobile channel, and other hard constraints. The most important challenge, it seems to me, is to accommodate and adapt our basic service concepts to the LBRV channel. In particular, the definition of a suitable wireless version of ISDN service offerings is needed.

Also, of course, once the equivalency issues have been suitably addressed, we can have dessert—that is, we can begin to develop the totally new and unique services that will be based on radio's unique characteristics as a transmission and access vehicle. That will be the fun part.

8.1.2 Quality of Service

Measuring circuit quality is one of the ancient arts of telephony [27]. Vast bodies of practice are built on the simple need to ensure some sort of standard performance. The subject is especially complex for analog transmission over a wireline network [28]. The many infirmities of the analog copper circuit—crosstalk, singing, echo, impedance mismatches, noise of all sorts—elicit a battalion of countermeasures and test procedures:

> For many years, test centers were staffed with skilled technicians who sat at large wooden consoles, called "test desks," which were equipped with telephone dials, meters, and an array of switches. When a customer called in a complaint, his/her line record was accessed . . . and a trouble ticket was issued to be distributed to the test center. By manipulating the switches in standard sequences, the tester could selectively apply dc and ac voltages to one or both of the conductors in the loop and observe the results of these actions on the meters. Meter responses could help determine the nature of a loop fault. [29]

Loop testing has been in many ways the heart of the telephone engineering.

The transition to digital circuits simplifies the assessment of circuit quality, for two reasons. First, it becomes possible—in principle at least—to specify a loss-less or error-free channel where a digital circuit is utilized, because of the ability to regenerate a digital signal [30]. Second, because even where regeneration is not possible (as in radio loops) or for whatever reasons the signal is degraded, most forms of degradation tend to become *homogenized* as bit-errors. Whether a particular bit-error is created by the energy from a crosstalk signal, or by a Rayleigh fade, or from the interference generated by a bolt of lightning, the impact on the circuit may be substantially the same. Nevertheless, even if the causes of bit-errors are thus amalgamated, the effects of bit-errors on different types of LBRV wireless access circuits can be very different. Indeed, the advent of wireless access shifts the focus of our interest from the causes of quality impairments to their effects. In other words, the definition of voice quality (for example) becomes crucial.

8.1.2.1 Quality of Voice Services

Wireless access brings LBRV coding into the telecommunications mainstream for the first time. This poses the difficult question of defining telephone-quality voice transmission in terms that are sufficiently objective to allow the evaluation of alternative coders. Moreover, these sophisticated coders respond to high BERs in different and often unusual ways. Testing procedures for these new LBRV circuits must also be refined.

Forms of Impairment: Digital versus Analog. Most of the impairments we are concerned with have to do with the effects of different *rates* and, to a lesser degree, different *patterns* of bit-errors. The BER, and certain other measures of the digital error rate, are the most basic measures of circuit quality for a wireless access system. This cuts two ways. On the one hand, it simplifies the testing procedures. For example, the site engineer need not be concerned with fading patterns or interference *per se,* unless he first detects a high BER that needs explanation. On the other hand, it may also complicate the diagnosis, because all "diseases" produce the same "symptom."

It should also be kept in mind that all wireless access terminals (and some network interface units) provide an analog circuit to the CPE or to the user's ear. In a PCN device, this analog segment may be practically negligible. But the subscriber line interface circuitry (SLIC) in a BETRS terminal is normally designed to drive a drop-line that may be several miles in length. Thus, there are still forms of analog impairment that can occur on this portion of the circuit. For example, idle channel noise (measured in dBrnc0) is still a useful measurement in many applications.

Measures of Codec Quality. In the absence of any channel errors, the use of an LBRV codec may alter the subjectively perceived voice quality compared to a standard 64 kb/s circuit. But a change in the tonal coloration of the voice circuit may not mean a reduction in quality. Such a judgment must be subjectively referenced. As anyone who has ever shopped for stereo speakers can attest, there are many subtle variations in sound coloration that do not easily equate to a better/worse quality assessment. "Voice quality" is clearly multidimensional and different LBRV coders represent different types of trade-offs and will produce outputs that sound different in often very subtle ways. The evaluation of baseline codec quality can be very difficult and expensive, because it must rely ultimately on controlled subjective testing by panels of representative subjects [31]. Nevertheless, the advent of wireless access and LBRV is bringing this issue to the fore. In Chapter 6 we discussed some of the different procedures and measures for making voice quality evaluations.

To date, the only standard as such in this area is the requirement promulgated by Bellcore proposing to limit the LBRV codec to 4.0 quantization distortion units (QDUs) or less [32]. As we noted in Chapter 6, while this may suggest a general benchmark, it is not adequate as the basis for coder comparisons or system engineering for wireless access systems, because of the failure to specify a realistic channel error-rate. In fact, not even a 64 kb/s codec will achieve 4.0 QDU when exposed to BER levels that are common in a real radio channel [33]. A properly designed LBRV coder should outperform a 64 kb/s PCM coder when realistic BERs are incorporated in the test.

Modes of Codec Failure. In evaluating an LBRV codec, once we have specified the baseline quality (at, say, a BER of 10^{-4}), it is also important to consider the likely failure modes of the particular codec under expected extreme conditions. Different types of failure may themselves produce different subjective reactions that must be assessed comparatively.

In an analog channel noise is perceived, roughly speaking, as *additive*—it overlays and ultimately obscures the desired signal (although there are many other failure modes in the analog radio channel—for example, the "picket fence" warbling effect for an FM channel received by a moving vehicle). The point is that analog noise is something we are to some degree accustomed to, from our everyday experience in a noisy acoustic environment. By comparison, digital channels experiencing high BERs may exhibit acoustically peculiar effects, such as clicking or popping, clipping (in a DSI channel), or more unusual deformations of tonal qualities (when an LPC-type codec takes an uncorrected hit on its predictive coefficients, for example).

The typical failure mode is the product of the entire coding system (codec, equalizer, error correction, and even channel assignment algorithms in the case of DSI). Is it more acceptable for a digital channel to break up in clicks and pops or to maintain some degree of intelligibility but lose the fidelity of tone and color? I am not personally aware of any studies that delve scientifically into this area, but my own practical experience with LBRV on radio channels is that the subjective response to the typical failure mode is almost as important as the baseline codec quality itself.

Measures of Codec Complexity. The issue of codec complexity is another enormously difficult and multidimensional problem with which the industry has only begun to come to grips. Some of the dimensions of the complexity question include:

1. Delay
2. Power consumption
3. Cost—one specification stipulates that "the complete codec should be implementable using a single fixed-point digital signal processor of moderate processing power" [34]
4. Echo cancellation requirements [35].

The measurement of complexity is not yet a science. Easy statements about the number of "mips" required to implement a particular algorithm have little usefulness in a fast-changing environment for DSP hardware. Nevertheless, the fact that there is almost always a trade-off between codec quality and complexity, as well as sometimes between different dimensions of "complexity" (e.g., power versus delay), indicates why we need a rigorous approach to the problem.

Grade of Service. When we expand the assessment of voice quality beyond the single-circuit level and begin to look at the statistics of circuit quality in a reasonably large system, we enter into what the traditional telephony world refers to as the *grade of service.* A standard definition:

> Transmission grade of service is a calculated measure of the expected percentage of telephone users who would rate the quality of telephone connections as excellent, good, fair, poor, or unsatisfactory when the connections include the effects of given transmission impairments. [36]

The calculation methods, models, and objectives for grade of service have been described in many transmission texts [37]. In general, these should be the same for an analog copper channel and for a LBRV digital radio channel.

8.1.2.2 *Quality of Nonvoice Services*

The evaluation of the quality of a digital data circuit is, mercifully, far more straightforward and quantitative. The standard measure is, of course, the BER. Other measures have been developed, mostly in response to the fact that the sources of radio channel degradations may not be uncorrelated at the bit level, but instead grouped in "bursts" (where a sequence of successive bits is affected by a single, longer external event, such as a lightning bolt). These alternative measures include:

1. Block error ratio (BLER)
2. Error-free seconds (EFS)
3. Severely errored seconds (SES) [38]
4. Word error-rate (WER) [39].

The specific standards for digital transmission (e.g., the BER requirement) should, in general, be referenced to comparable wireline requirements after the application of FEC and ARQ. The transmission rate may be allowed to drop, or the RF bandwidth increased, in compensation for difficult channel conditions, but a wireless access data circuit should provide the same error performance as that required by the external network.

8.1.3 Service Availability

"Do I get dial tone when I pick up the phone?" In somewhat oversimplified form, this is the question of service availability. (Oversimplified because we must also assume that a certain quality objective is continuously met for the duration of the call.)

In a wireless access system, service availability is the product of three things (see Figure 8.1):

1. *Signal availability:* Is there a radio signal level sufficient to support a communication link with the desired quality?

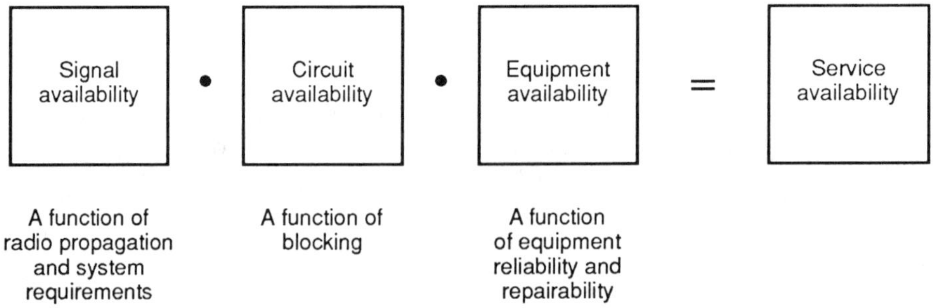

Figure 8.1 The components of service availability.

2. *Circuit availability:* Is there a trunked radio circuit available to handle the call?
3. *Equipment availability:* Is the equipment operating properly?

In other words, I may fail to get dial tone for one of three basic reasons. First, the radio signal may not be able to propagate from my terminal to the base station. Second, even if the RF link is adequate, there may not be an available radio channel or circuit to support my call. Third, there may be an equipment failure in the system that prevents the establishment of the call. From an engineering standpoint, each of these forms of availability must be handled separately. We shall discuss these in reverse order.

8.1.3.1 Equipment Availability

Equipment availability is a function of equipment reliability, ease of testing and diagnosis, and ease and speed of repair. Wireless access hardware is in this respect no different from any other sort of telecommunications gear, although the lack of a metallic circuit may pose some difficulties in adapting traditional techniques.

Reliability. The traditional measure of equipment reliability is *mean time between failures* (MTBF). The MTBF may be calculated for a component, an individual device (such as the subscriber terminal), or for the entire system.

The MTBF may be calculated (estimated) based on the circuit design, analyzed with the help of a computer model designed for this purpose. This is often the only way to obtain an MTBF in the early design and development phases of a new product. Such models are available in various military versions, where they may be integrated with the so-called "MIL-SPEC" [40], or from Bellcore [41]. In my experience, these models are useful but the results should not be interpreted literally. They may tend to understate the actual MTBF.

The other way to derive an MTBF is on the basis of actual field experience [42]. This is obviously preferable, although early failures ("infant mortality") may be higher in the initial phases of the product cycle, leading again to possible understatements of the MTBF.

A serious problem in this field, in my experience, is the difficulty that results from trying to derive useful MTBF objectives for a wireless system. It is easy for an operator to overspecify the MTBF for an individual component or device, which in a redundant system design may not be very meaningful. Another approach is to specify the number of seconds, minutes, or hours of *system outage* due to equipment failures. In an outage-oriented framework, the failure of a particular radio channel (for example) at the base station becomes less important, and properly so because the trunking across all radio channels means that no subscriber will be denied service based on such a failure. MTBF is more meaningful for the dedicated segments of the system, especially the subscriber terminal. Solid-state, sealed-unit designs can yield high reliability. Currently, manufacturers of thin-route subscriber radio and BETRS-type equipment are experiencing very good MTBFs on the terminal equipment (generally in the range of 15 years).

Testing and Diagnostics. In principle, testing of an intelligent wireless access terminal should be simple. Autodiagnostic procedures have been incorporated in BETRS and related telephony-oriented wireless systems. These tests can be initiated by remote instructions and should produce a diagnosis of possible trouble well before the customer experiences any difficulty. Unfortunately, that is not the end of the story.

One of the many ongoing revolutions in telephony is the development of automatic loop testing systems. Instead of the manual procedures and the test desks described above, systems are becoming available that offer not only a high degree of automation, but also integration with network management systems and remote provisioning centers. Loop testing is becoming a computer-controlled function at all levels.

This would be wonderful except for the fact that none of these procedures or systems ever envisioned the need to test a nonmetallic radio loop. Even the electronically derived pairs in a DLC system present difficulties:

> With a large investment in place optimized to test copper cable pairs, telephone companies were [and still are] reluctant to make significant changes in test equipment and methods to accommodate DLC systems, which could not be tested like copper loops. As a result, suppliers of test equipment have resorted to ingenious methods to allow the DLC channel and copper loops between the DLC RT and the customer to be tested with minimal impact on the embedded copper test systems. [43]

Ensuring backward compatibility of tomorrow's wireless access systems with today's loop testing systems will present an even greater set of challenges. For now, most operators are approaching the problem in a spirit of reasonableness. The real solution will come about when methods for testing wireless loops are appropriately codified.

Repair and Service. The standard measure of repairability is *mean time to repair* (MTTR). This is one area in which wireless access should be superior to wire-based loops. With a physical facility, such as a copper wire or an optical fiber, a failure in the circuit must usually be addressed head-on, so to speak. If there is a break in the cable, it must be found and spliced. The repair work must be performed where the problem is found—often knee deep in the mud, at night, in the rain, etc.

With a wireless access terminal, most field problems can be sidestepped by simply swapping out the bad terminal for a good one. The repairs can be performed off-line, at a service center or by the equipment supplier. The swap-out procedure for a BETRS installation normally takes about 20 minutes (excluding windshield time).

With portable applications of wireless access, the solution is even simpler: The customer brings the unit and may receive another unit on loan if extensive service is needed. In such cases, the MTTR (defined as the time required to restore service to the customer) may approach zero, from the operator's standpoint.

Nevertheless, certain procedures will need to be set up, including the maintenance of a small floating inventory of terminal devices. But, in general, by allowing the operator to avoid most field repair operations, wireless access should alleviate the engineering concerns about equipment availability.

8.1.3.2 Circuit Availability

A wireless access system comprises a set of trunked radio circuits. A given number of such circuits can serve a larger number of subscribers, based on the statistical nature of the offered traffic. The statistics of such systems have been thoroughly analyzed and detailed treatments are available from many sources [44]. We will not review these methods here. We shall note, however, that it is possible, using these methods, to determine the probability that a call attempt (e.g., from a wireless terminal) will be *blocked*—that is, the probability that no trunked radio circuit will be available for a particular subscriber when he requests it, due to the fact that all available trunked circuits are in use. This is what we mean by the *term circuit availability.*

It should be emphasized that circuit availability is a matter of *standard traffic engineering.* There is no difference between the traffic engineering for a trunk group in a transatlantic fiberoptic link and the engineering for a wireless access trunked radio system. There is some novelty about applying traffic engineering procedures *to the loop,* which has traditionally been based on dedicated plant (which does not require traffic analysis). But once this slight psychological barrier has been surmounted, the task is entirely familiar and mundane.

Traffic models can be simple or complicated. For the vast majority of actual wireless applications today, the simplest models are entirely adequate (in my experience). There are essentially only two parameters of concern: the *blocking rate* or *blocking probability,* and the *offered traffic* during the *busy hour.*

Blocking. The blocking probability is the percentage of call attempts that will be unsuccessful—i.e., where access to a circuit is not granted within some specific short interval of time. Normally, this number is given in the form of an external objective. For example, in telephony applications, a blocking rate of 0.5% ($P = 0.005$) is typical. This says simply that 1 out of every 200 call attempts will fail to get through. In the FCC rules governing cellular radio, a 2% blocking rate (1 failed call in 50 attempts) has been specified. It is clear that today the cellular blocking rates during the busiest periods are much higher. The reference to time periods points to the need to specify or qualify the blocking rate in terms of the busy hour (of which there are several definitions).

The blocking rate is engineerable; that is, the operator can select the blocking objective and design a trunked system to meet it. By adding or subtracting trunks to match the traffic generated by the customers, or by limiting the number of customers on the system, any desired level of blocking can be achieved. *There should be no inherent blocking rate for any true wireless access architecture.*

Offered Traffic. The offered traffic is the average amount of time that a subscriber uses the circuit during the busy hour. It is measured in Erlangs or hundreds of call-seconds (CCS). Traditionally, the telephone industry has used a nominal figure for offered traffic on loop circuits, usually 0.1 Erlangs per subscriber (3.6 CCS, or six minutes of circuit usage during the busy hour). Actual traffic figures are often considerably lower (0.02 to 0.05 Erlangs per customer). As noted in Chapter 7, the cost implications can be very significant. It is highly recommended that actual traffic be measured and used to size wireless access systems wherever possible.

More Complex Models. As wireless access systems and other types of trunked loop architectures become more commonplace, it may prove useful to expand the traffic model to address additional factors. These may include:

1. *Queuing:* Rather than blocking calls, it is entirely possible to set up reasonably short queues (with a maximum of, say, 60 seconds) that could allow even larger numbers of subscribers to be successfully served with a given number of radio trunks. Such systems may be quite acceptable to the consumer, especially if combined with some sort of priority override capability (which would be charged at a higher rate).
2. *Priority groups:* There is considerable latitude for the creative application of priority concepts in trunked radio systems. That is, different groups of subscribers (paying different rates) might be engineered for different blocking probabilities.
3. *Voice messaging:* The next generation of truly intelligent voice messaging systems may allow the system to slide a lot of low- to medium-priority traffic, especially one-way messages, into off-peak periods. The wireless access terminal could actually be integrated with mailbox features, to receive and store digitized messages (much as some fax machines today are capable of holding incoming data in memory).
4. *Digital speech interpolation:* As DSI becomes common, more sophisticated traffic models may take talk spurt statistics into account. The size of the DSI pools (which may be dynamic) will also be a consideration.
5. *Modeling the control circuit:* More sophisticated models may also perform a separate traffic analysis for the call setup and other control messages on the system.
6. *Bandwidth, or type of traffic:* As nonvoice traffic becomes quantitatively more important, it will become necessary to model the different characteristics of different forms of traffic, such as different holding times, channel utilization,

bidirectional symmetry, and bandwidth. We may even want to build error statistics into the traffic models (e.g., voice can tolerate higher BERs than data, and a CDMA system might be able to carry more voice circuits than equivalent bit-rate data circuits).

Summary. The payoff for the added complexity of future wireless access traffic models will be the opportunity to *gain effective capacity* by handling the traffic in a smarter, more flexible manner. The ultimate pragmatic concept of spectrum efficiency (alas unmeasurable) is the greatest number of messages carried on a given amount of spectrum. Because one message differs from another in potentially important ways—priority, urgency, tolerance for delay, circuit symmetry—we can look to exploit these differences to increase the carrying capacity of the system.

From an engineering standpoint, however, all these complexities will still be handled by pushing a button on a computer model. The challenge will be for the model builder, not the model user.

8.1.3.3 Signal Availability

Signal availability has to do with whether there is sufficient radio energy reaching the receiver, relative to the noise thresholds required by the modulation scheme, to produce an acceptable (low) BER in the demodulated signal. The actual received signal level (RSL) fluctuates over time as a function of changes in the environment, including atmospheric changes, foliation, and the behavior of reflectors that can alter the multipath propagation patterns. For a mobile terminal, all the Rayleigh fading phenomena become especially important. In short, the RSL cannot be predicted precisely, but its behavior follows certain statistical rules and can be modeled over time (Figure 8.2). We also know the threshold RSL required for a given BER output. We can therefore predict the percentage of time during which the RSL will be below the threshold, producing an unacceptable (i.e., unavailable) signal.

The discussion of signal availability naturally leads to the discussion of radio propagation and path modeling, which is the subject of Section 8.2. The important point here is that *signal availability is largely engineerable.* The RSL may be for all intents and purposes "given" by the environment, but the relationships between RSL, SNR, BER, and availability margins can all be engineered, at least potentially, except in the most extreme cases.

8.1.3.4 The Crux of Service Availability—Blocking

Service availability—the probability that a subscriber will obtain access on demand to a high-quality wireless access circuit—is the product of the three components described in this section: equipment availability, circuit availability, and signal avail-

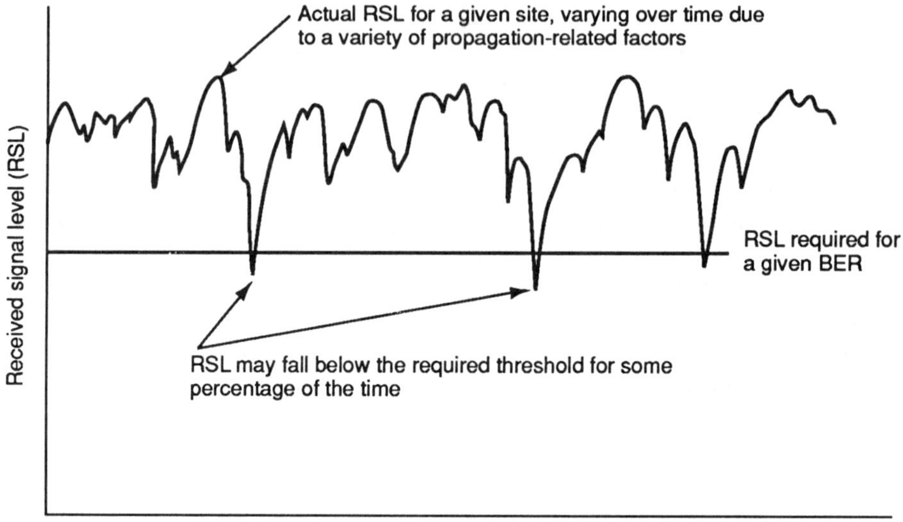

Figure 8.2 Signal availability, RSL and margin.

ability [45]. It should also be apparent that one of these components—circuit availability, the complement of the blocking rate—usually dominates the equation.

Consider a BETRS-type subscriber located 20 miles from the base station. Let us assume that equipment reliability is such that an MTBF of 15 years has been calculated for the dedicated subscriber terminal, with an MTTR of 20 minutes plus 4 hours of service outage prior to repair. All other system elements are inherently redundant. That yields an approximate equipment availability figure of 99.997%.

Let us assume that signal availability has been engineered at 99.99% (a typical requirement for a telephony application). Let us further assume that the system has been designed for a P.01 grade of service. In other words, circuit availability is 99% (i.e., the system has been engineered for 1% blocking).

We can see right away that the probability of failing to gain access due to blocking is at least 100 times higher than the probability of either an equipment failure or an RF link failure. Indeed, the product of these three numbers yields a *service availability* figure of 98.99%. If the blocking is reduced to 0.05% (increasing circuit availability to 99.5%), the service availability would be 99.49%. In other words, the combined effect of equipment and signal availability is only 1/100th of 1% reduction in service availability beyond the basic blocking rate.

The point is that in engineering such a system one should not be overzealous in setting requirements for the less important factors, especially for signal availability. In my experience, it is not uncommon for telephone company engineers, who are relatively unfamiliar with radio in the loop, to insist on unnecessarily high signal availability requirements such as 99.999% or even higher, when 99.9% signal availability would be more than adequate [46]. The difference may seem unimportant, but it can make a great difference in the nominal coverage area achieved by a wireless access system, as we will see in the following section. Indeed, if signal degradation events are plotted in time against calling traffic patterns, the effects of potential RF problems will often be seen to occur during periods of low usage. This means that such techniques as adaptive modulation (boosting the robustness of the signal in return for an expanded circuit bandwidth) can be used for off-peak periods of possible poorer signal conditions, without reducing peak system capacity.

8.2 RADIO ENGINEERING FOR WIRELESS ACCESS—AN OVERVIEW

As I have tried to indicate throughout this book, radio is in many ways a strangely forgiving medium, especially for digital transmission schemes at frequencies below 2 GHz or so. In most respects, loop engineering is easier with radio-based access. Once the radio coverage area is laid out, the planning process for providing service to a particular subscriber consists of little more than sticking a pin on a map, or, with the interactive computerized tools now becoming available, clicking the mouse on a screen.

Nevertheless, radio is still unfamiliar. For all its intractability, the copper loop still represents a tangible link between the subscriber and the network, whereas a radio wave travels through the mysterious ether, which nobody really owns or controls. It is the cousin of light, and yet it bends around corners and gets into crannies—except that sometimes it doesn't. It may penetrate yards of physical structure, and yet be defeated by a metallic window coating or by the coming of spring (and leaves on the trees).

Because of the uncanniness of radio transmission, the subject of RF engineering often preoccupies those approaching it for the first time. I have often remonstrated with telephone company engineers that this concern is largely unwarranted. Indeed, I firmly believe that if a few basic principles are followed, engineering a wireless access system should be a black box type of exercise. The plant engineer need not understand what radio is in any deep sense, nor be concerned with how many decibels can dance on the end of an antenna. *Wireless loop engineering should become basically a matter of reading maps* —and occasionally performing a diagnostic routine from a computer console. And reading maps lends itself very well to computerization. (Cochannel interference limited systems, such as cellular radio, are more challenging in one respect—siting the base stations and designing frequency plans. Nevertheless, even these more complex tasks should be reducible to fairly simple practice, supplemented by powerful computer software.)

I intend in this section to offer a very brief overview of the process. I shall be describing the practical steps for solving RF engineering problems, and in passing introduce a few basic concepts. For the details (which I am convinced are largely unnecessary for successful engineering), the reader is once again referred elsewhere [47]. And once again, the applications model I draw on is the wireless local loop or BETRS-type model. In one respect, this is a simpler application than most: The subscriber terminals tend to be mostly fixed, which allows the use of directional antennas for improved gain. On the other hand, mobile applications are usually engineered far more loosely than fixed applications. In part, this is due to the heritage of mobile radio in which the service and coverage expectations are certainly nowhere near the 99.99% availability level on which telephony applications usually insist. Traditional mobile radio coverage models have been far more catch-as-catch-can with respect to availability objectives [48]. Also, in part, it is because mobile services can afford to be less stringent, and therefore less precise, since the subscribers do have the capability to escape from small "holes" in the coverage. Broad geographical averages are therefore more acceptable. With fixed subscriber stations bolted to telephone poles, if the RF engineering has lacked sufficient precision to predict a small but deep shadow, the results may be very unsatisfactory and costly to correct. Therefore, I would contend that the extra decibels of the directional antenna and the less severe multipath notwithstanding, the fixed telephony subscriber is more challenging to the RF engineer than the mobile subscriber.

8.2.1 The RF Budget: Analyzing Circuit Requirements

The ultimate goal of RF engineering is to ensure that the wireless access customer answers "yes" to the following questions [49]:

> Does it sound okay?
>
> Is it reliable?

More precisely, the goal of engineering practice is to develop measures for translating these questions into fairly simple, quantitative terms suitable for field engineering purposes, as well as reasonably accurate predictive capabilities. As one source comments:

> Obviously, good engineering practice does not entail installing equipment, asking the subscriber these questions, and then modifying the system until the answer to both questions is "yes"; the engineer needs to understand the factors that will influence the response to these questions and then design the system accordingly. [50]

The more proximate goal of RF engineering, therefore, is to develop the ability to predict the answers to these questions from measured or estimated radio signal levels. In the following sections, we examine briefly some of the steps in the process.

8.2.1.1 *Circuit Quality, Reliability, and RSL*

Circuit quality is a function of several interrelated variables, including coder BER performance, the form of RF modulation, and the RSL, measured in dBm. These relationships will determine the *required RSL*—the signal level at the receiver necessary to ensure a good quality voice circuit. (We shall concern ourselves in this section only with voice circuit quality.)

Reliability, then, is a function of the difference between the *required RSL* and the *actual RSL* as measured at the receiver in an actual field installation. Normally, the actual RSL is higher than the required RSL, and the difference is known as the *fade margin.* The size of the fade margin can be related statistically to the fluctuations in the actual RSL (which is, of course, a sampled value in any actual field measurement) to determine signal availability.

If availability is being engineered to a certain level, then we can use these same relationships to specify how large a fade margin in dBm we must see in order to assure the required performance.

We will want to reduce both questions to a function of RSL, which is the best variable for field measurements and for modeling propagation paths. To do so, we will exploit the nicely chained relationship between audio quality, BER, SNR, and RSL (see Figure 8.3).

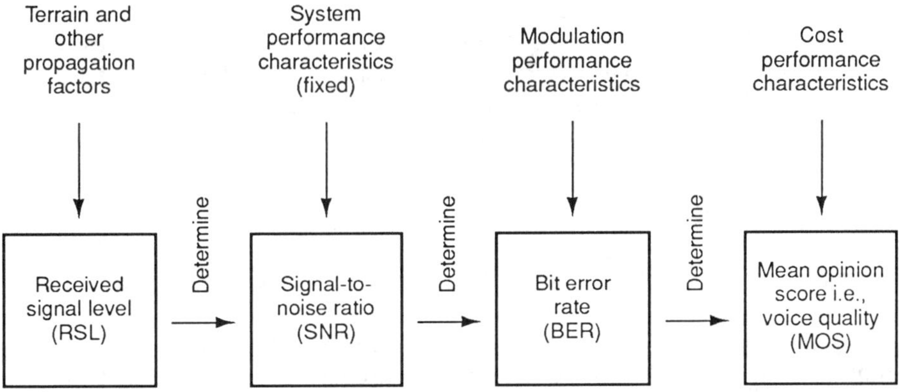

Figure 8.3 The RSL-SNR-BER-MOS chain.

8.2.1.2 *The BER Criterion*

For a digital voice circuit, and especially for an LBRV circuit, there is an important relationship between voice quality and BER. Essentially, this is a function of the voice coder (after equalization and error correction have been performed). Figure 8.4 shows a typical BER/quality curve for a particular coder [51]. There will be a unique curve for each particular coding algorithm.

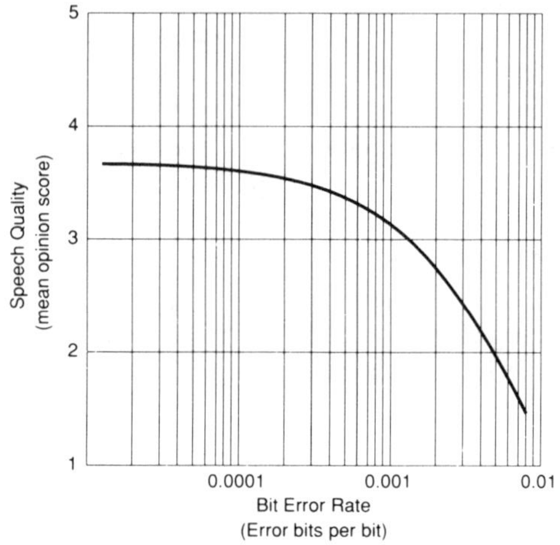

Figure 8.4 MOS-BER relationship.

Based on this relationship, we can select a BER criterion for circuit availability. The advantage of doing so is that BER can be easily measured, while voice quality is almost impossible to measure in the field in any objective way. For this particular coder and application, a BER criterion of 10^{-4} was selected.

8.2.1.3 The SNR Criterion

The next question becomes: What SNR is required by the demodulator to achieve the BER criterion?

Again, there is a predictable relationship between SNR and BER for each form of modulation. Figure 8.5 shows the SNR/BER curves for 4-level and 16-level differential PSK [52]. Thus we can now translate our quality objective into an SNR objective. In this case, approximately 18 dB is needed for a BER of 10^{-4} in 4-phase operation and 28 dB for 16-phase operation.

8.2.1.4 The RSL Criterion

Next question: What RSL is required at the receiving antenna in order to provide the necessary SNR to the demodulator?

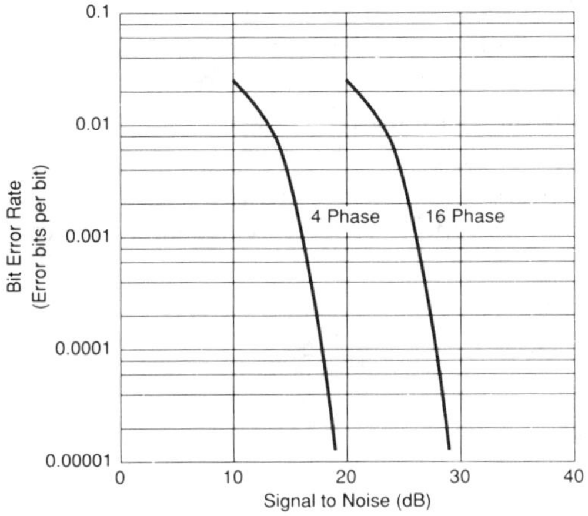

Figure 8.5 BER-SNR relationship for different modulation formats.

To answer this, a number of adjustments must be made to the SNR figure, the effect of which is basically to add a few dBm to the requirement to account for certain noise and/or loss factors associated with the radio equipment and antenna. For example, we may add about 9 dB to the 18-dB SNR required in our example for 4-phase operation at 10^{-4} BER [53]. The adjusted requirement is for a 27-dB SNR at the antenna. (The specific values depend on the equipment and modulation scheme employed.)

To derive the required RSL, we simply add this adjusted SNR requirement to the thermal noise floor—nominally −131 dBm [54]. Thus, for this example, the required RSL would be −104 dBm. In other words, we need to make sure that the received signal strength is at least −104 dBm in order to ensure at least 27 dBm of SNR with respect to the thermal noise floor.

8.2.1.5 Signal Variations and Margin Requirements

The actual RSL is subject to fluctuations over time, due to various forms of path loss (such as foliage changes on the vegetation or atmospheric effects). Thus, if we engineer a subscriber for an RSL of −104 dBm as measured on a given day, there will be some percentage of the time that the actual RSL will be lower than that, and the SNR and BER criteria will not be achieved. There are complex statistical relationships between a measured (or predicted) RSL at a given point in the field, the actual RSLs that will be experienced over time, and the required RSL for the desired BER performance. Unfortunately, this is an area of theoretical uncertainty. One of the standard models for RSL statistics is the so-called *Nakagami-Rice distribution func-*

tion [55] (see Figure 8.6). Nakagami-Rice provides a prediction of the cumulative distribution, that is, how often the RSL will be below a specified threshold. Figure 8.6 shows several different curves describing the relationship between the amount of fade margin and the signal availability. Without delving too deeply into the interpretations of this analysis, it may be observed that depending on what assumption we make about the gross characteristics of the propagation path (i.e., which curve we select), the fade margin required for 99.99% availability may range from 15 to 40 dB, a large range indeed.

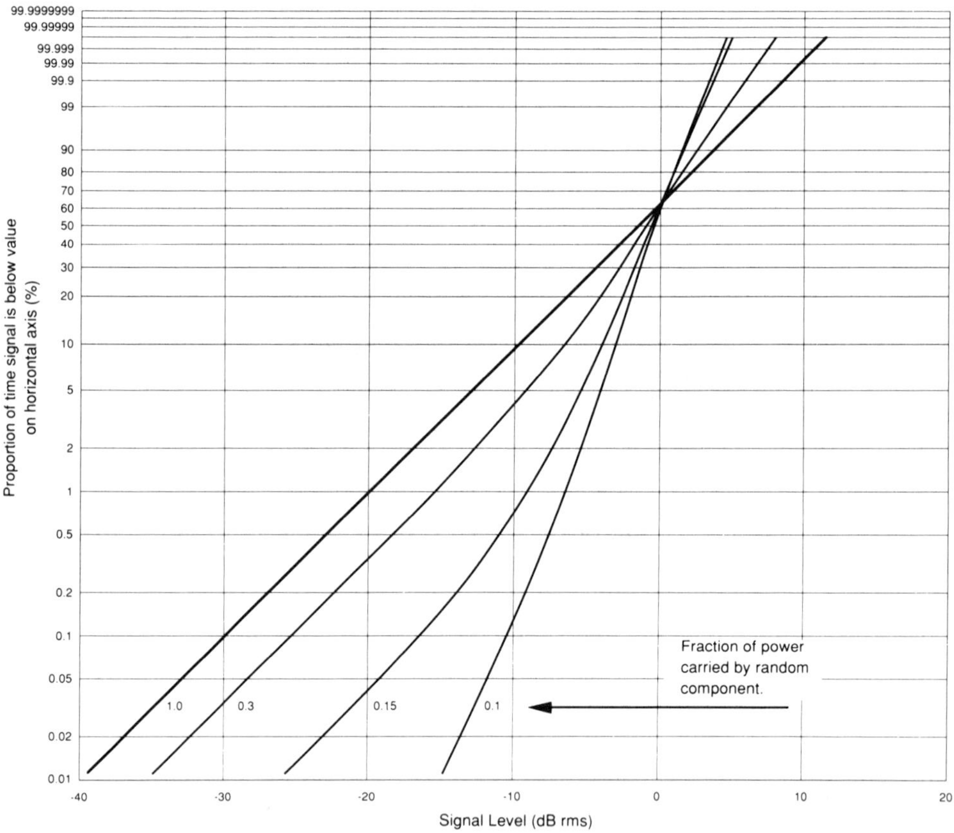

Figure 8.6 The Nakagami-Rice distribution.

In our experience with BETRS-type applications, the multipath components are usually not dominant, and a fade margin of 15 dB is often adequate for fixed service [56]. Thus, if a required RSL of −104 dBm is needed at the antenna, a measured RSL of −89 dBm should be adequate for ensuring that we have the desired SNR 99.99% of the time.

For a mobile application, however, where severe multipath may be anticipated, a fade margin of 30 to 40 dB may be more prudent, leading to an RSL criterion of −64 to −74 dBm.

Unfortunately, at the end of the day, we must still admit that these numbers are very crude. It is not clear that models like Nakagami-Rice are adequate for the large-scale deployment of wireless access, especially where they may lead to excessive conservatism in engineering. It is the easiest thing in the world to throw a few more dB onto the margin requirement. I have dealt with telephone company engineers who insisted on using a 40-dB fade margin for short (e.g., seven miles) essentially unobstructed line-of-sight paths to fixed BETRS subscribers. The result is to reduce coverage and often to render radio quite a bit harder to prove-in. With more experience with wireless access, we should be able to model the fade margin requirements and availability statistics on more of an empirical basis.

8.2.2 Path Analysis: Predicting RF Circuit Characteristics

Knowing the RSL requirement (with margin) is half the battle. The other half is predicting the actual RSL for a specific subscriber or for a large area that may require wireless access service. Of course, for an individual subscriber at a known location we can measure the RSL directly in the field. But it is highly desirable to have a reasonably good idea of the likely RSL ahead of time. This will indicate the coverage we may expect from a given base station and alert us to any specific subscriber engineering requirements. As for planning whole systems, it is impossible to make enough field measurements to feed a coverage model with any reasonable degree of resolution. A predictive model is essential.

8.2.2.1 Components of Path Loss

The radio propagation between two points is predicted by means of a *path loss model.* Path loss is the amount of power lost by the radio signal as it travels between the transmitter and the receiver, or, to put it another way, the difference in power at the receiver (RSL) compared with the transmitted power. It is measured in dBm.

The path loss in between a transmitter and a receiver is a function of frequency, distance, and terrain (including vegetation). Experience with radio systems has led to the development of a number of mathematical formulae that describe, often crudely, the relationships between these factors and signal level. We examine each of these factors briefly below.

Path loss can be calculated for a single path or for a variety of contiguous paths to achieve a feeling for the radio propagation over a larger area (such as a cell). Point-to-point path loss calculations are the basis for area coverage models, where the path is calculated iteratively. Although path losses can be calculated by hand for a single point-to-point analysis, it is obvious that area coverage analyses—which may involve hundreds of thousands of point-to-point calculations—must be computerized. We will also discuss briefly some of the computerized tools now becoming available for this function.

8.2.2.2 Frequency-Related Effects

Frequency is an overriding factor in predicting radio propagation. It determines the way in which the radio wave interacts with its environment. Some of the basic frequency-related effects are:

1. The higher the frequency, the more "optical" the transmission path and resulting coverage area. In other words, as we move higher in frequency, the photons behave more like light. Propagation becomes more line of sight with less of the "bending around corners" found at lower frequencies. Shadowing becomes sharper, and penetration through intervening physical matter decreases.
2. The higher the frequency, the more power required for a given coverage area. Or, said another way, less range is achieved for a given transmitter power. This effect may be pronounced at 2 GHz and higher frequencies now being considered for wireless access applications. Operations at these frequencies may be confined to smaller cells.
3. The higher the frequency, the more susceptible the transmission becomes to absorption by, for example, water in its various forms (rain, fog, humidity). Incidentally, this also means that higher frequencies engender greater potential health concerns.

Within any given application, however, the frequency is fixed, and can usually be handled as a distance/loss formula. For today's common radio-telephone frequency bands, such as the 150- and 450-MHz regions, and more recently the 800- to 900-MHz bands, there is a substantial body of field data on propagation characteristics. It is important to note that such relationships (e.g., see Figure 8.7 for the distance/loss function for 460 MHz) must be established empirically. Thus, many propagation characteristics of some of the higher frequency bands now being proposed for broad-based wireless access applications (such as the 2-GHz region) are still comparatively unknown. The next few years will undoubtedly witness a great deal of propagation testing at these frequencies to help establish the empirical basis for path loss and area coverage calculations.

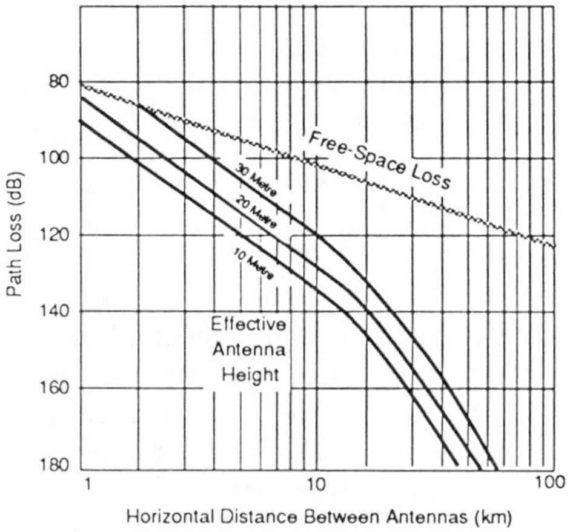

Figure 8.7 Distance/loss function for 460 MHz.

8.2.2.3 Distance-Related Effects

For a given frequency, the signal loss as a function of distance is given by a curve similar to those shown in Figure 8.7. Antenna height is one important variable. Note also the "knees" in the curves at 20 km or so where the horizon will normally begin to obstruct the transmission.

Another way that the distance/loss function is sometimes described is in terms of the R factor, where the factor is the exponent of the inverse of the radius used to describe the loss of signal power as a function of distance. In free space, this factor is equal to 2, so that the decline in signal level is given as $1/R^2$. For real-world situations, the R exponent may be set at anywhere from 3 to 6. Its value reflects a generalization of unpredictable terrain and horizon effects. Sometimes, path loss may be calculated with a changing R exponent; for example, $1/R^3$ may be used for the first ten miles, then $1/R^4$ for the next ten miles, and $1/R^6$ for the rest of the path.

8.2.2.4 Terrain-Related Effects

The best prediction is achieved when we can base the prediction on a model of the actual terrain intervening between the transmitter and receiver. Of course, if there is a clear line-of-sight path between both ends, then the specific terrain features do not matter and distance factors alone can be used to calculate the predicted RSL. But if there are obstacles in the way, such as hills, then a terrain-sensitive model should be employed.

The mere existence of such obstacles, however, does not mean that an acceptable signal cannot be obtained. At least at lower frequencies (such as 450 MHz), the interaction of the radio wave with the terrain may take a number of forms, including reflection, diffraction, and refraction. All of these should be incorporated into the path loss algorithm.

8.2.2.5 Path Loss Models

A path loss model is a mathematical formula that combines the various factors mentioned above, especially terrain and distance, and produces a predicted RSL for a given transmitter over a given path. The mathematical treatment of the various forms of terrain interaction in particular is complex, and at least so far it is not an exact science subject to a definitive solution. Different researchers or engineering firms may use different path loss models. Some models are publicly available, such as the Terrain Integrated Rough Earth Model (TIREM) developed by the U.S. Department of Defense. On the other hand, some of the better models are still proprietary [57]. The results of any model must be validated in the field, by means of a reasonable number of actual versus predicted RSL comparisons.

8.2.3 Area-Coverage Analysis

As we have stressed in the previous chapters, one of the most important aspects of wireless access is the added dimension of area coverage that it opens up for loop plant planners, economists, and engineers. To take full advantage of this added dimension, we need a model that is capable of predicting RSLs with a high degree of accuracy and detailed resolution over a large area. The best way to do this is to set up an iterative algorithm that analyzes a large number of propagation paths to contiguous points in the terrain. The only really good way to do this is with a computer.

Area coverage analysis (ACA) is really just beginning to come into its own. The simplistic models used for cellular propagation analysis are being supplanted by far more precise models needed for high availability systems such as BETRS [58]. These newer models are more precise in two senses. First, they tend to employ more sophisticated path loss formulas, which give better accuracy for specific locations than has normally been the case for mobile system analyses (where after all, it is argued, the moving vehicle effectively averages its location anyway). Second, the newer models use the latest available digitized terrain data, with a resolution of 3 seconds of arc, or about 200–300 feet between data points. All in all, for a typical BETRS-type ACA, the model may encompass more than 135,000 specific points and path loss calculations.

The results of such a model are displayed in a detailed ACA overlay. The overlay, which may be color-coded for easy field application, becomes in effect a permanent part of the record for the wireless access system. It shows the areas where service can be provided without any special engineering, as well as those areas where some additional subscriber site engineering may be required (see Section 8.3). Since general terrain features may be expected not to change very much, the ACA can serve as the basis for long-range and crisis-response planning. With an accurate ACA, it becomes possible to realize the benefits of area coverage to streamline the planning process for access facilities using radio.

Cochannel interference analysis is also possible with these sorts of models. In analyzing C/I ratios from two or more transmitters, it is necessary to run multiple ACAs and form the ratios at each point for the RSLs from the various transmitters.

8.3 SUBSCRIBER SITE ENGINEERING

Engineering for the base station equipment for a wireless access system is similar to engineering many other types of central office or microwave radio equipment. Normally it does not fall within the province of outside plant engineering. We shall not cover this topic here. The extension of some base stations into the field environment, as *radio ports* proposed by the Bellcore PCS advisory and others [59], will challenge outside plant engineers somewhat; however, since few if any of these sorts of systems have yet been deployed, it is rather speculative to discuss possible engineering practices for such devices.

On the other hand, subscriber site engineering is a new field for the outside plant engineer, which has become important as thin-route microwave and BETRS-type wireless access systems have entered the commercial marketplace. Engineering a wireless access subscriber station involves a number of novel methods and procedures, to address steps such as:

1. Subscriber transceiver installation, especially in an outdoor environment
2. Grounding
3. Provision of main and backup power
4. Interface to inside wiring and CPE
5. Test and cut-over to service.

For a detailed treatment of these issues, the reader is referred to the technical manuals available from the manufacturers of wireless access equipment, which address these topics in a very thorough manner [60].

In addition to these basic procedures, there are a number of special requirements for wireless access site engineering related to the goal of assuring that a specified signal availability criterion is met. These measures do not need to be applied within the standard coverage zone as defined by the area coverage analysis. But beyond this zone, in the fringes of the coverage area or in any section of the coverage area that experiences low RSLs (because of terrain blockages or heavy foliage, for example), it may be necessary to implement one or more of the following measures.

8.3.1 Modulation Level Selection

In some BETRS systems, it is possible to select the modulation level—i.e., 4-phase or 16-phase—on an adaptive basis, depending on link conditions. The effect is to gain dB in the link (and fatten the fade margin for a given RSL) by reducing the SNR criterion by 10 to 12 dB. The result is to expand the coverage zone, often substantially. This is one method for engineering improved availability for fringe subscribers that is relatively painless, since it does not affect subscriber hardware. Other software-based capabilities, including FEC and equalization, could in principle be handled in the same manner.

8.3.2 Subscriber Antenna—Gain

Another relatively easy engineering approach is to increase the gain at the subscriber by using a more powerful antenna. If the standard is a single yagi (9 dB of gain), a dual yagi (12 dB of gain) or a quad yagi (16 dB) can improve the fade margin and raise availability for a given RSL. Once again, the practical effect is to expand the coverage zone for a given transmitter. The ACA should be color-coded to indicate unambiguously to field engineers where these optional subscriber antennas would have to be used.

8.3.3 Subscriber Antenna—Height

Subscriber antenna height is another engineerable factor that can be used to improve availability for a given RSL.

8.3.4 Subscriber Drop-Line

Still another factor to be considered in field engineering wireless access systems is the ability of many subscriber terminals to drive a lengthy drop-line between the transceiver and the CPE (Figure 8.8). In some equipment specifications, the drop may be powered at up to 1300 ohms or more, which is sufficient to extend a copper drop several miles or more. What this means is that if a customer is located in a troublesome radio "hole" or "shadow," it is often possible to move the subscriber station and antenna a mile or two to one side to get out of the shadow and obtain an adequate RSL. A good ACA tool can help plan this type of installation.

(a)

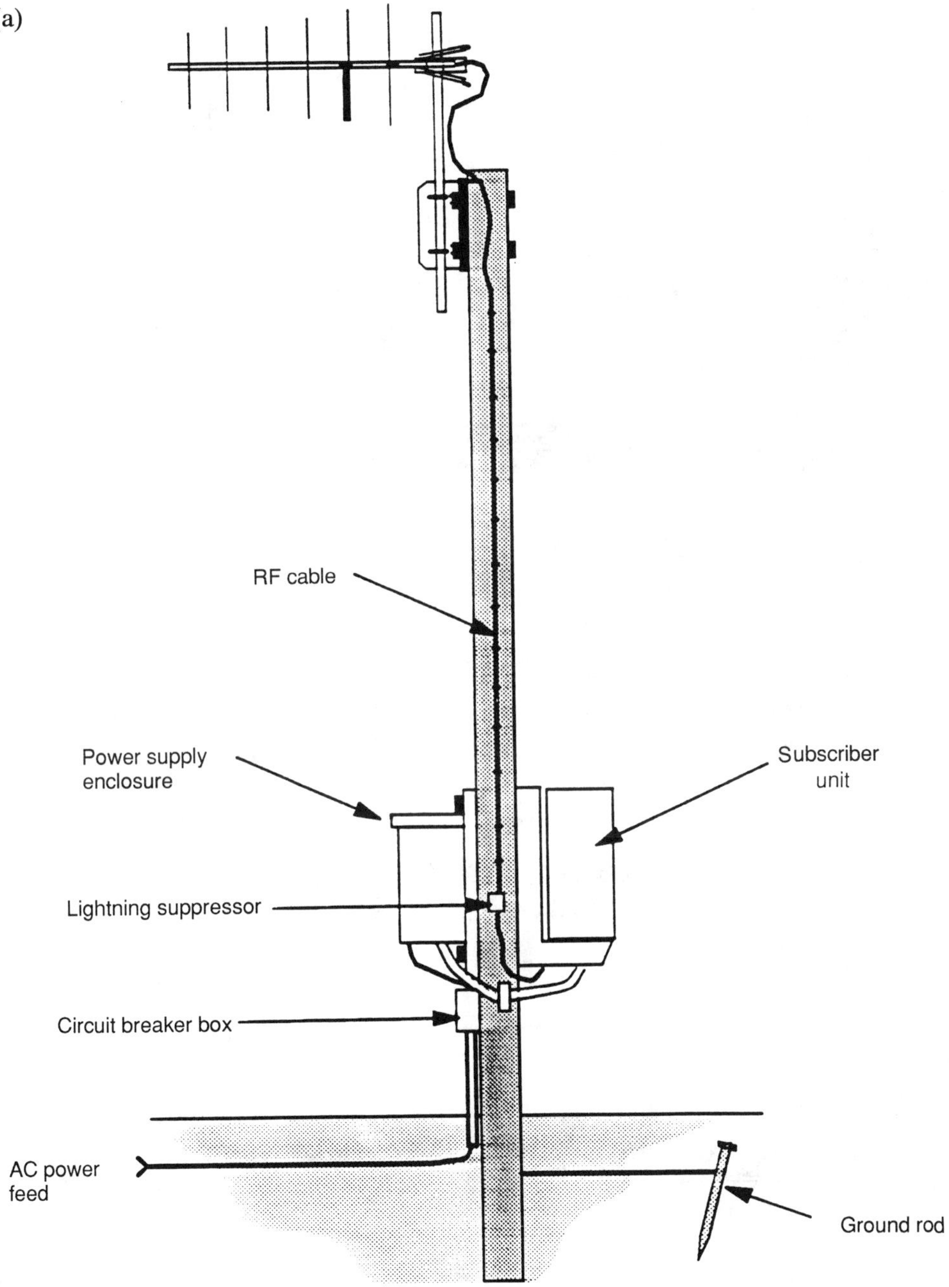

Figure 8.8 Schematic of lengthy drop line from subscriber terminal; (a) outdoor installation, (b) subscriber station cabling. (From *Ultraphone* manuals.)

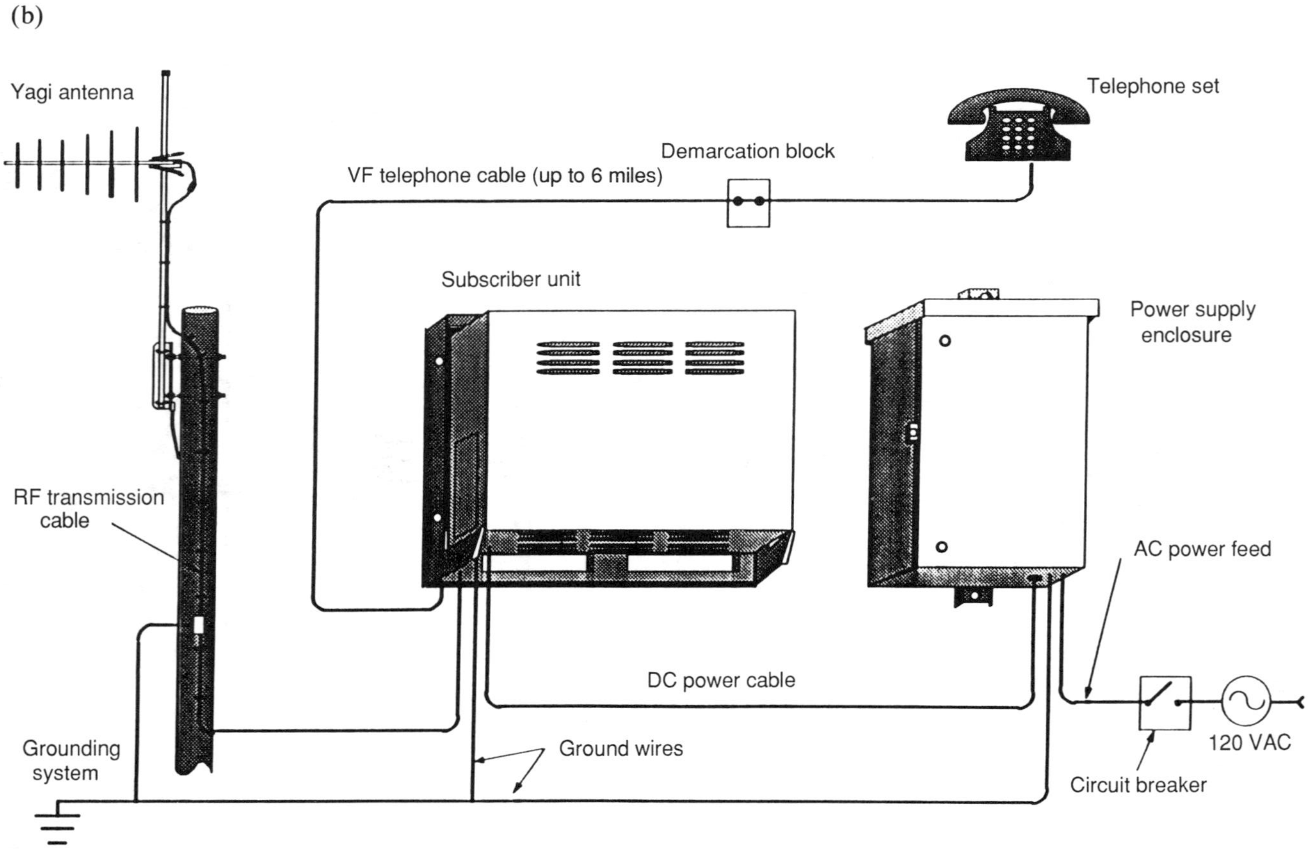

Figure 8.8 continued

NOTES

1. Many wireless access systems also provide for a rather wider window via over-the-air diagnostics that are available to the system operator (but often not directly available to field maintenance personnel). Still, the total amount of information available is quite limited, by design, in accordance with the black-box philosophy.
2. TOPS stands for Task Oriented Procedures, a Bellcore-developed and highly specified approach to organizing the documentation for many engineering procedures.
3. See Equipment Manuals for the Ultraphone 100 system available from IMM Corporation.
4. U.S. Department of Agriculture, Rural Electrification Administration, "BETRS Radio Systems," Bulletin 1751H-703, June 20, 1991, draft.
5. For example, Bellcore, *Generic Requirements for Basic Exchange Radio Systems,* TA-TSY-000911, Issue 2, July 1989.
6. For example, the statement that a cell outline represents a "90/90 coverage zone"—where 90% of the users will have a "good" signal 90% of the time—is so vague as to be practically meaningless, as many cellular engineers will at least privately admit. At IMM, as we began to have to model RF coverage and propagation for our BETRS customers to a 99.99% availability criterion, we found that many of the commercially available propagation modeling services did not provide reliable information to this level of accuracy.
7. AT&T, *Engineering and Operations in the Bell System,* Bell Laboratories, AT&T, 1977, Chap. 4; also Bellcore, *Telecommunications Transmission Engineering, Volume 3: Networks and Services,* 3rd ed., Bellcore Technical Publications, 1990.
8. Surveys of voice services are provided in AT&T, *Engineering and Operations in the Bell System,* Bell Laboratories, AT&T, 1977; Bellcore, *Telecommunications Transmission Engineering, Volume 2: Facilities,* 3rd ed., Bellcore Technical Publications, 1990; and Bellcore, *Telecommunications Transmission Engineering, Volume 3: Networks and Services,* 3rd ed., Bellcore Technical Publications, 1990. More detail is to be found in the many Bellcore technical references and technical advisories that are available for almost every segment of the network.
9. Bellcore, *Telecommunications Transmission Engineering, Volume 1: Principles,* 3rd ed., Bellcore Technical Publications, 1990, p. 666; also see Chap. 12 for a general description of speech signals.
10. *Ibid.,* Chap. 13.
11. *Ibid.,* p. 667.
12. There is a partial exception to this in the use of delta modulation in early DLC systems. See G.T. Hawley, "Historical Perspectives on the U.S. Telephone Loop," *IEEE Communications Magazine,* March 1991, pp. 24–28. However, low-bit-rate DLC did not become a widespread application.
13. See Bellcore, *Technical Analysis Report: Ultraphone 100 Codec Subjective Voice Quality Analysis,* AR-NPL-001041, Issue 1, January 3, 1989. Note: The distribution of this document may be limited to Bellcore and authorized clients.
14. See discussion of acceptable coder processing delays in the IS-54 digital cellular specifications and the GSM specifications, cited in notes 4 and 6 of Chap. 5.
15. Echo cancellation is performed in software by the BETRS subscriber terminals manufactured by IMM.
16. See Bellcore, *Telecommunications Transmission Engineering, Volume 3: Networks and Services,* 3rd ed., Bellcore Technical Publications, 1990.
17. For example, in early applications of BETRS with GTE of the Southwest, following the guidelines set out in the *Product Specification for Rural Subscriber Radio,* April 20, 1983, prepared by the GTSW Network Planning Department, the emphasis was on "good transmission quality" with implicit reference to voice services only, as discussed in Section 5.3.4. However, for later applications, facsimile transmission and 2400 b/s data modem transmission became firm POTS requirements.

18. E.S.K. Chien, D.J. Goodman, and J.E. Russell, "Cellular Access Digital Networks," *Proceedings of the International Conference on Digital Land Mobile Radio Communications,* Venice, June 30–July 2, 1987, pp. 84–93.
19. Bellcore, *Telecommunications Transmission Engineering, Volume 3: Networks and Services,* 3rd ed., Bellcore Technical Publications, 1990, p. 226.
20. *Ibid.,* p. 227.
21. *Ibid.,* p. 227.
22. *Ibid.,* p. 225.
23. Bellcore, *Telecommunications Transmission Engineering, Volume 1: Principles,* 3rd ed., Bellcore Technical Publications, 1990, Chapter 13.
24. Bellcore, *Telecommunications Transmission Engineering, Volume 3: Networks and Services,* 3rd ed., Bellcore Technical Publications, 1990.
25. Bellcore, *Generic Framework Criteria for Universal Digital Personal Communications Systems (PCS),* FA-NWT-001013, Issue 2, December 1990, Sec. 2.2.6.
26. *Ibid.,* Sec. 2.2.
27. See Kenzo Takahashi, "Trânsmission Quality of Evolving Telephone Services," *IEEE Communications Magazine,* October 1988, pp. 24–35.
28. See the illuminating first chapter of Bellcore, *Telecommunications Transmission Engineering, Volume 3: Networks and Services,* 3rd ed., Bellcore Technical Publications, 1990, for a discussion of the requirements for designing an analog network. Chapter 5 of the same work focuses more specifically on loop circuit engineering in the analog environment.
29. G. T. Hawley, "Historical Perspectives on the U.S. Telephone Loop," *IEEE Communications Magazine,* March 1991, p. 27.
30. See Chap. 2 of Bellcore, *Telecommunications Transmission Engineering, Volume 3: Networks and Services,* 3rd ed., Bellcore Technical Publications, 1990.
31. From a practical standpoint, subjective testing of coders should probably be done by recognized impartial organizations such as Bellcore, or by appropriate "third-party" institutions (e.g., the Rutgers WIN Labs), rather than by the individual carriers.
32. See, for example Bellcore, Bellcore, *Generic Framework Criteria for Universal Digital Personal Communications Systems (PCS),* FA-NWT-001013, Issue 2, December 1990, paragraph 3.5; also N. Kitawaki and H. Nagabuchi, "Quality Assessment of Speech Coding and Speech Synthesis Systems," *IEEE Communications Magazine,* October 1988, pp. 36–44.
33. See Bellcore, *Technical Analysis Report: Ultraphone 100 Codec Subjective Voice Quality Analysis,* AR-NPL-001041, Issue 1, January 3, 1989. Note: The distribution of this document may be limited to Bellcore and authorized clients.
34. Bellcore, *Generic Framework Criteria for Universal Digital Personal Communications Systems (PCS),* FA-NWT-001013, Issue 2, December 1990, p. 19.
35. *Ibid.,* p. 21.
36. Bellcore, *Telecommunications Transmission Engineering, Volume 1: Principles,* 3rd ed., Bellcore Technical Publications, 1990, p. 625.
37. *Ibid.,* see Chapter 25.
38. *Ibid.,* pp. 644 ff.
39. Bellcore, *Generic Framework Criteria for Universal Digital Personal Communications Systems (PCS),* FA-NWT-001013, Issue 2, December 1990, paragraph 4, p. 33.
40. MIL-SPEC models are available in Defense Department publications.
41. Available from Bellcore.
42. Field MTBF is based upon extrapolation from failure rates.
43. G. T. Hawley, "Historical Perspectives on the U.S. Telephone Loop," *IEEE Communications Magazine,* March 1991, pp. 27–28.
44. Many telephony textbooks include a chapter on traffic engineering, with further references for those with a special interest in the subject. The reader is referred to the following: J.C. Bellamy,

Digital Telephony, 1st ed., New York: John Wiley & Sons, 1982, Chap 9; J.C. McDonald, ed., *Fundamentals of Digital Switching,* New York: Plenum Press, 1983, Chap. 2; and with an emphasis on data traffic, M. Schwartz, *Telecommunications Networks: Protocols, Modeling and Analysis,* Reading, Massachusetts: Addison-Wesley, 1987, Chap. 2.

45. See also the discussion of availability in M.J. Miller and S.V. Ahamed, *Digital Transmission Systems and Networks, Volume 2: Applications,* Rockville, Maryland: Computer Science Press, 1987, pp. 10–11.
46. Note that the standard Bellcore reference on fiber to the home indicates a 99.99% availability requirement, which they admit may not always be met; see Bellcore, *A Compilation of Results from Recent Bellcore Studies on Fiber to the Home (FTTH),* SR-TSY-001468, Issue 1, December 1989, p. 18.
47. See, for example, R.L. Freeman, *Reference Manual for Telecommunications Engineering,* New York: John Wiley & Sons, 1985, Chaps. 12 and 13 and the references therein.
48. See George Calhoun, *Digital Cellular Radio,* Boston: Artech House, 1988, Chap. 1, for a discussion of the historical and philosophical differences between the Way of Radio and the Way of Telephony.
49. This entire section and the one that follows it draw extensively on the work of Al Sacuta and Paul Clements.
50. Al Sacuta and Paul Clements, "Speech Quality and System Availability," IMM Technical Note Number 1, IMM Corporation, 1988.
51. In this case, it is an early version of the RELP coder used by IMM Corporation in its BETRS system. The coding rate is approximately 14.6 kb/s net of TDMA overhead.
52. Again, this example is drawn from the IMM BETRS system.
53. Al Sacuta and Paul Clements, "Speech Quality and System Availability," IMM Technical Note Number 1, IMM Corporation, 1988, p. 8.
54. *Ibid.,* p. 7
55. Lucien Boithias, *Radio Wave Propagation,* New York: McGraw-Hill, 1987, p. 302.
56. See Note 3.
57. I cannot refrain from mentioning that one of the most comprehensive and most thoroughly validated models is that which has been developed by Dr. Craig Miller at IMM Corporation, primarily for BETRS-type applications. It has been used for tens of thousands of path loss calculations for specific BETRS subscribers, as well as hundreds of area coverage analyses. To date, unfortunately, the algorithm remains proprietary.
58. For example, the Miller model described in the previous note.
59. Bellcore, *Generic Framework Criteria for Universal Digital Personal Communications Systems (PCS),* FA-NWT-001013, Issue 2, December 1990.
60. See especially the applications literature available from SR Telecom and IMM Corporation.

Part V
THE LAMINAR NETWORK

Fiber optics and digital radio are emerging as the basis for enhanced network access for the twenty-first century. Both technologies have been deployed extensively in long-distance transmission applications, but they are just beginning to be used as access technologies, and considerable uncertainty exists as to their respective roles in the overall access picture. Despite our historical tendency to think in terms of singular solutions propagated across the entire network—the ultimate version of "seamless" communications—many observers are coming to the recognition that the most likely future is what is here dubbed the *Laminar Network:* a series of access "fabrics" overlaying one another, some on a small scale (e.g., microcell radio), some of global proportions (VSAT), some optimized for narrowband transmissions, others for broadband, some for vehicular communications, others for fixed or portable, some for data-dominant traffic, others for voice-dominant traffic, and still others for the transmission of image-based traffic—all rather imperfectly stitched together. In this section, we will examine in two brief chapters the complementary roles of radio and fiber, and the likely trajectory of the post-cellular (or trans-cellular) radio access industry.

Chapter 9
FIBER OPTICS VERSUS DIGITAL RADIO

9.1 ADVANCED NETWORK STRUCTURES: GEODESIC OR LAMINAR?

In the Beginning—The Archipelago

Before we examine the fiber/radio equation, let us first step back for a broad structural look at the whole eco-system of telecommunications. As we noted in Chapter 2, in the early days of telephony the structure of the network was very simple: A local exchange consisted of a relatively small number of customers connected in the simplest manner possible (i.e., a star network) to a local switch. The exchange was like an isolated island: 98 to 99% of all the traffic stayed within its boundaries and a few lines to the outside were enough to handle the occasional toll call. Indeed, many exchanges had no external connections at all, acting more like community intercom systems. The available census data would indicate that in fact the vast majority of all electrical communications systems were not interconnected, at least through the first 50 years of the industry's life [1]. The "Network" was little more than an archipelago of these largely self-contained, self-sufficient communities (see Figure 9.1). Network access meant a handful of trunk lines to the nearest city, patched into the local switchboard and connected by human operators on those rare occasions.

Stage 2: Hierarchy

In time these islands tended to grow together, which brought an end to this relative simplicity. Managing an exchange was one thing; managing a system of densely interconnected exchanges called for a very different network architecture. The past 40 years in particular have seen the emergence of complex network structures in response to the sheer quantitative expansion of the network and the need to interconnect millions of access points and billions of messages. The first response was hierarchy: By grouping exchanges and their traffic flows in a pyramid of subsets and

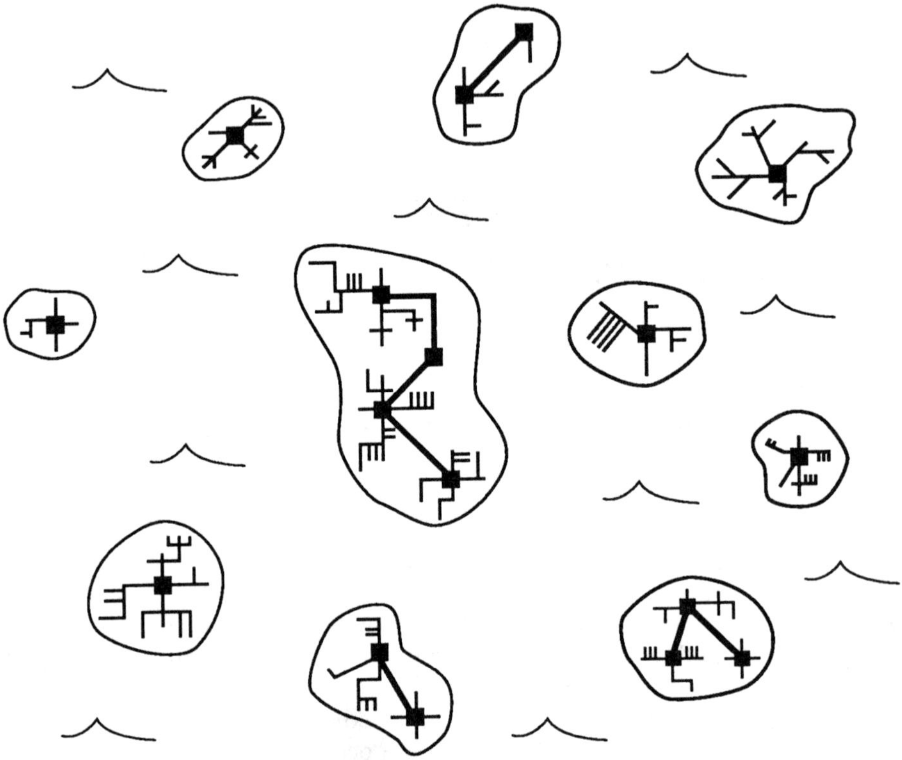

Figure 9.1 Stage 1 in the growth of the network structure—the Archipelago.

supersets and assigning toll switches to the various groupings, a hierarchical network structure emerged that is still the standard described in most texts (see Figure 9.2).

Stage 3: Geodesy

It might seem that these hierarchical structures could be elaborated indefinitely. Yet it turns out that strict hierarchy represents an early and passing phase for the evolution of most complex systems. In the design of telecommunications networks, nonhierarchical structures have already begun to emerge in many of the most technologically advanced sectors. By using intelligent processing at the various nodes and even within the transmission channels themselves, it is possible to build a more robust, cheaper, and more efficient network without piling more layers onto the pyramid [2]. For example, the long-distance network is moving towards nonhierarchical configurations based on the Signaling System No. 7 control fabric [3]. Many new

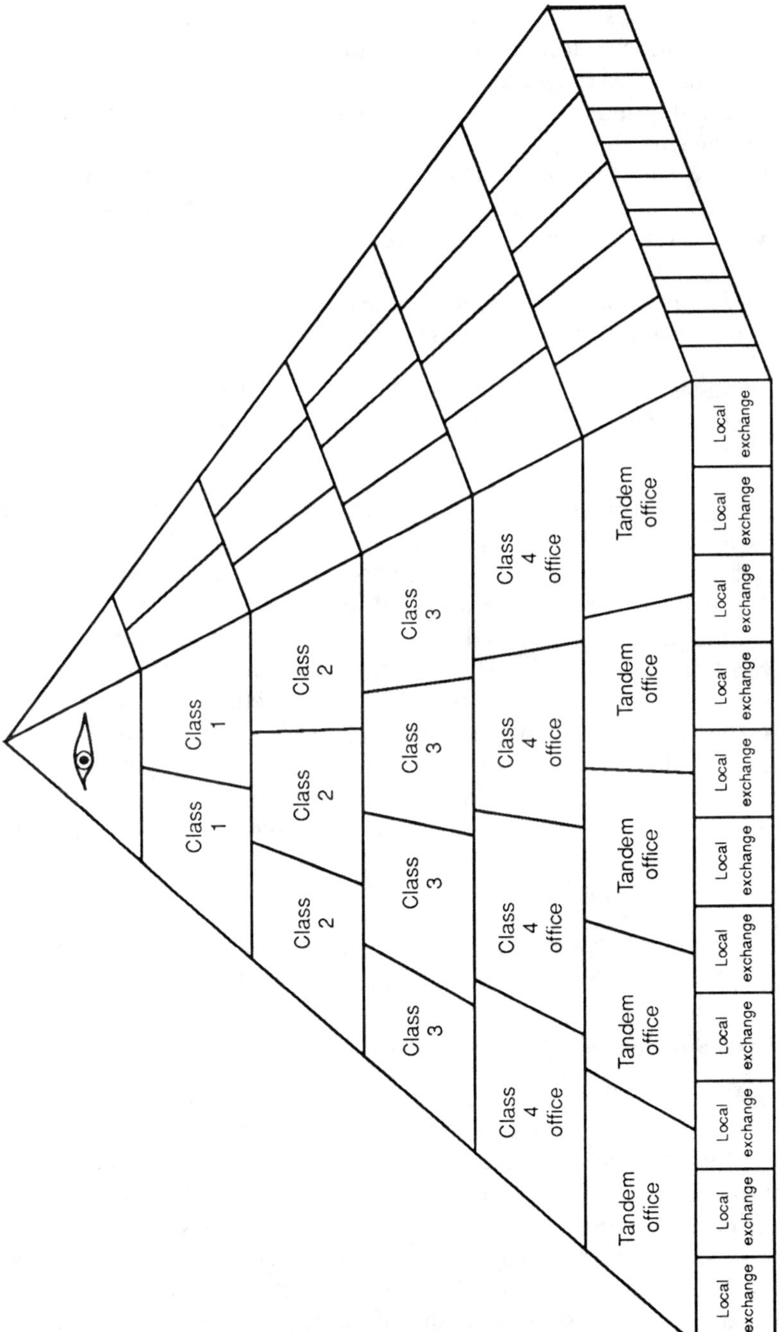

Figure 9.2 Stage 2 in the growth of the network structure--the zenith of hierarchy. (From Standard Bell system.)

network topologies—grids, buses, and rings—have been developed to replace the simple star configuration in local-access-type applications (although so far mainly for data-oriented local area networks) [4].

One of the first writers to call attention to this trend and to think through its significance for the overall evolution of the telephone industry was Peter Huber. In his influential 1987 report for the Justice Department on the postdivestiture evolution of competition, Huber highlighted this development in the title of the report itself, *The Geodesic Network,* one of the more stimulating neologisms of the past decade [5].

Essentially, Huber's conception of a geodesic structure is one where the nodes are connected together by the shortest possible transmission path:

> When switching is cheap and transmission is expensive the efficient network is a ring. . . . The nodes (switches or computers) are connected along a "geodesic"—a path of minimum length. . . . [6]

In the "old days" expensive switching gear had to be centrally located, connected by a star network to remote subscribers, and managed in a hierarchical pyramid. But now the availability of cheap digital electronic components is leading to the decentralization of switching and other information-processing functions. In principle, every node can become a "switch" and the total transmission mileage in the system can be reduced tremendously. Smart nodes thus alter the switching/transmission equilibrium, which determines the optimal network structure. In place of a star, we see a ring. In place of hierarchy, we see geodesy (see Figure 9.3). Looked at slightly differently, instead of concentrating network intelligence in large central facilities, geodesy means that the intelligence can be pushed out toward the periphery, toward the user: "Network functionality will therefore continue to rush out to the edges [7]."

Geodesy also evens out the competitive playing field:

> The geodesic network is structurally competitive in the most fundamental sense of the phrase. . . . Network dispersion guarantees the survival of competition, by providing the next-generation MCIs with countless network gateways at which to peddle their services and wares. [8]

The Geodesic Network offers a striking insight into the trajectories of trends in network evolution that are still in their early stages. In fact, today the network is still predominantly hierarchical; yet the abandonment of hierarchical structures is gathering momentum, especially in the core public network and in specialized computer networks. The packet-oriented, "switchless" local-area network is perhaps the prototype for the geodesic future. And if it holds for local networks, then perhaps it will pertain to larger networks as well. The trade-off between processing power and copper or glass is real, at least for some segments of the network, and some applications.

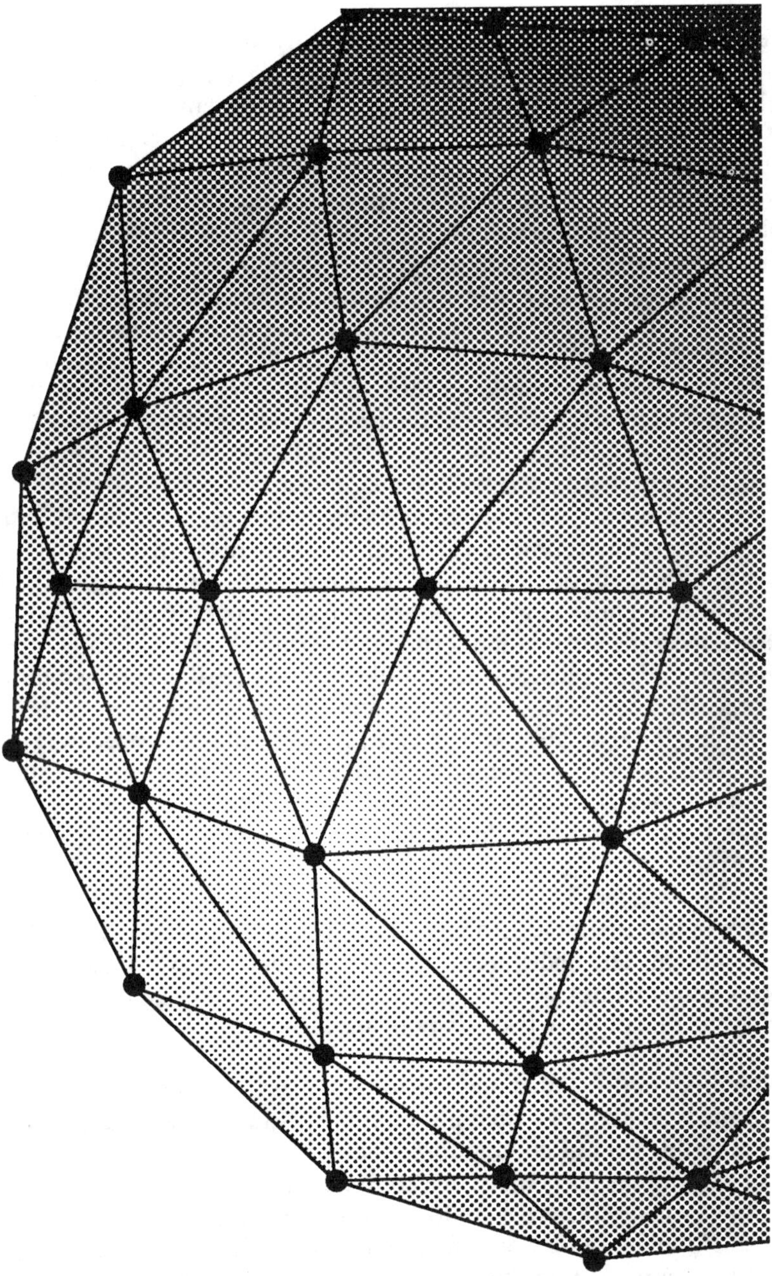

Figure 9.3 The geodesic network.

Geodesic Access?

Planners are even beginning to consider the possibility of ring-like structures in the access environment [9]. Yet as in many other things, we find a different set of conditions in the world of access. For one thing, the access environment is almost entirely based on analog voice (or voiceband) traffic. The marvelous processing engines that are driving the geodesic trend require digital input signals. Until we can guarantee the digitalization at the customer's premises, the local loop will remain an impermeable barrier to the dispersal of the network of which Huber writes. This is not to say that intelligent systems may not grow on the loop side of the central office; indeed, as noted above, local-area networks are in some respects leading the way. But most LANs are still cut off from the public network, isolated by the analog telephone loop. In effect, they are islands in a new archipelago. To connect them into a true digital network, we will need a true digital loop, as the long-range planners at Bellcore and elsewhere have argued for years. Yet the proper architecture for a digital access network is very much in question. Star, ring, or bus? Fiber to the home? Fiber to the curb? Radio tails? "Alternative access?" ISDN, or what?

Moreover, regardless of which *physical* configuration prevails, the efficient digitalization of the access environment will require equipment based on standards that are still under development or, at best, just beginning to be commercialized. These range from fairly conventional fiberoptic transmission systems (e.g., SONET), to ISDN and B-ISDN transmission and terminal equipment, and ultimately to such exotic architectures as packetized voice on fiber rings, ATM switching, etc. Most technology forecasters do not foresee the commercial availability many of these elements until the year 2000 or later. And, of course, there will be the staggering cost of upgrading the access network—several hundred billion dollars by current estimates (as discussed in Chapter 2).

Moreover, as Huber readily admits, there is a strong streak of optimization in the geodesic vision:

> Networks are in fact assembled from two basic building blocks: nodes and the links between them. The nodes are switches or, increasingly often, computers. The links are transmission systems. . . . Efficient network design is the fine art of balancing the mix of these two components. [10]

"Balancing the mix" in this global sense assumes the existence of a "guiding hand" that can blend the right alloy of switching and transmission, software and hardware. Yet it is precisely this central guiding force that we have now jettisoned with the breakup of AT&T and the acceptance of a competitive market structure in the telecommunications industry. There is no longer a single planning authority that can implement a single coherent network plan. Indeed, one of the greatest obstacles facing the new network architectures, from Signaling System No. 7 to ISDN, is that they must now win their way in the marketplace. The central decision-maker of the old AT&T pyramid has been replaced by an increasingly chaotic Darwinian process.

There are those who believe that "natural selection" will somehow produce the same sort of optimization, through a harsh winnowing of alternatives until the one "fittest" option emerges and propagates itself. I believe this is a misreading of evolutionary experience. Single-outcome processes—i.e., where a single species comes to dominate the eco-system—are extremely rare in biological evolution. As we will comment further in Chapter 10, the tendency toward single-species dominance is found only in the most severe environments. The more benign the environment, the more abundant, the more variegated and heterogeneous the eco-system, the larger the number of species, and the more "alternative" or "parallel" outcomes emerge. The fish of the coral reef or the birds of a tropical jungle show this more typical result in all its glory.

I would argue that the successive qualitative revolutions and rapid cost reductions in digital electronics are creating precisely the sort of abundant, "tropical" style of eco-system for telecommunications that will foster diversity, redundancy, multiple-outcome processes, and will tend to defeat any Darwinistic tendency toward tight optimality. Thus, while the trend toward "dispersal of the network" that Huber has identified is undoubtedly real, the actual outcome is probably not geodesy—the true "shortest" line between every node and every other node—but, more simply, *a proliferation of "gateways" or access points into the network,* as Huber himself also recognizes. While in the hierarchical network, the typical subscriber had only a single pathway into the public network—the fateful "bottleneck" of the local loop and the local exchange switch—in the geodesic network, more customers will enjoy multiple ways of accessing the network. For example, today, setting aside regulatory constraints, a customer in a typical urban setting probably has at least seven access gateways:

1. The AT&T "point of presence" (POP)
2. The MCI POP
3. The Sprint POP
4. The cellular A carrier
5. The cellular B carrier
6. The local fiber-based alternative access provider
7. The local loop to the telephone company's nearest central office.

Of course, the quality and kind of access provided may differ considerably. The long-distance carriers are unable today to offer access for purposes of local calls. The nature of the interconnections available to the cellular carrier may be inferior to those offered by the local wireline telephone company. But, in principle, all of these carriers can be reached by a customer, and all can offer some form of access to the wider shared network. Moreover, the customer might also, depending on size, enjoy the additional option of building his own network and managing his own access—

for example, by means of a corporate VSAT system (see Section 5.6.4 for descriptions of such systems).

The Laminar Network

This heterogeneity—which already exists—begins to suggest a somewhat different vision of the network of the future. To clarify it, let us extrapolate. Imagine that, say, 10 years hence, there are no more restrictions on competition or interconnect arrangements in the local exchange. Let us further imagine that, for the sake of round number theory, there are now 10 carriers providing services to customers in a given metropolitan area (e.g., New York): three long-distance carriers, two cellular operators, two PCN operators, the local cable TV company, which has installed a two-way capability on its network, a fiber-based alternative access company, and the remnant of the local Bell operating company. Each of these 10 carriers is serving customers in the New York metropolitan area, and each of them has found a technical alternative for providing access to and from its customers that does not rely on the Bell operating company's copper local loops. The cellular and PCN carriers use radio, of course, to connect their customers to their own switches. The fiber-based carrier has laid his own fiber infrastructure to reach his customers, who are mainly large corporations. The long-distance carriers have built out from their points of presence to reach their largest customers, using a mixture of technologies (cable, fiber, and radio), and so forth.

What emerges thus is what I would call a *laminated access network,* where multiple carriers provide competing, redundant (duplicative, if you like) access facilities that, in effect, overlay one another. The kinds and qualities of the access offered are not necessarily identical. For example, let us assume that one of the PCN carriers has concentrated almost entirely on public microcell applications (e.g., telepoint), with very short range transmitters, while the other PCN carrier has focused on serving industrial parks and business and college campuses. The two cellular providers have also differentiated themselves somewhat; the A carrier focuses on the consumer market with service packages and pricing oriented toward the Ford and Chevy segments, while the B carrier has implemented a higher quality and higher cost package of digital services is marketed chiefly to corporate fleet managers. The three long-distance carriers are still going at it head to head, although the enhanced digital services available from each differ slightly. In short, although all offer the same generic type of service, the market is now *segmented* much as, say, the automobile market is broken into various segments with different manufacturers addressing different segments with somewhat different strategies. And indeed, just as most companies today subscribe to more than one long-distance carrier, we may assume that many customers will subscribe to several of these local access services, either to make sure they have backups or to avail themselves of different types of service offerings. For example, the local cable TV carrier will probably not try to duplicate ordinary residential telephone service *per se,* but may offer enhanced videotext or home-shopping services that take advantage of the customer's TV screen (which the phone company cannot access as easily).

In short, the laminar network is a series of partly competing, partly complementary, somewhat differentiated, overlapping access fabrics (see Figure 9.4). It is not, as the reader may have already surmised, the "seamless network" envisioned by some. There is not a single telecommunicating terminal that the user carries from home to highway to office to airplane. Instead, by the year 2001, the average upscale consumer probably still has a wireline phone at home, which is used mainly for family telephone calls. He or she also has a cellular phone factory-installed in the dashboard of the new car; plus an interactive videotext option (with ordinary telephony as a "feature") through his cable TV company; plus a PCN-type device for use in the office and possibly in some public places instead of a pay phone. In addition, he/she has recently received an offer from AT&T to install a special telephone terminal (similar to the special AT&T pay phones we see today) right in the home that can be used for more direct access to AT&T worldwide services. Finally, the local fiber-based carrier is marketing a separate hook-up to support a new ultra-high-speed fax machine.

Is this a disaster scenario? Certainly Theodore Vail would be twirling like a neutron star. Consider the duplication of networks and access facilities. In this simple scenario, there are (prospectively) up to four separate wireline connections (copper or fiber) and two radio terminals for our average household. Each access "line" connects to a separate switch, managed by a separate company, carrying redundant overhead. The monthly billing statements are a nightmare for the check-writer (if we still have checks to write).

But is it much different from any other competitive market? In my town of Philadelphia, there are more than 100 car dealers, all addressing the same market. Imagine the waste, the duplication of facilities, the costs of advertising, and the many other forms of competitive friction. As anyone who goes out to buy a car well knows, the system is not a very efficient way to deliver an automobile from the manufacturer to the consumer. Nevertheless, it works, we accept it. We understand the many fundamental efficiencies that the competitive marketplace fosters, even though a Soviet-style (or Henry Ford-style) approach to streamlined production and delivery could eliminate a lot of waste in the short-run.

The truth is that *all competitive markets are laminar.* They all involve multiple overlapping distribution structures aimed at the same population base. In telecommunications, the core network is already laminar, with duplicate, overlapping, competing facilities. Laminar growth must now occur in the access network. This is, I believe, an imperative of the competitive framework. (The only other alternative—Open Network Architecture—is based on the odd idea that somehow laminarity and competition can exist in the core network, but must be avoided in the access network

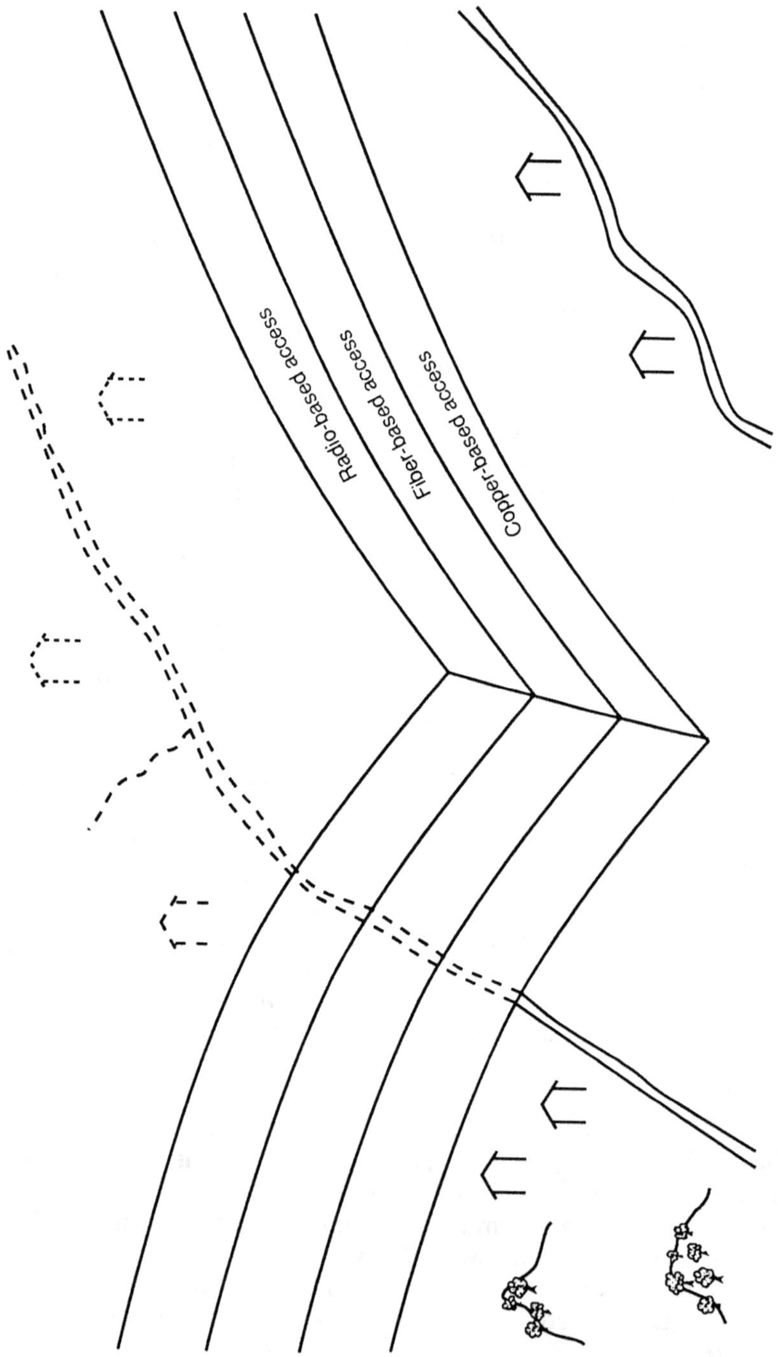

Figure 9.4 The laminar network.

by the promulgation of strict and ever-more-complex rules to ensure equal access [11]. It will die stillborn.)

Overlapping Fabrics

From the standpoint of the access facilities, the laminar network will develop as a series of overlapping access fabrics—a patchwork quilt of many layers and different sized pieces in each layer.

Copper (Two-wire). At the "bottom" (if historical age is the scale of "depth") will be the frayed but still vital copper fabric that we have discussed in Chapter 1. It may be that at some point in the distant future there will be no more copper wire in the telephone network, but with more than 130 million access lines and hundreds of millions of circuit-miles already installed on copper, that fabric is likely to continue to play a useful role for a long time to come.

Copper (Coaxial). There is already another large base of transmission fabric based on coaxial cable, used to provide cable TV access to more than half of the households in the United States. The cable TV industry is moving swiftly to embrace fiber optics, but its actual application is still in the early stages. Probably, as with the copper two-wire fabric, fiber will first replace the feeder portions of the cable TV networks, and coaxial cable will continue to be used for the drop segment for a longer period of time. Although most of this network is now one-way only, much of it could be adapted for a two-way service.

Optical Fiber. Fiber is moving into both copper access networks, working out from the center. In the telephone (two-wire) fabric, fiber is beginning to replace the feeder links. Clearly, fiber will become more prevalent as wider bandwidth services are popularized.

Digital Radio. Radio is the perfect overlay. As we have discussed in Chapters 5 and 7, the technical and economic performance of radio is optimal for area-coverage (compared to any of the three physical media identified above, which are constrained to follow actual physical routes).

The laminar network, then, will consist of multiple layers of transmission facilities for accessing the core network at an increasing number of gateways (see Figure 9.5). The lowest levels will still be copper-based fabrics, the vast installed base of wireline telephony and coaxial cable TV plant that will continue is use for decades. Growing over these there will be several new layers of fiberoptics plant—and, because of its nature, even more layers of digital radio. Even within a given "fabric layer" there will almost certainly be a great deal of technical diversity. For example, fiber systems will emerge from at least three sources:

1. The telephony environment, with generally synchronous formats like SONET

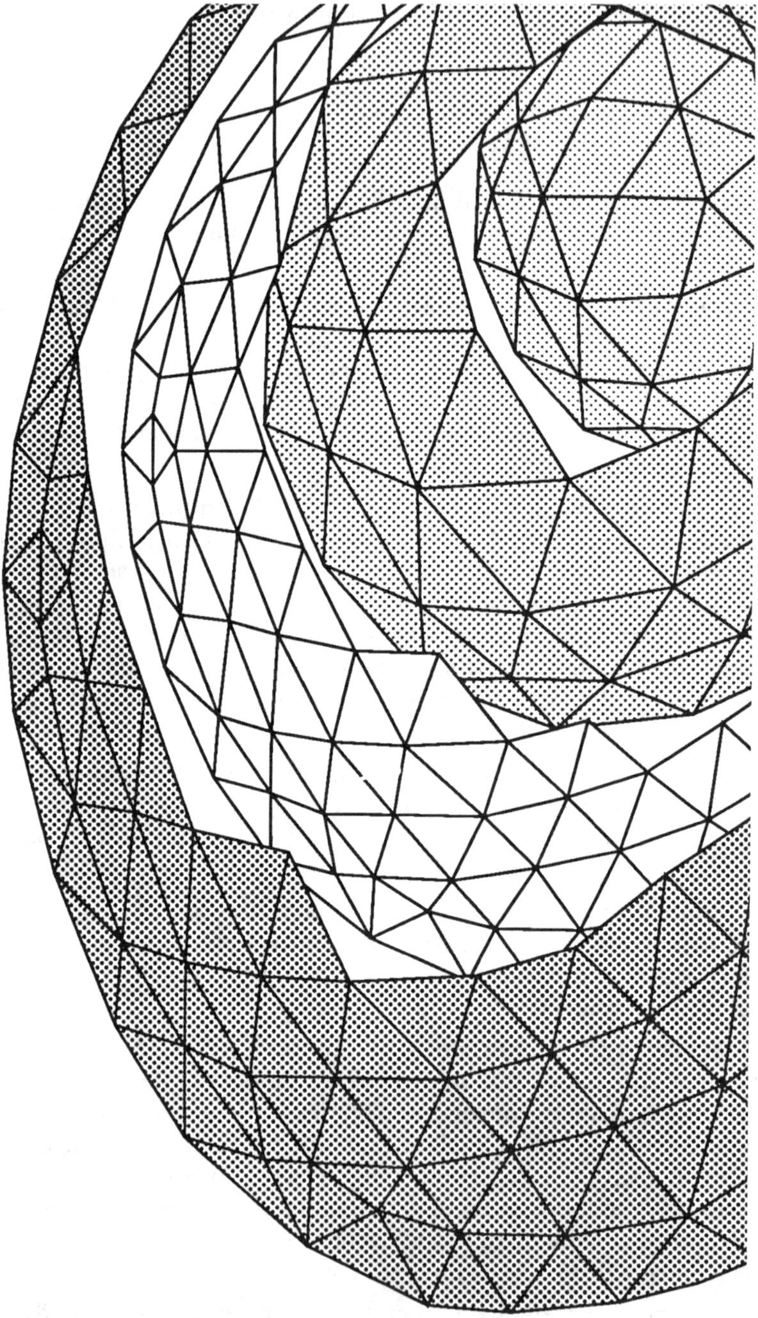

Figure 9.5 The multiple fabrics of the laminar network.

2. The data communications environment, with ring and bus-type packet architectures in a general asynchronous approach
3. The cable TV environment, which may evolve both analog and digital formats—undoubtedly the imminent sunburst of high-definition television will drive its own transmission standards for fiber-based access.

There is a forlorn hope in some quarters (mainly in the telephony industry) that somehow all these technical trends could be merged in a single "seamless" fiber-optic format—that is what B-ISDN concepts are often said to represent. Whether this ever comes about, it seems certain that at least for the next decade or two there will continue to be different technical versions of the fiber fabric [12].

The same is true of the radio fabric. We have surveyed many of the applications of wireless access in Chapter 5—and I at least have reached the conclusion that there is likely to be considerable diversity in radio system architectures, driven mainly by the need to optimize for specific applications requirements. We already have analog FDMA (cellular radio) in several forms worldwide; there are at least three forms of digital TDMA cellular radio, and possibly for noncellular mobile radio in some applications. Standards in some mobile markets, on the other hand, will probably go toward digital FDMA. In-building, telepoint, and other emerging wireless architectures show little centripetal tendency—on the contrary, almost every new application, as it reaches the stage of commercialization, discovers the need to develop its own radio format. We shall return to the question of this diversity, and its management, in Chapter 10.

In any case, the laminar network provides the framework within which the relative roles of fiber and radio must be assessed.

9.2 THE ROLES OF FIBER AND RADIO

In discussions of wireless access, there usually comes a point where the question is inevitably raised: *What about fiber?* It is a valid question; the case for radio is made primarily against copper, which may be irrelevant if copper is to be rapidly substituted by a far superior access technology based on optical fiber. Does not the "fiber revolution"—especially in its most aggressive scenarios for "fiber to the home"—somehow blunt or even forestall the "wireless access revolution"?

Fiberworld?

In the old centrally managed Bell System hierarchy, network planners could evaluate the alternatives, select the best available technology for a particular requirement, and then standardize that solution throughout the network. Many of those in the telephone industry still implicitly adopt this perspective when thinking about new technology. They assume that, for example, the local exchange is and will remain a monopoly, and that they (as managers of the monopoly) will have a stable business environment within which to deploy new access technology. Since they are more familiar with the merits of fiberoptic transmission—and especially as it opens up the possibility for still-forbidden but nevertheless tempting lines of new business, such as video delivery—telephone company planners tend to assume that the "fiberization" of the long-distance network (which has been overstated, but let that be) will now extend itself into the loop environment. They can already see its beginnings with SONET systems creeping into certain feeder applications, and they are bombarded with forecasts about the approaching "fiber/copper crossover point"—beyond which it will be cheaper to put fiber in new loops, all the way to the house, than to use the age-old copper solutions.

Indeed, fiber optics has been sold so assiduously, for so long, as a nearly universal solution that for many in the telephone industry "Fiberworld" is already here. I remember speaking to one gathering of telephone executives some years ago who received most of what I had to say with the aplomb and lack of emotion that are sometimes characteristic of our industry, until I broached the idea that digital radio might become an important mainstream access vehicle. This apparently triggered some autonomic response, for they suddenly became quite animated. "What about fiber in the loop?" they demanded in a somewhat less than friendly manner. I must have had a less enthusiastic answer than they expected, and the next thing I knew several of them had jumped up from the table and were chirping "Fiber! Fiber! Fiber to the home! Fiber everywhere!" It could be costly, I ventured. "Cheap, cheap," came the chorus, "They're making fiber out of plastic, it'll be so cheap. Plastic fiber!" When I persisted in suggesting there might be some applications where radio might be preferable, this set off a new chorus: "Bandwidth! Bandwidth! Gigabits! Video!" They went on for several minutes trading reassurances that in the near future fiber would surely triumph, and the meeting ended without my being able to get in another word.

Will fiber triumph? Much of the argument for Fiberworld is bandwidth-driven, so to speak. The carrying capacity of optical fiber transmission systems is so enormous that it can bedazzle one into thinking that fiber possesses a true technological imperative. We even hear about bandwidth "too cheap to meter" (reminiscent, of course, of the claims for nuclear power in the early 1950s). If bandwidth is good, then it would seem that fiber must prevail. And who would argue that lots of bandwidth is not an unalloyed blessing for mankind. (On the other hand, the need for spectrum efficiency constrains every radio application in terms of its bandwidth and message-carrying capacity.) In a monopoly environment, setting aside for a moment the question of economics, the all-fiber scenario garners an air of plausibility.

But in a laminar network, where there are many providers and many solutions, and no true central optimization process, the outcomes are likely to be rather different. The roles of fiber and radio will be determined not by some perceived abstract superiority but by specific, and fairly near-term, competitive advantages that they offer to system operators and, ultimately, to their customers. In a laminar marketplace, the sources of valid competitive advantage are sometimes different from what we might project on the basis of a purely technical analysis. In a sense, this is as simple as saying that a jet engine may be technically superior to a two-stroke gasoline engine, but that doesn't mean that you want a jet engine to power your lawn mower. To understand the roles of fiber and radio, we have to look at their likely competitive benefits in an open market situation.

9.2.1 The Competitive Advantages of Fiber-Based Access

In our brave new laminar world, then, what actually are the selling points of fiber optics as an access vehicle? (We will continue to set aside the question of costs until the next section.)

The obvious answer is fiber's huge bandwidth. The advantage of fiber transmission in interexchange trunking and long-haul applications has been clearly proven: one hair-thin fiber circuit can replace thousands of copper wires, depopulate stuffed conduits, eliminate electrical components from the outside plant environment, and generate substantial savings. According to one source, during the 1980s fiber optics had replaced copper on 41% of the "interoffice equivalent voice circuits" [13]. This is in many respects an amazing achievement.

But where, in today's *access* network, does mega-bandwidth provide added value? Because an access circuit, *by definition,* is a low-usage circuit. In today's industry, it is not clear where the mega-bandwidth is needed for ordinary access applications. The vast majority of access lines requires only a single voice telephone circuit, for which fiber is really no better suited than copper. Using fiber optics for ordinary telephony is indeed similar to mowing the lawn with a jet engine. It is vast overkill for a very ordinary job that low-tech (copper-based) solutions can handle quite well. It is true that fiber provides relief from some of the difficulties of electrical interference and the like; but fiber loops also raise the troublesome question of powering the loop from the subscriber's premises (see Chapter 5), which copper loops do not have to be concerned with. Finally, there is the matter of the difficulty of economically proving-in fiber over conventional copper for ordinary POTS services, which we shall address in greater detail in Section 9.3.

What about tomorrow's telecommunications industry? Won't we need fiber's bandwidth to deliver the broadband Information Age services that we have alluded to in earlier chapters? The answer is yes—at some point. But let us consider what those likely high-bandwidth services are and who will be interested in them.

Video. The obvious candidate for a high-bandwidth service for which there is an established demand is television. The success of the cable TV industry over the last

10 years seems to point the way toward a huge opportunity for the telephone carriers. Most of the major telephone companies have run numbers that show that if they were able to deliver video services, and reap the revenue from them comparable to what today's cable TV companies bring in, they could instantly justify fibering half the homes in America.

There are a few problems with this scenario. For one thing, it is still (as of this writing) illegal. Telephone companies are essentially prohibited by law from competing with cable TV companies in video services to end-consumers [14]. This will probably change, but until it does TV revenues will not be there to fuel the fiberization of America.

Second, the cable TV industry itself is well-entrenched. They have a sunk investment in coaxial cable, which I suspect might prove hard to compete against for some period of time, even for well-endowed telephone companies. Moreover, cable TV companies are aggressively deploying fiber in their own networks, readying themselves for the day when they will have to compete. There are signs that cable TV is approaching the stage of a "cash cow" business—it is fairly mature in the sense that it has expanded to fill most of its important markets, and it generates large cash flows—the better to defend those markets, attack other markets (e.g., telephony?), and to support upgrades to new technology.

Third, as we will examine further in Section 9.3 , the economics of video delivery by telephone companies may not be that good.

LAN/MAN/WAN Access. Another prospective need for bandwidth in the access environment will come as local-area networks begin to grow out and interconnect—as the Data Archipelago begins to become the Data Network. This is not a market that I am particularly familiar with, so I am reluctant to forecast its evolution. But it seems to me that it will be rather gradual, at least measured in terms of the numbers of lines that might require a broadband access for packet data communications.

B-ISDN. What of the broadband ISDN architectures we read so much about? Once again we may wonder how quickly they will become a reality, on a mass scale. Judging from the status of the discussions in today's literature, my guess is that we are still at least a decade away from true commercial introduction, and several decades before B-ISDN becomes even reasonably ubiquitous [15].

In short, all of the projectable requirements for true broadband capabilities that might give an edge to fiber in the access environment are still beyond the immediate planning horizon, even for telephone company strategic think tanks. As we enter the 1990s, I think this fact has sunk in for quite a few telephone managers and has led to an appreciable cooling of the fiber fever, at least as far as true access applications are concerned. This is not to say that these broadband requirements will never develop; I believe they will, reaching the critical mass for commercialization probably by the end of this decade. Nevertheless, in the access environment (as opposed to the core network environment), fiber is still very much a technology in search of a mission. As one article on fiber to the home concluded:

> Many researchers and technologists have dreamed of using optical fiber with its low loss and immense bandwidth for improving the capability and productivity of businesses and the quality and enjoyment of people's lives. "Fiber-to-the-Home" appears to be the rallying cry for making this dream a reality. The ultimate implications of the FTTH concept in laying the foundation for the new "Information Age" are far-reaching and can only be imagined at this time. [16]

In other words, it is such a technological marvel that it is bound to be useful to mankind, somehow. Meanwhile, we run experiments to prove we can mow the lawn with a turbine engine.

9.2.2 The Competitive Advantages of Radio-Based Access

As noted above, the chief selling point for fiber is bandwidth. By comparison, radio brings three major selling points to the market, at least two of which are immediately useful to some of the established carriers:

1. Radio can provide basic access (i.e., POTS or POTS plus near-term narrowband digital services) for less cost, per customer, than wire (or fiber) in some applications.
2. Radio can offer truly new services (e.g., mobile) based on its unique properties as a transmission facility.
3. Radio can allow rapid entry by a competitor into an established market segment.

Low Cost Access. This is the application we have dwelt on throughout most of this book. In the United States it is BETRS, or BEXR; elsewhere, it is, more simply, the wireless local loop. In Chapters 5 and 7 especially, we have looked at the many situations in which radio can immediately offer a cheaper vehicle for an ordinary voice-grade telephone line to an ordinary customer. In general, many of these applications are found in rural or nonmetropolitan areas where long copper loops mean high costs for conventional plant (and fiber costs are certainly no better). As our discussion has made clear (I hope), quite a few applications are also found in urban areas.

Unique Capabilities. Radio can allow communications with moving vehicles, airplanes, and ambling pedestrians. It can offer tetherless communications in homes and offices. It can offer rapid deployment possibilities in military, emergency, and temporary communications situations. In most cases, radio is uniquely suited for the application; wire or fiber simply cannot do it. Thus, as we see, radio is penetrating rapidly into many communications segments where these features are required.

Facilitation of Competitive Entry. The advent of microwave radio in the 1950s and 1960s made it possible for a small, lightly capitalized entrepreneurial company such as MCI (i.e., Microwave Communications, Inc.) to enter successfully the long-distance telephone market that had been monopolized for decades by AT&T. The real attractiveness of radio in many applications may not be that it is necessarily the optimal solution for a particular problem, but simply that the low overhead and rapid startup that radio systems make possible can allow a competitor to get into business quickly and move more nimbly than the wirebound giants he confronts. I believe this is one of the most important aspects of the appeal of PCN—it points to a path by which a potential competitor could enter the telephone business (for example) without mounting a gargantuan construction program to string wire and cable. It must be admitted, of course, that the ability to use radio as a competitive entry vehicle is still being reined in by the federal and state regulators.

Thus, I believe that radio-based access is already experiencing a stronger pull from the market than fiber-based access. As we shall comment in Section 9.3, the economics for radio are also more favorable today. In time, with the unshackling of broadband services, it is likely that fiber optics will find its proper and probably quite broad-based applications in the access network. But I do not believe that fiber is likely to eliminate wireless access as a key part of the twenty-first-century network.

9.2.3 Complementarties

I was once asked, semi-hypothetically, whether I would advise "the CEO of a major telephone company" to "go with fiber or with radio for the local loop of the future." I answered by drawing an analogy. In the computer business today, there are supercomputers, conventional mainframes, minicomputers, microcomputers, workstations, laptops. No single species of computer has driven out all its competition—indeed the all market segments continue to grow. Some jobs require a mainframe, others a laptop. The hypothetical question posed to me, therefore, was a little like being asked to advise the head of IBM, or the CEO of one of IBM's large corporate customers, as to whether he should "go with mainframes or microcomputers." The answer, at least for a major company, is that both, all, have to be pursued.

I believe that fiber will undoubtedly play an important role in twenty-first-century access, wherever its enormous bandwidth is required. But there will be many other applications—cellular mobile radio is but one example—where the radio access option will prevail, due to its competitive advantages. Moreover, the rapid deployment capability of radio will allow it to be used as a precursor to fiber in many situations. Radio may ultimately be replaced by fiber in certain applications—but the length of that "ultimately" could extend beyond our working lifetimes. For example, knowing what I know about the economics of rural wireline telephony, and witnessing the deepening commitment of the United States to market mechanisms in telecommunications, I think it is likely that fiber optics will take many decades to reach some segments of the population, under almost any conceivable scenario for its deployment. Radio will be the gap-filler for many people, for a long time. Even if the ultimate mix is 95% fiber and 5% radio—there will still be tens of millions of lines of radio-based access worldwide under even the most fiber-friendly scenarios.

In short, in considering radio and fiber as alternatives, we need to see beyond the current, rather flat and commodity-like telephone market, where voice is the dominant service. For basic voice service, copper wire is fine in many situations; radio and fiber begin to stand out—and to differentiate themselves from each other—as we look beyond basic exchange service to enhanced services, whether B-ISDN or mobile telephony. The further we march into the future, the more the fundamental complementarity of fiber and radio emerges.

9.3 THE COMPARATIVE ECONOMICS OF FIBER AND RADIO

9.3.1 Cost Estimates

We have examined the economics of radio and copper in Chapters 2 and 7. To recapitulate, copper loop first costs appear to *average* between $1200 and $2000 per access line (depending on the study), but these costs also exhibit a very wide range due to distance-sensitivity of wireline cost factors—from perhaps $300 to $400 at the low end to $10,000 a line in extreme cases in remote areas. Even if we ignore the extremes, we find that rural loops often show costs in the range of $3000 to $5000—so that there is a ratio of at least 10:1 between high and low ends of the cost spectrum. The costs for wireless access, by comparison, are much less variable, and today (1991) probably range between $2000 and $4000 per access line for a full telephony type application. While wireline costs are fairly mature, radio-based access costs can reasonably be expected to drop by at least 50% or so over the next five years.

What are the costs of fiber in the loop (FITL)? We have been hearing and reading for years about the imminent crossover in which fiber access costs would drop below copper-based access costs [17]. In light of the high variability of copper-based access costs, it has never been entirely clear what the crossover actually implied. Enough experience has been developed with FITL that a somewhat more sober assessment is beginning to develop. For one thing, the current average (?) costs for FTTH are apparently somewhat more than twice the average for copper—$3000 to $3100 per access line [18]. However, other sources suggest that those costs may be "about $5000" for a three-mile fiber loop [19]. Once again, it is not entirely clear to me what these numbers mean, beyond the fact that fiber is still quite a bit more expensive than copper for access applications.

Some of the most recent studies begin to reflect a decidedly more pessimistic tone, regarding *both* the difficulty of understanding the true costs of fiber:

> The cost of copper line technology is very dependent upon geographical location. . . . The line cost is found to be inversely proportional to customer density. Consequently, when fiber deployment is targeted at areas of high customer density and churn, *its cost effectiveness compared to copper can be very difficult to substantiate.* [Emphasis added] [20]

and the apparent implications of the data, however imperfect:

> If ubiquity is actually a requirement for viable new services [i.e., if we have to deal with average costs for mass markets], *then FITL is not likely in itself to be the basis of new services for many years,* particularly since fiber rehabilitation of deteriorated copper plant is expected to remain elusive *owing to the low cost of copper rehabilitation.* [Emphasis added] [21]

Even Bellcore's assessment is rather guarded:

> The initial thrust is to promote the near-term universal usage of fiber in lieu of copper in the loop distribution to support traditional telephone service. Overall, the results of these studies indicate that the deployment of fiber links dedicated to each customer . . . *will not soon be economic when compared with copper.* [Emphasis added] [22]

Moreover, it is not clear that these costs are relevant for a broadband service over fiber. If my analysis in the preceding section is correct, then fiber will not penetrate rapidly as a simple substitute for copper for voice access only. It will be "pulled" by broadband services. Yet broadband access will involve considerably more expense than POF (POTS on fiber). One of the Bellcore studies notes that "a true broadband connection, which is not yet fully defined, *could run between $5000 and $10,000 per residence today.*" [Emphasis added] [23]

It begins to seem as though we do not understand the economics of fiber-based access very well.

9.3.2 Revenue Estimates

And, of course, there is the other half of the equation: revenues. Even assuming that we can deliver some sort of defined broadband service, and the public wants it, what sort of revenue potential does this hold and how does it compare to the cost of the rewiring? John McLaughlin has addressed this question in a rather irreverent way, under the heading of "We will make money by 'Glassing America.'" It is worth quoting him at some length:

> This brings us to the technocratic visionaries and their solution to all of your problems: Fiber optics to every home. The Electronic Highway. Information Infrastructure. The Information Utility . . .

> Crank a few numbers. The only tangible market for broadband services to the home is for video entertainment. In 1988 all of the cable TV system operators in the United States generated $13.4 billion in revenues from all sources . . . A lot of analysts see future growth of cable TV revenues as highly constrained. With cable now passing 85% of U.S. homes, there are fewer new potential subscribers being added to the base every year . . .
>
> Assuming that the telephone industry were allowed to compete for this market through deployment of fiber optics, the estimated entry prices range from $150 billion up to $400 billion. And that is the price to compete with cable operators who have political friends, who know their business and who have plant in place today. If we assume that the capital investment costs are at the lowest point in the range—$150 billion—and if we assume that [the telephone companies] eventually captured half of the total cable TV revenues—about $7 billion in 1988 dollars—we are talking about a gross return on invested capital of less than 5% per year. And that totally ignores operating costs. [24]

9.3.3 Cost Structures

Finally, there is the question of the cost structure of alternative forms of access. As we have seen, the early stage of the fiber-in-the-loop application means that our knowledge of actual costs is rather hazy. Yet regardless of the values of the variables, the set of relationships between cost components, the service requirements, and the environment of the application is important in determining the relative long-term economic advantages and disadvantages of each access technology.

Table 9.1 summarizes some of the aspects of the cost structure for the three access alternatives. Perhaps the essential point is that copper and fiber are both quite similar. They are both based on wireline physical plant; both incur construction and materials costs that are roughly proportional to the distance or length of the loop; both show a wide range between the lowest cost customers and the highest cost customers. It is often claimed that fiber will have much lower maintenance than copper. In the interexchange application, where fiber eliminates the need for electronics in the outside plant, this may well be true. I wonder whether there will be much difference in the access application, where the amount of outside plant electronics is, if anything, somewhat higher for fiber because of the terminal required at the customer's premises. Knowing the practical limitations and the unavoidable hazards of the access environment, I tend to doubt that the long-term maintenance experience will be very different. Both copper and fiber are likewise rather inflexible investments. They are subject to "stranding" if the population moves around very much or if the forecasts are inaccurate (although for POTS the capacity of the fiber is so great that probably there will never be a need for upgrading a fiber feeder-route simply because of subscriber growth). In short, without minimizing the tremendous technological leap forward that fiber represents, what is striking about the access environment (at least for POTS access) is how similar the two are in their overall cost structure.

Table 9.1
The Comparative Economics of Copper, Fiber, and Radio-Based Access Facilities.

Category	*Copper*	*Fiber*	*Radio*
Avg. first cost (1990)	$1500–1800	$3000	$2000–3000
Avg. first cost (1998)	$2000	$1500–2000	$800–1200
Ratio of highest cost to lowest cost	10:1*	Same as copper	2:1
Distance sensitive	Yes	Yes	No (as a rule)
Favorable population densities	Higher	Higher	Lower
Risk of stranded investment	Yes	Yes	No (as a rule)
Salvage value	Zero or negative	Same as copper	Positive
Cost per bit	Highest	Lowest	Low
Support for digital services	Limited	Unlimited	OK up to T1**
Maintenance	5%***	Half of copper(?)	Half of copper

*This excludes the extreme rural loops (over $5000). If these are considered, the actual range of costs for copper-based access is probably more like 50 to 1.
**Certainly the ISDN BRI (144 kb/s) should be readily supportable; higher rates up to several megatbits are probably feasible for next generation TDMA and/or CDMA architectures, although there are many questions about the pricing and application for such services.
***A typical value for rual copper loops; urban loop maintenance is probably lower.

By contrast, the cost structure for radio is very different. It is distance-insensitive and density-insensitive. The range between the highest cost and lowest cost customers is much smaller. The maintenance advantage is demonstrable. In Chapter 7 we drew a more detailed comparison between the cost structures for copper and radio access, and I suspect that many of the generalizations would hold for a fiber versus radio comparison as well.

9.4 HYBRID FIBER/RADIO ARCHITECTURES

But then fiber *versus* radio is probably the wrong way to look at it. Increasingly, network strategists are thinking in terms of how fiber and radio cannot simply coexist, but may truly complement one another in the broad modernization of the access network.

In part, this has been brought about by the difficulty of making the numbers work for an all-fiber solution (for reasons already discussed). In the 1980s, the rallying cry for fiber advocates was FTTH. This "concept" (for it can hardly be called yet an "architecture") calls for an optical link from end to end, from my photonic terminal in my home or office all the way through to the telephone switch (and perhaps beyond). FTTH is certainly the most elegant solution, and it still has its hardline adherents.

Nevertheless, the troublesome economics of FTTH have caused many people to embrace a back-off scenario—Fiber to the curb (FTTC) [25]—as a less costly alternative to full FTTH. Essentially, FTTC brings the fiber interface to a point close to a number of subscribers, and uses standard metallic facilities (that is, twisted-pairs of copper wire) to connect the drop segment to the customer's premises. The argument is made that good quality copper facilities can deliver large bandwidths over short distances, so the FTTC approach does not represent an abandonment of the overall broadband architecture, although in some cases FTTC is differentiated from FTTH as a perhaps qualitatively different type of access. Nevertheless, the motivation for FTTC is lower cost [26].

It may become necessary to back off somewhat further. There are now discussions of "fiber to the pedestal," which presumably means the distribution frame or cross-connect point interfacing the drop lines to the distribution cable, which is normally a little further from the customer than "the curb." Then there is "fiber to the hub" or "fiber through the hub," in which large corporate customers may enjoy fiberized access but distributed customers still connect via standard distribution cable [27]. Once again, the uncertainties over actual costs mean that the economically optimal fiber/copper interface point is, in fact, still uncertain. "Fiber to the Whatever" was the rather cheerful expression coined recently by Bell Labs guru Robert Lucky [28].

In any case, the back-off from FTTH implies the need for a hybrid loop. It is usually assumed that this must mean fiber and copper, but a number of people are beginning to consider the marriage of fiber and radio as well. In one respect, fiber/radio hybrids would be at a disadvantage: It would be difficult to deliver the true broadband capabilities over a radio link. However, for a range of interim or lower rate forms of access, digital radio could serve quite well. There is also the possibility that radio systems may seek out fiber as the backbone technology for supporting unique radio-based access services, such as neighborhood cordless communications. A recent article by the Donald Cox group at Bellcore underscores this point:

> While digital radio will allow significantly higher transmission rates (several hundred kilobits per second) than currently supported by analog wireline loops, it falls considerably short of the gigabit-per-second rates that can be transmitted over optical fiber. Although high-rate fiber transmission can provide many new and exciting services, it still requires a physical fixed connection. Low-power digital radio, when integrated with network intelligence, could provide tetherless personal portable communications. . . . In fact the near-term recommended deployment architecture of placing fiber to the curb

> provides an economic solution to installing radio ports. That is, the radio port can be attached at the Optical Network Unit (ONU) at the curb. [29]

As of this writing, this is on the leading edge of network planning, so it is not possible to speak definitively about fiber/radio hybrids. In general terms, however, I think we can identify several different approaches that may develop.

9.4.1 DLC-Type Architectures

The simplest hybrid configuration involves the substitution of fiber for the feeder portion of the loop and radio for the distribution and drop portion (see Figure 9.6). This is a structure that telephone engineers are very familiar with, and since the loop is already organized in this manner it would allow for fiber and radio (and copper) to be deployed essentially independently, which could have many advantages in implementation.

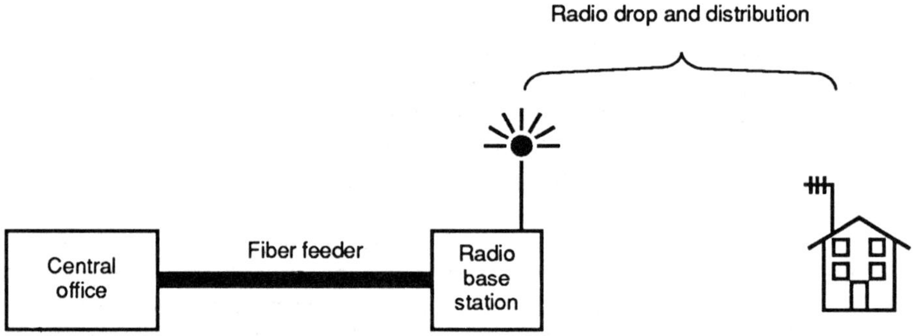

Figure 9.6 DLC-type radio/fiber hybrid.

The disadvantage is that it tends to constrain radio to the existing feeder plan, rather than exploiting the area-coverage concepts discussed in Chapter 7.

9.4.2 Radio Tails

The second approach is somewhat more vague, but would involve using radio for the last segment from the shared facilities (e.g., distribution cable) to the end-user:

> Both fiber and radio will be used to implement the local loops of tomorrow. Fiber in the local loop is only cost effective, however, for clusters of lines . . . [and] for large and medium sized businesses. For residential users, however, it

> only makes sense to take fiber to the street end and connect a number of customers to a box at the end of the street. . . . For this last 100 meters to 200 meters [one] could use copper or . . . radio tails. Well over half the costs of a typical local telephone operation are associated with the maintenance of copper cables. . . . Radio tails would avoid these costs. [30]

In essence, this is FTTC with a radio drop (see Figure 9.7). It would require rethinking the way the network is currently laid out from a number of perspectives (e.g., power, maintenance, CPE and demarcation point, transport formats). The biggest problem is that by replacing only the drop, we do not gain as much of the potential savings over copper as in other architectures (see Section 7.2.3). My feeling is that this will be used only where portability or some other radio-unique service is to be offered.

9.4.3 Cable TV-Based Architectures

The cable TV industry starts with both an advantage and a disadvantage in one—its existing tree-and-branch one-way transmission architecture. In one respect, it is a leg-up on the problem: Cable TV operators have a lot of plant out there, and much of it could be adapted, in principle, to support radio-based two-way access. As fiber comes into the cable TV feeder plant, it may make this transition all the easier by eliminating some of the obstacles to efficient upstream communications [31]. Radio ports could allow cable TV operators to leverage their existing plant and offer two-way telephony-type services (Figure 9.8). (We are ignoring the regulatory obstacles here.)

On the other hand, the cable TV network was never designed for such uses. The architecture of a radio system that could be compatible with today's television broadcasts is still open to many questions. Moreover, the cable TV network in any given franchise area was not designed with radio propagation in mind. Easy solutions such as placing radio ports at amplifier points do not prove to be very efficient in terms of coverage. Some analysts in the cable industry are coming to the conclusion that a new backbone (fiber-based) will be needed to support two-way communications [32] (see Figure 9.9).

9.4.4 Nontraditional Loop Architectures

Down the road a few years there will undoubtedly be many interesting futuristic loop architectures, many of which will involve fiber/radio hybrids. The FITL thinkers are actively evaluating quite a few alternative backbones, including rings, buses, and other LAN-like configurations [33]. It may well be that 10 or 15 years from now both the standard star network and the standard circuit-switched loop will have gone the

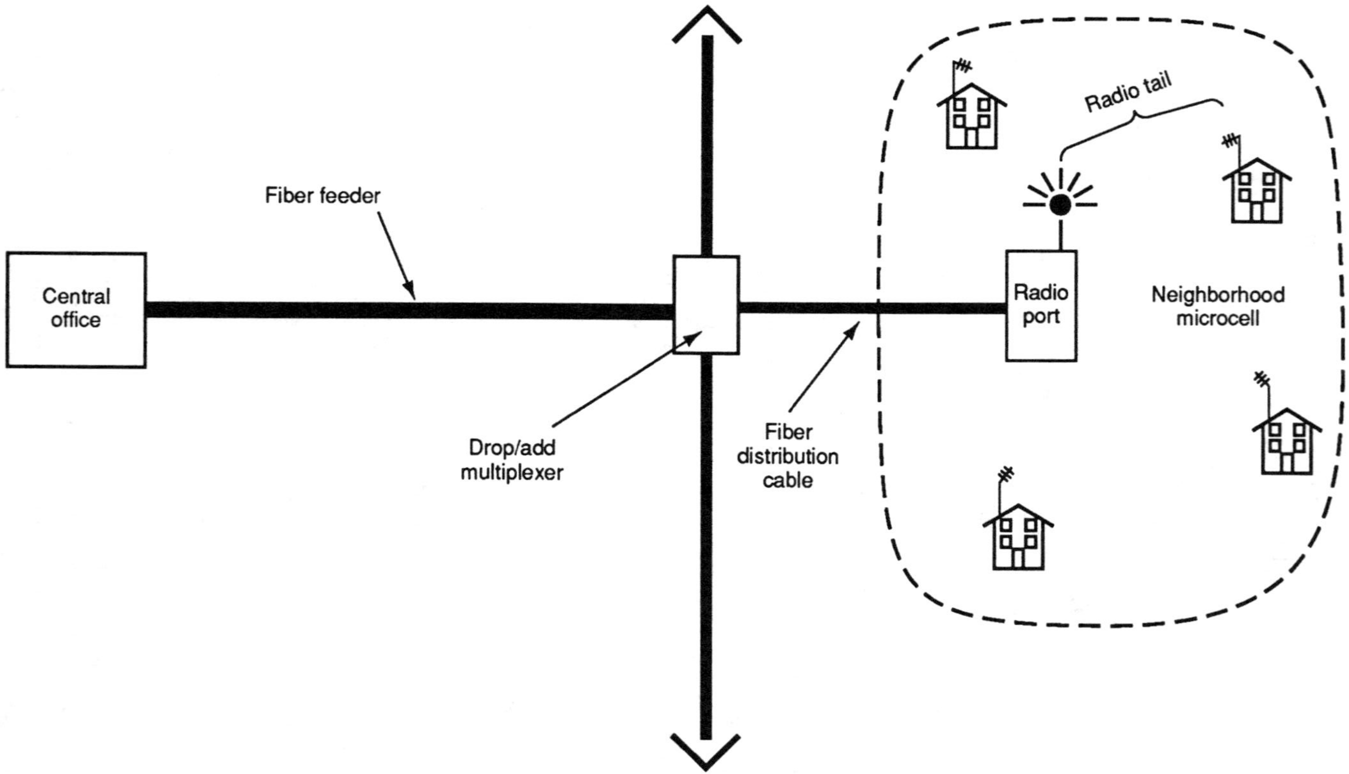

Figure 9.7 The radio tail.

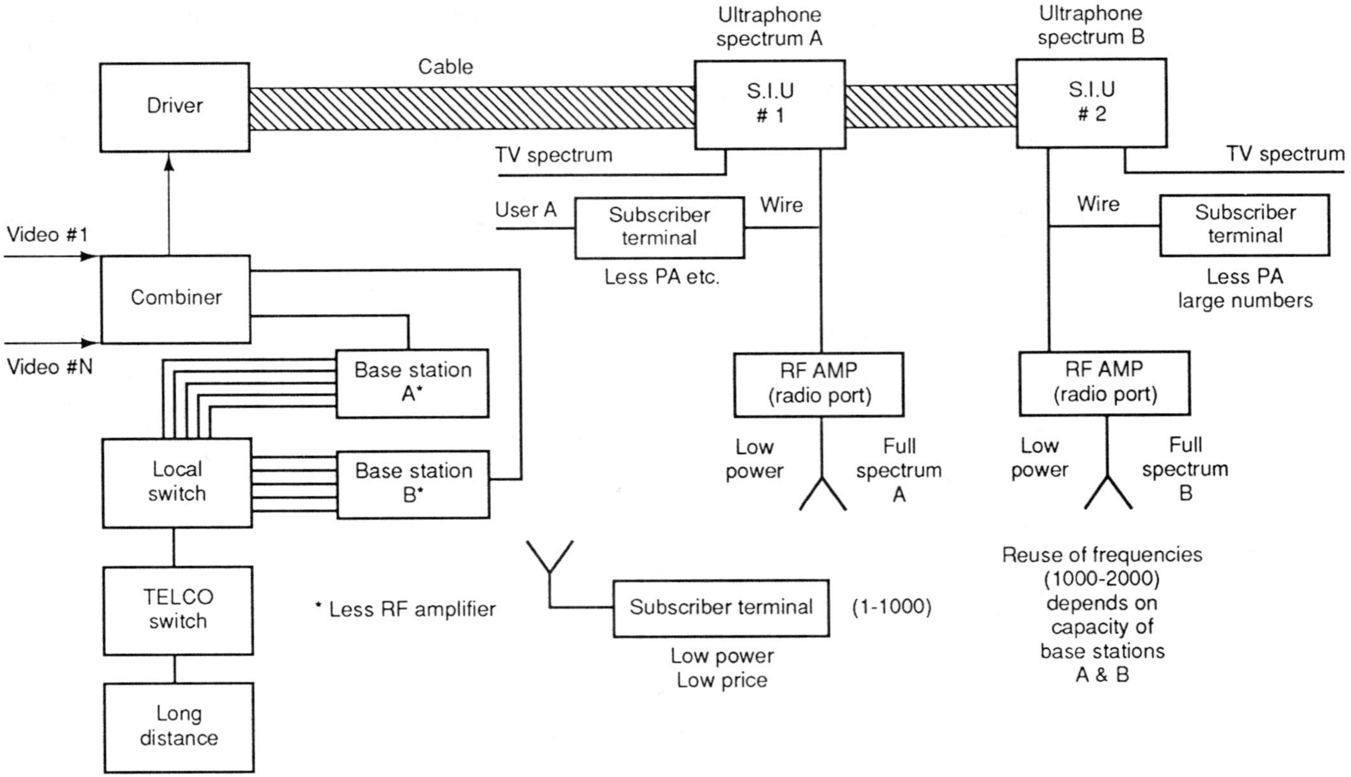

Figure 9.8 The cable TV-based radio hybrid. (From IMM Corporation.)

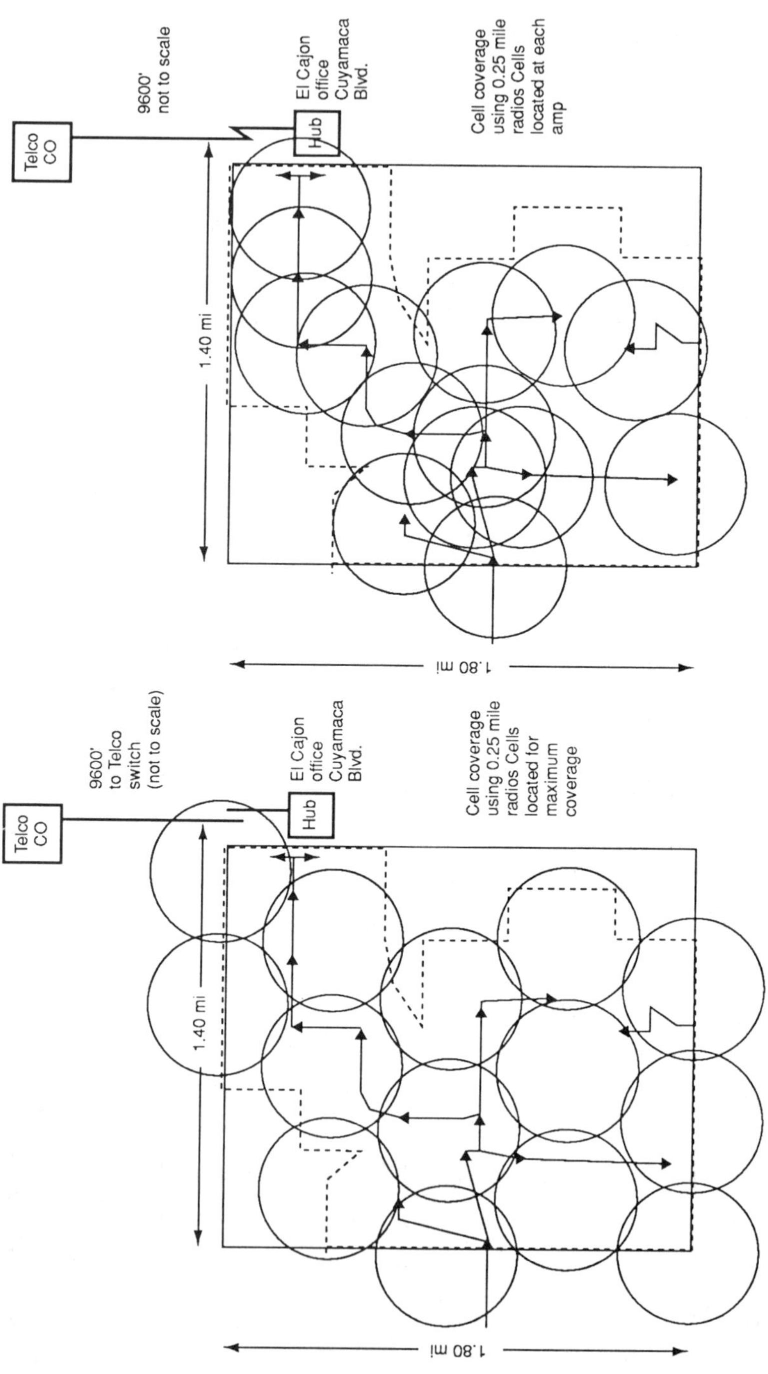

Figure 9.9 Two scenarios for fiber/radio deployment.

way of the buffalo. ATM and B-ISDN networks may drive the development of novel loop access arrangements, which may or may not rewrite the economics of today's access disequilibrium. It is too early to say very much about such exotic access structures, but look for this to be an upcoming field over the next few years.

I was somewhat amused recently to see two strong advocates of FITL come forth with what must be acknowledged as an ingenious pro-fiber argument.

> Virtually all studies comparing fiber to copper on a life-cycle basis assume a 20-30 year useful life for copper. When one considers wireless technology, including Personal Communications Services (PCS), competing broadband technologies, and the emergence of broadband ISDN around the mid-1990s, even a 20-year useful life of copper becomes suspect. Few if any published studies have

Table 9.2 (a)
Spectrally Efficient Wireless Access.

Site#	*Location*	*Telephone Company*	*Date of First Service*	*Number of Homes*
1	Glendo, Wyoming	U S West	September 1986	19
2	Allen, Kansas	S&A Telephone	October 1986	25
3	Sonora, Texas	GTSW	April 1987	160
4	Wahsington, D.C.	WHCA #1	August 1987	14
5	Ripley, Missouri	South Central Bell	August 1987	15
6	Silver Lake, Oregon	PTI	September 1987	156
7	Etzikom, Alaska	AGT	September 1987	39
8	Espanola, Texas	Contel	October 1987	30
9	Eldorado, Texas	GTSW	December 1987	30
10	Kodiak, Alaska	PTI	December 1987	65
11	Dubois, Wyoming	Dubois Telephone	June 1988	39
12	Washington, D.C.	WHCA #2	July 1988	10
13	San Clemente, California	SONGS	October 1988	10
14	Ft. Davis, Texas	Southwestern Bell	November 1988	13
15	Sitka, Alaska	PTI	November 1988	33
16	Chinle, Arizona	Navajo Com.	November 1988	210
17	Victorville, California	Contel of CA	January 1989	40
18	Hardy, Arkansas	Liberty Telephone	April 1989	44
19	Breckenridge, Texas	Southwestern Bell	June 1989	28
20	Selma, Alabama	BellSouth	August 1989	17
21	Clarksdale, Missouri	BellSouth	September 1989	20
22	Kingman, Arizona	Citizens Utilities	October 1989	75
23	Ely, Minnesota	Contel of Minnesota	October 1989	25
24	Quartzite, California	Contel of California	November 1989	40
25	Dot Lake, Alaska	AL Pwr. & Tele.	November 1989	11
26	Pine Mtn., Oregon	PTI	December 1989	70
27	Harding Lake, Alaska	PTI	December 1989	35
28	Caballo, New Mexico	Contel of West		59
29	Telluride, Colorado	U S West		76
	Total number of homes served			1408

Table 9.2 (b)
Fiber Optics Transmission to the Home Trials.

Site#	*Location*	*Telephone Company*	*Date of First Service*	*Number of Homes*
1	Hunters Creek, Florida	Southern Bell	November 1986	95
2	Heathrow, Florida	Southern Bell	June 1988	256
3	Princeton Gate, New Jersey	New Jersey Bell	August 1988	51
4	Leawood, Kansas	Southwestern Bell	October 1988	64
5	Memphis, Tennessee	South Central Bell	November 1988	43
6	Ridgecrest, California	Contel	December 1988	100
7	Perryopolis, Pennsylvania	Bell of Pennsylvania	January 1989	80
8	Mendota Heights, Minnesota	U S West	3Q 1989	100
9	Sidney, New York	Contel	2Q 1989	—
10	Cerritos, California	GTE	2Q 1989	705
11	Coco Plum, Florida	Southern Bell	2Q 1989	12
12	Governors Island, N.C.	Southern Bell	2Q 1989	4–8
13	Mira Vista, Texas	Southwestern Bell	3Q 1989	80
14	Lynnfield, Massachusetts	New England Tel	4Q 1989	100
15	Savannah, Georgia	BellSouth	2Q 1989	192
16	Orlando, Florida	BellSouth	2Q 1989	55
17	Lake View Terr., Florida	BellSouth	2Q 1989	100
18	Scottsdale, Arizona	U S West	3Q 1989	96
19	Norcross, Georgia	BellSouth	1990	135
20		Ameritech	—	100
21	Ft. Myers, Florida	United	—	—
22	Olathe, Kansas	Southwestern Bell	October 1989	—
	Total number of homes served			2370

> considered the impact of PCS or other wireless/copper comparisons on the remaining life of copper. [34]

Thus, we may even find that "radio in the loop" actually *facilitates* the acceptance of FITL. There is something satisfying about the thought that perhaps radio and fiber may actually reinforce each other's advantages in their respective loop applications.

It is interesting to note, in closing, that as of this writing (more or less), the actual penetration of the two technologies into the access environment is very similar in quantitative terms. Table 9.2 shows the respective lists of trials and first applications for FTTH and wireless digital access. In each case, as the 1990s began, there were several dozen sites and several thousand customers. As a partisan, I may observe, however, that the FTTH trials were all run as technology showcases—to my knowledge, not a single one of these sites could have been justified on a first-cost basis against conventional copper-based access. On the other hand, most of the wireless access sites were proven-in economically against copper in rigorous analysis by conventional industry costing methods, and these sites are in service today precisely because radio-based access proved to be less expensive.

NOTES

1, See the three major census surveys for 1912, 1922, and 1937: Department of Commerce, Bureau of the Census, *Telephones & Telegraphs, and Municipal Electric Fire-Alarm and Police Patrol Signaling Systems—1912,* Washington, D.C.: U.S. Government Printing Office, 1915; Department of Commerce, Bureau of the Census, Census of Electrical Industries: *Telephones—1922,* Washington, D.C.: U.S. Government Printing Office, 1924; and Department of Commerce, Bureau of the Census, Census of Electrical Industries: *Telephones and Telegraphs—1937,* Washington, D.C.: U.S. Government Printing Office, 1939.

2. M. Schwartz, *Telecommunications Networks: Protocols, Modeling and Analysis,* Reading, Massachusetts: Addison-Wesley, 1987, Chap. 12. Schwartz summarizes the logic in favor of nonhierarchical networks as follows: "A primary reason for the cost effectiveness of nonhierarchical routing, particularly in a relatively large country, is that traffic patterns tend to change in different offices and different parts of the network as a function of the time of day or season of the year. Hierarchical routing is essentially static and must be designed for peak traffic in the network. Some network capacity will thus remain idle during nonpeak intervals. Adaptive routing, implicit in nonhierarchical routing, can be adjusted to changing traffic patterns with time as traffic at various offices peaks at different hours. A given load can thus be accommodated by fewer resources." (p. 632)

3. *Ibid.,* pp. 645 ff.

4. A.E. Joel, "Circuit-Switching Fundamentals," in J.C. McDonald, ed., *Fundamentals of Digital Switching,* New York: Plenum Press, 1983, pp. 1-39; see also Mario Gerla and Joseph Bannister, "High Speed Local Area Networks," IEEE Spectrum, August 1991, pp. 26-31. 5 P. W. Huber, The Geodesic Network: 1987 Report on Competition in the Telephone Industry, Washington, D.C.: U.S. Government Printing Office, 1987.

6. *Ibid.,* p. 1.3.

7. *Ibid.,* pp. 1.3, 1.13–14.

8. *Ibid.,* pp. 1.20–1.21.

9. Bellcore, *A Compilation of Results from Recent Bellcore Studies on Fiber to the Home (FTTH),* SR-TSY-001468, Issue 1, December 1989, Bellcore, Chap. 5.

10. P. W. Huber, *The Geodesic Network: 1987 Report on Competition in the Telephone Industry,* Washington, D.C.: U.S. Government Printing Office, 1987, p. 2.3.

11. See A.D. Lipman, *Telecom Deregulation, Telephony,* 1987; also U.S. Congressional Budget Office, *The Changing Telephone Industry: Access Charges, Universal Service, and Local Rates,* Washington D.C.: U.S. Government Printing Office, 1984.

12. My favorite one-sentence description of the problem is the following: "This infrastructure will be a combination of FITL, HDSL (vis-a-vis the variation described herein), SMDS, and BISDN, all using ATM transport" from P.D. Lattner, R.L. Fike, and G.A. Nelson, "Business and Residential Services for the Evolving Subscriber Loop," *IEEE Communications Magazine,* March 1991, p. 114.

13. L.K. Vanston, R.C. Lenz, and R.S. Wolff, "How Fast is New Technology Coming?," *Telephony,* September 18, 1989, p. 50.

14. See R.M. Pepper, "Through the Looking Glass: Integrated Broadband Networks, Regulatory Policy and Institutional Change," Federal Communications Commission OPP Working Paper Series, No. 24, November, 1988, pp. 21 ff., including the somewhat pessimistic conclusion: "Because of the crossownership and MFJ restrictions, if local exchange carriers—especially the BOCs—have to

wait until they provide video programming themselves before justifying building broadband fiber networks—or before upgrading narrowband fiber networks to broadband capacity—it may be a long time before switched broadband networks reach the majority of U.S. homes. . . . "

15. I am well aware of the forecasts from Bellcore and elsewhere that predict a much more rapid uptake of B-ISDN by the mass market. See, for example, L.K. Vanston, R.C. Lenz, and R.S. Wolff, "How Fast is New Technology Coming?," *Telephony*, September 18 1989, pp. 49–52. We shall see.
16. P. Kaiser and P. W. Shumate, "Fiber-to-the-Home," *Optics News*, October 1989, p. 20.
17. I first read these predictions in the mid-1980s, at which time the crossover was anticipated before the end of that decade. There was a heavy component of wishful thinking in a lot of this: "What I am calling for is nothing less than a massive priority effort to achieve cost effective fibering of the distribution loop to the residences with 1990 as the worse [sic] case target" from R.G. Dewitt, "Fiber Frontier: The Local Loop," *Telephone Engineer & Management*, July 1 1987, p. 104; see also M. Gawdun, "Lightwave Systems in the Subscriber Loop," *Telecommunications*, May 1987, pp. 65–74, 85, where the cost goals for a fiber loop are targeted at $1500 by 1990—which clearly did not happen. Today, the elusive goal is usually said to be sometime in the "mid-1990s." One of the more bullish fiber prophets is R. M. Pepper, "Through the Looking Glass: Integrated Broadband Networks, Regulatory Policy and Institutional Change," Federal Communications Commission OPP Working Paper Series, No. 24, November, 1988, p. 7: "Within two to five years, the cost of providing 'plain old telephone service' (POTS) using fiber in the local loop will fall below the cost of providing POTS using today's copper technology" and, p. 16, "8 to 18% of U.S. households will be provided telephone service over a fiber optic network extending to the subscriber's premise by the year 2000."
18. P.W. Shumate, "Cost Projections for Fiber in the Loop," *IEEE LCS Magazine*, February 1990, pp. 73–76; also K. Lu, R. Wolff and F. Gratzer, "Installed First Cost Economics of Fiber/Broadband Access to the Home," *Proceedings of the IEEE Global Telecommunications Conference*, November 28, 1988, pp. 48.4.1–48.4.7.
19. N. Janus, ed., *The Demand for Fiber in the Local Loop*, Potomac, Maryland: Phillips Publishing, 1988, p. 13.
20. T.R. Rowbotham, "Local Loop Developments in the U.K.," *IEEE Communications Magazine*, March 1991, p. 57.
21. P.D. Lattner, R.L. Fike, and G.A. Nelson, "Business and Residential Services for the Evolving Subscriber Loop," *IEEE Communications Magazine*, March 1991, p. 109.
22. Bellcore, *A Compilation of Results from Recent Bellcore Studies on Fiber to the Home (FTTH)*, SR-TSY-001468, Issue 1, December 1989, p. 34.
23. P.W. Shumate, "Optical Fibers Reach into Homes," *IEEE Spectrum*, February 1989, p. 44; a slightly more optimistic forecast is offered in P. Kaiser and P.W. Shumate, "Fiber-to-the-Home," *Optics News*, October 1989, pp. 14–20, suggesting perhaps $2000 to $3000 per line costs for broadband access by 1998.
24. J.F. McLaughlin, Executive Director of the Program on Information Resources Policy, Harvard University, "Making Money as a Telephone Company: Risks, Opportunities, and Likely Failures," prepared for the United States Telephone Association Capital Recovery Seminar, October 23, 1989.
25. Bellcore, *A Compilation of Results from Recent Bellcore Studies on Fiber to the Home (FTTH)*, SR-TSY-001468, Issue 1, December 1989, p. 12.
26. "Significantly lower costs-per-customer can be realized by providing 'Fiber-to-the-Curb' access only, with wire pairs completing the connection to the customer" from P. Kaiser and P.W. Shumate, "Fiber-to-the-Home," *Optics News*, October 1989, p. 15.
27. R. Lamerdin, "Pacific Bell's Plan for Fiber in the Local Loop," *Telephone Engineer & Management*, February 1, 1988, pp. 61–67.
28. Quoted in *Report on AT&T*, industry newsletter, May 13, 1991.

29. D.C. Cox, W.S. Gifford, and H. Sherry, "Low-Power Digital Radio as a Ubiquitous Subscriber Loop," *IEEE Communications Magazine,* March 1991, p. 94.
30. Peter Purton, "Radio over Fiber?" *Telephony,* November 1989.
31. Bellcore, *A Compilation of Results from Recent Bellcore Studies on Fiber to the Home (FTTH),* SR-TSY-001468, Issue 1, December 1989, pp. 13 ff.
32. At the early stages of this analysis, I can only reference as-yet unpublished internal studies by the staff of IMM Corporation, and also unpublished studies conducted or sponsored by CableLabs, Inc., of Boulder, Colorado.
33. See K. Lu, R. Wolff, and F. Gratzer, "Installed First Cost Economics of Fiber/Broadband Access to the Home," *Proceedings of the IEEE Global Telecommunications Conference,* November 28, 1988, pp. 48.4.1–48.4.7, Sec. 4.3 on "Topologies"; see also Bellcore, *A Compilation of Results from Recent Bellcore Studies on Fiber to the Home (FTTH),* SR-TSY-001468, Issue 1, December 1989.
34. P.W. Shumate and R.K. Snelling, "Evolution of Fiber in the Residential Loop Plant," *IEEE Communications Magazine,* March 1991, p. 72.

Chapter 10
THE PCN TRAJECTORY

10.1 DOWN FROM THE MOUNTAIN—ANOTHER PARABLE

In 1986—not that long ago—the first wireless digital access system in the United States was installed, as a field prototype, on a Wyoming hilltop overlooking the North Platte River. In this remote corner of the Old West, where the map carries names like Poison Spider Creek, Rattlesnake Range, Medicine Bow Mountain, Lost Springs, Hanging Woman Creek, and (looming over all of them) the Little Bighorn, a handful of homesteaders had staked out their ranches east of the Platte in the 1920s and 1930s, about 25 miles from the tiny town of Glendo. Later the river was dammed several miles below the town, and as the reservoir filled up, it flooded out old roads, a few homes down on the bottomland, and the open-wire telephone line that had connected those eastern ranches to the network. For years, the ranchers were cut off. The phone company experimented with submarine cable under the lake, but it never worked well. Periodically, the planners reassessed other service options, such as running a line from Glendo south to below the dam, through the Platte canyon and then back up the east side of the lake—well over 100 kilofeet even before the customer drop was calculated. The last estimate in 1984 was about $40,000 per ranch.

So when the idea of trying out a new wireless system was put forward, no one in Glendo had to think too hard about saying yes. No need for a detailed economic analysis; whatever it cost, radio had to be cheaper than wire. And since those ranchers had not had any telephone service for decades, there was no real risk in asking them to be guinea pigs for a new technology. Nobody was going to quibble too much about, say, the voice quality from a new low-bit-rate coder. If the system worked even half-decently, the ranchers would be delighted and the telephone company would score points with a lot of important people.

The biggest challenge was still the terrain. A pine-covered hilltop located about 12 air-miles across the reservoir from Glendo was chosen as the site for the radio base station, because of the favorable radio coverage it offered. There was no road to the top at first, and our initial survey was conducted by helicopter, although later a path was surveyed and the summit was reachable, in dry weather, by four-wheel drive. It was advisable to travel with at least two vehicles, however. In spring and summer, the chief hazard was the mud that collected in poorly drained dips in the trail; some of these "pits" were big enough to mire even a four-wheel drive jeep and the second vehicle might find it necessary to pull the first one out. (In winter, there was snow and bitter cold to 40 below, and nobody went up.) It took about an hour ascending this way, not counting any delays in getting through Mrs. Hammer's gate. (Mrs. Hammer had not agreed to pay the $4000 "aid to construction" that the telephone company insisted on charging each rancher, so she wasn't getting a new wireless phone, but the road crossed her property and sometimes she left the gate locked. Sometimes she came out to meet the vehicles coming up, and at least once she is said to have carried a shotgun with her.) In all, counting installation, troubleshooting, and demonstration visits, the teams probably made 50 trips up and down that summer and fall.

The chief compensation for all this was the stunning scenery. I made four trips up myself and at various times I saw herds of pronghorn antelope, deer, and, once, a bald eagle. The drivable slope up the backside of the hill came to a sheer 800-foot precipice on the front side, with a view all the way west to the Laramie range 60 miles away. The phone company cleared a small space, poured a concrete pad, and set up the all-weather hut for the radio base station equipment. An 18-GHz microwave system (also experimental) was used to connect the T1 link back to the switch in Glendo. There was also a radio relay about a mile from town where the microwave path just grazed a small ridge.

The project gathered momentum as the turn-up date approached. Subscriber installation began in August—there was some urgency to get everything in before the weather turned bad—and by September 13 the system was ready to go on the air. The official "first phone call" was to be placed from the then-head of the FCC, Mark Fowler, to Mrs. Doris Leonard (the official "first customer"). The FCC wanted to promote the BETRS concept, then heading into a rule-making proceeding, and the Mountain Bell telephone company wanted to demonstrate its commitment to rural service. As a result, the call became a minor media event, broadcast over a speakerphone in the FCC offices in Washington with a fair number of reporters listening in while Fowler chatted with Mrs. Leonard. In the end, it was deemed a success. Nine days later, the telephone company invited the entire Wyoming State Public Utility Commission out to tour the site, including the trek up to the base station. As it became clear that the equipment worked well and reliably, that the customers were satisfied, there was a brief burst of wireless enthusiasm. The BETRS concept was created and authorized by the FCC. Seminars and conference sessions on "digital rural radio" began to appear. Bellcore issued its first study of the BETRS application. Mountain Bell even began running ads in the local Wyoming papers, accompanied by 60-second radio spot advertisements touting the Glendo trial.

In some ways, the follow-up was slower than a lot of us had hoped. US West did not install its second wireless access system until 1989, although in the meantime quite a few other companies picked up on the trend. But in another respect, the subsequent course of events developed much faster than any of us anticipated.

The Genie Escapes Again

I have already used the metaphor of Aladdin's genie to describe the unforeseen consequences of the introduction of long-haul microwave radio back in the 1950s (Chapter 3). Looking back on the developments in the field of wireless access since 1986, it seems to me that the same process is under way once again. A new technological "force" has been unleashed that is proving difficult to contain within a particular application or industry niche.

For example, it is interesting that the Glendo trial also saw the (quite unplanned) first mobile use of digital TDMA radio. It occurred during the installation of the subscriber terminals. The individual ranches are very large, thousands of acres in some cases, and the installation program took the installer teams over hundreds of square miles of open country, during which time they were often out of contact for hours at a stretch. Coordinating their activities and troubleshooting the system turn-up was especially hard. Testing often involved three separate groups—one at the Glendo central office, one on top of the mountain at the base station, and one out at a subscriber location. But the installers soon found that they could hook up the subscriber units in the back of the van and stay in communications with the other team members. At first they only made these calls when they had stopped the vehicle and aimed the yagi antenna toward the mountain. Soon, inevitably, they found that although the terminals were "for fixed use only," with a whip antenna they would work just fine while the van was moving. Before long, the Doppler effect was being painstakingly researched at all feasible vehicle speeds up to 110 mph or so.

These first prototype terminals were rather large metal boxes designed to mount on a pole or in the basement next to a fuse box. Each terminal weighed nearly 35 pounds, and the power supply (with car battery backup) weighed even more. It drew nearly 60 W of power in the idle mode and more than 100 W while transmitting (which meant that you kept the engine running while you made your calls). In short, it was probably the least convenient mobile communications package ever fielded. But none of this daunted the installers who found it the antidote to long hours of driving incommunicado. Moreover, they perceived immediately the digital clarity of the transmission, quite superior to the standard static-wracked analog mobile radio call. For all its physical cumbersomeness, the system embodied a very sophisticated TDMA radio, with an adaptive 4/16-ary DPSK modulation scheme and a sub-16 kb/s voice coder (the first ever used in any commercial telephone system, which had to pass the "Mrs. Leonard test" for quality), and it delivered excellent performance even in mobile operations. From the installers' standpoint, the question was simple: If it works, why not use it? [1]

That was the beginning. On one of the next installations, no sooner had the system been turned up for *fixed* service than we heard that the customer [2] had installed a couple in vehicles for actual ongoing service, and had even tested one in an airplane! It worked.

The next step grew out of an inquiry from the director of communications for a nuclear power plant in Southern California; his problem was that the conduits in the concrete structure of the containment facility were full, and he needed a more flexible system of communications for special situations such as the refuelings that occurred every so often. It was the complete opposite of rural—yet at his persistence a system was installed in the plant, using leaky coaxial cable for distribution of the radio signal. It worked.

The following summer there was a fire in Yellowstone National Park. Part of the fire zone lay within the coverage area of a BETRS system that had been installed about nine months before. The fire fighters were more than willing to hoist around the heavy fixed station subscriber terminals, which became their chief means of communication for a couple weeks. It worked.

Next came "Operation Last Mile"—a government- sponsored simulation of a response to a large-scale natural disaster, carried out in Alaska in 1989. The scenario assumed that all communications facilities had been knocked out by a major earthquake, and the purpose of the exercise was to restore all the elements of the total system—switch, satellite link, local loop—as quickly as possible. Wireless BETRS equipment was adapted for the "last mile" and service restored within 12 hours.

The following month a nonsimulated hurricane (Hugo) came off the Atlantic and tore up the wireline plant all across the eastern end of Puerto Rico, including the communications at the Roosevelt Roads Naval Station. The "last mile" system was airlifted into Puerto Rico and for three weeks wireless access served as the local loop for many vital services until wireline facilities could be restored.

Six months later a Colombian airliner crashed in the woods of Long Island after running out of fuel on the approach to Kennedy airport. Although within sight of Manhattan's towers and situated in the middle of one of the world's densest population and communications centers, the crash site was a mile from the nearest wireline link. The cellular system was quickly swamped by normal rush hour traffic. Within six hours a BETRS-type system had been set up to coordinate the rescue and crash investigation teams.

That fall, a small telephone company in West Texas installed the same equipment—for an entirely mobile communications system in a rural area where cellular may not make it for years.

Six months later, in March 1991, hours after the fighting stopped in the Persian Gulf, several wireless digital access systems were on their way to Kuwait, where they became the basic local link for most government communications. The devastation of the Kuwaiti telecommunications infrastructure was almost total, and these systems remain in place as of this writing, serving (among others) the fire fighters working in the burning oil fields. Most of the units are used in a mobile mode.

The trend is becoming clear. Wireless digital access technology is robust and highly flexible with respect to the application. People will take a radio system into almost any situation and try to make it work, and it will in fact work in a lot of applications that its designers may not have anticipated. Five years ago (as of this writing), that first test of digital wireless access technology in the "wilds" of Wyoming was an obscure experiment, strictly a rural application, a "last resort" for an extreme situation. At that time only a handful of people in the entire industry had given any thought to "wireless access" as described in this book.

Five years later, wireless is on almost everyone's agenda in one form or another. The major telephone companies all have cellular subsidiaries, and most of them also have a strategy group for wireless technology on the (wireline) telephony side of the house as well. Most equipment manufacturers are working on a number of wireless access systems. We even have technical standards for applications such as BETRS, cellular (vehicular), cordless, wireless in-building, dispatch. We have proposals large and small for new radio-based communications systems, ranging from Motorola's Iridium project to the proliferating "Part 15" markets. There are more than 100 wireless experiments under way as of this writing. There is a new conference on wireless every couple of weeks.

We even have a name for the genie—"PCN."

10.2 THE FOUR NO'S OF PCN

PCN has stormed into the industry consciousness with a ferocity that has startled a lot of us. In Britain, where the acronym was coined, it is a code word for nothing less than a top-down revolution for both cellular and wireline operators, a wedge for competition. The vision expressed in the "Phones on the Move" report is marked by an unbridled optimism regarding the role of radio in the network. Under the banner of PCN, there are three new operating consortia, preparing to farm to local exchange. Suddenly, from having been a "last resort" in Wyoming and Alaska, wireless access is on the verge of soliciting subscribers in the Midlands. In one breathtaking step, the British seem to have leapt to the end game. And so casually executed.

Here across the Inland Sea, the appropriately ambiguous term *personal communications* has indeed come to be the *phrase* of last resort, to throw over any and every new wireless concept that pops up. We are not slow, and there was already quite a bit of "PCN-ing" going on by June 1990, when, following the British lead and with an eye to the global competition in all corners, the FCC opened an inquiry into what they called PCS, for *personal communications service.* Thus, the idea achieved legitimacy, which enhances its marketability. But the focus was not yet sharp. For one thing, in taking up the British model, Americans have paid less attention to the institutional revolution—local exchange competition—and more to the vaguely charismatic marketing notions of "portability" and "seamlessness," as well as broad libertarian proclamations about the "freedom" that personal communications will somehow bring. For example, in the *Notice of Inquiry,* PCS is defined as the use of radio to "free individuals from the constraints of the wireline public switched network" [3]. This might mean many things—for example, the freedom from the problems caused by the difficult economics of copper-based access plant (see Chapters 3 and 4) which could radically alter the rural/urban disequilibrium in telecommunications opportunities. Apparently, however, the FCC's concept of "freedom" follows more along the lines of traditional consumer marketing psychology; PCN will convey "freedom" in the same sense that microwave ovens or lawncare services do—through convenience and timesaving. In this blander view, the "constraints of the wire network" are the extension cords that string across our kitchens and over our work desks, as well as the nuisance of having to manage several telephone numbers per person. After noting that there are a number of "basic forms" of PCS, from cordless phones to pagers (!), the *Notice of Inquiry* zeros in on "even more advanced forms of PCSs":

> The most significant feature of the next generation of these services appears to be a movement towards person-to-person, instead of station-to-station, communications. Existing PCSs require the user to have a different telephone instrument depending upon whether the user is at home, in the office, or in the car. In addition, each of these devices requires its own telephone number. . . . Thus, future PCSs are expected to permit individuals to use the same device in several different environments, including in the home, in the office, perhaps in conjunction with a wireless PBX, or for mobile public telephone service. [4]

What does this really mean? What is the significance of "person-to-person" communications—direct inward dialing? (We have that already.) Consolidating telephone numbers? Seamless transitions for urban businessmen and businesswomen? What really are the implications of "personal communications" for this industry?

Let me begin this final meditation with what I have come to call the *Four No's of PCN:* no spectrum, no service, no licenses, no standards. In all the enthusiasm for what might be, it is wise to take stock first of what we still lack.

No Spectrum. Unlike the United Kingdom and most other countries, there is no central authority in the United States with the power to restructure the spectrum map quickly. There is no easy path to obtaining large amounts of spectrum for PCS or any other new service. Instead, we must exercise the democratic (i.e., political) process in several rounds. The question of spectrum politics is complex beyond the scope of this epilogue, and will undoubtedly occupy many of the best minds in Washington for many years to come. But consider:

1. For roughly 40 years (1946–1983), mobile telephony was limited to less than 5 MHz of total spectrum. In New York City, there were only a couple dozen channels available. The "great leap forward" called cellular radio managed to extract 50 MHz from the TV broadcasters—but it took 15 years to do it [5].

How soon can we expect to see large new blocks of spectrum for a new radio service?

2. From a technical standpoint (principally for reasons of radio propagation characteristics), the prime spectrum for this type of application (i.e., point to multipoint, with mobile or portable subscribers) is between 1 MHz and 1 GHz. But there is virtually no available spectrum remaining below 1 GHz, as the *Notice of Inquiry* acknowledges [6].
3. Even above 1 GHz, finding sufficient new spectrum in the United States will almost undoubtedly mean displacing some existing users. These users fall into several categories, including the federal government (principally the Defense Department), the broadcasters, and the private users of radio communications systems (mainly large corporations but also the incredibly powerful ham radio community). Past experience has shown that all these groups are tenacious of their spectrum rights.
4. Other regions of the spectrum that could offer some relief appear to be still out of reach for the near term. The hoped-for reallocation of UHF TV spectrum between 470 and 800 MHz (which, by my estimate, is about 90% idle nationwide [7]) is currently frozen pending decisions about HDTV. Meanwhile, the Defense Department continues to hold huge tracts of spectrum that by all accounts are used rather inefficiently [8]; the proposed "peace dividend" in radio spectrum for commercial users has been stalled, at least temporarily, by renewed military urgency [9].

No Service Definition. What exactly is PCS, or what should it be? The outcome of the FCC's inquiry and rule-making will ultimately have to answer this question in some detail. In particular, it will be necessary (at least given past experience with the spectrum allocation process) to define very carefully what PCS is not. Therein lies much of the difficulty:

1. PCS is presumably not "simply" more spectrum for cordless extension phones (of which there may be as many as 50 million units in the United States). Therefore, one key question becomes: What service requirements will differentiate PCS from cordless phones? Presumably the ability to make calls from locations other than the coverage area of the immediate residential base station will be important, but how rigorously will this requirement be defined? And what happens to today's cordless telephone business?
2. Cellular operators figure that whatever PCS is, they either provide it already or can do so within the framework of the new "Flexible Cellular" rules [10]—and basically they are right. With a few exceptions, cellular operators are now free to redefine their service offerings as they see fit, even to the point of reorganizing their own spectrum. Several of them are moving in this direction, at least as a defensive measure [11]. I will comment further below upon the cellular

PCS initiatives, but the question raised for any PCS/PCN rule-making is: How will the new service be different from today's "flexible cellular"?

3. Wireless in-building systems, including wireless PBXs, seem poised to follow their own path to commercialization. There is in fact little technical rationale for forcing such systems to adopt the considerable system overhead associated with some of the "seamless network" schemes. My reading of the marketplace dynamics of this segment is that it will develop very well on its own, quite possibly with manufacturer-proprietary technologies, as it "grows out" from existing PBX systems (rather than "growing in" from the public network, cellular, or wherever). Once again, therefore, the question of the value-added of PCS/PCN over a non-PCS wireless PBX must be raised in the course of defining the personal communications service.
4. Having spent a number of years working in and thinking about product design and market development, one of my conclusions (and certainly it is not original) is that the more tightly focused the application that is designed for, the more likely the product is to succeed. This is not to deny the kind of "Aladdin's genie" process I have just described; serendipity, and innovations by users, do occur and one must be ready for it. But I find it is best to build out from one established, focused application. Cellular has done that with vehicular telephony. Vague, all-encompassing market concepts often find difficulty gaining that first foothold. Consider the still unfinished saga (as of this writing) of CT-2 in the United Kingdom and elsewhere. What began as a fairly focused proposal for a better cordless extension phone (obviously a well-established market with certain distinct needs) was later enhanced to include a variety of much hazier applications, including Telepoint and even a version of wireless PBX functionality. The uncertain applications focus of ISDN is perhaps another example of the dangers of trying to be all things to all customers. I will return to this point below.

No Licensing Procedure. Surely this is the most difficult question of all. How is the FCC going to avoid a repetition of the cellular licensing fiasco [12]?

1. The traditional FCC method for awarding licenses has been through *comparative hearings,* especially in the broadcast arena where hearings have worked fairly well. But comparative hearings in the first round of cellular licensing quickly showed their administrative infeasibility—the process was simply too cumbersome for the FCC machinery, and that was with only half a dozen or so applicants per market. A PCS license for, say, New York City, would probably attract hundreds of applications, which is far more than the comparative process could ever deal with.
2. The "solution" for later cellular licensing rounds was a *lottery.* It is generally agreed that the effects were disastrous. All pretense of rational evaluation was eliminated, as winning became a matter of pure chance. The number of appli-

cations skyrocketed to thousands per franchise, as "applications mills" began peddling cellular "opportunities" to ordinary citizens. There was very limited screening of these applications. Speculators and con artists jumped into the fray, and fraud became a serious problem. The breakdown caused many qualified potential operators to abandon cellular altogether, and led to the effective concentration of the industry in the hands of the major telephone companies (which is precisely what competitive licensing was supposed to prevent). Some people did benefit: There were financial windfalls for dentists and truck drivers and housewives who happened to hold winning lottery tickets. But overall the industry still staggers under the debt load it took on in large part to buy out the lucky winners.

3. The third licensing approach often discussed is *spectrum auctioning.* The idea is to sell or lease the electromagnetic real estate to the high bidder, like offshore oil leases. This concept has a long history and is encrusted with many pros and cons. Suffice it here to say that, like the lottery concept 10 years ago, the spectrum auction has never been really put into practice and the dynamics of such a mechanism are difficult to predict [13]. And—aside from theoretical considerations—the courts have so far held that the current charter of the FCC does not encompass the authority to auction the spectrum. This means that it will probably take an act of Congress before this option could be exercised [14].

No Standards. There is a case to be made for industry-wide standards for certain system interfaces for any radio system, perhaps including elements of the air interface. For one thing, standards encourage more manufacturers to make product investments. But where do these standards come from? In the past, standards were sometimes written into the FCC rules, and carried thus the force of law. This may make for orderly, if highly commoditized, markets. Nevertheless, there are many difficulties with government-set standards for high-tech industries, some of which I have summarized elsewhere [15]. In any case the "flexible cellular" rule-making took the FCC out of the standards-setting business for mobile radio, at least for the time being. This puts the spotlight on the industry, especially the equipment manufacturers, to come up with a standard. There are quite a few obstacles to this process, however.

1. In the United States, assuming the FCC does not mandate a technical standard, a manufacturer's adherence to an industry-developed standard is strictly voluntary. Antitrust law militates against industry agreements that may be seen as a "restraint of trade." There is often an antitrust lawyer present at meetings of standards committees. This means that while a standard may be agreed to by a majority of the industry, there is no legal barrier to any manufacturer or operator who wishes to bring out a nonstandard system. In the past, for the mobile radio industry, this potential for standards diversity was tempered somewhat by the lack of technological alternatives. With digital radio, how-

ever, there are many technical options. The ultimate unenforceability of voluntary standards makes it unlikely that we shall see a return to the monolithic technical environment of the FM epoch.

2. Even setting aside the question of the enforceability of standards, it seems clear that developing a PCN standard will be a lengthy process. Consider that it has taken three years to develop what is still basically the first draft of the North American digital cellular standard. The technical options for PCN/PCS are even more diverse, and, with the introduction of CDMA into contention, many of the questions are now fraught with uncertainties of a more fundamental nature than most standards committees have traditionally had to deal with [16]. If the 60 or 70 interested companies sat down tomorrow to begin working on PCN standards (which they have not yet done as of this writing), how many years would it be before a detailed standard emerged?
3. Unfortunately, this increases the risk for equipment manufacturers. The likely delays in the standards process present them with some difficult choices. They may choose to delay investments until a (voluntary) standard has been confirmed—and in the meantime non-U.S.-based companies who operate in the much more stable technological environments of Europe and Japan may be able to introduce their equipment here sooner, regardless of standards, because of the lead time they have gained over the more chaotic and hesitant market here. Indeed, this seems likely to happen in the early rounds of PCN-type applications (e.g., CT-2, DECT) where a solid European standard may coalesce far sooner than American standards processes are likely to bear fruit of any kind. Or they may take the risk of developing their own equipment now, bringing it to the market, and hoping it will be successful enough in commercial terms that it can become a *de facto* standard. Of course, this virtually assures a non-standardized PCN marketplace for at least some period of time, and probably means that some manufacturers and customers will ultimately take a loss on having backed a wrong horse. (The beta/VHS battle in video recorders comes to mind.)
4. Finally, the question of intellectual property rights (IPR) has raised its head. IPR encompasses patents, copyrighted software, trade secrets, and the like—those forms of knowledge or technology over which one company may exercise a proprietary claim to the possible exclusion of potential competitors. In many cases (i.e., patents and copyrights) this claim is legally enforceable. Traditionally, the telecommunications industry has not had to be too concerned about IPR. In the United States, most of the basic telecommunications inventions (including, by the way, the basic patents on cellular radio) were freely available through cross-licenses with AT&T under the 1956 Consent Decree. But that structure is no longer in place. Instead, patents are becoming competitive weapons, to be wielded, licensed, or withheld based on corporate strategies rather than a Court decree.

How does this affect standards development? The crucial question is: What happens when the industry sets a standard, or wants to, where one company owns some of the technology incorporated into that standard? One "easy answer" is that such IPR should be made available royalty-free to all users [17]. But then the companies that have borne the risk of developing new technology would not be compensated for their investment. The second not-so-easy answer is that the IPR owner should agree to license (for royalties) the necessary technology to all. But what if there are a number of such companies, with IPR of varying scope and importance [18]? How are royalties to be set in a fair and equitable manner? And what if one of the IPR owners refuses to license equitably, or at all? Ironically, the telecommunications industry is just beginning to understand Information Age property rights.

The purpose of reciting the Four No's of PCN is not to instill pessimism, but to emphasize the fact that the industry is at a starting point. In many ways, there is a clean slate; PCN comes into the world far less encumbered by either technical or regulatory presuppositions than was the case with cellular radio [19]. All of these issues (with the possible exception of the licensing procedure) can be resolved without undue agony. For the first time in decades, the prospect of obtaining large blocks of spectrum for a new service appear at least reasonably favorable. The marketplace will sort out many of the service-related questions, and experience with digital radio systems will strengthen the standards process [20]. As far as licensing it concerned, it may well be that technology will help solve these problems in a way that we will examine in the final section of this chapter.

To continue this summary analysis of the PCN phenomenon, we now turn to three sets of questions that underlie much of the current debate over the meaning and the future of PCN services: What are the applications for PCN? What are the technologies? And who, if anyone, has a special claim to the PCN franchise?

10.3 APPLICATIONS

We cannot escape the centrifugal tendencies of radio. In Chapter 5, we browsed through a partial catalog of wireless access applications, including those in what we labeled the *PCN set,* which itself is a mesocosm of different concepts. Almost every new PCN proposal expands the set somewhat, and now that it has acquired a certain cachet there is a tendency to attach "PCN" to almost any new idea involving radio-based communications. There is little agreement except as to what PCN is *not:* It is not cellular, not BETRS, not SMR, and so forth. Perhaps it is "more spectrum for something else."

But to turn from negative characterizations to positive ones, let us start with the truism that every application and every product must provide something of value to the telecommunications user. And there are, I believe, only three fundamental ways in which a radio-based communications system can provide this value:

1. *Portability* or, if you will, "tetherlessness": Radio can allow the user to carry his access capability with him, whether walking, riding in a car, visiting the neighbors, or just sitting in the backyard.
2. *Economics:* Radio can allow a user to obtain access for less cost than wireline alternatives—although as we have seen in Chapter 7, the economics are multidimensional and encompass everything from lower first-costs, to greater deployment flexibility in time and space, to the avoidance of stranded investments.
3. *Digital services:* For the reasons discussed in Chapters 4, 5 and 7, radio can in some cases provide quicker access to enhanced digital services.

Looking at PCN/PCS through this prism, we can see essentially meta-applications:

1. *PCN as a better cordless phone:* This is the prevalent applications idea; PCN is seen as an enhanced cordless phone that is capable of operating over a much wider area. The scope of the coverage ranges from the Telepoint idea for certain (semi-isolated) "phone zones" in train stations, airports, restaurants, and hotel lobbies, to Bellcore's "neighborhood" cordless phone, to the grandiose seamless portable network concepts espoused by some. The emphasis is strictly on portability or tetherlessness—that is the added-value that the customer buys. The majority of PCN proposals falls into this category.
2. *PCN as a cheaper way to provide telephone access:* We have argued throughout this book that radio should be fundamentally less expensive than wire for many ordinary access lines. This is what underlies the recent NYNEX proposal for a "wireless local loop" with portable CPE [21]. It is also the basic rationale for BETRS and the rest of the telephony applications set (see Chapter 5). It is explicitly a part of the definition of PCN service in the United Kingdom: "The UK has more than 20 million telephone subscribers who access the public network through a copper wire link," write the authors of the "Phones on the Move" report. "There is no inherent technical reason why these subscribers should not access public telephone networks through mobile radio means." [22]
3. *PCN as a digital access system:* This alternative has received the least emphasis so far, although the some recent proposals perhaps point in this direction [23]. PCN could be sold as a digital access vehicle, with portability as a feature, especially as the local exchange bypass market begins to develop.

Which of these concepts will ultimately prevail? Throughout this book I have expressed my view that the case for universal portability is not clear. Highly portable access instruments are already available—from analog cordless extension phones to analog cellular portables and even simplex radios—at reasonable prices. Some people buy them, many do not. How much cheaper, how much smaller, how much wider coverage—before everyone carries a portable access device and the copper wires wither away?

On the other hand, we have examined in considerable detail the economic case for wireless access. Unfortunately, the PCN version of the economic case is the weakest, because it involves only the replacement of the drop wire (see Section 7.2.3.1). And the Bellcore residential wireless access proposal, for example, would compete not only against drop wire but also against conventional cordless phones—and the competition could prove rather difficult for PCN in this environment (see Section 5.5.2).

My view is that the emerging market for digital services could be the key to the PCN equation. By adding a digital capability to the portability inherent in the PCN idea, we could significantly expand the value-added of PCN service. Moreover, the copper network is not really equipped to compete in many areas. The portability phenomenon in the computer business has already gone quite a way, and many observers believe that during the 1990s there will be further breakthroughs in laptop design and technology that will turn that segment into the dominant consumer market for computing. Adding a wireless, portable and (especially) digital communications capability to these systems could be the key to PCN. Time will tell.

10.4 TECHNOLOGIES

What technologies will become the basis for PCN/PCS?

In today's environment, this question becomes contentious almost as soon as it is posed. The TDMA/CDMA debate has spilled over from the cellular arena to other applications. PCN will probably operate at new frequencies that hold many unknowns for engineers and product designers, and new system architectures will require extensive field experience before their performance parameters are truly known. Without recapping the discussions in Chapter 6 (especially Section 6.4.8), let me offer a few brief observations about the matter of PCN technologies.

1. There probably is no clear single winner. In some applications, TDMA will prevail; in others, CDMA may prove superior. Much of the question of CDMA's potential applicability will depend on the timing of the development of the particular market; there is a lot of work ahead to establish CDMA's true capabilities.
2. "Overlays" probably won't work. The idea that somehow technology (e.g., CDMA) can solve the licensing problem for the FCC by allowing everyone to share the same spectrum will prove illusory. Interference is interference, and

we must remember that high-capacity CDMA systems are designed to manage interference, not to make it somehow go away. The Part 15 experience in the next couple years will show us a great deal about the limits of the sharing idea.

3. Smaller cells, with fixed or slow-moving users, will make a lot of things easier (compared to either the BETRS or cellular-type environments). Smaller cells mean lower power, less delay spread, less concern about multipath, the possibility of higher symbol rates with better infrastructure economics, easier hand-off. I have a lot of optimism in particular about indoor and very short-range outdoor PCN applications. These should lead to cheap, simple implementations compared to, for example, vehicular mobile systems.
4. On the other hand, if we add back the capability to operate a PCN device in a moving car, we will add back a lot of cost and complexity. If we add back long-range capabilities, we will add back cost and power requirements. For these reasons, I tend to be skeptical of the merging of PCN and vehicular (cellular) applications.
5. The most critical aspect of systems design is probably not the radio (that is to say, the air interface) but the network that controls it. The intelligent features that many PCN advocates speak of are almost all network issues. This includes call-in versus call-out capability, numbering plans, location-finding capabilities, hand-off, and so forth. The "seamlessness" of a PCN/PCS has more to do with the network than with the radio technology.
6. The propagation characteristics for systems operating at 2 GHz or higher are still fundamentally unknown, and will only be understood after extensive field experience has been gathered. Recall that Bell Labs studied 800-MHz propagation for the better part of two decades (including the large-scale field trial in Chicago)—and still there were nasty surprises in store for cellular operators. This will be especially true for CDMA systems, where many aspects of performance may not be known until really large systems have been tested (with thousands of users and dozens of cells).
7. The economics for mass production of 2-GHz hardware are still unknown. We may have faith that in time all things will cost $1.98, in quantity—but I am still not convinced that 2-GHz equipment can be made more cheaply than 800-MHz equipment under any reasonable set of assumptions. If there is in fact a persistent cost penalty for 2-GHz systems, then it will greatly affect our assessment of the viability of potential PCN applications.

10.5 OPERATING CONCEPTS

Who owns PCN? Who should have the right, or the preference, to become a PCN/PCS operator?

Three answers have been put forward to this question, leading to three different operating concepts:

1. *PCN as a cellular adjunct:* For the past several years, the cellular operators have found themselves in something of a dilemma. They have enjoyed the financial success associated with a bullish stock-market appreciation of their corporate values; at the same time, this has placed them in a position of having to defend the basis of much of this perceived value—the semi-monopoly, or scarcity value of their spectrum. Especially in the major markets, there would appear to be room for additional competitors, were the spectrum to be made available. The eye which they cast over the PCN phenomenon is therefore, understandably, rather jaundiced. Indeed, competition for cellular was precisely the motivation for the creation of the PCN licenses in the United Kingdom [24].

   The excellence of the capitalist system notwithstanding, nobody in their right mind wants competition—so the cellular operators have by and large attacked the very idea of PCN as a separate service. Some of them have spent considerable time and money to demonstrate that they can do everything that goes by the name of PCN within the current cellular framework. For the cellular community, the panglossian scenario would be for the FCC to authorize additional spectrum for them to use for new PCN-type services. Their investors would be protected, the public would be well-served by experienced operators, and the FCC would sidestep the impossible job of having to grant new licenses.

   This has no chance of happening. I can understand why the cellular operators must defend their economic turf, but none of their arguments is valid. The duopoly is already a creaking obsolescence which, with the advent of vastly more spectrum-efficient digital radio technology, is less and less viable. Beyond these political considerations, the proposition that a cellular operator can behave like a supermarket vendor of all sorts of communications services is inherently unlikely. I am afraid that if spectrum remains scarce, there will be an inevitable "Gresham's Law" type of effect, with more profitable services driving out less profitable ones. Unless we expect that all PCN services will be priced the same per minute of airtime, this is bound to happen. It is not feasible for a minute of Telepoint service (for which there is always the 25-cent pay phone alternative) to be priced the same as a minute of car phone service (there may be no alternative out on the crowded highway). For economic, technical, and regulatory reasons, we can expect that the new PCN spectrum will not simply be handed to today's cellular operators.

2. *PCN as a wireline adjunct:* The wireline telephone companies would like to utilize radio as a natural extension of their basic local exchange network. The Bellcore PCS scenario, among others, is a reflection of this assumption. By

inference, the telephone companies would like to view this as a continuation of the local monopoly—and there's the rub. For as we have said throughout this book, radio does not lend itself to the "natural monopoly" argument.

It seems to me that this dilemma is a little more difficult for the regulators to deal with. There is a case to be made, on the grounds of public convenience and social utility as well as those of private profit-seeking, for allowing the telephone companies to use radio as a substitute, extension, or adjunct for wire plant—we have rehearsed these arguments *in extenso* in Chapters 3, 4, 5, 7, and 9. On the other hand, there is no good reason to constrain the use of radio only to the telephone company. Once it is admitted into the local exchange, it clears the way inevitably for competition, for alternative operators.

3. *PCN as a new competitor:* The third answer is the best in every respect except for the implementation of it. Instead of allowing either of the established telecommunications carriers to capture PCN for its own, the new service should be bartered out to the entrepreneurial community for which America is so famous. Open competition, open entry, come one, come all! This is the path marked out by the strongest historical precedents and the one that most accords with our secular ideologies of government.

   The problem of implementation is essentially the problem of licensing: If "anyone" can have a license, still we cannot be sure that just "anyone" can build and operate a system. And while the FCC is not in the business of guaranteeing the success of its licensees, it recognizes that without certain precautions it is all to easy to create conditions that will lead to unstable markets, speculation, fraud, windfall profits, and deficient service. We can reasonably expect that PCS licensing could look a lot like the later rounds of the cellular licensing, with which almost no one was happy.

10.6 TOWARD SPECTRUM ABUNDANCE

The path toward PCN is fraught with difficulties—that would seem to be the pessimistic message of this final chapter. But much of the uncertainty will be resolved in the normal course of events over the next several years. Spectrum will be found, pried loose, commandeered, dividended; the wide range of ongoing PCN experiments in many countries, as well as the experience with operational PCN systems in Britain, will clarify the applications scenario; the standards-setters will begin their work in earnest. It will take longer than we might hope—although I was recently struck by a comment from a friend in the business, in response to my question about a particular PCN experiment, to the effect that he thought it was "before its time" [25]. There is much wisdom in this phrase. For unlike cellular radio, which was built on a foundation of decades of mobile radio experience and large, measurable reservoirs of pent-up demand, the markets for most PCN services are still inchoate, still in the embryo. How many consumers have ever really wished for a wireless pay phone (i.e., Telepoint) or given serious thought to the possibilities of a wireless PBX? There is a strong element of "technology-push" in many of the PCN scenarios, and still relatively weak "market-pull." It is not uncommon for the telecommunications industry to spawn a service "before its time"—from PicturePhone in the 1960s to ISDN in the 1980s we have seen this pattern many times.

In this context, then, I would hope that the reader does not interpret the discussion of the Four No's of PCN as mere pessimism. The charisma of portable communications is real enough, even with today's relatively imperfect instruments, and I am sure that as the answers to the question of the value added are refined and focused, portability will become a feature that may penetrate many segments of the communications market. If I differ from some of the PCN enthusiasts, it is perhaps in that I see portability more as a *feature* than as a primary functionality. People will buy a portable computer when the computer part of it is adequate—that is why the Timex Sinclair portable of the early 1980s failed and the laptops of the early 1990s are succeeding—the difference is not in the degree of portability but in the computer capabilities. Similarly, I can today purchase a portable telephone in several forms, and if all I want to do is talk on it and I am willing to put up with a diminished voice quality (compared to wireline, especially for cordless extension phones), I will probably be relatively satisfied. And these "primitive" portables can be fairly inexpensive. How will PCN take me beyond these capabilities, especially if it asks me to pay a premium? In a sense, therefore, the real market question about PCN is: *What does it offer me besides portability?*

But my final observation has to do with the coming sea-change in the overall environment of radio communications. Ever since the beginning of the mobile radio business, the industry has been straitjacketed by the real and perceived scarcity of spectrum. There seemed to be far more potential users of the radio services than the finite electromagnetic spectrum could accommodate. Although technologists continued to push out the spectrum frontier, the growth in the potential demand continued apace. This scarcity mindset has distorted the industry structure in a number of ways:

1. Scarcity imparts an added value to any license. Since the number of licenses is necessarily limited, the license acquires the patent of monopoly. This appears to be empirically true even in markets where there is more than one licensee (e.g., the cellular duopoly) as long as there are not very many of them and the entry of new competitors is restricted or prohibited.
2. The monopoly value in turn engenders a speculative interest in spectrum licenses. The value of a nonmonopolistic business is essentially based on its earning ability, i.e., the substance of the business. The value of a monopoly franchise is inflated by the investors' perception of the potential value of any bottleneck. To create or expand value in a nonmonopolistic business, I have

to work at it. To realize value in a monopoly franchise, I may do little more than take title to it and immediately resell it (as was the case with many winners of the cellular franchises [26]), or sit tight for a while the market develops and then sell out. Controlling speculation has bedeviled the FCC from the very beginning.

3. Scarcity actually strengthens the argument against competition and open entry. In the original rule-makings for determining the number and nature of cellular radio licenses, the potential damage from allowing too many competitors to share the spectrum was precisely calculated, in terms of the reduced overall capacity due to loss of trunking efficiency [27]. The Solomonic result was the creation of the cellular duopoly, which allowed nominal competition under what has proved to be (in most cases) a rather monopolistic framework.
4. Scarcity also leads to hoarding (this is as true for spectrum as it was for rice in China during the famines). In this case, both the FCC and the operators hoard spectrum in various ways. The operators hoard spectrum individually (by literally holding title to unused or lightly used channels, which I have found to be a very widespread practice [28]) and collectively (by defending underutilized spectrum allocations that they happen in some cases to have inherited [29]).

   The FCC calls its hoards "reserves"—which are presumably intended for future use and are doled out with excruciating caution. (For example, the FCC managed to pry loose 110 MHz of upper UHF TV spectrum from the broadcasters in 1970—from which it created the cellular, SMR, and several other allocations. There are still several MHz of reserve from this almost unprecedented bounty that are unallocated more than 20 years later. Such conservatism, while understandable in the context of perceived spectrum scarcity, nevertheless casts a pall over the entire community of prospective users and service developers, and leads to a predictable fight over each scrap as it is tossed out.)
5. The scarcity mindset thus increases the friction in the allocation and licensing process, raising the cost and lengthening the time associated with bringing any new service to market. It also politicizes the process; it becomes a case of not just company versus company, but interest group versus interest group and industry versus industry. The result is delay. Cellular radio is again the best case in point. From a technical standpoint, we could have had cellular radio much earlier (especially since it did not involve any new radio technology)—certainly by the mid-1970s. But the politicization of the process delayed cellular by 10 years or so, by which point the obsolescence of the technical framework was beginning to become apparent [30].
6. These procedural costs, as well as the surplus value of the monopoly that so often accrues as a windfall to the speculator, must be borne in the end by the

consumers. It is difficult to calculate just how much higher the rates for cellular radio are today because of these factors but I believe it could be substantial in many cases [31].

It has been this way since the beginning, and now it is all about to change. The perpetual state of spectrum scarcity is going to give way to a new era of "spectrum abundance," and almost all of these distortions will begin to disappear. Not overnight, but certainly by the end of 1990s, we will have left scarcity behind us.

The basis for this change is technological. The digital radio systems now coming on the scene will increase traffic-carrying capacity per MHz of spectrum by a factor of up to 15 or so (see Chapter 6). Capacity baselines will probably double again by the second half of the 1990s. Furthermore, these numbers reflect crude calculations based on *equivalent voice circuits;* in terms of message-carrying capacity, as digital services and packet services begin to come onto the networks, the practical capacity increases will be much greater due to improved control architectures. For example, instead of using an entire voice channel (say, 16 kb/s with overhead) to transmit a short burst of packet data (e.g., for a telemetering application), the packets will be handled in a specialized channel or possibly in a flexible bandwidth architecture. Finally, the perfection of microcells and picocells will push the reuse factor into the extremely favorable region where gains of 100 times capacity or more can be realized simply through more controlled cellularization [32].

Imagine, then, a capacity baseline increase of a factor of 20 by the year 2000 (which is conservative). Assuming reasonable cellularization plans, this would provide capacity for something more than 250,000 cellular users per MHz of spectrum in a given metropolitan area (e.g., Los Angeles). Or several times that many dispatch users per MHz of spectrum. Or 100,000 BETRS users per MHz [33]. Now imagine that we have 200 MHz of new spectrum to work with. Even a small slice—say 10 MHz—could support (with efficient digital radio and advanced cellularization) over a million access lines of almost any type—equivalent to one million households, or three million people. In other words, a small allocation would be enough to serve all the access requirements for a large city.

We can begin to envision the meaning of spectrum abundance. Obviously, there would not be anywhere near enough demand for communications in any city to saturate the 20 operators who could each provide service with 10 MHz of spectrum under these technological conditions. In fact, there probably would not be 20 operators. There might be 10, and each of them would be operating at well below the maximum capacity of his spectrum allocation. Indeed, there would be at that point no need to restrict formally entry—new companies would be able to come into the market if they can make a business go of it. To do so, they would have to do what almost every other kind of business does today: that is, win market share in a market where there is no pent-up demand and where the traditional equilibrium pricing of classical economics will hold.

There will thus be no intrinsic or monopoly value to holding a license *per se.* The only value will come from using the license to serve paying customers, and the value of the franchise will reflect the value of the discounted cash flows. There will be little if any speculative surplus.

There will be so much capacity that almost any number of operators can be permitted without a deleterious effect on trunking efficiencies.

The elimination of speculation value from spectrum licenses will mean that by and large only qualified, serious business applicants will bother to seek a license. If I can get a license pretty much any time I want, then the value to me of the license you hold today, above and beyond the value of the basic business, is zero. The FCC can quietly go back to a watchdog role, and get out of the casino business.

The instinct for spectrum hoarding will disappear.

I would not want to say that spectrum abundance, created by digital radio technology, cures all ills—but it comes close. It eliminates much of the need for regulation and allows for market mechanisms to function in the same way they do for most of the rest of the economy (at least for the PCN set of applications; the telephony set will still be troublesome). It eliminates the need for a mandated service definition; there will be enough capacity to allow experimentation to take its course. Once again, there is nothing really to fear in this: Experimentation thrives in the marketplace. It is only scarcity that so abhors failure that it becomes afraid to experiment.

If this technological imperative will indeed deliver us from so many of today's problems, then what should be the role of the FCC or any other regulatory body?

Aside from their constabulary functions, there is essentially one creative role for the regulators: to encourage and promote spectrum-efficient technology. For in some respects, efficiency is not natural. Waste is natural. When land was seemingly abundant in the early days of this country, it was used wastefully, plowed, exhausted, abandoned. In the Amazon today the jungle is given over to slash-and-burn cultivation and after a few years the "farmer" moves on. When, in the past, the FCC has granted licenses with no care for the efficiency with which the spectral real estate has been used, the predictable result has been that radio technology stagnated. We are still using today the same basic radio system demonstrated by Edwin Armstrong in 1935. On the other hand, it was only the FCC's tough stand when, less than two years after the beginning of cellular service, the cellular operators came pleading for more spectrum to continue using their slash-and-burn technology, that has propelled us forward toward the next technological chapter [34]. Had more spectrum been granted to the cellular operators at that time, we would certainly not have a digital cellular standard today. By refusing to grant additional spectrum, the FCC forced the industry to get serious about spectrum efficiency for the first time. For all that the FCC is often berated, it must be acknowledged that they have shown real leadership in the transition toward digital radio technology, and toward spectrum abundance.

The only "penalty" for spectrum abundance is that it will take us further from the ideal of a seamless, single-standard, universal wireless access network. Abundance fosters diversity, heterogeneity, redundancy, laminarity. The old metaphor would have been the Tower of Babel. The new metaphor is perhaps the Darwinian one of speciation.

Consider the Arctic eco-system. In a harsh environment, the biomass is concentrated into a few, "no-nonsense" species that become very successful and multiply into huge herds, flocks, or swarms. Anyone who has ever seen the seal rookeries or the great caribou herds or, for that matter, the unbelievably dense clouds of predator-less mosquitoes swarming over the tundra, has seen this "arctic logic" at work. Uniformity, basic designs, few colors or frills, little waste of any kind, and huge numbers.

The tropical eco-system—where food and energy are abundant and nature benign—is quite different. In the tropical rain forest, the species multiply and differentiate themselves endlessly. Almost every hillside may have its own species of frog or finch. Almost every plant has a specialized caterpillar that feeds on nothing else. The energy flowing through the system is so abundant that much of it can be diverted to goals other than raw survival—and the jungle blossoms in a riot of colors and "features" for mating, for defense, for symbiosis. Surprisingly, there is not that great a difference in the total biomass for certain non-arid arctic and tropic environments. But the distribution of that living stuff among different species is completely different. Where the arctic system may have only a handful of species, each uniquely occupying its special niche, in the tropics there are hundreds of thousands of different living creatures and every niche is filled with dozens or hundreds of species.

The world of FM and spectrum scarcity is like the arctic eco-system. It exhibits a relative paucity of radio applications—mainly voice communications from moving vehicles—and it fosters technological uniformity. The world of digital radio, in contrast, will be more like the tropical eco-system: marked by differentiation, the elaboration of new services and "features," a wide variety of technical architectures all jostling with each other, and irreducible diversity. I say "irreducible" because I do not believe that once the genie has escaped from the bottle we shall ever be able to get him back in. Radio will continue to find its way into new applications, and the environmental demands of each application will tend to encourage application-specific technical solutions. The users and manufacturers will endlessly bemoan this "lack of standards" and the standards committees will work overtime to try to bring more order to things. But the entrepreneurs, large and small, who have come into the industry in the last 10 years will continuously thwart the standards-setters by coming up with new and better technologies, systems, applications. In time we will learn to live with it, primarily by codifying certain basic interfaces (probably *not* including the entire air interface) that will allow much more complex digital systems to interwork to a modest degree. We already know the name for this from the computer industry—*format conversion.* If the experience of the computer folks is any guide, even the interfaces will not be all that easy to manage. In any case, welcome to the jungle.

NOTES

1. I should note that this system has remained in service at least to the date of this writing, and continues to deliver good service. This is a credit to many people, of course those at Mountain Bell (especially Dick Wunder and Dennis Barp, as well as Kris Shelton) who made a go of it, and those at M/A-COM (now Hughes Network Systems) who built the prototype system. The M/A-COM team was led by Jack Fassel, Pete Vena, Graham and Wendy Avis, Tom Bernard, and Jim Gandolfi. Had their hardware not survived that first winter on the mountain, I would not be writing this book now. At IMM, those who played the largest roles in the trial included Dick Saunders, Jud Kenney, Paul Levitsky, Brian Kiernan, Bill Vanderslice, Thad McGarrah, and Ted Kunko. I believe it was Ted Kunko who placed history's first TDMA digital mobile telephone call, in early September 1986.
2. In this case, the customer was the White House Communications Agency.
3. Federal Communications Commission, "In the Matter of Amendment of the Commission's Rules to Establish New Personal Communications Services," Docket 90-314, adopted June 14, 1990, released June 28, 1990, p. 1.
4. *Ibid.*, pp. 1–2.
5. See the discussion of the cellular spectrum struggle in my first book, *Digital Cellular Radio,* Boston: Artech House, 1988, Chap. 3.
6. Federal Communications Commission, "In the Matter of Amendment of the Commission's Rules to Establish New Personal Communications Services," Docket 90-314, adopted June 14, 1990, released June 28, 1990.
7. This exceedingly high figure is primarily due to the antiquated rules regarding UHF taboo channels. Today, each active UHF channel automatically removes from service anywhere from 60 to 90 MHz of spectrum in that coverage area. These are the "taboo" channels which, under the FCC rules, are set aside to prevent interference. Almost everyone in the industry acknowledges that such huge set-asides are now unnecessary, from a technical standpoint.
8. The DOD spectrum situation seems to range from glut to famine. In the central European military theatre (admittedly not a part of the U.S. spectrum map, but illustrative nevertheless), there are severe spectrum shortages for many military needs. However, there are other large allocations (e.g., certain radar frequencies) that are very lightly occupied, on a geographic basis.
9. For at least the past two years (as of this writing) there have been legislative initiatives in the direction of releasing some of this electromagnetic hoard for civilian use. It is to be hoped that such steps may soon be taken; however, such a release would only be the first step toward a full reallocation. The idea of an electromagnetic "peace dividend"—a relatively painless and essentially cost-free benefit from the end of the Cold War—comes from Jack Taylor.
10. Federal Communications Commission, *Notice of Proposed Rule Making,* "In the Matter of Amendment of Parts 2 and 22 of the Commission's Rules to Permit Liberalization of Technology and Auxiliary Service Offerings in the Domestic Public Cellular Radio Telecommunications Service," Docket 87- 390, released October 15, 1987.
11. NYNEX's cellular subsidiary has made the most noise about cellularized PCS, with their proposal for a CDMA-based PTS in New York City.
12. See the final section of Chap. 4 of my previous book, *Digital Cellular Radio,* Boston: Artech House, 1988, pp. 120–134, for a more flavorful recounting of this recent history.
13. My personal feeling has always been that auctioning is an idea that looks better at first glance than it does when it is examined in detail. At the least, one can say that implementation would be tricky (i.e., hardly a relief for the overburdened FCC), prone to many of the same abuses as the other methods, and probably ripe for porkbarrel politics (the special interest groups are already mobilizing to exempt various user categories from the auction requirement). In the end, it boils down to a

user tax, levied rather inequitably on those who communicate by means of radio. This may seem okay for stock brokers chatting on their BMW car phones—but what about residential users of cordless phones, or rural telephone subscribers for whom BETRS is a lifeline service? Is there a purpose served by increasing the costs of their services (perhaps substantially, if the experience with cellular "pop" values is any indicator)?

14. As of this writing, there is a bill before Congress that would authorize spectrum auctions.
15. *Ibid.*, pp. 426 ff.
16. Indeed, most technical standards are developed or formalized after the fact, that is, after the technology has been developed and is already commercially accepted. With the selection of TDMA for digital cellular, although the application was new, there was a basis of commercial experience with TDMA systems dating back two decades. With CDMA, we are dealing with a communications technique that is still very poorly understood and has never been deployed in any commercial telephony system, or indeed any system even remotely similar to the PCN applications now being contemplated. Standards bodies are used to dealing with a fairly high level of certainty about the technical alternatives—the uncertainties they wrestle with have to do with market or business or applications questions. With CDMA, such bodies are being asked to explore deep technical uncertainties as well. By the way, the same issue confronts the FCC should they choose to enter into the PCN/PCS standards business. The FCC is not technically qualified to resolve many of the technical conundrums that CDMA will present.
17. This was the initial position taken by many of the operators in the very difficult IPR discussions regarding the GSM digital cellular standard in Europe. It has been suggested in other contexts as well, but it now generally recognized as being unacceptable and unworkable.
18. In the case of the IS-54 digital cellular standard, there were three companies asserting relevant IPR at the time the standard was adopted.
19. With cellular, the drawn-out process meant that by the time cellular arrived on the scene in the mid-1980s, it was pre-committed to both a technical and a regulatory framework that were essentially 10 years out of date.
20. I believe that the digital cellular standards processes in both Europe and the United States have taught the industry a great deal about how to deal with Information Age technology standards.
21. NYNEX Enterprises, "Request for Information on Radio Loop Access," RFI No. 546-NSC-10B, October 12, 1990; see also NYNEX press release, "NYNEX to Test Wireless Loop Technologies," August 2, 1990, and associated filing with the FCC.
22. British Department of Trade and Industry, "Phones on the Move," January 1989, pp. 3–4.
23. See the recent Apple Computer proposal, for example.
24. See Section 5.5 in Chap. 5.
25. Credit to Pete Campanella of NYNEX.
26. See the discussion in Chap. 4 of *Digital Cellular Radio,* Boston: Artech House, 1988.
27. The "logic" of the various positions and other details of this discussion are recorded in my previous book. *Ibid.*, p. 54–61.
28. For example, we once needed to operate a BETRS-type system in a major metropolitan area where all the 450-MHz BETRS channels were already held by existing licensees and were supposedly in use carrying very heavy traffic. After discussions with the predominant licensee (holding 12 channels in all) led to the conclusion on his part that allowing this particular BETRS application to proceed would be to his benefit, we were informed that there were actually 8 channels that would be completely available to us! All of these channels were supposed to be occupied by IMTS traffic in an extremely congested mode. I have since seen this type of situation so often that I am tempted to conclude that for some allocations (such as the 450-MHz channels) it is nearly universal.
29. The classic case of this is the UHF television spectrum between about 470 and 800 MHz. This is a massive amount of prime spectrum (more than six times the size of the entire cellular radio allocation) that by my calculation is utilized to less than 10% of its capacity. Indeed, the rules for this

spectrum are marked by a number of prohibitions, which are very extensive, and have the effect of taking large numbers of channels permanently out of service. The argument has been made that these spectrum reserves are actually necessary for the future growth of the UHF television industry, notwithstanding the fact that these reserves have been in place since the 1960s and have never been used. The latest defense of this sort is advanced under the banner of HDTV—which under some scenarios might use a small amount of this fallow spectrum. Of course, in the meantime, this fallowing policy increases the scarcity value of today's UHF licenses—which is at least a not unwelcome side-effect for the broadcasters.

30. All in all, the cellular process stretches from the 1950s (the first formal proposals), through 1968–1970 (the main enabling rulings by the FCC which authorized cellular service and allocated spectrum) to 1984–1990 (when the licenses were issued and systems constructed).

31. I have never rigorously attempted this calculation. However, I do know that the value of today's cellular franchises per "pop" is several times higher than can be substantiated on the basis of even optimistic cashflow projections. In some cases (e.g., some RSAs), the cash flow-based valuation of the franchise is actually negative—the systems never become money-making. This surplus value is attributable to the monopoly value, which is based on scarcity.

32. The sensitivity of the reuse gain factor to cell radius is startling. For example, medium-sized cells (in the "cellular radio" context) of 5 to 8 miles radius yield gain factors (measured in terms of Erlangs per MHz) of one to five times over single-cell systems, depending on the reuse pattern. By contrast, half-mile radius cells yield gains of several hundred times the single-cell capacity. See the calculations in Chap. 4 of *Digital Cellular Radio,* Boston: Artech House, 1988, pp. 113–120.

33. The calculation is not precise, because the inputs to the model are uncertain over a fairly wide range. In particular, the numbers are sensitive to our assumptions about traffic levels per customer. Will mass market cellular continue to generate about 0.026 Erlangs per customer at the busy hour or will lower prices lead to higher usage? Will BETRS be able to use higher levels of modulation and directional antennas to pack cells and circuits even more densely than mobile systems? The reader is invited to use the back of his own envelope for his calculations; however, I am confident that the order of magnitude cited here is appropriate.

34 G. Calhoun, *Digital Cellular Radio,* Boston: Artech House, 1988, p. 114.

INDEX

The Artech House Telecommunications Library

Vinton G. Cerf, Series Editor

Advances in Computer Systems Security, Rein Turn, ed.

A Bibliography of Telecommunications and Socio-Economic Development, Heather E. Hudson

Codes for Error Control and Synchronization, Djimitri Wiggert

Communication Satellites in the Geostationary Orbit, Donald M. Jansky and Michel C. Jeruchim

Current Advances in LANs, MANs, and ISDN, B.G. Kim, ed.

Design and Prospects for the ISDN, G. Dicenet

Digital Cellular Radio, George Calhoun

Digital Signal Processing, Murat Kunt

Digital Switching Control Architectures, Giuseppe Fantauzzi

Digital Transmission Design and Jitter Analysis, Yoshitaka Takasaki

Disaster Recovery Planning for Telecommunications, Leo A. Wrobel

E-Mail, Stephen A. Caswell

Expert Systems Applications in Integrated Network Management, E.C. Ericson, L.T. Ericson, and D. Minoli, eds.

FAX: Digital Facsimile Technology and Applications, Second Edition, Dennis Bodson, Kenneth McConnell, and Richard Schaphorst

A Guide to Fractional T1, J.E. Trulove

Handbook of Satellite Telecommunications and Broadcasting, L.Ya. Kantor, ed.

Information Superhighways: The Economics of Advanced Public Communication Networks, Bruce Egan

Integrated Broadband Networks, Amit Bhargava

Integrated Services Digital Networks, Anthony M. Rutkowski

International Telecommunications Management, Bruce R. Elbert

International Telecommunication Standards Organizations, Andrew Macpherson

Introduction to Satellite Communication, Burce R. Elbert

Introduction to Telecommunication Electronics, A. Michael Noll

Introduction to Telephones and Telephone Systems, Second Edition, A. Michael Noll

The ITU in a Changing World, George A. Codding, Jr. and Anthony M. Rutkowski

Jitter in Digital Transmission Systems, Patrick R. Trischitta and Eve L. Varma

LANs to WANs: Network Management in the 1990s, Nathan J. Muller and Robert P. Davidson

The Law and Regulation of International Space Communication, Harold M. White, Jr. and Rita Lauria White

Long Distance Services: A Buyer's Guide, Daniel D. Briere

Low-Angle Microwave Propagation: Physics and Modeling, Adolf Giger

Mathematical Methods of Information Transmission, K. Arbenz and J.C. Martin

Measurement of Optical Fibers and Devices, G. Cancellieri and U. Ravaioli

Meteor Burst Communication, Jacob Z. Schanker

Minimum Risk Strategy for Acquiring Communications Equipment and Services, Nathan J. Muller

Mobile Information Systems, John Walker

Networking Strategies for Information Technolgy, Bruce Elbert

Optical Fiber Transmission Systems, Siegried Geckeler

The PP and QUIPU Implementation of X.400 and X.500, Stephen Kille

Packet Switching Evolution from Narrowband to Broadband ISDN, M. Smouts

Principles of Signals and Systems: Deterministic Signals, B. Picinbono

Private Telecommunication Networks, Bruce Elbert

Radiodetermination Satellite Services and Standards, Martin Rothblatt

Residential Fiber Optic Networks: An Engineering and Economic Analysis, David Reed

Setting Global Telecommunication Standards: The Stakes, The Players, and The Process, Gerd Wallenstein

The Telecommunications Deregulation Sourcebook, Stuart N. Brotman, ed.

Television Technology: Fundamentals and Future Prospects, A. Michael Noll

Telecommunications Technology Handbook, Daniel Minoli

Telephone Company and Cable Television Competition, Stuart N. Brotman

Terrestrial Digital Microwave Communciations, Ferdo Ivanek, ed.

Troposcatter Radio Linds, G. Roda

Voice Processing, Walt Tetschner

Voice Teletraffic System Engineering, James R. Boucher